Logical Connectors

$p \vee q$ or

$p \wedge q$ and

$\neg p$ not

$p \rightarrow q$ conditional

$p \leftrightarrow q$ biconditional

$p \Rightarrow q$ implication

$p \Leftrightarrow q$ logical equivalence

Symbols Used with Combinatorics and Probability

$P(n,r)$ number of r-permutations

$C(n,r) = \binom{n}{r}$ number of r-combinations

$n!$ n factorial

$P(n; n_1, n_2, \ldots, n_k)$ number of ordered selections from n elements with $n_1, \ldots, n_k$ duplicates

$p(A)$ probability of an event

$p(A \mid B)$ conditional probability

Other Symbols

$p, q, r, \ldots$ propositions and predicates

$x, y, z, \ldots$ real numbers

$i, j, k, \ldots$ integers

$\sum\limits_{i=1}^{n} a_i$ = the sum $a_1 + a_2 + \cdots + a_n$

$\prod\limits_{i=1}^{n} a_i$ = the product $a_1 \cdot a_2 \cdots \cdot a_n$

$\exists$ there exists

$\forall$ for all

π pi ($3.14159 \ldots$)

e the natural log base ($2.71828 \ldots$)

$O(f(x))$ the set of functions asymptotically dominated by $f(x)$

$\{x \mid p(x)\}$ set-builder notation

(a,b) an ordered pair; also used to denote the GCD (greatest common divisor)

$a \, R \, b$ a is related to b in the relation R

$[a]$ equivalence class

K_n complete graph on n vertices

$K_{n,m}$ complete bipartite graph with $m + n$ vertices

Rules of Inference

$$\frac{\begin{array}{l} p \Rightarrow q \\ p \end{array}}{q} \qquad \text{modus ponens}$$

$$\frac{\begin{array}{l} p \Rightarrow q \\ q \Rightarrow r \end{array}}{p \Rightarrow r} \qquad \text{syllogism}$$

$$\frac{p \Rightarrow q}{\neg q \Rightarrow \neg p} \qquad \text{contrapositive}$$

$$\frac{\begin{array}{l} p \Rightarrow q \\ \neg q \end{array}}{\neg p} \qquad \text{modus tollens}$$

$$\frac{p \Rightarrow (q \text{ and } \neg q)}{\neg p} \qquad \text{contradiction}$$

Logical Structures and Operations

if .. then

if .. then .. else

while .. do

for .. do

repeat .. until

case .. of

:=   becomes

Introduction to Discrete Mathematics

Introduction to Discrete Mathematics

James Bradley

Calvin College

Addison-Wesley Publishing Company
Reading, Massachusetts · Menlo Park, California · New York
Don Mills, Ontario · Wokingham, England · Amsterdam · Bonn
Sydney · Singapore · Tokyo · Madrid · Bogotá · Santiago · San Juan

Sponsoring Editor: Thomas N. Taylor
Production Supervisor: Marion E. Howe
Art Consultant: Loretta Bailey
Copy Editor: Helen Greenberg
Illustrator: KATERPRINT Company, Ltd.
Manufacturing Supervisor: Roy Logan
Cover Design: Marshall Henrichs
Text Designer: Marie McAdam

Library of Congress Cataloging-in-Publication Data

Bradley, James, 1943–
   Introduction to discrete mathematics.

   Includes index.
   1. Mathematics—1961– . 2. Electronic data
processing—Mathematics. I. Title.
QA39.2.B725 1987   004'.01'51   87-27067
ISBN 0-201-10628-0

To my children:
Jeanette and Peter

Preface

In recent years there has been a growing recognition of the potential value of a freshman–sophomore level course in discrete mathematics that is not calculus based. The primary impetus for such a course has come from computer science where faculty have recognized the existence of a broad collection of discrete mathematics concepts underlying even the beginning courses in the computer science curriculum. But faculty in other disciplines also have seen the value of such a course, notably faculty in mathematics, the social sciences, engineering, and the natural sciences. This book is written as a text for such a course.

Goals

Shortly before I started writing this book, a group of leading mathematicians held a conference to discuss the future of college mathematics, especially the role of discrete mathematics in the first two years of the curriculum. Many expressed goals they felt should characterize discrete mathematics courses at that level. Here is a list gleaned from their proceedings:

Students should

- develop in mathematical maturity—modeling and reasoning skills plus the ability to estimate, generalize, simplify and detect sloppy reasoning;
- master the basic concepts, results, methods, vocabulary, and notation associated with contemporary discrete mathematics;
- become acquainted with historically and culturally significant problems in the field such as the traveling salesman problem, the knapsack problem, and the Hamilton cycle problem;
- appreciate the need for proof and be able both to read and to do elementary proofs;
- appreciate the place of creativity in mathematics and have some experience applying their own creativity to discrete mathematics;
- be provided with a mathematics corequisite for the introductory computer science courses and a prerequisite for subsequent courses in data

structures, theory of computation, algorithms, and advanced discrete mathematics;

- be able to apply standard algorithms to common discrete structures and modify these algorithms when necessary;
- be able to analyze the efficiency of such algorithms and compare the quality of algorithms in terms of elegance and efficiency;
- understand the notion of discrete mathematical model and be able to use basic discrete mathematics tools to model appropriate situations;
- enjoy the subject.

Besides accomplishing the above goals, a text in discrete mathematics is expected to be at an intellectual level comparable to a Calculus text and have a good measure of unity.

Features

I believe it is possible to write a discrete mathematics text that can help students significantly in accomplishing the above goals and at the same time have unity and be at a level comparable to Calculus. I have used the following features in trying to do this:

- An algorithmic approach. Over forty algorithms are included in the text. These are written in a Pascal-like pseudo-code with frequent comments to make them as readable as possible. Also the complexity of almost every algorithm presented is computed and its correctness verified. The concepts of complexity and verification of algorithms are of critical importance in computer science, and they also have significant mathematical content. The approach is based on the conviction that a discrete mathematics course is the proper place to give these topics the careful mathematical treatment they need, and thus they are treated here with some care. This has the fringe benefit of strengthening in students' minds the linkage between mathematics and computer science.
- Content has been selected for consistency with recommendations of MAA panels, articles published in the *Communications of the ACM,* and other studies of appropriate topic selections for a freshman–sophomore level discrete mathematics course.
- Motivational discussions, applications, and examples are included throughout to make the book as readable and accessible as possible for freshmen and sophomores who have not had calculus.
- The text has been class tested over a period of three years and much of its development is based on that experience. Students have found it highly readable and feel that they are able to learn from it.
- The book begins with a discussion of the traveling salesman problem and the problem is used throughout to illustrate concepts. This is a very important research problem in contemporary mathematics and computer science and yet the statement of the problem can be easily

grasped by novices. Thus students are brought immediately into contact with the research frontiers in the two disciplines. But also, almost every discrete mathematics topic discussed in this text is applicable to the study of the traveling salesman problem. Thus it serves as a unifying theme for the material.

- The history of many topics is explained and historically important problems are presented and discussed at a number of points.

- The proofs of theorems are included throughout with only a few exceptions. These are written to be accessible to freshmen and sophomores, but without compromising mathematical rigor.

- Many exercises are included with each section and these are graded in difficulty from routine to quite challenging. Many exercises requiring proofs are included. In Section 3.6, where rules of inference are presented, a number of exercises are included that are especially designed to introduce students to the concept of proof.

- The notion of mathematical model is presented in the first chapter. Models are discussed throughout the book and a number of exercises requiring students to form elementary models themselves are included.

- A number of features are included to help integrate the text; the most prominent of these are algorithms, discrete models, and the traveling salesman problem. Recursion is also a very important integrating theme in the second half of the book.

- The text includes many examples and applications. The most frequent ones are drawn from computer science; however, many are drawn from the social and natural sciences as well.

- Each section concludes with a list of terminology introduced in the section and a summary of its main points. Students have frequently commented that they found the terminology list and summaries helpful.

- A glossary is included. There is quite a bit of new terminology introduced in a discrete mathematics course, and the glossary provides an easy way for students to look up terms whose meaning they may have forgotten.

Organization

Chapter 1 is primarily motivational but it serves two other valuable purposes—it introduces the notion of computational complexity in an informal way and it introduces the concept of the mathematical model. Thus the unifying themes of algorithms and discrete models are introduced right at the beginning. Chapter 2 is background material in sets, functions, and finite series. Much of this material will be review for many students, so it should be covered as quickly as possible; it is included for completeness. Chapter 3 deals with logic; the material on propositions and predicates in Sections 1 through 4 is also likely to be review for many students and should be covered quickly. Sections 5 through 8 contain extremely important material, dealing with quantifiers, rules of inference, mathematical induction, and combinational circuits. Chapter 4 is on algorithms. It begins with a discussion of the algorithmic language used here and then proceeds to a thorough, mathematically grounded discussion of O-notation, an important topic that is often given superficial treatment. Both the analysis and verification of algorithms and the notions of efficient and inefficient algorithms are introduced here. Chapter 5 deals with basic concepts in number theory. Chapter 6 deals with recursion. I have found that recursion is best explained against a background of mathematical induction. Also students are generally more familiar with equations than algorithms, so recurrence equations are done first, then recursive algorithms. This allows the use of recurrence equations in the analysis of the complexity of recursive algorithms, and induction in their verification. Chapter 7 addresses combinatorics and discrete probability. Chapter 8 deals with relations and introduces digraphs. Rather than introducing relations with functions in Chapter 2, I find it more effective to save relations until students have more examples to build their understanding on. Also, digraphs provide a nice tool to illustrate relations and this way I can introduce digraphs and graphs in subsequent chapters. Chapter 9 deals with graphs and Chapter 10 with trees. Several appendixes are also included; the first two include essential material on arrays and matrices. These topics are included in appendixes to give instructors more flexibility as to when to present them or, with students who have had them before, to skip them. In any case, students will need the material in Appendix A before Section 4.5, if that section is done, and Appendixes A and B before Chapter 8. The other appendixes provide additional depth on some topics that I wanted to include but felt were inappropriate for the body of the text.

Dependency Chart The following chart illustrates the interdependency of sections:

Chapter 1

Chapter 2

Chapter 3

Appendix A   Sections 4.1–4.4   Section 5.1   Sections 5.2–5.5

Section 4.5   Chapter 6.   Appendices A and B

Chapter 7   Chapter 8

Chapter 9

Chapter 10

Sections 3.8, 4.5, 6.4, 6.6, 8.3, 8.4, and 9.4 and portions of some other sections could be omitted without loss of continuity.

Acknowledgments

Above all else, I want to express my thanks to God both for the gift of the ability to write this text in the first place and for the stamina to complete it. I also want to thank my wife, Hope, and my family for their patient endurance of evenings and Saturdays I spent in front of a keyboard rather than with them. My colleagues, both at Nazareth College of Rochester, NY, and at Calvin College, deserve a great deal of credit for their encouragement and help, notably Janet Elmore (who suggested the problem that introduces Chapter 6), Judith Rose, Herbert Elliott, Mary Harrigan, Richard DelVecchio, Joseph Kelly, John Edelman, and many others. I thank my students who not only patiently endured drafts and revisions but also offered me the benefit of their considerable wisdom as to what was presented effectively and what was not. Of particular note is John Freckleton, who strongly encouraged me to write. But I would also like to thank the following students from Nazareth College: Joe Arieno, Pat Assel, Richard Barth, Gina Cecala, Kristine Clauss, Sylvia Cooney, Brenda Dupee, Tim Freed, Lawana Jones, Wendy Marsden, Sabrina May, David Munson, Judy Olivieri, James Palamar, Kathy Pinckey, Jim Porter, Trisha Post, Dina Rice, Phyllis Roberts, Lori Schmidt, Teresa Snyder, and Keith Turner; and from Calvin College: Bruce Abernathy, Judy Arnett, John Brewer, Derek Brouwer, Bennet Bush, Rick Conklin, Joel DeBruin, Harmen DeJong, Deb DeRose, Alan DeVries, Dave Dorner, Dave Dreyer, Dan Fletcher, Kevin Hoag, Carl Hordyk, Steve Klaasen, Steve Kroese, Rich Manni, Richard McClain, Nancy Morrow, Paul Mulder, Jong Myung, Patrick Nagle, Joel Oakes, Stephern Pase, Priya Ramchandran, Debbie Smith, David Stevens, Rick Stiles, Sean Stroub, Alice VandeHeide, Jackie VandenBurg, Kevin VanderMeulen, Mark VanGorp, Ron Vanlwaarden, John Verbrugge, Ken VerHulst, Jonathan Youngsma, Brent Zomerlei, and Steve Zuidema.

In conclusion, I would like to thank the following reviewers for their valuable help and suggestions:

Robert Earles, St. Cloud University

Michael D. Grady, Loyola Marymount College

Denny Gulick, University of Maryland

Georgiana Klein, Grand Valley State College

Joseph B. Klerlein, Western Carolina University

Thomas Koshy, Framingham State University

Richard S. Palais, Brandeis University

Dana Richards, University of Virigina

Diane M. Spresser, James Madison University

Michael Stecher, Texas A&M University

Don Thompson, Pepperdine University

Keith Yale, University of Montana

Grand Rapids, Michigan   J.B.

Contents

Discrete
Models

1

Let us start with a problem. Suppose you are a traveling salesman[†] and there are five cities, including your own, that you would like to visit. Suppose also that the cities are called Amherst, Big Forks, Cattaragus, Devon, and Eden (*A, B, C, D,* and *E* for short). Figure 1.1 is a map of these cities. Starting from *A,* you would like to decide in what order to visit the cities so that the total distance (or time) you must travel to visit each city and return to *A* is minimal. What order should you follow? Before reading further, take a few minutes to see if you can figure out the answer.

Figure 1.1 Map of five cities.

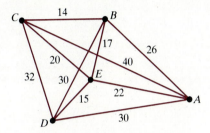

This problem and others like it have come to be called the "traveling salesman problem" and are among the most widely studied problems in contemporary mathematics and computer science. There are several solution techniques, but none that is altogether satisfactory.

The Brute Force Algorithm

The most obvious way to solve the problem is by a **brute force** technique:

1. List all of the possible routes.
2. Compute the length of each.
3. Select the shortest route.

Table 1.1 is a list of each of the routes starting at *A* and the distance associated with each. From this list, it is obvious that the best routes are *ABCEDA* and *ADECBA.* In fact, both routes are the same, merely traveled in opposite directions. So here is a method that works. But it is not a very good method because the process of listing all possible routes and computing the length of each is so time-consuming. With five cities there are 24 (i.e., $4 \cdot 3 \cdot 2 \cdot 1$) routes, starting at *A*. With six cities, there are 120 ($5 \cdot 4 \cdot 3 \cdot 2 \cdot 1$), with seven cities 720 ($6 \cdot 5 \cdot 4 \cdot 3 \cdot 2 \cdot 1$), etc.

[†] Although "traveling salesperson" has the advantage of being sex-neutral, we will keep the traditional and less awkward term, "traveling salesman."

TABLE 1.1   Distances (in Miles) for All Routes in Figure 1.1 that Both Originate and Terminate at *A*

ABCDEA	109	*ADBCEA*	116
ABCEDA	105	*ADBECA*	137
ABDECA	131	*ADCBEA*	115
ABDCEA	130	*ADCEBA*	125
ABEDCA	130	*ADEBCA*	116
ABECDA	125	*ADECBA*	105
ACBDEA	121	*AEBCDA*	115
ACBEDA	116	*AEBDCA*	141
ACDBEA	141	*AECBDA*	116
ACDEBA	130	*AECDBA*	130
ACEBDA	137	*AEDBCA*	121
ACEDBA	131	*AEDCBA*	109

Suppose a very fast computer program could list a route and find its length in 10 microseconds (10^{-5} second). With 20 cities this approach would take about 38,573 years ($19 \cdot 18 \cdot 17 \ldots 3 \cdot 2 \cdot 1 \cdot 10^{-5}$ seconds) to find the shortest route. Adding one more city would multiply this time by 20. We describe this result by saying that the time it takes is proportional to $(n-1)!$ ("$n-1$ factorial," a condensed way of saying $(n-1) \cdot (n-2) \cdot (n-3) \ldots 3 \cdot 2 \cdot 1$).

The Nearest Neighbor Algorithm

While brute force may be very effective in solving small problems, it is unworkable with larger ones. Two things should be obvious from the list of routes given in Table 1.1. First, every route was listed twice, for instance, *ABCDEA* and *AEDCBA*. So we could conceivably cut the processing time in half. This is helpful; it would reduce our computer program time by 19,287 years. However, it would still take 19,287 years to solve the problem! So we have to find something that will make a bigger improvement. Second, many of the routes are obviously inappropriate; for instance, going from *A* directly to *C* without going through *E* is only 2 miles less than going from *A* to *E* and then to *C*, which seems foolish. One method that holds out the hope of avoiding such foolish routes is this:

1. Start at *A*.

2. Go to *A*'s nearest neighbor.

3. Move from there to its nearest neighbor that has not yet been visited.

4. Continue this process until all cities have been visited.

5. Return to A

This approach is sometimes called the **nearest neighbor algorithm** and is an example of a type of approach known as a "greedy algorithm"—get the best "deal" you can at each step but never look further ahead than that step. Apply it to our problem before reading further. You should get the sequence AEDBCA, with a total distance of 121 miles.

To determine how long this approach would take, we note that at each city the four other cities need to be checked to see which is closest and whether it has already been visited. So $5 \cdot 4$ checks need to be made. If each check takes 10^{-5} second, the time required is $20 \cdot 10^{-5}$ seconds. For 20 cities this would mean $20 \cdot 19 = 380$ checks. This comes to $380 \cdot 10^{-5}$ seconds, or $3.8 \cdot 10^{-3}$ seconds, far less than 38,573 years. With n cities, then, we say that the time required is proportional to $n \cdot (n - 1)$.

Thus the nearest neighbor algorithm does not give us the shortest route, but 121 miles may not be a bad solution. Also, this algorithm certainly doesn't take very long to perform. When compared to the optimal answer of 105 miles, it may be close enough to decide that finding the best solution isn't worth the extra effort.

Why Do We Need Mathematical Models?

Is there always a solution that allows us to visit each city only once? The answer is no; consider the map in Figure 1.2.

Sometimes there is a solution that visits each city only once and sometimes there is not. Is there, then, any way to tell from the problem itself when there is a solution that passes through each city only once? There are a couple of approaches one could use to answer this question. One is the method of an experimental scientist: Try out many different examples, note which have solutions that pass through each city only once and which do not, and try to observe some patterns common to one group of examples or another. This method is often very helpful and is the one mathema-

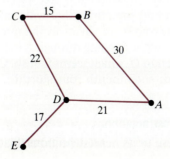

Figure 1.2 Map of five cities where not all are connected.

ticians frequently use when faced with a new problem. However, it faces a limitation common to all experimental work: Even when we have discovered a pattern that works in every case we have examined, there is no way to know whether a case we have not yet thought of could show our conclusions to be completely false.

Recall geometry for a moment. Basic terms like "point," "line," "plane," and "congruent" are assumed, and then axioms such as "Equals when added to equals give equal sums" and "Things congruent to the same thing are congruent to each other" are added. Other definitions such as those of a triangle and a circle are included, and significant theorems about triangles and circles can be proven. This same approach can be applied to the traveling salesman problem (and to many other problems as well). Such a collection of undefined terms and axioms is called a **mathematical system**. When a mathematical system is developed to represent features of a particular situation, it is called a **mathematical model**. The process of developing such terms and axioms is called **mathematical modeling**.

Example of a Mathematical Model

For instance, the traveling salesman problem can be modeled using the idea of a **weighted graph**. (Note that "graph" here means something quite different from the kinds of figures one draws on graph paper when studying functions in algebra.) One way to define a weighted graph is to say that it consists of a finite, nonempty collection of objects called **vertices** and a finite collection of **edges**, each associated with a pair of vertices. For Fig. 1.1, the vertices are *A, B, C, D,* and *E*. The edges are the line segments *AB, AC, AD, BD, BC,* etc. Each edge has a number called a **weight** associated with it. Thus the edge *AE* has a weight of 22, the edge *AD* has a weight of 30, etc. Comparing to geometry, the terms "vertex" and "edge" are basic in the same way that the terms "point" and "line" are basic to geometry. We can also add axioms. For instance, let's add the following statements:

Every distinct pair of vertices has exactly one edge associated with it.

For any pair of vertices (say, *A* and *C*) and any other vertex (say, *B*), the weight of edge *AC* is no more than the weight of *AB* plus the weight of *BC*.

Compare these statements to Figures 1.1 and 1.2. Note that Figure 1.1 satisfies both of these statements but Figure 1.2 does not. It is possible to prove that any weighted graph satisfying these statements has a solution to the traveling salesman problem that visits each city only once. This is a theorem in the same sense as geometric theorems: It is true—at least within the framework of the axioms and definitions that have been set down. Thus mathematical modeling provides the means to answer the kinds of questions we are asking and is the principal subject of this book.

In a nutshell, then, a mathematical model is a precise, but idealized, description of a situation (or several situations) using symbols for the entities and relationships involved. In our case, the entities are the cities, symbolized by letters, and the relationships are the roads joining them, symbolized by pairs of letters. A model is idealized in that the modeler tries to focus on the most essential features of a situation and symbolize them while leaving out the less important features. Thus the names of the cities, the state in which they are located, and their population are regarded as inessential and are dropped from the model. On the other hand, the distance between the cities is considered essential and is kept. In this way, the modeler examines the basic features of a situation in a form that can be manipulated on paper, in one's head, or in a computer without being distracted by peripheral information. One can then observe patterns and find properties that are true of the original situation being modeled and of other situations as well.

The process of developing, studying, and applying models is sometimes described in terms of the **mathematical modeling life cycle** (see Maki and Thompson, 1973, section 1.1). Figure 1.3 depicts the cycle. In our case the *real world* is the actual problem our salesman faced, with all the bumps in the road, the red lights, and the little towns that we omitted from Figure 1.1. Alternatively, it could be a neuron net in a living creature, an electrical circuit, or a pattern of social relations observed by a sociologist. The *real model* is Figure 1.1 without all of the messy details just mentioned. We are left with just the five cities, idealized straight or gently curving roads, no red lights, etc. The *mathematical model* is the definition of the weighted graph given previously and the two axioms we added to that definition. Note that at this point, all reference to cities and their names have been omitted and the situation has been described in terms of symbols, a new vocabulary (edges, vertices), and verbal statements (the axioms) that use numerical concepts ("pairs of vertices," "one edge," "no more than"). The model is abstract; that is, although it can be applied to a particular concrete situation

Figure 1.3 Mathematical modeling life cycle. (From Daniel P. Maki/Maynard Thompson, *Mathematical Models and Applications,* p. 4, © 1973. Reprinted by permission of Prentice-Hall, Inc., Englewood Cliffs, New Jersey.)

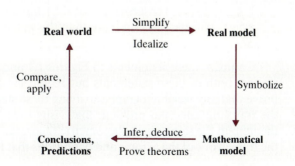

(the traveling salesman problem), it doesn't depend on that situation. For example, for the sociologist, the vertices could be interpreted as people. The edges could indicate that two individuals know each other, and the weights could be a measure of the closeness of their relationship.

Once we have selected the mathematical model, we can add the powerful tool of deductive logic to our intuition and observations and prove theorems. These theorems, in turn, can be applied to any situation that is consistent with the assumptions of the model. Thus the theorem cited earlier about a solution to the traveling salesman problem that passes through each city only once applies to the problem of Figure 1.1 but not to the problem of Figure 1.2. To deal with Figure 1.2, we would need to go back, alter our assumptions, form new axioms, and prove new theorems. This process can be repeated many times until we find a suitable model. It can be one of the most creative and exciting aspects of mathematics.

Types of Models

There are two principal ways of classifying mathematical models—discrete versus continuous and stochastic versus deterministic. Thus models can be of any of types shown in Figure 1.4.

The weighted graph concept we previously examined is an example of a discrete model. Newton's law of gravitation—that the gravitational force between two objects is proportional to the product of their masses and inversely proportional to the square of the distance between them—is a continuous model. **Discrete models** are typically based on the set of natural numbers or on finite sets, while **continuous models** are based on the set of real numbers. Thus, in the traveling salesman problem, there is a finite set of cities and a finite set of edges connecting those cities. But an object in motion can occupy any position along a continuum of points. Since our focus here is on discrete models, this book will emphasize concepts growing out of a study of finite sets and the set of natural numbers, as well as a study of some common discrete models.

The distinction between **deterministic** and **stochastic models** depends on whether probability is involved. If there are no chance elements, the model is deterministic. Both the weighted graph model and Newton's laws of motion are deterministic models. On the other hand, models of rolling a die or of the arrival of cars at a turnpike toll booth are stochastic. Although most of the models we will examine are deterministic, Chapter 7 will

Figure 1.4 Types of mathematical models.

Continuous deterministic	Continuous stochastic
Discrete deterministic	Discrete stochastic

address stochastic models. Chapters 1 to 4 present the mathematical background necessary to deal with discrete models, and the remaining chapters will discuss some of those models and their applications.

TERMINOLOGY

brute force algorithm	edge
nearest neighbor algorithm	weight
mathematical system	mathematical modeling life cycle
mathematical model	discrete model
mathematical modeling	continuous model
weighted graph	deterministic model
vertex	stochastic model

SUMMARY

1. The traveling salesman problem can be "solved" by the brute force algorithm or by the nearest neighbor algorithm. The brute force algorithm is accurate but inefficient. The nearest neighbor algorithm is efficient but not accurate.

2. The weighted graph is an example of a mathematical model. Traveling salesman problems can be modeled by the concept of the weighted graph.

3. Mathematical models are symbolic representations of a simplified, idealized aspect of the real world. They are typically developed and refined through the mathematical modeling life cycle.

4. Mathematical models are classified as discrete vs. continuous and as deterministic vs. stochastic. This book is about discrete models. Its primary focus will be on discrete deterministic models.

EXERCISES 1.1

Apply the brute force algorithm to solve the traveling salesman problem in Exercises 1–4.

1.

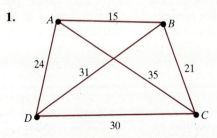

2.

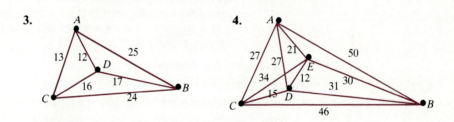

3.

4.

Apply the nearest neighbor algorithm to solve the following problems.

5. Exercise 1   **6.** Exercise 2

7. Exercise 3   **8.** Exercise 4

9. Verify that the brute force algorithm really takes 38,573 years to solve in the case of 20 cities.

10. To compare the efficiency of the nearest neighbor algorithm to that of the brute force algorithm, complete the following table. Recall that the number of steps in the brute force algorithm was proportional to $(n-1)!$ and the number of steps the nearest neighbor algorithm took was $n(n-1)$.

n	$(n-1)!$	$n(n-1)$
2	1	2
3	2	6
4	6	12
5		
6		
7		
8		
9		
10		

11. Recall that the nearest neighbor algorithm requires that pairs of cities be examined $n(n-1)$ times. Suppose we take into account that at the last city no examinations are necessary—the salesman can simply return to the starting city. How would this change the number of examinations needed?

12. Suppose we compute the number of steps required by the nearest neighbor algorithm in a different way: At A, the $n-1$ remaining cities all need to be checked to see which is closest. Once that city is found, move to it. Then examine only the $n-2$ cities remaining, etc. Find an expression for the number of steps that this variation on the algorithm would take.

The following description refers to Exercises 13–18.

Suppose that we modify the nearest neighbor algorithm so that instead of looking only one step ahead, it looks two steps ahead. For instance, if ABD turned out to be the shortest two-city route from A, we would move from A to B. But once at B, we would not go on to D; rather, we would find the shortest two-city route from B, etc.

13. Apply this modified algorithm to Exercise 1.

14. Apply this modified algorithm to Exercise 2.

15. Apply this modified algorithm to Exercise 3.

16. Apply this modified algorithm to Exercise 4.

17. For n cities, how many examinations does this involve, assuming that at each step you must check every other city to see if it has been visited, as well as the distance to those cities that have not been visited?

18. Modify your answer to Exercise 17 to take into account the fact that when the last city is reached, no checks are needed.

The following questions relate to Figure 1.2.

19. Show that there is no route from any city that visits all others only once and returns to the original city.

20. Modify Figure 1.2 by adding edges and weights in such a way that the problem does have a solution that visits each city only once.

21. Find the solution of your modified version by brute force.

22. Find the solution of your modified version by the nearest neighbor algorithm.

For Exercises 23–26, identify the vertices, edges, and weights.

23. Exercise 1   24. Exercise 2

25. Exercise 3   26. Exercise 4

27. Consider another situation to which the definition of a weighted graph could be applied: airline routes. What would the vertices, edges, and weights be in this situation?

28. What would the vertices, edges, and weights be in a corporate hierarchy chart?

29. What would the vertices, edges, and weights be in a family tree?

30. What would the vertices, edges, and weights be in an electrical circuit?

For Exercises 31–38, state whether the specified weighted graph satisfies the two axioms presented in the text and explain why.

31. Exercise 1

33. Exercise 3

32. Exercise 2

34. Exercise 4

35.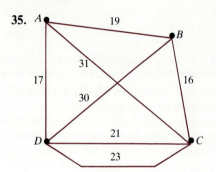

36.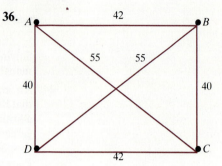

37.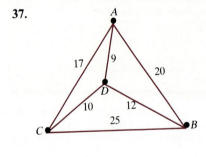

38.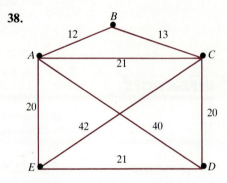

For each of the following situations, state whether you think that a mathematical model of the situation is more likely to be deterministic or stochastic and whether it is more likely to be discrete or continuous. In each case, give reasons for your answer.

39. The motion of a rocket carrying an earth satellite.

40. The outcome of the spin of a roulette wheel.

41. The growth in a rabbit population in a certain geographical area.

42. The concentration of penicillin in the bloodstream for several hours after taking a penicillin pill.

In the following two problems and their solutions, all the elements of the mathematical modeling life cycle are present. Identify which aspects of the problem and solution correspond to the different phases of the cycle.

43. *Problem:* A small grocer has an opportunity to sell some tropical fruit and wants to make as much profit as possible. She has $120 to spend; she can buy mangoes for 30 cents each and pineapples for 45 cents each. From past experience, she knows that she can usually sell twice as many pineapples as mangoes if the profit on each is the same.

Solution: Let m denote the number of mangoes, and p the number of pineapples bought. Then

$$0.30m + 0.45p = 120$$

and

$$p = 2m$$

Solving these equations simultaneously, she concludes that she should buy 100 mangoes and 200 pineapples. She does this and sells them all for a tidy profit.

44. An individual is dropping stones off a bridge into a river below. He observes that the higher up he is, the longer the stones take to reach the water. Ignoring all other factors, he decides to see how the time depends on the height. After gathering some data, he concludes that the time is proportional to the square root of the height ($t = k\sqrt{h}$). He then goes to several bridges whose height he knows, computes from his formula how long it should take the stones to reach the water, and compares his predictions with the actual times. He concludes that his formula is accurate.

REFERENCES

Bellmore, M., and G.L. Nemhauser, "The Traveling Salesman Problem: A Survey." *Operations Research* 16:538–558, 1968.

Knuth, Donald, *The Art of Computer Programming,* Vol. 1., *Fundamental Algorithms.* Reading, Mass.: Addison-Wesley, 1973.

Lawler, Lenstra, Rinnooy Kan, and Shmoys, *The Traveling Salesman Problem.* New York: Wiley, 1985.

Maki, Daniel, and Maynard Thompson, *Mathematical Models and Applications.* Englewood Cliffs, N.J.: Prentice-Hall, 1973.

Foundations

2

In our discussion of the traveling salesman problem in Chapter 1, the weighted graph model started with a collection of vertices. Collections of objects of one sort or another occur in every discrete model. Set theory, which provides us with a vocabulary and tools to deal with these collections, will be our starting point.

2.1
BASIC CONCEPTS OF SETS

Perhaps the most basic of all intellectual skills is the process of classifying and grouping similar things. Since mathematics is abstract, i.e., since it seeks to find and describe similar patterns among diverse situations, the process of forming and describing categories is fundamental to all of mathematics. Set theory is the tool that mathematicians use for this purpose. Thus the vocabulary and ideas presented in this section will be used in almost every chapter of this book.

Basic vocabulary[†]   A **set** is a well-defined collection of objects. An **element** (or **member**) is one of those objects. An element of a set is said to **belong to** or **be a member of** the set.

Sets and Set Notation

EXAMPLE 1   Here are some typical examples of sets:

a) the positive, even integers
b) the letters in the English alphabet
c) the solutions of the equation $x^2 - 6x + 5 = 0$
d) the capitals of the New England states ∎

Notation   There are two principal ways to denote a set: the **tabular form** —a list of all of the elements in the set—and the **set-builder form**—a statement of the property that the elements have in common. Braces, {and}, are used to mark the beginning and end of the description of a set. The symbol " | " means "such that." Capital letters will generally be used to denote sets

---

† Note that this is "basic vocabulary" and not "definitions." To define "set" and "element," we would have to use other words, which in turn would require other words to define them, etc. In order not to regress indefinitely, we take "set," "element," and "is a member of" as starting points, give many examples to illustrate them, and define subsequent vocabulary using these terms.

and lowercase letters to denote elements of sets. The Greek letter $\in$ will be used to denote set membership, i.e., if x is a member of set A, we write $x \in A$. Thus $\{x \in \mathbf{N} \mid x > 6\}$ is read "the set of all x in the natural numbers such that x is greater than 6." If x is not a member of A, we write $x \notin A$.

EXAMPLE 2   The sets of Example 1 can be written in either form:

Tabular Form	Set-Builder Form
a) $\{2, 4, 6, 8, \ldots\}$	$\{x \mid x$ is a positive, even integer$\}$
b) $\{a, b, c, d, \ldots, z\}$	$\{x \mid x$ is a letter in the English alphabet$\}$
c) $\{1, 5\}$	$\{x \mid x^2 - 6x + 5 = 0\}$
d) $\{$Hartford, Providence, Boston, Augusta, Concord, Montpelier$\}$	$\{c \mid c$ is the capital city of a New England state$\}$

Notice the use of the **ellipses,** i.e., the "$\ldots$" that appears in Examples 2a and 2b. This means "continue on in the same pattern." In some cases (Example 2a, for instance), the pattern continues forever. In others it does not. Also, note that the set-builder form requires the use of a variable at the beginning. For large sets, only the set-builder form is practical. ∎

Definition   There are a number of numerical systems that are important in discrete mathematics.

$\mathbf{N} = \{1, 2, 3, 4, \ldots\}$, the **natural numbers** or **positive integers**

$\mathbf{Z} = \{\ldots, -3, -2, -1, 0, 1, 2, 3, \ldots\}$, the **integers**

$\mathbf{R} = \{x \mid -\infty < x < +\infty\}$, the **real numbers**

$\mathbf{Z}_n = \{0, 1, 2, \ldots, n - 1\}$, the **integers mod n**

$\mathbf{Q} = \{a/b \mid a, b \in \mathbf{Z}, b \neq 0\}$, the **rational numbers**

$\mathbf{I} = \{r \in \mathbf{R} \mid r \neq a/b$ for any $a,b \in \mathbf{Z}\}$, the **irrational numbers**

The natural numbers are the foundation of discrete models. As such, discrete mathematics could be characterized as the study of models based on the natural number system. The real numbers are the basis of most continuous models and the basis of calculus, although they have applications in discrete mathematics as well.

EXAMPLE 3   Write the set of all natural numbers greater than 7 in set notation.

SOLUTION   The solution, in set-builder form, is $\{n \in \mathbf{N} \mid n > 7\}$. ∎

Not all sets of interest are numerical, however. The following is an important class of nonnumerical sets.

Definition   Let Σ be a set of symbols, e.g., $\Sigma = \{a, b, c\}$. In this case, Σ is called an **alphabet.** If characters from Σ are **concatenated,** i.e., written next to each other in a list, the resulting lists are called **words** or **strings.** The set of all words that can be formed from an alphabet Σ is denoted Σ^*.

↳ Together with the empty string
λ , is denoted by Σ^*

EXAMPLE 4   Let $\Sigma = \{0, 1\}$. Find the set, S_3, of all strings over Σ of length 3 or less.

SOLUTION   In tabular form,

$$S_3 = \{0, 1, 00, 01, 10, 11, 000, 001, 010, 011, 100, 101, 110, 111\}.$$

In set-builder form,

$$S_3 = \{w \in \Sigma^* \mid \text{length of } w \text{ is less than or equal to 3}\}. \quad \blacksquare$$

Relationships Between Sets

Sets provide more than just a way to list the basic elements from which discrete models will be built. They provide a number of tools with which to compare those lists and perform various operations on them. The remainder of this section deals with those tools.

Definition   Two sets are **equal** if they have the same elements.

EXAMPLE 5   Let $A = \{1, 2, 3\}$ and $B = \{1, 2, 3, 1\}$. Does A equal B?

SOLUTION   Yes. A and B have the same elements. B is, in fact, improperly written. Duplicate elements should not be included in the tabular form for a set. $\blacksquare$

It often happens that all of the elements of one set are also members of another set. For instance, if

$$A = \{1, 2, 3, 4, 5, 6\}$$

and

$$B = \{2, 3, 4, 5\}$$

then A contains B.

Definitions   If every element of B is also an element of A, we say that B is a **subset** of A and denote this relationship as $B \subset A$ or $A \supset B$. $A \not\subset B$ means that A is not a subset of B. B is a **superset** of A if A is a subset of B. If the elements of a set or collection of sets are all selected from some common set, the common set is called a **universal set.**

EXAMPLE 6 $\mathbf{N} \subset \mathbf{Z} \subset \mathbf{Q} \subset \mathbf{R}$. We could look upon $\mathbf{R}$ as a universal set from which $\mathbf{N}$, $\mathbf{Z}$, and $\mathbf{Q}$ are selected.   ∎

EXAMPLE 7   Using Σ^* as defined in Example 4, if $S_n = \{w \in \Sigma^* \mid \text{length of } w \le n\}$, then

$$S_1 \subset S_2 \subset S_3 \subset \ldots \quad ∎$$

Note carefully the difference between $\in$ and $\subset$. For instance, if $A = \{1, 2, 3, 4, 5\}$, $\{1, 2\} \subset A$ but $1 \in A$. Also, $\{1\}$ denotes the set containing 1 and is usually read "singleton one," whereas 1 denotes the element itself. Hence $\{1\} \subset A$ but $1 \in A$. Similarly, $\{1\}$ is not a member of A.

Definition   A set B is a **proper subset** of A if B is a subset of A and B does not equal A.

EXAMPLE 8   Let $A = \{x \mid x + 5 = 10\}$ and let $B = \{x \mid x - 5 = 0\}$. Then A is a subset of B, since all of the elements of A (namely, the element 5) are elements of B. But A is not a proper subset since $A = B$. Note also that $B \subset A$. $A = B$ can also be defined as meaning that $A \subset B$ and $B \subset A$.   ∎

Some authors use the symbol $\subseteq$ to denote subset and $\subset$ to denote proper subset. We will follow a different convention, using $\subset$ to denote both proper and improper subsets.

One important way to represent sets is with **Venn diagrams.** For instance, to represent a universal set **U** and two subsets A and B of **U**, one can draw a diagram like the one shown in Figure 2.1. Note that drawing it

Figure 2.1   Venn diagram representing two sets.

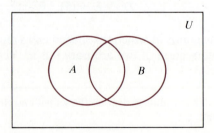

Figure 2.2 *A* is a subset of *B*.

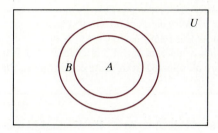

this way appears to assume that *A* is not a subset of *B*, *B* is not a subset of *A*, there are some elements common to both, and some outside both. The Venn diagram is drawn as in Figure 2.1 when we do not know which of these statements is true. It is possible, however, that some parts of the diagram may have no elements. If we do know something about the relationship between *A* and *B*, we can draw a Venn diagram that represents that relationship. For instance, if *A* is a proper subset of *B*, Figure 2.2 applies.

Set Operations

Definition   Let *A* and *B* be sets. The **union** of *A* and *B* is $\{x \mid x \in A \text{ or } x \in B\}$ and is denoted $A \cup B$. The **intersection** of *A* and *B* is $\{x \mid x \in A \text{ and } x \in B\}$ and is denoted $A \cap B$.

EXAMPLE 9   Let $C = \{1, 2, 3\}$, $D = \{4, 5\}$, and $E = \{5, 6, 7\}$.

Then

$$C \cup D = \{1, 2, 3, 4, 5\}.$$

Also

$$D \cup E = \{4, 5, 6, 7\}$$

and

$$D \cap E = \{5\}. \blacksquare$$

Venn diagrams in which the shaded areas represent first the union and then the intersection of two arbitrary sets are given in Figure 2.3.

Definition   The symbol ∅ denotes the set with no elements; it is called the **empty set** or **null set**.

Figure 2.3 Shaded areas represent the union and the intersection of two sets.

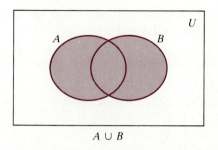 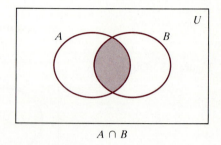

$$A \cup B \qquad\qquad A \cap B$$

In set theory, $\emptyset$ is the counterpart of zero in the integers. If we think of integers as measuring quantities, 0 is the number with no quantity, just as $\emptyset$ is the set with no elements. It is also important as a convenience, allowing us to write "$C \cap D$" and treat $C \cap D$ as a set even if C and D have no elements in common.

Definition   The **difference** between two sets, A and B, is denoted $A \backslash B$ and means $\{x \mid x \in A \text{ and } x \notin B\}$. $A \backslash B$ is also called the **relative complement** of B.

EXAMPLE 10   Let $A = \{0, 1, 2, 3, 4\}$, let $B = \{1, 2\}$, and let $C = \{3, 4, 5\}$. Then

$$A \backslash B = \{0, 3, 4\},$$

and

$$A \backslash C = \{0, 1, 2\}.$$

Note that $A \backslash C$ is still meaningful even though C is not a subset of A. ∎

The Venn diagram for the difference, $A \backslash B$, is given in Figure 2.4.

Figure 2.4 $A \backslash B$.

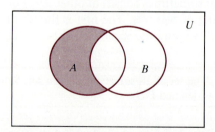

Definition   Two sets are **disjoint** if their intersection is the empty set. The **complement** of a set S (relative to some specified universal set) is the set of all elements not in S and is denoted S'.

Venn diagrams representing disjoint sets and the complement of a set are given in Figures 2.5 and 2.6.

Figure 2.5   Disjoint sets.

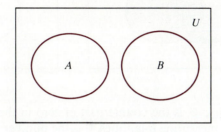

Figure 2.6 A and its complement, A'.

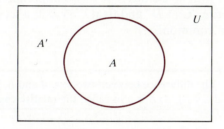

EXAMPLE 11   Let $A = \{000, 001, 010, 011, 100, 101, 110, 111\}$ and $B = \{0, 1, 00, 01, 10, 11\}$. Then A and B are disjoint. Also, note that relative to S_3, $A = B'$. ∎

EXAMPLE 12   Find the complement of $G = \{1, 2, 3, 4, 5\}$ relative to **N** and to **Z**.

SOLUTION   Relative to **N**, $G' = \{n \mid n \geq 6\}$.
Relative to **Z**, $G' = \{\ldots, -3, -2, -1, 0, 6, 7, 8, \ldots\}$ ∎

Sets can sometimes consist of other sets. For instance $\{\{1, 2\}, \{2, 3\}\}$ is a set consisting of two sets. The first is the set of solutions of the equation $x^2 - 3x + 2 = 0$ and the second is the set of solutions of $x^2 - 5x + 6 = 0$.

Definition   The set of all subsets of a given set, S, is called the **power set** of the given set. It is denoted **P(S)**.

EXAMPLE 13   Let $N = \{1, 2, 3\}$. Find the power set of N.

SOLUTION   The power set is the set containing all of the subsets of N. Thus

$$P(N) = \{\varnothing, \{1\}, \{2\}, \{3\}, \{1, 2\}, \{1, 3\}, \{2, 3\}, \{1, 2, 3\}\}.$$ ∎

At this point, we have not formally defined what we mean by the number of elements in a set. This is best done after we introduce some additional concepts in Section 2.4. However, approaching Example 13 intuitively, it is obvious that N has three elements, $P(N)$ has eight elements, and $2^3 = 8$. In a later section, we shall show that the number of elements in $P(S)$ is always two raised to the power n where n is the number of elements in S.

We can summarize the main properties of the set operations as follows:

Laws of the Algebra of Sets
<hr>

Idempotent Laws

1a. $A \cap A = A$ 1b. $A \cup A = A$

Associative Laws

2a. $(A \cap B) \cap C = A \cap (B \cap C)$ 2b. $(A \cup B) \cup C = A \cup (B \cup C)$

Commutative Laws

3a. $A \cap B = B \cap A$ 3b. $A \cup B = B \cup A$

Distributive Laws

4a. $A \cap (B \cup C) =$ 4b. $A \cup (B \cap C) =$
    $(A \cap B) \cup (A \cap C)$ $(A \cup B) \cap (A \cup C)$

Identity Laws

5a. $A \cap U = A$ 5b. $A \cup \emptyset = A$

6a. $A \cap \emptyset = \emptyset$ 6b. $A \cup U = U$

Complement Laws

7a. $A \cap A' = \emptyset$ 7b. $A \cup A' = U$

8a. $(A')' = A$ 8b. $U' = \emptyset, \ \emptyset' = U$

DeMorgan's Laws

9a. $(A \cap B)' = A' \cup B'$ 9b. $(A \cup B)' = A' \cap B'$

<hr>

Venn diagrams can also be used to illustrate these properties and to test whether other relationships are true.

EXAMPLE 14 Use Venn diagrams to illustrate Law 9b,

$$A' \cap B' = (A \cup B)',$$

the second of DeMorgan's laws.

SOLUTION   See Figure 2.7.

Figure 2.7   The cross-hatched areas are the same for both expressions.

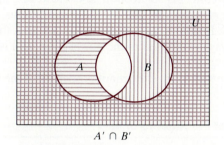
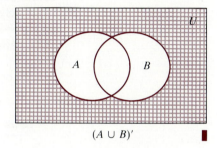

$A' \cap B'$ $(A \cup B)'$

EXAMPLE 15   Draw Venn diagrams to show that

$$(A \cap B) \cup C \neq (A \cup B) \cap C.$$

SOLUTION   See Figure 2.8.

Figure 2.8 $(A \cup B) \cap C$ is on the left; $(A \cap B) \cup C$ is on the right.

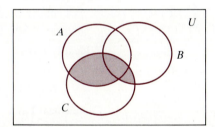
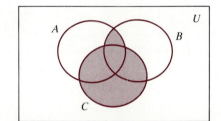

TERMINOLOGY

set	integers	string	intersection
element	real numbers	equal sets	empty set
member	integers mod n	subset	null set
belongs to	rational numbers	superset	difference
is a member of	irrational numbers	universal set	relative complement
tabular form	alphabet	proper subset	disjoint
set-builder form	concatenate	Venn diagram	complement
ellipses	word	union	power set
natural numbers			

SUMMARY

1. The words "set" and "element" are undefined starting points for mathematical thought. However, a set can be described as a collection of objects; the objects are the elements.

2. One way to denote a set is by listing its elements, using the tabular notation— {list of elements}. Another is by describing a common property of the members with the set-builder notation—(variable | property the members share}.

3. Some important sets are **N, Z, R,** and $\mathbf{Z}_n$.

4. A set all of whose elements are contained in another is a subset.

5. The principal operations on sets are union, intersection, difference, and complement.

6. The empty set is the set with no elements.

7. Other important concepts involving sets are disjointness, complement, and the power set of a set.

8. The laws of the algebra of sets summarize the principal properties of the set operations.

EXERCISES 2.1

Write in set builder form only.

1. The real numbers between 0 and 1 inclusive of 0 and 1.

2. The real numbers between 0 and 1 exclusive of 0 and 1.

3. All strings over {0, 1} of length 4.

4. All strings over {a, b} of length 2 or less.

Write in tabular form only.

5. All strings over {0, 1} of length 4.

6. All strings over {a, b} of length 2 or less.

7. The elements of $\mathbf{Z}_6$.

8. The elements of $\mathbf{Z}_{11}$.

Write the sets of solutions of the following equations in both set-builder and tabular form.

9. $x^2 - 4 = 0$

10. $x^2 = x$

11. $x^3 - x = 0$

12. $x^2 = 0$

In each case, determine whether the sets listed are equal or unequal.

13. $A = \{1, 3, 5, 7\}$ and $B = \{5, 1, 7, 3\}$

14. $\{x \in \mathbf{N} \mid x^2 - 3x + 2 = 0\}$ and $\{1, 2\}$

15. Σ_1^* and Σ_2^* where $\Sigma_1 = \{0, 1\}$ and $\Sigma_2 = \{a, b\}$

16. $\{w \in \{0, 1\}^* \mid \text{length } w \leq 1\}$ and $\{0, 1\}$ itself

For Exercises 17–24, let $U = \mathbf{Z}_{12}$, $A = \{0, 1, 2, 3, 4, 5, 6, 7\}$ and $B = \{4, 5, 6, 7, 8, 9, 10, 11\}$. List the elements.

17. A' **18.** B'

19. $A \cap B$ **20.** $A \cup B$

21. $A \backslash B$ **22.** $B \backslash A$

23. $A' \cup B'$ **24.** $B' \cup A$

25. Are A and $A \backslash B$ disjoint?   **26.** Are A and $B \backslash A$ disjoint?

27. Consider again the traveling salesman we discussed at the beginning of Chapter 1. Suppose there are four cities (B, C, D, E) that can be visited on any given day. The salesman may visit none or all or any number in between. List all the possible sets of cities that could be visited. (*Hint:* There are 16.)

Exercises 28–33 refer to the list of sets of cities developed in Exercise 27. In each case, determine how many of the sets contain the following members.

28. B **29.** C

30. B and C **31.** D and E

32. All except E **33.** All except B

Let $U =$ the set of all animals, $M =$ the set of all mammals, $D =$ the set of all dogs, $C =$ the set of all cats, and $L =$ the set of all collies. Indicate whether each of the following is true or false, and give a reason for your answer.

34. $U \supset M \supset D \supset L$ **35.** $U \supset M \supset D \supset C$

36. $C \cap D = \emptyset$ **37.** $D \backslash L \subset C$

38. $U \backslash M \subset D$ **39.** $D \backslash C = D$

40. $L \backslash M \neq \emptyset$ **41.** $U \cap M = M$

Use the Venn diagram shown in Figure 2.9 to label the areas indicated in Exercises 42–49.

Figure 2.9 Venn diagram representing two sets.

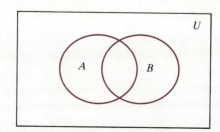

42. $A \cup B$ **43.** $A \cap B$

44. A' **45.** B'

46. $A \backslash B$ **47.** $B \backslash A$

48. $(A \cup B)'$ **49.** $(A \cap B)'$

Suppose $U = \{0, 1\}^*$, $M = \{0, 1, 00, 01, 10, 11\}$, $N = \{00, 01, 10, 11, 000, 001, 010, 011\}$, $P = \{000, 001, 010, 011, 100, 101, 110, 111\}$, and $R \{00, 01, 10, 11\}$. Illustrate the relationship between each of the following pairs of sets using Venn diagrams.

50. M and N **51.** M and P

52. M and R **53.** N and $M \cap R$

Illustrate each of the following laws of the algebra of sets by drawing Venn diagrams and shading the appropriate parts.

54. $(A \cap B)' = (A' \cup B')$

55. $A \cup (B \cap C) = (A \cup B) \cap (A \cup C)$

56. $A \cap (B \cup C) = (A \cap B) \cup (A \cap C)$

57. $A \cup A' = U$

Show that each of the following relationships is false by means of Venn diagrams.

58. $A \cup B = A \cap B$ **59.** $A \backslash B = B \backslash A$

60. $(A \cup B) \cap C = (A \cap B) \cup C$

Exercises 61–66 are based on the following table of beverage preferences:

	Milk, No Coffee	No Milk, No Coffee	Milk, Coffee	No Milk Coffee
Teachers	5	1	9	40
College Students	20	10	30	6
High School Students	24	4	13	2
Elementary Students	30	5	1	0

Let T = teachers, C = college students, H = high school students, E = elementary school students, M = milk drinkers, and F = coffee drinkers. Indicate how many people are in each of the following sets.

61. $T \cap M \cap F$ **62.** $T \cap M \cap F'$

63. $(C \cup H) \cap (M \cup F)$ **64.** $E \cap F$

65. $C \cap F'$ **66.** $(C \cup H \cup E) \cap (M \cup F)$

2.2

RUSSELL'S PARADOX

In Section 2.1 we showed that we often want to form sets of sets. In other words, we want to look at sets whose elements are themselves sets. Normally this is fine, but there are situations in which it can lead to problems. The most famous of these is a mathematical paradox known as **Russell's paradox.** It was originally formulated by the mathematician and logician Bertrand Russell around 1900. A philosopher, Gottlob Frege, had just completed a major work on set theory, a work that he felt would uniquely strengthen and clarify the foundations of mathematics. Before the book was even published, Russell pointed out what appeared to be a logical contradiction in Frege's theory. Frege included Russell's paradox as an appendix to his book.[†] This event subsequently turned out to be just the first of many dramatic discoveries that challenged the foundations of mathematics in the first half of the twentieth century.

Russell's paradox is similar in some ways to the logical contradiction in the statement "The sentence you are now reading is false." To see the contradiction, suppose the sentence is true. If it is true, then it tells us itself that it is false. Alternatively, suppose it is false. It says it is false, so if that assertion of falsity is itself false, then the sentence is true. Either way we get a contradiction. One way to resolve the paradox is to say that such a sentence is nonsense and is meaningless, i.e., to omit it from what we consider to be acceptable sentences by saying that it is inherently inconsistent. In other words, we say that a sentence's grammatical correctness is not sufficient reason to accept it as a meaningful sentence in a language.

This idea arises in set theory in the following way. First, let's consider a set whose members are all sets. It is possible for such a set **to contain itself** as a *member.* Recall that any set contains itself as a *subset* (an improper subset, to be sure, but still a subset). But this time we want it to contain itself as a member. For instance, let S be the set of all nonempty sets. S is itself nonempty; its members consist of many sets, such as $\{1\}$, $\{2, 3\}$, **N**, etc. Since S is nonempty and contains all nonempty sets, it certainly contains itself. There are also **sets (of sets) that do not contain themselves** as members. For

---

[†] Some years later, Russell commented on Frege's response to what Russell had observed: "As I think about acts of integrity and grace, I realise that there is nothing in my knowledge to compare with Frege's dedication to truth. His entire life's work was on the verge of completion, much of his work had been ignored to the benefit of men infinitely less capable, his second volume was about to be published, and upon finding that his fundamental assumption was in error, he responded with intellectual pleasure clearly submerging any feelings of personal disappointment. It was almost superhuman, and a telling indication of that of which men are capable if their dedication is to creative work and knowledge instead of cruder efforts to dominate and be known" (Heijenoort (ed.) 1967, p. 127).

instance, consider the set, F, of all finite sets. F is itself infinite, since there are infinitely many finite sets. To understand this, note that {1}, {2}, {3}, etc., are all finite sets but that there are clearly infinitely many such sets.

The paradox arises when we let S be the **set of all sets that do not contain themselves** as members. Suppose S is a member of S. S consists only of sets that do not contain themselves as members. So, if S is a member of S, it is not a member of S. Alternatively, suppose S is not a member of S. But S consists of all such sets—sets that do not contain themselves as members. Therefore, it is a member of itself. Once again, we get a contradiction either way.

One way to resolve this paradox is to do the same thing we did with the sentence "The sentence you are now reading is false." We must realize that just as not every grammatically correct sentence one can formulate is acceptable, neither is every set. This should not be surprising. If set theory is really modeling of some basic processes in human thought, it should also have some of the same limitations as human language. Russell's paradox provides a glimpse of some of the limitations of human thought. One of the main difficulties, though, is that we do not have a rule that tells us which sets are meaningful and which are not. That is, just as the grammar of a language does not provide sufficient rules to guarantee that all of its sentences are meaningful, the rules of set theory are not sufficient to guarantee that every set we attempt to formulate can be formulated. This is a serious problem for both philosophy and mathematics. Russell's paradox also turns out to have an important analogy in the theory of computation in the halting problem. This is an important problem that points up some of the limits of computation. We will not go into it further here, but almost any text on the theory of computation will treat it in more detail.

TERMINOLOGY

Russell's paradox	set of sets
set that contains itself	set of all sets that do not contain themselves
set that does not contain itself	

SUMMARY

1. Paradoxes arise in language, mathematics, and computer science. Besides being fun to explore, they reveal some of the limits of thought.

2. The most important paradox in modern mathematics is Russell's paradox, which deals with the set of all sets that are not members of themselves. The computer science analogy of this paradox is found in the halting problem, which deals with our inability ever to write a program that would decide if an arbitrary program would halt in a finite amount of time.

EXERCISES 2.2

Exercises 1–8 are based on Eves and Newsom, *The Foundations and Fundamental Concepts of Mathematics,* © 1965, and used with permission of Holt, Rinehart, and Winston, New York. Each of the situations contains a paradox. Show that this is so.

1. The barber of Seville shaves only those residents who do not shave themselves but shaves all of them. Does the barber shave himself?

2. Every city in a certain country must have a mayor, and no two cities may have the same mayor. Some mayors do not reside in the city they govern. A law is passed requiring nonresident mayors to live by themselves in a special area, A. There are so many nonresident mayors that A is itself declared a city. Where shall the mayor of A reside?

3. Suppose a librarian decides to compile, for inclusion in a library, a bibliography of all those bibliographies that do not list themselves. Is this possible?

4. A crocodile has stolen a child, but promises to return the child, provided that the mother guesses whether the child will be returned or not. What should the crocodile do if the mother guesses that the child will not be returned?

5. A missionary has been captured by the cannibals. The cannibals offer the missionary an opportunity to make a last statement on the condition that if his statement is true, he will be boiled, and if it is false, he will be roasted. What should the cannibals do if he says "I will be roasted"? (*Hint:* the missionary is a friar.)

6. Show that the statement "All generalizations are false" is self-contradictory.

7. What would happen if an irresistible force collided with an immovable object?

8. Suppose the statement "God is omnipotent" means "God can do anything." Can God make a stone so heavy that He cannot lift it?

2.3

FUNCTIONS—BASIC CONCEPTS

Why Study Functions?

The search for patterns is at the heart of every intellectual endeavor—in art history and music, as much as in computer science or mathematics. One of the principal ways of modeling patterns mathematically is through the concept of the function. In essence, a function consists of a set of conditions or inputs that can occur, a set of outcomes or outputs, and a rule that tells us what output to expect for what input. We will make these notions more precise shortly. Thus functions model diverse situations. For example, Newton's laws of motion can give us a function that tells us precisely how far an idealized baseball (with no air resistance) will fall in a given number of seconds when dropped from an idealized cliff (with no rocks or ledges to block its path). More complex versions of the same function may include air resistance, collisions with other objects, etc. A different function entirely will tell us how much of a given chemical will be produced by a chemical reaction involving different quantities of the reagents. Still another function will tell us how much time it takes to solve the traveling salesman problem for any number of cities using the brute force algorithm.

This book assumes that you have had experience with functions, so that most of the material in this section will consist of a review. Thus the main purpose of this section is to summarize the principal aspects of functions we will be using.

EXAMPLE 1 Fill in the next number in the following pattern without looking at the solution. Also, try to discover a rule for finding y from x.

x	y
1	0
2	3
3	8
4	15
5	

SOLUTION The next number is 24 and the rule is $y = x^2 - 1$. Another way to find the number is to note that the differences on the right are 3, 5, and 7. Hence one would expect a next difference of 9. Also, adding the current and previous x's and the previous y will give the new y (e.g., $24 = 5 + 4 + 15$). ∎

EXAMPLE 2 Try to predict the next y for these x's:

x	y
1	2
2	6
3	12
4	20
1	−7
5	

SOLUTION One would expect the next number to be 30 except for the −7 next to the second 1. The presence of two different outputs for a single input makes it impossible to discern a consistent pattern and hence impossible to predict what the next y will be. ∎

These two elementary examples illustrate the essence of the idea of a function. Example 1 is a function because for any single input we have one and only one output. Example 2 is not a function; one input may produce more than one output. Figure 2.10 illustrates a function. Note that each X goes to only one Y, although one Y may come from two or more X's.

Figure 2.10   A function from X to Y.

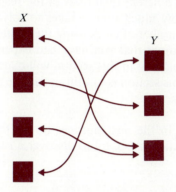

Cross Product Sets

Definition   Let X and Y be two sets. The set $X \times Y$ (the **cross product of X and Y**) is the set of all **ordered pairs (x, y)** where $x \in X$ and $y \in Y$. That is,

$$X \times Y = \{(x, y) \mid x \in X, y \in Y\}$$

The idea is that we are considering all possible pairs of elements where the first element of the pair has to be from X and the second has to be from Y.

EXAMPLE 3   Let $X = \{1, 2, 3\}$ and $Y = \{1, 2\}$. Find $X \times Y$.

SOLUTION $X \times Y = \{(1, 1), (1, 2), (2, 1), (2, 2), (3, 1), (3, 2)\}$ ∎

EXAMPLE 4   If $X = \mathbf{R}$ and $Y = \mathbf{R}$, what is $X \times Y$?

SOLUTION   In this case, the set $X \times Y$ is the ordinary Cartesian plane (the x-y coordinate system), familiar from coordinate geometry, repre-

Figure 2.11 **R** × **R**—the Cartesian plane.

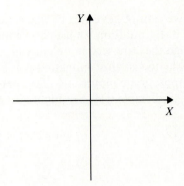

sented in Figure 2.11. The elements of this set are the ordinary points (x, y) in the plane. ∎

EXAMPLE 5 If $X = \mathbf{Z}$ and $Y = \mathbf{Z}$, find $X \times Y$.

SOLUTION $X \times Y$ consists of all possible pairs (x, y) of integers, for instance, $(1, 1)$, $(2, 3)$, $(-1, 7)$. Graphically, it can be represented by the intersections of the horizontal and vertical lines in Figure 2.12.

Figure 2.12 **Z** × **Z**. The intersections represent the elements of the set.

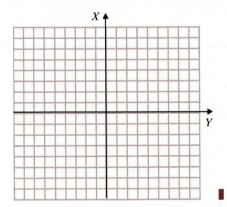

Functions

Definition Let X and Y be two sets. A **function** is a subset, f, of $X \times Y$ with the property that each $x \in X$ has one and only one $y \in Y$ associated with it.

In other words, we cannot have two *y*'s associated with the same *x*, as we had in Example 3. By tradition we usually denote functions by lowercase letters even though they are sets, not elements.

A common way to visualize a function is as a "black box"—a machine that transforms appropriate inputs into corresponding appropriate outputs in an absolutely predictable way. See Figure 2.13.

Figure 2.13 One way to visualize a function.

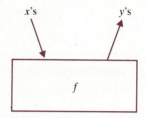

EXAMPLE 6 Let $X = Y = \{1, 2, 3, 4\}$. Is the set $\{(1, 2), (2, 3), (3, 4), (4, 1)\}$ a function?

SOLUTION   Yes. Each *x* has one and only one *y* associated with it. ∎

EXAMPLE 7 Let $X = Y = \{1, 2, 3, 4, 5, 6\}$. Is the set $f = \{(1, 3), (2, 5), (4, 2), (1, 5), (5, 6), (6, 3)\}$ a function?

SOLUTION   It is not, for two reasons. There is no *y* associated with an *x* of 3, and there are two distinct *y*'s associated with 1. ∎

EXAMPLE 8 Suppose $x = \mathbf{R}$ and $Y = \mathbf{R}$. Let $y = 3x + 2$. Is this a function?

SOLUTION   This is a function, since both conditions are met. First, there is a *y* for every $x \in \mathbf{R}$, namely, $3x + 2$. Second, $3x + 2$ establishes a rule that specifies a unique *y* value associated with every *x*. ∎

The function in Example 8 establishes a set $F = \{(x, y) \mid y = 3x + 2\}$. This is usually abbreviated as $f(x) = 3x + 2$. This specifies the value of *y* for any *x*. If we want to find the value of *y* corresponding to some particular *x*, say $x = 2$, we use **substitution**, i.e., we write $f(2) = 3 \cdot 2 + 2 = 8$. Similarly, if $x = 4$, we write $f(4) = 3 \cdot 4 + 2$.

EXAMPLE 9 Let $\Sigma = \{0, 1\}$, $X = \Sigma^*$, and $Y = \mathbf{N}$. If $w \in \Sigma^*$, let $n = L(w)$, where $L(w)$ is the number of symbols in *w*. Is *L* a function?

SOLUTION L is a function, since both conditions are met. First, there is an n for every w in Σ, namely, its length. For instance, $L(001) = 3$ and $L(0101) = 4$. Second, the rule specifies a unique n value for every w, since any string has only one length. ∎

We will reserve notations such as $L(w)$ or $f(x)$ to refer only to the value of a function at a particular w or x, not to the entire function; L, f, and other such symbols will refer to the entire function.

Definitions If $f \subset X \times Y$ is a function, we call X the **domain** of the function and Y the **codomain**. $\{ f(x) \in Y \mid x \in X \}$ is called the **range**.

That is, the codomain is all the possible values for $f(x)$; the range is the ones that actually occur. Some books do not make the distinction between codomain and range; hence they use the word "range" for what we are calling the "codomain."

Definitions If $y = f(x)$ where f is a function, a formula relating each x and its corresponding y is sometimes called the **rule**. Also, x is called the **independent variable** and y is called the **dependent variable**. We will often write $f: X \rightarrow Y$ to denote a function, its domain, and its codomain. $f: X \rightarrow Y$ is read "f is a function from X to Y" or "f maps X to Y." If X and Y are both $\mathbf{R}$ (or $\mathbf{Z}$ or other sets that can be drawn in a coordinate system), the **graph** of a function is $\{(x, f(x)) \mid x \in X\}$.

Other names used in place of "function" are **transformation, mapping,** or **map.** For our purposes, they mean the same thing, although in more advanced mathematics "transformations" often refers to a special type of function.

EXAMPLE 10 Identify the domain, codomain, independent variables and dependent variables for Examples 6, 8, and 9.

SOLUTION For Example 6, the domain, codomain, and range are the same, namely, $[1, 2, 3, 4]$. The independent variable is x and the dependent variable is y. For Example 8, the domain, codomain, and range are again the same, namely, $\mathbf{R}$. The independent variable is again x and the dependent variable is again y. In Example 9, the domain is Σ^* and the codomain and range are both $\mathbf{N}$. The independent variable is w and the dependent variable is n. ∎

Vertical Line Test

An important way to test whether a particular subset of $\mathbf{R} \times \mathbf{R}$ is a function is via the **vertical line test.** The idea is to draw the graph of the function. If every vertical line that can be drawn through points in the domain meets the graph at one and only one point, the graph represents a function.

EXAMPLE 11   Let $X = Y = \mathbf{R}$. Consider the function given by the rule $y = 3x - 2$. Sketch its graph and show that it is a function by the vertical line test.

SOLUTION   See Figure 2.14.

Figure 2.14   The vertical line test applied to the function $y = 3x - 2$.

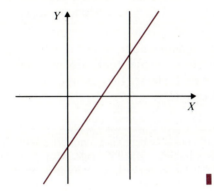

Some Common Functions

There are a number of familiar functions that are of particular importance in the analysis of algorithms, a topic we will begin to consider in Chapter 4. We will assume that you have had previous experience with these functions. The functions and their graphs are summarized in Table 2.1, assuming that both the domain and the codomain are $\mathbf{R}$. Two other classes of functions, exponential and logarithmic functions, are especially important and will be discussed in Section 2.5.

If the domain and codomain of the functions listed in Table 2.1 had been $\mathbf{Z}$ instead of $\mathbf{R}$, the graphs would have been quite different. Table 2.2 illustrates two of them. Note that the graph consists only of the points indicated by the heavy dots. Such functions are called **discrete functions,** unlike those in Table 2.1, which are **continuous functions.** The function from Σ^* to $\mathbf{N}$ in Example 9 is also an example of a discrete function.

TABLE 2.1 Some Important Functions

Name	Equation	Graph
Constant function	$f(x) = c$	
Identity function	$f(x) = x$	
Linear function	$f(x) = ax + b$	
Quadratic function (or parabola)	$f(x) = ax^2 + bx + c$	
Cubic function	$f(x) = x^3$	
Square root function	$f(x) = \sqrt{x}$	
Cube root function	$f(x) = \sqrt[3]{x}$	
Reciprocal function	$f(x) = \dfrac{1}{x}$	

TABLE 2.2   Some Important Functions on Z

Name	Equation	Graph
Constant function	$f(x) = c$	
Identity function	$f(x) = x$	

Piecewise Definition

Three functions require some extra attention.

EXAMPLE 12   The **absolute value function**, $y = |x|$, is defined as follows:

$$|x| = \begin{cases} x & x \geq 0, \\ -x & x \leq 0. \end{cases}$$

Sketch its graph.

SOLUTION   This is our first example of a **piecewise definition of a graph**, i.e., the rule defining how to obtain $f(x)$ from x is different on different pieces of the domain. Such functions occur frequently. The graph of $|x|$ is in Figure 2.15.

Figure 2.15 $y = |x|$

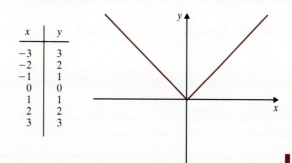

x	y
-3	3
-2	2
-1	1
0	0
1	1
2	2
3	3

EXAMPLE 13   The **floor function,** $y = \lfloor x \rfloor$, (also called the **greatest integer function**) is defined as the greatest integer less than or equal to x. Thus if a positive, real number x is written in decimal form:

$$x = n \cdot d_1 d_2 d_3 \ldots,$$

then

$$\lfloor x \rfloor = n.$$

That is, for positive real numbers, $\lfloor x \rfloor$ is the result of dropping the decimal part of x. For negative real numbers that are not integers, it is the result of dropping the decimal part and then subtracting 1. Sketch the graph of the floor function.

SOLUTION   See Figure 2.16.

Figure 2.16 $y = \lfloor x \rfloor$

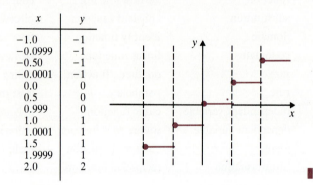

x	y
-1.0	-1
-0.0999	-1
-0.50	-1
-0.0001	-1
0.0	0
0.5	0
0.999	0
1.0	1
1.0001	1
1.5	1
1.9999	1
2.0	2

EXAMPLE 14   Let $x \in \mathbf{R}$. The **ceiling function** (or **least integer function**), denoted $y = \lceil x \rceil$ is defined as the least integer greater than or equal to x. Thus $\lceil 3.5 \rceil = 4$ and $\lceil 3.00001 \rceil = 4$. Note that we cannot always find the ceiling of x by adding 1 to the floor of x. If x is an integer, $\lfloor x \rfloor = x = \lceil x \rceil$, whereas if x is not an integer, $\lceil x \rceil = 1 + \lfloor x \rfloor$. Graph this function.

SOLUTION   (Left for the exercises.)

Two other classes of functions that come up on occasion are **polynomial functions**—functions of the form $a_n x^n + a_{n-1} x^{n-1} + \cdots + a_1 x + a_0$—and **rational functions**—functions that are a ratio of polynomial functions. We will treat these functions when the need arises.

Multivariate Functions

So far we have looked only at functions with one input and one output. Most realistic situations, though, depend on more than one factor. Functions can be used to model such multi-input and multi-output situations.

For instance, $z = f(w_1, w_2)$ may denote the concatenation function, and $f(w_1, w_2)$ is $w_1 w_2$, the string consisting of the symbols in w_1 followed by the symbols in w_2. Its domain is $\Sigma^* \times \Sigma^*$, its codomain is Σ^*, and for each pair of strings, (w_1, w_2), there is one and only one z. An extensive coverage of **multivariate functions** on the set of real numbers can be found in most calculus texts.

TERMINOLOGY

cross product	mapping	discrete function
ordered pair	map	absolute value function
function	vertical line test	continuous function
substitution	constant function	piecewise definition of a function
domain	identity function	floor function
codomain	linear function	greatest integer function
range	quadratic function	ceiling function
rule	parabola	least integer function
independent variable	cubic function	polynomial function
dependent variable	square root function	rational function
graph	cube root function	multivariate function
transformation	reciprocal function	

SUMMARY

1. Let X and Y be two sets. The set $X \times Y$ is the cross product of X and Y and is the set of all ordered pairs of elements (x, y) where $x \in X$ and $y \in Y$. That is,

$$X \times Y = \{(x, y) \mid x \in X \text{ and } y \in Y\}.$$

2. A function is a subset, f, of $X \times Y$ with the property that each $x \in X$ has one and only one $y \in Y$ associated with it.

3. If $X = \mathbf{R}$ and $Y = \mathbf{R}$, a subset of $X \times Y$ may be checked by the vertical line test to see if it is a function.

4. Some important functions are the constant function, the identity function, the linear function, the quadratic function, the cubic function, the polynomial function, square and cube root functions, the reciprocal function, the absolute value function, and the floor and ceiling functions.

5. Although most functions we draw here will be drawn as continuous functions, most functions we deal with have domains that are discrete sets.

EXERCISES 2.3

1. Find the next value of y for this function. Find a rule relating y to x.

x	y
0	-1
1	0
2	7
3	26
4	

2. Find the next two values of y for this function. Again, find a rule relating y to x.

x	y
0	0
1	1
2	3
3	7
4	
5	

In Exercises 3–6, find the cross product of sets A and B.

3. $A = \{1, 2, 3\}$, $B = \{1, 2\}$

4. $A = \{s, t\}$, $B = \{a, e, i, o, u\}$

5. $A = \{1, 2, 3\}$, $B = \varnothing$

6. $A = \varnothing$, $B = \{a, b\}$

Let $A = B = \{1, 2, 3, 4\}$. Tell whether each of the following subsets of $A \times B$ is a function and why.

7. $\{(1, 1), (2, 2), (3, 3), (4, 4)\}$

8. $\{(1, 4), (2, 3), (2, 2), (4, 1)\}$

9. $\{(2, 3), (2, 4), (3, 4), (4, 4)\}$

10. $\{(1, 1), (2, 1), (3, 1), (4, 1)\}$

For each of the following functions, let $A = B = \mathbf{Z}$.

11. If $f(x) = x^2 - 1$, find

   a) $f(0)$ b) $f(1)$ c) $f(-1)$ d) $f(a)$

   e) $f(x + 1)$ f) $f(x^2)$ g) $f(x) + f(y)$ h) $f(2x^2 - 1)$

12. If $f(n) = 2n - 3$, find

   a) $f(0)$ b) $f(1)$ c) $f(-1)$ d) $f(a)$

   e) $f(x + 1)$ f) $f(x^2)$ g) $f(x) + f(y)$ h) $f(2x^2 - 1)$

For each of the following rules, assume that the domain and codomain are $\mathbf{R}$. Identify the range, the independent variable, and the dependent variable.

13. $y = 3x + 1$ 14. $g = r^2$ 15. $h = u^2 - 3$

In Exercises 16–19, a string function that operates on strings of 0's and 1's is given. Find its domain, codomain, and range.

16. L shifts a string one place to the left. The leftmost symbol is lost.

17. R shifts strings one place right. The leftmost symbol becomes a zero.

18. P acts on strings that begin with the symbols 101. It removes these symbols.

19. S adds the symbols 11011 to the end of any string.

For each of the following relationships, set up a table of values, sketch the curve, and decide whether it is a function by the vertical line test.

20. $y = 2x - 1$

21. $x = y^2$

22. $y = x^3/3$

For each of the following, sketch the curve and state the domain and range.

23. $y = 2 - x$

24. $y = -x^2$

25. $y = 100$

In the following exercises, you are given some experimental data that contain errors. Plot the data and infer a function that approximately fits them.

26. $(-3, -6.8)$, $(-3, -6.6)$, $(-2, -5.0)$, $(-2, -5.2)$, $(-1, -2.6)$, $(-1, -3.1)$, $(0, -1.9)$, $(0, -0.8)$, $(1, 1.1)$, $(1, 1.2)$, $(2, 2.9)$, $(2, 3.5)$, $(3, 3.9)$, $(3, 2.9)$

27. $(-3, 3)$, $(-3, 2.5)$, $(-3, 3.1)$, $(-2, 1.8)$, $(-2, 2.5)$, $(-1, 1)$, $(-1, 1)$, $(-1, 0.5)$, $(0, 0.5)$, $(0, 0.1)$, $(0, -0.2)$, $(1, 1.8)$, $(1, 0.6)$, $(2, 2.0)$, $(2, 2.1)$, $(3, 3.2)$, $(3, 2.9)$

28. $(-3, 4.1)$, $(-2, 4.3)$, $(-1, 3.9)$, $(0, 4.0)$, $(1, 3.7)$, $(2, 4.1)$, $(3, 4.1)$

In the following situations, find a function that fits the observations.

29. The distance an objects falls under the influence of gravity is observed to be proportional to the square of the time passed since it was released.

30. A computer program processes a large quantity of data. It is observed that the time it takes to run is directly proportional to the amount of data it is given.

For each of the following exercises, graph the given data and infer a function that fits them.

31.

x	y
2	Undefined
3	0
4	1
5	1.41
6	1.73
7	2

32.

x	y
0	0
0.4	0
0.8	0
1.2	1
1.6	1
2.0	2
2.4	2

33.

x	y
−27	−1.00
−8	−.667
−1	−.333
0	0.00
1	.333
8	.667
27	1.00

34.

x	y
−3	−.667
−2	−1.00
−1	−2.00
0	Undefined
1	2.00
2	1.00
3	.667

35.

x	y
−3	4
−2	3
−1	2
0	1
1	0
2	1
3	2

36.

x	y
−3	−3
−2	−2
−1	−1
0	0
1	3
2	2
3	1
4	0

37. Make a table of values for the floor function and sketch its graph.

In Exercises 38–46, find the range and domain of the given function and sketch the graph of each.

38. $f(x) = \sqrt{(x-1)}$

39. $P(x) = \sqrt[3]{(2x+1)}$

40. $r(x) = \dfrac{1}{3x}$

41. $b(z) = |z^2 - 1|$

42. $f(x) = \begin{cases} x^2 & x < 0 \\ -x^2 & x \geq 0 \end{cases}$

43. $k(m) = \lfloor 2m \rfloor$

44. $c(x) = \lceil \dfrac{x}{2} \rceil$

45. $f(x) = 1 + x^4$

46. $r(x) = \dfrac{x}{x-1}$

2.4
FURTHER PROPERTIES OF FUNCTIONS

So far we have looked at functions primarily in terms of finding the outputs that correspond to given inputs. But we often want to ask questions that look at the situation differently: Given the output, did it come from a single input? If so, what? To answer this question, we need to understand the concept of the inverse of a function. First, we need to know some preliminary concepts. These concepts will also enable us to clarify the notion of the size of a set.

Composite Functions

Definition   Suppose $f\colon X \to Y$ and $g\colon Y \to Z$. We define the **composite** as the function $g \circ f\colon X \to Z$, which is defined by the rule $(g \circ f)(x) = g(f(x))$.

This can be visualized as shown in Figure 2.17.

Figure 2.17 $g \circ f$ is the composite of g and f.

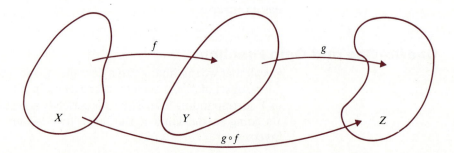

EXAMPLE 1   Let $X = Y = Z = \mathbf{R}$, let $f(x) = x^2$, and let $g(y) = \sqrt{(y-1)}$. Find $(g \circ f)(x)$.

SOLUTION $g(f(x)) = \sqrt{(x^2 - 1)}$.   ∎

EXAMPLE 2   Again, let $X = Y = Z = \mathbf{R}$, let $f(x) = \sqrt{x}$, and let $g(y) = y^2 + y + 1$. Find $(g \circ f)(x)$.

SOLUTION $g(f(x)) = (\sqrt{(x)})^2 + \sqrt{x} + 1 = x + \sqrt{x} + 1$.   ∎

Inverse Functions

Definition   Let $f: X \to Y$. By the **inverse of f** we mean a function, denoted $f^{-1}: Y \to X$, such that $f^{-1}(f(x)) = x$ for all $x \in X$ and such that $f(f^{-1}(y)) = y$ for all $y \in Y$.

In other words, if f takes x into y, f^{-1} brings that same y back to that same x. See Figure 2.18.

Figure 2.18 f and its inverse.

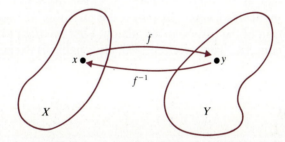

Unfortunately not every function has an inverse. For instance, if $f(x) = x^2 - 1$, $f(1) = 0$ and $f(-1) = 0$. Thus an inverse function of f (if there were one) would have to take 0 to both 1 and −1. But then it would not be a function. So some functions have inverses and others do not. The following concepts will enable us to decide whether a particular function has an inverse.

One-to-One and Onto Functions

Recall that we said that given a domain, X, and a codomain, Y, a function, f, is a subset of $X \times Y$ such that for every x in X there is one and only one y in Y corresponding to it. Thus it is perfectly acceptable to have two or more x's going to the same y. But such "two-to-one" functions do not have inverses.

EXAMPLE 3 Suppose $X = Y = \{1, 2, 3, 4\}$ and let f be defined by the rule $f(1) = 2, f(2) = 4, f(3) = 1$, and $f(4) = 2$. Then f is a function since each x goes to only one y, but it does not have an inverse since both $f(1)$ and $f(4)$ equal 2. See Figure 2.19. ∎

Figure 2.19 A simple function that has no inverse.

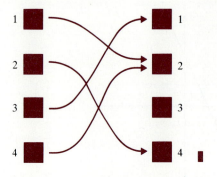

Definition A function, f, from domain X into codomain Y is **one-to-one** if for each y in Y there is at most one x such that $f(x) = y$. A function is **onto** if for each y in Y there is at least one x in X such that $f(x) = y$. That is, a function is onto if its codomain equals its range. If a function is both one-to-one and onto, it is a **one-to-one correspondence.**

The easiest way to check for one-to-one and onto for functions with domain **R** and codomain **R** is by the **horizontal line test.** If a horizontal line drawn through any point in the codomain meets the graph of the function in at most, one point, the function is one-to-one. If any such line meets the graph in at least one point, the function is onto. See Figure 2.20.

Figure 2.20 Use of the horizontal line test to show that f is one-to-one and onto.

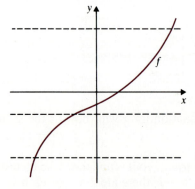

EXAMPLE 4   Let $X = Y = \mathbf{R}$. Let $f(x) = 2x + 3$. Is f one-to-one? Is it onto?

SOLUTION   By the horizontal line test, it is both one-to-one and onto; hence it is a one-to-one correspondence. See Figure 2.21.

Figure 2.21 $y = 2x + 3$ is one-to-one and onto.

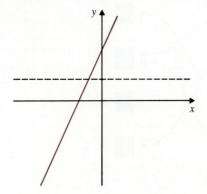

EXAMPLE 5   Let $X = \mathbf{R}$. Is the function $f(x) = x^2 - 1$ one-to-one and onto?

SOLUTION   This time, it is clear that the function is neither one-to-one nor onto. Values of y below -1 have no x's that correspond to them and values above -1 have two. See Figure 2.22.

Figure 2.22 $y = x^2 - 1$ is neither one-to-one nor onto.

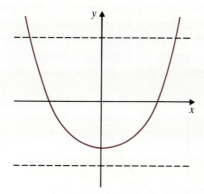

Solving the equation for x will show us the same thing:

$$x = \pm\sqrt{(y + 1)}$$

If $y < -1$, the square root is undefined, so there are no x's corresponding to those y's. If $y > -1$, there are 2 x's for each y. ∎

EXAMPLE 6 Is $y = x^3$ one-to-one and onto?

SOLUTION Yes. Again, we can apply the horizontal line test. See Figure 2.23.

Figure 2.23 $y = x^3$ is one-to-one and onto.

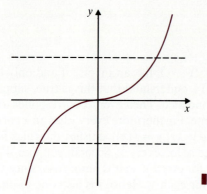

The notion of one-to-one correspondence provides a way for us to define formally the size of a set, a notion mentioned intuitively in Section 2.1.

Definition Let S be a set. Then S **has n elements** or has **cardinality** n if there exists a one-to-one correspondence between S and the set $\{1, \ldots, n\}$ for some $n \in \mathbf{N}$. If such a one-to-one correspondence exists, S is **finite;** if S is not finite, it is **infinite.** The cardinality of a set is denoted $|S|$.

EXAMPLE 7 Let $\Sigma = \{0, 1\}$, $S_2 = \{0, 1, 00, 01, 10, 11\}$, $S_3 = \{0, 1, 00, 01, 10, 11, 000, 001, 010, 011, 100, 101, 110, 111\}$. Then

$$|\Sigma| = 2,$$
$$|S_2| = 6,$$
$$|S_3| = 14,$$
$$\Sigma^* \text{ is infinite.} \quad \blacksquare$$

Note that **N, Z, Q,** and **R** are all infinite. However, it can be shown that there are one-to-one correspondences between **N** and **Z, N** and **Q,** and **Z** and **Q,** but *not* between **R** and any of the other three. Thus we cannot simply use the symbol ∞ to denote the cardinality of any infinite set. The subject of infinite cardinal numbers is an important and interesting subject in its own right, but we will not pursue it further here. (See, for instance, Lewis and Papadimitriou, 1981, sections 1.6 and 1.7.)

Note also that if $|A| = n$ and $|B| = n$, then both A and B are in one-to-one correspondence with $\{1, \ldots, n\}$. Thus there is a one-to-one correspondence between A and B. That is, two finite sets have the same number of elements if and only if there is a one-to-one correspondence between them.

Inverse Functions

A function $f: X \to Y$ has an inverse if and only if it is a one-to-one correspondence. To understand that this is true, suppose first that f is a one-to-one correspondence. Then every x corresponds to *one and only one y,* since it is one-to-one. Furthermore every y has an x corresponding to it, since it is onto. Thus $\{(y, x) \mid y = f(x)\}$ satisfies the rules for being a function—one and only one x for every y. Second, suppose that f has an inverse. Then there is an x for every y, so it is onto. Also, since f^{-1} is a function, there is at most one x for each y. Hence f is also one-to-one.

EXAMPLE 8 If $f(x) = 2x + 3$, find f^{-1} and verify that it is the inverse of f.

SOLUTION Let $y = 2x + 3$. Then $x = f^{-1}(y) = (y - 3)/2$. We can verify that this function is the inverse of f by substitution:

$$f(f^{-1}(y)) = 2f^{-1}(y) + 3 = 2((y - 3)/2) + 3 = y - 3 + 3 = y$$

and

$$f^{-1}(f(x)) = (f(x) - 3)/2 = (2x + 3 - 3)/2 = 2x/2 = x. \quad \blacksquare$$

EXAMPLE 9 If $f(x) = x^3$, find f^{-1} and verify the answer.

SOLUTION If $y = x^3$, $x = f^{-1}(y) = \sqrt[3]{y}$. Again, verify by substitution:

$$f^{-1}(f(x)) = \sqrt[3]{f(x)} = \sqrt[3]{(x^3)} = x$$

and

$$f(f^{-1}(y)) = (f^{-1}(y))^3 = (\sqrt[3]{y})^3 = y. \quad \blacksquare$$

EXAMPLE 10 Let $f: \mathbf{Z}_5 \to \mathbf{Z}_5$ be defined as follows:

$$f(0) = 1, f(1) = 2, f(2) = 3, f(3) = 4, f(4) = 0.$$

Show that f has an inverse and find it.

SOLUTION f is one-to-one and onto, since each element in $\mathbf{Z}_5$ has one and only one element mapped to it. Thus f has an inverse. It is defined as follows:

$$f^{-1}(0) = 4, f^{-1}(1) = 0, f^{-1}(2) = 1, f^{-1}(3) = 2, f^{-1}(4) = 3. \quad \blacksquare$$

Conclusion

Why have we done all this? Functions are powerful tools partly because there are many familiar functions that arise frequently. If a function is one-to-one and onto, everything we know about functions is available to help us understand its inverse. If it is not one-to-one or not onto, we cannot invert it and get a function. Although this reduces considerably what we can do in such situations, we do have some mathematical tools that will help us. The principal one is the concept of the relation, which we will be examining in Chapter 8.

TERMINOLOGY

composite of g and f horizontal line test

inverse of f has n elements

one-to-one function   cardinality

onto function   finite set

one-to-one correspondence   infinite set

SUMMARY

1. A function, f^{-1}, is called the inverse of a function f if it has the property that $f^{-1}(f(x)) = x$ for all x's in the domain of f and such that $f(f^{-1}(y)) = y$ for all y's in the codomain of f.

2. Not all functions have inverses; a function has an inverse if and only if it is a one-to-one correspondence.

3. A set is said to have n elements if it can be put into one-to-one correspondence with the set $\{1, \ldots, n\}$ for some $n \in \mathbf{N}$.

EXERCISES 2.4

For each of the following pairs of functions, find $g \circ f$ and $f \circ g$.

1. $f(x) = x - 3$, $g(x) = |x|$
2. $f(x) = 2x$, $g(x) = |x|$
3. $f(x) = x/2$, $g(x) = x^2$
4. $f(x) = x^2$, $g(x) = \sqrt{x}$
5. $f(x) = \lfloor x \rfloor$, $g(x) = 2x$
6. $f(x) = \sqrt{(x - 1)}$, $g(x) = 1/x$

For each of the following problems, sketch f, g, and $g \circ f$.

7. Exercise 1   8. Exercise 2

9. Exercise 3   10. Exercise 4

11. Exercise 5   12. Exercise 6

Sketch a graph of each of the following functions and use the horizontal line test to tell whether it is a one-to-one correspondence. If not, explain why not.

13. $\{(1, 1), (2, 2) (3, 3), (4, 4)\}$ 14. $\{(1, 3), (2, 3), (3, 3), 4,3)\}$

15. $f(x) = x^2 + 1$, $x \in \mathbf{R}$ 16. $f(x) = 6$, $x \in \mathbf{R}$

17. $g(x) = |x|$, $x \in \mathbf{Z}$

18. $n(z) = \begin{cases} z & z < 0, z \in \mathbf{Z} \\ 2z + 1 & z \geq 0 \end{cases}$

For the next four exercises, assume that both the domain and the codomain of f are $\{x \in \mathbf{Z} \,|\, x > 0\}$. Decide if these are one-to-one correspondences.

19. $f(x) = 1/x$ 20. $f(x) = \sqrt{x}$

21. $f(x) = x^2$ 22. $f(x) = |x|$

For each of the following functions, decide if it has an inverse, and if it does, find it. If not, explain why not. Assume the domain in $\mathbf{R}$.

23. $h(x) = x^3/3$

24. $n(x) = |x - 1|$

25. $f(x) = \begin{cases} x^2 & x < 0, \\ -x & x \geq 0 \end{cases}$

26. $g(x) = \lceil x + 1 \rceil$

27. $f \colon \mathbf{Z}_6 \to \mathbf{Z}_6$ such that $f(z) = \begin{cases} z + 3 & z < 3, \\ z - 3 & z \geq 3 \end{cases}$

28. $f \colon \{1, 2, 3, 4, 5\} \to \{a, b, c, d, e\}$
    where $f(1) = c, f(2) = a, f(3) = c, f(4) = b, f(5) = d$

For each of the following, find the inverse and verify that it is the inverse.

29. Exercise 19   30. Exercise 20

31. $b(x) = 2x - 3$ 32. $s(x) = 3\sqrt[3]{x}$

Find the cardinality of the following sets.

33. $S = \{w \in \Sigma^* \,|\, \text{length of } w \leq 2\}$, $\Sigma = \{0, 1, 2\}$

34. $S = \{w \in \Sigma^* \,|\, \text{length of } w \leq 3\}$, $\Sigma = \{0, 1, 2\}$

35. $P(S)$ where $S = \{0, 1\}$

36. $P(S)$ where $S = \{0, 1, 2\}$

2.5

EXPONENTIAL AND LOGARITHM FUNCTIONS

There are two functions that arise particularly often in models—the exponential and logarithm functions. The exponential function is used to model processes of growth and decay, which are widely studied in all branches of science. It is also extremely important in computer science because of its role in the analysis of algorithms. The logarithm function is the inverse of the exponential function; hence it is present whenever the exponential occurs. Because of their special importance and because students so often have difficulty with them, we will examine them in this section.

The Exponential Function

Definition Let $X = Y = \mathbf{R}$ and let $f(x) = a^x$ for some real number $a > 0$. Then $f(x)$ is called the **exponential function with base a.**

Before looking at some graphs of the exponential function, there are two properties of it that we need to examine:

E1: $a^0 = 1$

E2: $a^{-n} = (1/a^n)$

These are not theorems to be proven but notational conventions that were adopted because they are convenient. Consider E1. The following relationships are well known.

$$a^1 = a$$
$$a^2 = a \cdot a$$
$$a^3 = a \cdot a \cdot a$$
$$\text{etc.}$$

That is, a^n means n a's multiplied by each other. But it is not at all obvious what zero a's multiplied times each other would mean. One way to define it is to notice that $a^1 = a^2/a$, $a^2 = a^3/a$, etc. Following the same pattern, one would expect the relationship

$$a^0 = a^1/a.$$

But $a^1 = a$ and hence, for consistency, a^0 is most naturally defined to be 1. Similarly

$$a^{-1} = a^0/a = 1/a,$$
$$a^{-2} = a^{-1}/a = 1/(a^2).$$

Continuing in this same pattern, we can also understand the rationale for E2.

EXAMPLE 1 Suppose $f(x) = 2^x$. Set up a table of values for $f(x)$ and sketch its graph.

SOLUTION   See Figure 2.24.

Figure 2.24   Graph of $y = 2^x$.

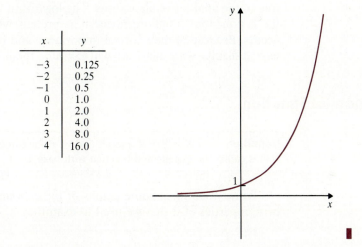

x	y
-3	0.125
-2	0.25
-1	0.5
0	1.0
1	2.0
2	4.0
3	8.0
4	16.0

EXAMPLE 2   Suppose $f(x) = 3^x$. Set up a table for f and sketch its graph.

SOLUTION   See Figure 2.25.

Figure 2.25   Graph of $y = 3^x$.

x	y
-3	0.037
-2	0.111
-1	0.333
0	1.0
1	3.0
2	9.0
3	27.0
4	81.0

EXAMPLE 3 Suppose $f(x) = (1/2)^x$. Set up a table for f and sketch its graph.

SOLUTION   See Figure 2.26.

Figure 2.26 Graph of $y = \left(\dfrac{1}{2}\right)^x$.

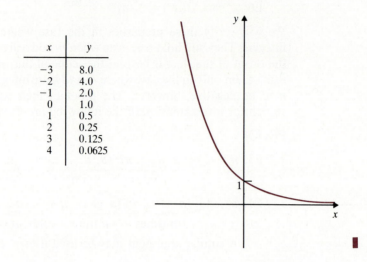

x	y
-3	8.0
-2	4.0
-1	2.0
0	1.0
1	0.5
2	0.25
3	0.125
4	0.0625

EXAMPLE 4 Sketch graphs of $f_0(x) = (1/2)^x$, $f_1(x) = 1^x$, $f_2(x) = 2^x$, $f_3(x) = 3^x$, and $f_4(x) = 4^x$ on the same set of axes.

SOLUTION   Note first that $1^x = 1$ for any x. See Figure 2.27.

Figure 2.27 Comparison of graphs of exponential functions.

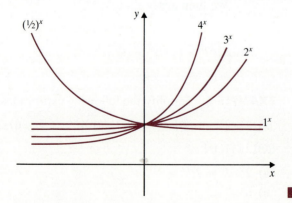

Further Properties of the Exponential Function

The following properties of the exponential function are used frequently.

E3: $a^m \cdot a^n = a^{m+n}$

E4: $a^m/a^n = a^{m-n}$

E5: $(a^m)^n = a^{mn}$

E6: $a^{m/n} = \sqrt[n]{(a^m)} = (\sqrt[n]{a})^m$

We will verify these properties in the case where m and n are positive integers. They still hold true even when m and n are real numbers, although the proof of that fact is beyond the scope of this book. In fact, we will be using them only in the case where m and n are integers; we will allow m and n to be negative, however. The proof of their validity in this case is a straightforward extension of the following proofs using property E2.

PROOF

E3: $a^m \cdot a^n = \underbrace{(a \cdot a \ldots a)}_{m \text{ times}} \cdot \underbrace{(a \cdot a \ldots a)}_{n \text{ times}} = \underbrace{a \cdot a \ldots a}_{m + n \text{ times}} = a^{m+n}$

E4: $a^m/a^n = \underbrace{(a \cdot a \ldots a)}_{m \text{ times}}/\underbrace{(a \cdot a \ldots a)}_{n \text{ times}} = \underbrace{a \cdot a \ldots a}_{m - n \text{ times}} = a^{m-n}$ if $m > n$

A similar argument may be used if $m \leq n$.

E5: $(a^m)^n = \underbrace{a^m \cdot a^m \ldots a^m}_{n \text{ times}} = \underbrace{\underbrace{(a \ldots a)}_{m \text{ times}} \cdot \underbrace{(a \ldots a)}_{m \text{ times}} \ldots \underbrace{(a \ldots a)}_{m \text{ times}}}_{n \text{ times}} = a^{mn}$

E6: First, we define $a^{1/n}$ to be $\sqrt[n]{a}$. We can justify this definition as follows. E5 was proven for integers. But we want a definition of $a^{1/n}$ that allows us to use E5 for rational numbers as well. Thus we must have $(a^{1/n})^n = a^{(1/n)n} = a^1 = a$. But if $x^n = a$, then $x = \sqrt[n]{a}$. Thus $a^{1/n} = \sqrt[n]{a}$.

We then apply E5 to $a^{m/n}$:

$$a^{m/n} = (a^m)^{1/n} = \sqrt[n]{(a^m)}.$$

Also, $a^{m/n} = (a^{1/n})^m = (\sqrt[n]{a})^m$. ☐

EXAMPLE 5 Simplify the following expressions.

a) $2^5 \cdot 2^2$ **b)** 6^0 **c)** $4^{3/2}$ **d)** $3^{3x}/3^x$ **e)** $(2^x)^x$

SOLUTION

a) $2^5 \cdot 2^2 = 2^{5+2} = 2^7$

b) $6^0 = 1$

c) $4^{3/2} = (\sqrt{4})^3 = 2^3 = 8$

d) $3^{3x}/3^x = 3^{3x-x} = 3^{2x}$

e) $(2^x)^x = 2^{x \cdot x} = 2^{x^2}$ ∎

EXAMPLE 6 Show that $(1/2)^n = 2^{-n}$ and that $(a^2)^{3/2} = a^3$.

SOLUTION

$$(1/2)^n = (2^{-1})^n = 2^{-n}$$
$$(a^2)^{3/2} = a^{2(3/2)} = a^3 \quad ∎$$

EXAMPLE 7 Here's an ancient problem that illustrates how rapidly exponential functions grow.

A certain young man pleased his king greatly, and the king asked him what reward he would like to receive. He replied that he would be more than satisfied if the king took a chess board and on the first day placed one grain of rice on one of the squares. On the second day, the king would place two grains on another square. On the third day, he would place four grains on a third square, the fourth day eight on a fourth square, etc., until he had given him rice for each square on the board, that is, for 64 days. Thinking that the young man had asked very little, the king readily agreed. Estimate the amount of rice the young man would receive on the 64th day if the king were able to keep his end of the bargain.

SOLUTION   Let $f(n)$ denote the number of grains the young man is to receive on day n. Then $f(1) = 1, f(2) = 2, f(3) = 4, f(4) = 8$, etc. That is,

$$f(2) = 2 \cdot 1 = 2^1$$
$$f(3) = 2 \cdot f(2) = 2 \cdot 2 \cdot 1 = 2^2$$
$$f(4) = 2 \cdot f(3) = 2 \cdot 2 \cdot 2 \cdot 1 = 2^3$$
$$\text{etc.}$$

Since $f(n)$ continues to double each time, we can infer that

$$f(n) = 2^{n-1}$$

and hence $f(64) = 2^{63}$.

Now, to evaluate 2^{63}, we can use properties E1–E6 to make some reasonable estimates quite quickly:

$$2^{63} = 2^{60} \cdot 2^3 = (2^{10})^6 \cdot 2^3$$
$$2^{10} = 1024, \quad 1000 = 10^3$$

Hence

$$2^{63} \approx (10^3)^6 \cdot 8 = 8 \cdot 10^{18}.$$

Using an estimate of 50 grains of rice per gram, this is roughly equal to 1.6×10^{11} metric tons of rice, an amount that substantially exceeds the entire annual world production of rice. ∎

The Logarithm Function

Applying the horizontal line test to the preceding graphs, we can see that the exponential function is one-to-one as long as a is not 1. Also, if we look at it as a function from **R** to **R**⁺ it is onto, since with this codomain every y value has a corresponding x value. Thus, using domain **R** and codomain **R**⁺, the exponential function is a one-to-one correspondence and hence has an inverse. The inverse of the exponential function with base a is the **logarithm function with base a,** denoted $y = \log_a x$. We will usually refer to it simply as the *log function*. Recall that for any inverse function $f(f^{-1}(x)) = x$ and $f^{-1}(f(x)) = x$. These give us the following two properties of the exponential and log functions.

I1: $a^{(\log_a x)} = x$

I2: $\log_a(a^x) = x$

EXAMPLE 8   Find the following.

a) $\log_2 8$

b) $\log_3 81$

c) $\log_a(a)$

d) $2^{(\log_2 3)}$

e) $5^{(\log_5(x+1))}$

SOLUTION

a) $\log_2 8 = \log_2(2^3) = 3$

b) $\log_3 81 = \log_3(3^4) = 4$

c) $\log_a(a) = \log_a(a^1) = 1$

d) $2^{(\log_2 3)} = 3$

e) $5^{(\log_5(x+1))} = x + 1$ ∎

EXAMPLE 9   Sketch $y = \log_a x$ for several values of a.

SOLUTION   All logarithms we will see have $a > 1$. So we will sketch only for $a = 2$, 3, and 4. See Figure 2.28.

Figure 2.28 Comparison of log functions.

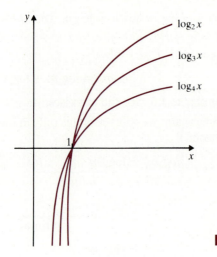

Just as the principal characteristic of the exponential function is that it grows very rapidly, the principal characteristic of the log function is that it grows very slowly. Another way of looking at the difference between the exponential and log functions is this: $y = a^n$ means that 1 must be multiplied by a n times to get y. Similarly, if n is a power of a, $y = \log_a n$ means that n must be divided by a y times to get 1. If n is not a power of a, $\lceil y \rceil$ is the number of times n must be divided by a to get a number less than 1. For instance, if $n = 8$, $\log_2 n = 3$, since $8/2^3 = 1$. If $n = 7$, $\lceil \log_2 7 \rceil = 3$, since $7/2^3 < 1$.

Properties of Log Functions

Following are the basic properties of the log function.

L1: $\log_a 1 = 0$
L2: $\log_a a = 1$
L3: $\log_a mn = \log_a m + \log_a n$
L4: $\log_a(m/n) = \log_a m - \log_a n$
L5: $\log_a(m^n) = n \log_a m$

The arguments for verifying these properties depend on applying the concept that the log function is the inverse of the exponential function and using an appropriate property of the exponential function.

PROOF

L1: Let $\log_a 1 = x$. Then $a^x = 1$. Hence $x = 0$.
L2: Already done in Example 4.

L3: Let $x = \log_a m$ and $y = \log_a n$. Then $m = a^x$ and $n = a^y$.

Hence $mn = a^{x+y}$ and therefore $x + y = \log_a(mn)$. Combining these terms, we get

$$\log_a mn = \log_a m + \log_a n.$$

L4: Similar to L3. See the exercises.

L5: Once again, we will verify this only in the case where n is a positive integer.

$$log_a m^n = \underbrace{\log_a(m \cdot m \ldots m)}_{n \text{ times}}$$

$$= \underbrace{\log_a m + \log_a m + \cdots + \log_a m}_{n \text{ times}}$$

$$= n\log_a m \quad \square$$

EXAMPLE 10   Suppose $\log_2 3 = 1.585$ and $\log_2 5 = 2.322$. Find the following.

a) $\log_2 6$ b) $\log_2 1.6$ c) $\log_2 0.5$

SOLUTION

a) $\log_2 6 = \log_2 2 \cdot 3 = \log_2 2 + \log_2 3 = 1 + 1.585 = 2.585$
b) $\log_2 1.6 = \log_2(8/5) = \log_2 8 - \log_2 5 = 3 - 2.322 = 0.678$
c) $\log_2 0.5 = \log_2(1/2) = \log_2 2^{-1} = -1$ ∎

EXAMPLE 11   Using $\log_2 3 = 1.585$ and $\log_2 5 = 2.322$, find $\log_3 2$ and $\log_3 5$.

SOLUTION

Let $x = \log_3 2$. Then $3^x = 2$.
$x\log_2 3 = \log_2 2 = 1$
$x = 1/(\log_2 3) = 1/1.585 = 0.631$
Now let $x = \log_3 5$. Then
$3^x = 5$
$x\log_2 3 = \log_2 5$
$x = (\log_2 5)/(\log_2 3) = 2.322/1.585 = 1.465$ ∎

EXAMPLE 12   Many calculators have a button marked LN for computing the natural logarithm, $\log_e(x)$, where $e = 2.71828 \ldots$ Show how to compute $\log_2(x)$ on a calculator that has an LN button but no other way to compute logarithms.

SOLUTION   Let x be a number for which we want to find $\log_2(x)$. Let y denote $\log_2(x)$. Then

$$y = \log_2(x)$$
$$x = 2^y$$
$$\log_e(x) = \log_e(2^y)$$
$$\log_e(x) = y \log_e 2$$
$$y = \log_e(x)/\log_e 2$$
$$\log_2(x) = \frac{LN(x)}{LN(2)} \quad \blacksquare$$

TERMINOLOGY

exponential function with base a

log function with base a

SUMMARY

1. The properties of the exponential function with base a are

   E1: $a^0 = 1$ E4: $a^m/a^n = a^{m-n}$

   E2: $a^{-n} = 1/a^n$ E5: $(a^m)^n = a^{mn}$

   E3: $a^m \cdot a^n = a^{m+n}$ E6: $a^{m/n} = (\sqrt[n]{a})^m = \sqrt[n]{(a^m)}$

2. The properties of the function $\log_a x$ are

   L1: $\log_a 1 = 0$ L4: $\log_a(m/n) = \log_a m - \log_a n$

   L2: $\log_a a = 1$ L5: $\log_a(m^n) = n \log_a m$

   L3: $\log_a mn = \log_a m + \log_a n$

3. There are two properties relating the exponential and log functions:

   I1: $a^{(\log_a x)} = x$ I2: $\log_a(a^x) = x$

EXERCISES 2.5

For each of the following functions, set up tables and then sketch a graph.

1. $y = 4^x$ **2.** $y = 3^{2x}$ **3.** $y = (1/3)^x$ **4.** $y = 2^{(x/2)}$

Simplify the following.

5. $2^6 \cdot 2^2$ **6.** 2^{-3} **7.** $3^a \cdot 3^{-a}$ **8.** $5^7 \cdot 5^{-4}$

9. $(2^2)^3$ **10.** $27^{x/3}$ **11.** $2^3/2^4$

12. Verify property L4 of the log function.

13. Sketch $y = \log_5 x$ and $y = \log_2 x$ on the same set of axes.

Simplify each of the following expressions.

14. $\log_2 4$ **15.** $\log_3(1/27)$ **16.** $\log_2 2^n$ **17.** $3^{\log_3(x+1)}$

Using the facts that $\log_2 7 = 2.807$ and $\log_2 5 = 2.322$, find the following.

18. $\log_2 35$ **19.** $\log_2 14$ **20.** $\log_2 625$ **21.** $\log_2 175$

Using the LN button on a calculator, compute the following.

22. $\log_2(13)$ **23.** $\log_2(2048)$ **24.** $\log_3(50)$ **25.** $\log_{10}(1025)$

26. The following is an ancient French puzzle: A farmer has a pond that he is very fond of. He notices that the area of the pond covered by lily pads is doubling every day. He knows that if the lily pads ever cover his entire pond, they will kill it. He checks the pond after 29 days and discovers that it is half covered. How many days does he have to save his pond?

27. The authors of the book *Limits to Growth* use the puzzle in Exercise 26 to illustrate the concept that a natural resource that has been in abundant supply for many generations and still seems to be plentiful could be completely consumed in one generation. Explain how this situation could occur.

28. Money invested at interest rate r will double in value in roughly $70/r$ years. Suppose Tina invests $1000 at 10% and JoAnne invests $1000 at 14%. Compute the approximate value of each investment after 35 years and after 70 years.

29. Show that $\log_a(x) = LN(x)/LN(a)$ for any $a > 0$.

2.6

FINITE SERIES

Cumulative processes often arise in discrete mathematics. For instance, a computer program may include many repetitions of a particular calculation. Each repetition requires a certain amount of time and may consume additional memory as well. The programmer would like to be able to estimate the amount of time and space the repetition will take before actually running the program.

Such calculations frequently involve the addition of fairly long lists of numbers or expressions. To facilitate such calculations, a special notation called **sigma notation** or **summation notation** has been developed. It uses the Greek letter Σ (sigma) to stand for "sum." Here are some examples.

EXAMPLE 1

a) $\displaystyle\sum_{i=1}^{n} i = 1 + 2 + 3 + \cdots + (n-1) + n$

b) $\displaystyle\sum_{i=0}^{n} 3i = 0 + 3 + 6 + 9 \cdots + 3(n-1) + 3n$

c) $\displaystyle\sum_{i=1}^{n} (3i + 2) = 5 + 8 + 11 + \cdots + (3n-1) + (3n+2)$

d) $\displaystyle\sum_{i=1}^{n-1} i = 1 + 2 + 3 + \cdots + (n-2) + (n-1)$

e) $\displaystyle\sum_{i=2}^{n} i = 2 + 3 + 4 + \cdots + (n-1) + n$

f) $\displaystyle\sum_{j=1}^{n} j = 1 + 2 + 3 + \cdots + (n-1) + n$

g) $\displaystyle\sum_{i=1}^{n} i(3i + 2) = 5 + 16 + 33 + \cdots + (n-1)(3n+1) + n(3n+2)$ ∎

Consider Example 1g. The n above the Σ is called the **upper limit.** The 1 below the Σ is called the **lower limit** and the i below the Σ is called the **index.** Note that the index is a **dummy variable,** i.e., what we call it does not matter. For instance, Examples 1a and 1f are the same series, although one has i as an index and one has j. The index is often omitted below the Σ; thus, you will frequently see sums like 1a written as

$$\sum_{1}^{n} i$$

More General Ways to Write Series

Series are often written with functions and variables as well, as shown below.

EXAMPLE 2

a) $\displaystyle\sum_{i=m}^{n} f(i) = f(m) + f(m+1) + \cdots + f(n)$

b) $\displaystyle\sum_{i=1}^{n} a_i = a_1 + a_2 + \cdots + a_n$ ∎

Note the use of the **subscripts** in Example 2b. These are often used to distinguish members of a list, a_1 being the first element in the list, a_2, the second, etc.

Manipulating Series

Following are some basic properties of sums written using sigma notation. One of the situations in which people often get confused when first working with sigma notation is where the traditional "+" notation and the new sigma notation are intermingled. Understanding these properties thoroughly should overcome much of that confusion.

$$R1: \sum_{i=0}^{0} a_i = a_0$$

$$R2: \sum_{i=0}^{n} a_i = a_0 + \sum_{i=1}^{n} a_i$$

$$R3: \sum_{i=0}^{n} a_i = a_n + \sum_{i=0}^{n-1} a_i$$

$$R4: \sum_{i=0}^{n} a_i = \sum_{i=0}^{m} a_i + \sum_{i=m+1}^{n} a_i \qquad \text{if } 0 < m < n$$

PROOF

R1: When the upper and lower limits are the same, there can be only one term, namely, the one corresponding to the common value of the limits.

$$R2: \sum_{i=0}^{n} a_i = a_0 + a_1 + \cdots + a_n = a_0 + \sum_{i=1}^{n} a_i$$

$$R3: \sum_{i=0}^{n} a_i = a_0 + a_1 + \cdots + a_n = a_n + \sum_{i=1}^{n-1} a_i$$

$$R4: \sum_{i=0}^{n} a_i = a_0 + a_1 + \cdots + a_m + a_{m+1} + \cdots + a_n = \sum_{i=0}^{m} a_i + \sum_{i=m+1}^{n} a_i \quad \square$$

Properties of Finite Sums

Properties of P1 through P5 summarize some additional properties of finite sums.

$$P1: \sum_{i=1}^{n} c = cn$$

$$P2: \sum_{i=1}^{n} cf(i) = c\sum_{i=1}^{n} f(i)$$

P3: $\displaystyle\sum_{i=1}^{n}(f(i) + g(i)) = \sum_{i=1}^{n}f(i) + \sum_{i=1}^{n}g(i)$

P4: $\displaystyle\sum_{i=1}^{n}(f(i) - g(i)) = \sum_{i=1}^{n}f(i) - \sum_{i=1}^{n}g(i)$

P5: $\displaystyle\sum_{i=1}^{n}(f(i + 1) - f(i)) = f(n + 1) - f(1)$

PROOF

P1: $\displaystyle\sum_{i=1}^{n}c = \underbrace{c + c + \cdots + c}_{n \text{ times}} = cn$

P2: $\displaystyle\sum_{i=1}^{n}cf(i) = cf(1) + cf(2) + \cdots + cf(n)$
$$= c\,(f(1) + f(2) + \cdots + f(n)) = c\sum_{i=1}^{n}f(i)$$

P3: $\displaystyle\sum_{i=1}^{n}(f(i) + g(i)) = f(1) + g(1) + f(2) + g(2) + \cdots + f(n) + g(n)$
$$= f(1) + f(2) + \cdots + f(n) + g(1) + g(2)$$
$$+ \cdots + g(n)$$

$$= \sum_{i=1}^{n}f(i) + \sum_{i=1}^{n}g(i)$$

P4: See the exercises.

P5: $\displaystyle\sum_{i=1}^{n}(f(i + 1) - f(i)) = f(2) - f(1) + f(3) - f(2) + f(4) - f(3)$
$$+ \cdots + f(n + 1) - f(n)$$
$$= f(n + 1) - f(1) \quad \square$$

Some Important Series

It's not enough to be able to model a cumulative process as a finite sum using sigma notation. Usually we would like a formula for the sum in terms of the upper and lower limits. There are a number of common series for which the formulas are well established and well worth committing to memory. These formulas can be modified to accommodate different limits.

One finite series that arises quite often in discrete mathematics is the **arithmetic series,** also called the **arithmetic progression:**

$$1 + 2 + 3 + \cdots + (n - 1) + n$$

There is an interesting anecdote about this series. Carl Friedrich Gauss (1777–1855) was one of the greatest mathematicians of the nineteenth century. As a boy, he was sent to a boarding school known more for its strict discipline than its academic quality. As a punishment, he and his class-mates were given the exercise of adding up all of the integers from 1 to 100. The teacher had barely assigned the problem when Gauss raised his hand with the correct answer. Gauss had discovered a clever way to sum the arithmetic series with $n = 100$. He wrote the sum horizontally, wrote it again horizontally in the opposite order, and then added term by term:

$$\begin{array}{r} 1 + \quad 2 + \quad 3 + \cdots + \quad 99 + 100 \\ 100 + \quad 99 + \quad 98 + \cdots + \quad 2 + \quad 1 \\ \hline 101 + 101 + 101 + \cdots + 101 + 101 \end{array}$$

Hence he could see that the answer was half of 101×100, or 5050. Gauss's teacher recognized that the boy possessed an unusual talent for mathemat-ics and recommended that his father send him to a school that would be better suited to a student of his ability, which his father did.

We can use Gauss's method to sum the arithmetic series:

$$\begin{array}{r} 1 + \quad\quad 2 + \quad\quad 3 + \cdots + (n-1) + n \\ n + (n-1) + (n-2) + \cdots + \quad\quad 2 + 1 \\ \hline (n+1) + (n+1) + (n+1) + \cdots + (n+1) + (n+1) \end{array}$$

The last line totals to $n(n+1)$; hence the sum of the arithmetic series is $n(n+1)/2$. The sums of this series and three others are tabulated as proper-ties F1 through F4.

$*$ F1: $\displaystyle\sum_{i=1}^{n} i = \frac{n(n+1)}{2}$

  F2: $\displaystyle\sum_{i=1}^{n} i^2 = \frac{n(n+1)(2n+1)}{6}$

  F3: $\displaystyle\sum_{i=1}^{n} i^3 = \frac{n^2(n+1)^2}{4}$

$*$ F4: $\displaystyle\sum_{i=0}^{n} ar^i = a\frac{1 - r^{n+1}}{1 - r} \quad r \neq 1$

F1, F2, and F3 are called **p-series**, since they are all of the form

$$\sum_{i=1}^{n} i^p$$

for some power p. F4 is called the **geometric series** or **geometric progres-sion**. F1 was verified in our discussion of Gauss. We will now verify F4; the others will be covered in Section 3.7.

PROOF OF F4

First, by property P2,

$$\sum_{i=0}^{n} ar^i = a\sum_{i=0}^{n} r^i.$$

Next,

$$\sum_{i=0}^{n} r^i = 1 + r + r^2 + \cdots + r^n.$$

Let s denote this sum, multiply both sides of the resulting equation by r, and subtract. We get

$$s = 1 + r + r^2 + \ldots + r^n$$
$$rs = r + r^2 + r^3 + \cdots + r^n + r^{n+1}$$
$$\overline{s - rs = 1 - r^{n+1}}$$

or

$$s = \frac{1 - r^{n+1}}{1 - r}. \quad \square$$

EXAMPLE 3   Find

a) $\displaystyle\sum_{i=1}^{15} i$ b) $\displaystyle\sum_{i=7}^{20} i$

SOLUTION

$$\sum_{i=1}^{15} i = \frac{15(15 + 1)}{2} = 120$$

$$\sum_{i=7}^{20} i = \sum_{i=1}^{20} i - \sum_{i=1}^{6} i = \frac{20(21)}{2} - \frac{6(7)}{2} = 210 - 21 = 189 \quad\blacksquare$$

EXAMPLE 4   Consider again Example 7 of Section 2.5. What is the total number of grains of rice that would be accumulated in 64 days?

SOLUTION

$$\sum_{n=1}^{64} 2^{n-1} = \sum_{n=0}^{63} 2^n = \frac{1 - 2^{64}}{1 - 2} = 2^{64} - 1$$

Note that $2^{64} = 2 \cdot 2^{63} = 2^{63} + 2^{63}$. Hence $2^{64} - 1 = 2^{63} + 2^{63} - 1$.

Recall that the number of grains of rice received on the 64th day was 2^{63}. Hence the amount of rice received on the day 64 is one more than the number of grains received on all previous days added together! In general, the geometric sum follows this pattern:

$$1 + 2 + 4 = 7, \text{ 1 less than 8}$$
$$1 + 2 + 4 + 8 = 15, \text{ 1 less than 16}$$

etc.

This provides a different look at why exponential functions grow so rapidly. ∎

TERMINOLOGY

sigma notation	subscript
summation notation	arithmetic series
upper limit	arithmetic progression
lower limit	p-series
index	geometric series
dummy variable	geometric progression

SUMMARY

1. The following relationships are true for all finite series:

R1: $\displaystyle\sum_{i=0}^{0} a_i = a_0$

R2: $\displaystyle\sum_{i=0}^{n} a_i = a_0 + \sum_{i=1}^{n} a_i$

R3: $\displaystyle\sum_{i=0}^{n} a_i = a_n + \sum_{i=0}^{n-1} a_i$

R4: $\displaystyle\sum_{i=0}^{n} a_i = \sum_{i=0}^{m} a_i + \sum_{i=m+1}^{n} a_i \quad$ if $0 \le m < n$

2. The following properties of finite series are also useful.

P1: $\displaystyle\sum_{i=1}^{n} c = cn$

P2: $\displaystyle\sum_{i=1}^{n} cf(i) = c\sum_{i=1}^{n} f(i)$

P3: $\displaystyle\sum_{i=1}^{n} (f(i) + g(i)) = \sum_{i=1}^{n} f(i) + \sum_{i=1}^{n} g(i)$

P4: $\displaystyle\sum_{i=1}^{n} f(i) - g(i)) = \sum_{i=1}^{n} f(i) - \sum_{i=1}^{n} (g)i)$

P5: $\displaystyle\sum_{i=1}^{n} (f(i+1) - f(i)) = f(n+1) - f(1)$

3. A few series come up frequently enough to deserve special attention. These series and their sums are as follows.

F1: $\displaystyle\sum_{i=1}^{n} i = \frac{n(n+1)}{2}$

F2: $\displaystyle\sum_{i=1}^{n} i^2 = \frac{n(n+1)(2n+1)}{6}$

F3: $\displaystyle\sum_{i=1}^{n} i^3 = \frac{n^2(n+1)^2}{4}$

F4: $\displaystyle\sum_{i=0}^{n} ar^i = a\frac{1-r^{n+1}}{1-r} \qquad r \neq 1$

F1, F2 and F3 are called *p-series*. F1 is also called the *arithmetic series;* F4 is called the *geometric series.*

EXERCISES 2.6

Expand the following sums.

1. $\displaystyle\sum_{i=1}^{n}(i+1)$

2. $\displaystyle\sum_{i=0}^{n-1}(i+1)$

3. $\displaystyle\sum_{i=2}^{9} i^2$

4. $\displaystyle\sum_{i=1}^{n} 2^i$

5. $\displaystyle\sum_{i=0}^{n-1} 3$

6. $\displaystyle\sum_{i=1}^{6} (-1)^i$

Write the following sums using sigma notation.

7. $1 + 3 + 5 + \cdots + (2n - 1)$

8. $1 + 4 + 9 + \cdots + 144$

9. $2 + 4 + 8 + \cdots + 2^n$

10. $(1/2) + (1/4) + \cdots + (1/64)$

Verify the following expressions.

11. $\displaystyle\sum_{i=1}^{n}(i+1) = \sum_{i=2}^{n+1} i$

12. $\displaystyle\sum_{i=1}^{10}(2i+1) = 10 + 2\sum_{i=1}^{10} i$

13. $\displaystyle\sum_{i=3}^{13}(i+1) = \sum_{j=4}^{14} j$

14. Property P4 in the text

In the following exercises, find the given sum using the properties P1–P5 and the sums F1–F4. Express your answer as a function of n or, if possible, as a number.

15. $\sum_{i=1}^{n} 4(i + 1)$

16. $\sum_{i=1}^{n} (i^2 - 1)$

17. $\sum_{i=1}^{n} (2^i - 1)$

18. $\sum_{i=0}^{n-1} 2(i + 1)^2$

19. $\sum_{i=1}^{n} (i^3 + 2i - 1)$

20. $\sum_{i=7}^{21} (i - 1)$

21. $\sum_{i=1}^{n} (i + 1)$

22. $\sum_{i=1}^{9} (i + 1)$

23. $\sum_{i=1}^{n} ((i + 1)^2 - i^2)$

24. $\sum_{i=1}^{9} (2^{(i+1)} - 2^i)$

REVIEW EXERCISES—CHAPTER 2

Write each of the following sets in both set-builder form and tabular form.

1. The New England states.

2. The U.S. presidents since 1970.

3. The nonnegative integers less than 10.

4. The integers between −10 and 10.

5. The positive even numbers less than 19.

6. Find the set of all possible four-letter words that could be formed from the letters *TEAM*. "Words" are acceptable even if they don't make sense, e.g., "EATM" is a word. (*Hint:* your list should have 24 words.)

The following questions refer to the set of words developed in Exercise 6. In each case, write the specified subset of that set.

7. $M = \{$words starting with $A\}$

8. $N = \{$words ending in $E\}$

9. $P = \{$words with E before $A\}$

10. $R = \{$words that make sense$\}$

11. $M \cap N$

12. $N \cup P$

13. $R \setminus N$

14. $N \setminus R$

15. Annette has five coins in her desk drawer—a penny, a nickel, a dime, a quarter, and a 50-cent piece. She reaches in and randomly grabs for some coins. She may come up with any number from zero to all five. List all possible sets of coins she could come up with.

The next five questions refer to the sets of coins mentioned in Exercise 15. In each case, tell how many of the sets have the specified property.

16. The total value of the coins is less than 50 cents.

17. The value is less than 50 cents and there are no pennies.

18. The value is 75 cents or more.

19. The set has exactly one member.

20. The set has exactly two members.

Suppose $U = \mathbf{N}$, $M = \{1, 2, 3, 4, 5\}$, $N = \{3, 4, 5, 6, 7\}$, $P = \{6, 7, 8\}$, and $R = \{4, 5\}$. Illustrate the relationship between each of the following pairs of sets using Venn diagrams.

21. N and R 22. R and $M \cap N$

23. R and P'

Illustrate each of the following relationships by drawing Venn diagrams and shading the appropriate parts.

24. $A \cap A' = \emptyset$ 25. $A \cup B = (A \setminus B) \cup (A \cap B) \cup (B \setminus A)$

26. $A \cap B \subset A$ 27. $A \cap B' \subset A$

Show that each of the following relationships is false by means of Venn diagrams.

28. $(A \cap B) \cup C' = (A \cap C) \cup B'$ 29. $A \cup B = (A \setminus B) \cup (B \setminus A)$

30. $(B \setminus A)' \subset A$

Let $A = B = \{1, 2, 3, 4\}$. Tell whether each of the following subsets of $A \times B$ is a function and why.

31. $\{(1, 3), (2, 1), (2, 4), (3, 3) (4, 2)\}$ 32. $\{(1, 2), (2, 3), (3, 4)\}$

For each of the following functions, let $A = B = \mathbf{Z}$.

33. If $f(z) = 1 - z - z^2$, find

   a) $f(0)$ b) $f(1)$ c) $f(-1)$ d) $f(a)$

   e) $f(x + 1)$ f) $f(x^2)$ g) $f(x) + f(y)$ h) $f(2x^2 - 1)$

For each of the following, sketch the curve and state the domain and range.

34. $y = 3 - x.$ 35. $y = x^2$

In Exercises 36–38 you are given some experimental data that contain errors. Plot the data and infer a function that approximately fits them.

36. $(-3, 8.1)$, $(-3, 9.3)$, $(-2, 4.5)$, $(-2, 3.9)$, $(-1, 2.0)$, $(-1, 0)$, $(0, -.5)$, $(0, .4)$, $(1, 1.0)$, $(1, 1.5)$, $(2, 3.9)$, $(2, 4.7)$, $(3, 7.9)$, $(3, 10.1)$

37. $(2, 0)$, $(2, 0.1)$, $(2, 0.2)$, $(3, 1)$, $(3, 0.8)$, $(3, 1.1)$, $(4, 1.5)$, $(4, 1.3)$, $(5, 1.7)$, $(5, 1.9)$, $(6, 2.1)$, $(6, 1.7)$, $(6, 1.9)$, $(7, 2.3)$, $(7, 2.1)$

38. $(0, 1.01)$, $(0, 1.10)$, $(0, 1.07)$, $(1, 0.50)$, $(1, 0.46)$, $(2, 0.31)$, $(2, 0.31)$, $(2, 0.40)$, $(3, 0.30)$, $(3, 0.21)$, $(4, 0.21)$, $(4, 0.15)$, $(5, 0.16)$, $(5, 0.18)$, $(6, 0.11)$, $(6, 0.13)$

In the following situations, find a function that fits the observations.

39. A business is interested in modeling the profit it earns on an item as a function of its price. At \$1 per item the company breaks even, since manufacturing costs

are also $1. At $3 per item it also breaks even, since that is more than most people will pay. At $2 its profit is a maximum.

40. An integer variable x is between 0 and 50 inclusive. A particular X in the domain is selected for special study. Write an expression for the function $D(x)$, the distance from any x to X, and state its domain and range.

In Exercises 41–49, find the range and domain of the given function and sketch its graph.

41. $g(x) = \sqrt{(5 - 2x)}$ **42.** $h(x) = \sqrt[3]{(x - 1)}$

43. $s(n) = 1/(n - 1)$ **44.** $a(y) = |1 - y|$

45. $d(x) = \begin{cases} 1 - x & x < -2, \\ 4 - (x/2) & x \geq -2 \end{cases}$ **46.** $l(m) = \lfloor m/2 \rfloor$

47. $b(z) = \lceil 1 - x \rceil$ **48.** $p(x) = x^3 - 3x - 1$

49. $t(x) = 1/((x - 1)(x - 2))$

For each of the following pairs of functions, find $g \circ f$ and $f \circ g$.

50. $f(x) = \begin{cases} x - 2 & x < 2, \\ 3 - x & x \geq 2 \end{cases}$ $g(x) = 2x$,

51. $f(x) = \sqrt{(x - 1)}$, $g(x) = 1/x$

For each of the following problems, sketch f, g, and $g \circ f$.

52. Exercise 50   **53.** Exercise 51

Sketch a graph of each of the following functions and use the horizontal line test to tell whether it is a one-to-one correspondence. If not, explain why not.

54. $\{(1, 4), (2, 3), (3, 2), (4, 1)\}$ **55.** $\{(1, 4), (2, 3), (3, 3), (4, 4)\}$

56. $f(x) = (1/3)x^3$ **57.** $f(x) = 2x - 17$

58. $h(x) = \lceil x \rceil$ **59.** $q(r) = \begin{cases} r + 1 & r < 0, \\ r & r \geq 0 \end{cases}$

For each of the following functions, decide if it has an inverse, and if it does, find it. If not, explain why not.

60. $f(x) = 3x^2 - 1$ **61.** $g(x) = 10 - x$

For each of the following, find the inverse and verify that it is the inverse.

62. Exercise 54   **63.** Exersise 56

Expand the following sums.

64. $\sum_{i=1}^{n} 3$ **65.** $\sum_{i=-2}^{3} (i - 1)$

66. $\sum_{i=2}^{7} 2i/(i - 1)$ **67.** $\sum_{i=0}^{n-1} i(i + 1)$

Write the following sums using sigma notation.

68. $-2 + 0 + 2 + 4 + 6 + 8$ **69.** $1 + 2 + 5 + 10 + 17 + 26 + 37$

Verify the following expressions.

70. $\displaystyle\sum_{i=0}^{15}(i+1) = 100 + \sum_{i=0}^{9}i^2 + \sum_{i=11}^{15}i^2$ **71.** $\displaystyle\sum_{i=0}^{n-1}(i+1) = \sum_{i=1}^{n}i$

In the following exercises, find the given sum using the properties P1–P4 and the sums S1–S4.

72. $\displaystyle\sum_{i=6}^{12}3i^2$ **73.** $\displaystyle\sum_{i=0}^{13}2i(i+1)$

74. $\displaystyle\sum_{k=5}^{10}3^k$ **75.** $\displaystyle\sum_{i=0}^{n-1}(i+1)$

REFERENCES

Eves, Howard, and Carroll V. Newsom, *The Foundations and Fundamental Concepts of Mathematics,* rev. ed. New York: Holt, Rinehart & Winston, 1965.

Halmos, Paul, *Naive Set Theory.* Belmont, Calif.: D. Van Nostrand, 1960.

Heijenoort, J. van, ed., *From Frege to Gödel.* Cambridge, Mass.: Harvard University Press, 1967.

Kemeny, John G., J. Laurie Snell, and Gerald L. Thompson, *Introduction to Finite Mathematics,* 3rd ed. Englewood Cliffs, N.J.: Prentice-Hall, 1974.

Leithold, Louis, *The Calculus with Analytic Geometry.* 3rd ed. New York: Harper & Row, 1976.

Lewis, Harry R., and Christos Papadimitriou, *Elements of the Theory of Computation.* Englewood Cliffs, N.J.: Prentice-Hall, 1981.

Lipschutz, Seymour, *Set Theory and Related Topics.* New York: McGraw-Hill, 1964.

Swokowski, Earl W., *Calculus with Analytic Geometry,* 2nd ed. Belmont, Calif.: Prindle, Weber, and Schmidt, 1981.

Logic

3

3.1

INTRODUCTION TO LOGIC

In order to be useful, a model must be abstract. That is, it must identify basic patterns and principles independently of the particulars of a situation and enable one to deal with the patterns alone. In the same sense, the logical tools for dealing with models must be abstract. They must focus on patterns and principles of reasoning rather than particular instances of reasoning. Because deductive logic is such a critical part of the modeling life cycle, in this chapter we will set aside other aspects of discrete models and focus on logic and reasoning. Logic and reasoning have come to be seen in a specific way in the twentieth century. Perhaps the best way to introduce and explain them is historically. We will give a very brief outline here. An excellent work presenting these ideas in more detail is the book by Eves and Newsom listed in the References.

Babylonian and Egyptian Mathematics

Fortunately, due to the dry climate in their part of the world, we have a great deal of evidence on the mathematics of ancient Babylon and Egypt. We have clay tablets dating to 3000 B.C. and earlier. They indicate that the Babylonians and Egyptians were experimental mathematicians, i.e., they discovered many things by trial and error. For example, they developed a variety of mathematical tables such as multiplication tables, tables of squares and square roots and of cubes and cube roots, tables of reciprocals, exponential tables for finding compound interest, and others. They found a formula for solving quadratic equations and discovered the method of completing the square. But they also found a number of formulas and methods that they believed to be correct but that were not. For instance, one means of finding the area of a circle was to take the square of eight-ninths of the circle's diameter. This is actually equivalent to taking $\pi = (4/3)^4 = 3.1604\ldots$, an estimate that was probably close enough to the true value of $3.14159\ldots$ for their purposes. But it was not correct. Another error is their apparent use of the formula $K = (a + c)(b + d)/4$ for the area of a quadrilateral with sides a, b, c, and d. Where the quadrilateral is a rectangle with $a = c$ and $b = d$, this formula gives the correct answer, but in general it does not. So, although the experimental trial-and-error method is capable of leading to great discoveries, it lacks a means of verifying those discoveries. That is, what seems correct in every case that has been tested may not be true in other cases not yet tested. Notice an important result: The ancients' lack of a means to verify their results led them not only to incorrect theorems but also to incorrect computational procedures. This is part of the reason why it is important for computer scientists and others who are concerned with computation to know something about proof.

Greek Axiomatics

The Greeks, on the other hand, used a very different system. Their approach to the study of philosophy was deductive, depending on the establishment of first principles and paying careful attention to the use of reason. They applied the same approach to mathematics, and as a result developed the **axiomatic method.** This method had four components. First, all technical words (*line, point,* etc.) were carefully defined, whereas non-technical words were accepted as having their ordinary meanings. Second, **axioms** and **postulates** were set down. Axioms were statements that were regarded as universal truths; postulates were more narrow—statements that were true in the particular study at hand, but not necessarily in a broader setting. Thus, for instance, the statement "If equals are added to equals, the wholes are equal" was regarded as an axiom. The statement "A straight line can be drawn from any point to any point" was regarded as a postulate of geometry. Third, technical terms could be added as long as they were defined in terms of the ones already defined. Finally, further statements about the technical terms (theorems) could be added only if they could be derived from axioms, definitions, and/or previously proven theorems. The result was a system of mathematics that seemed to provide a solid foundation for truth, and, unlike the Babylonian and Egyptian experimental method, seemed to provide a means to verify the truth of assertions and the correctness of computational procedures. Also, Aristotle and others did a great deal to clarify the deductive process by which theorems could be legitimately derived from other axioms, definitions, and theorems so that the process as well as the vocabulary was laid on a solid foundation.

The Parallel Postulate

While a great deal of work was done in mathematics and logic in succeeding centuries, the next revolutionary breakthrough did not occur until the early nineteenth century. The last of Euclid's 10 axioms and postulates was the **parallel postulate:**

If a straight line falling on two straight lines makes the interior angles on the same side together less than two right angles, the two straight lines, if produced indefinitely, meet on that side on which the angles are together less than two right angles.

Gauss and other mathematicians showed that the parallel postulate could be replaced with either of the following two statements, and one would still have axioms for geometry that do not contradict each other.

1. Through a given point not on a given line, there are no parallels to the given line.
2. Through a given point not on a given line, there is more than one parallel that can be drawn to the given line.

This discovery was revolutionary in the mathematics of that time. Mathematics was supposed to be true. But if it was true, how could there be three different geometries, each consistent within itself but inconsistent with the others? For example, assuming Euclid's parallel postulate, triangles have 180°. If we assume no parallels, triangles have more than 180°, and if we assume more than one parallel, they have less than 180°. Thus mathematicians began very carefully to measure triangles. The best that anyone could say was that the angles of a triangle are *approximately* 180°, but since no one could measure perfectly, no one could be certain. So the attempt to decide which geometry was correct by measuring triangles did not succeed.

Effects of the Discovery of Non-Euclidean Geometry

This situation prompted a fresh look at the foundations of mathematics. One result was a recognition that there are difficulties in attempting to define all technical terms, as Euclid's geometry attempted to do. For instance, a point was defined as "that which has no part." A line was "length without breadth." But *part, length,* and *breadth* were not defined. If Euclid had defined those terms, he would have had to do so by using other terms, whose definitions would have had to use other terms, and so on forever. This kind of circularity is inevitable in dictionary definitions. Consider, for example, the following set of dictionary definitions, starting with the word *ambiguous.*

> Ambiguous—open to various interpretations, equivocal, doubtful, uncertain, obscure, indistinct
>
> Equivocal—uncertain, doubtful, dubious, suspicious, susceptible of double interpretation, ambiguous
>
> Uncertain—doubtful, vague, indistinct
>
> Doubtful—uncertain, ambiguous
>
> Obscure—unclear, indistinct, doubtful, dubious, ambiguous

These definitions tend to be circular. This posed a serious problem: Technical terms are supposed to be defined unambiguously, but the very process of definition is inherently circular. The resolution of this problem came with the use of **undefined terms,** also known as **primitives.** In geometry this means considering words like *point, line,* and *plane* as undefined. Other definitions are then built up using a minimum number of undefined terms. The undefined terms can then have various interpretations as long as those interpretations are consistent with the axioms in which the undefined terms are used. For instance, if *plane* is interpreted to mean the surface of a sphere, *line* to mean a great circle on that sphere, and *point* to mean what we normally think of as a point on the surface, we have a geometry in which there are no parallel lines and all triangles have between 180° and 360°. There is another surface that can be visualized as looking like the inside of a infinitely long trumpet. If this is interpreted as a plane, we have a geometry in which there is more than one parallel and all triangles have less than 180°.

The key point is that the undefined terms are freed from any particular concrete interpretation. David Hilbert, one of the leading mathematicians of the early twentieth century, expressed this situation by saying that "We must always be able, instead of talking of 'points, lines and planes,' to talk even of 'tables, chairs, and beer mugs'." In other words, our mathematics has to be independent of whatever meaning its primitive terms have in everyday speech. Thus a mathematical model can be applied to any situation in which the primitive terms can be matched to aspects of it in such a way that the axioms of the model are satisfied, but not to any other situation.

Axioms

Another consequence of the reexamination of the foundations of mathematics was that the distinction between axioms and postulates that the Greeks made was dropped; all such statements are now considered axioms. Axioms are no longer seen as universal truths, but rather as statements expressing relationships between the primitive terms. Theorems that are derived from axioms are then applicable to any situation that can be modeled by the primitive terms and axioms.

The idea that mathematical theorems were true in some ultimate metaphysical sense was also dropped. Instead theorems were seen as valid, i.e., as statements that were derived from the axioms, primitive terms, definitions, and previously proven theorems according to the legitimate rules of inference.

In this book, we will use a somewhat more informal approach than the axiomatic method actually requires. This is sometimes called the *naive* approach. In set theory, for instance, this means incorporating concepts that can be spelled out axiomatically in the definitions. We will use this approach because some of the issues raised by a formal axiomatic treatment of set theory are beyond the scope of this book.

Symbolic Logic

Simultaneously with these advances in our understanding of the axiomatic method, significant breakthroughs in logic occurred. The Greeks had outlined the valid laws of inference (the process by which one was able to make legitimate logical deductions) quite clearly. But in the mid-nineteenth century, George Boole succeeded in developing a symbolic system for use with logic. This system, which came to be known as **symbolic logic,** will be the main topic of the next several sections of this book since it is the principal tool mathematicians use to describe abstractly the patterns of reasoning accepted as valid and invalid. Just as undefined terms made it easier to see that mathematics depends on the abstract relationships that exist between entities, rather than on any particular interpretation of those entities, symbolic logic made it easier to see logic in terms of patterns of reasoning as distinguished from particular instances of reasoning. In fact, it was the very abstractness of this symbolic system that made it possible for Claude Shannon and others earlier in this century to develop the first electronic circuits that performed logical and arithmetic computations and that ultimately led to the development of the computer. The following sections, then, will focus on the foundational concepts in symbolic logic. We will close the chapter with some applications of logic to the design of combinational circuits.

TERMINOLOGY

axiomatic method   undefined terms

axioms   primitives

postulates   symbolic logic

parallel postulate

SUMMARY

1. Trial-and-error methods are capable of discovering much important mathematics; however, deductive methods are needed to verify conjectured theorems and computational methods.

2. Modern mathematics has been developed by the use of the axiomatic method, which includes the use of primitive or undefined terms, axioms, and deductive logic.

3. Symbolic logic allows us to study the patterns of valid and invalid reasoning apart from the content of particular arguments.

EXERCISES 3.1

1. In terms of the mathematical modeling life cycle, explain the role of logic in modeling.

2. What are the strengths and weaknesses of experimental mathematics?

3. Explain the distinction the Greeks made between axioms and postulates.

4. What is the parallel postulate and why was it controversial?

5. Explain what undefined terms are and why they are so important in contemporary mathematics.

6. What does it mean to say that a mathematical theorem is "true"?

7. Explain what Hilbert meant when he said that, in our geometry, we should be able to talk in terms of tables, chairs, and beer mugs, as well as points, lines, and planes.

3.2
PROPOSITIONAL LOGIC

. . . it is necessary to pass from natural languages to a formal language in order to study logic successfully. We could, therefore, start building up such a formal language at once in order to have a basis for our logical investigations. However, the formal languages have grown up against the background of the natural languages. Thus we shall be able to understand the structure of a formal language better if we know which characteristics of the natural language it reflects and how it does this. We can find this out best by practicing "translating" everyday statements into the formal language, or, in other words by symbolising *these statements. The symbolisation of statements is an art which can, to a large extent, be learned by practice. The symbolisation of arbitrary linguistic statements can be very difficult. However, the exercise is simpler if the statements which are to be translated are mathematical statements, since the traditional mathematical statements have a fairly transparent logical structure and since it is precisely this logical structure that we want to bring out in the symbolisation.*

H. Hermes, *Introduction to Mathematical Logic*

Propositions

In mathematics, we use constants, variables, functions, and relationships (such as $=$, $<$, $\geq$, etc.). In applying symbolic notation to logic, we do much the same thing. We start with some basic concepts.

Definition A **proposition** (sometimes called a **statement**) is a verbal or mathematical sentence that is either true or false. That is, it is unambiguous, and its truth can be definitely decided.

We use lowercase letters, primarily *p, q,* and *r,* to denote variables whose allowable values are elements of the set of all propositions. The **truth**

value of a proposition is the word *true* or the word *false,* depending on whether the proposition is true or false. We will consider **simple propositions,** simple declarative sentences whose truth can be decided, and **compound propositions,** combinations of simple propositions using various rules of combination that we will look at shortly.

EXAMPLE 1 Which of the following are propositions?

a) $2 > 1$ **b)** $39.3 < 61.057$ **c)** $\{1, 2, 3\} \cap \{3, 4, 5\} = \{3\}$

d) $f(x) = x^2$, $f: \mathbf{R} \to \mathbf{R}$, is a function

SOLUTION   All of them are propositions. Also, all of them are true. The first three involve comparisons using equality or inequality. The last involves verifying that a particular relationship satisfies the definition of function. It can also be considered as determining whether a particular element (the function $f(x) = x^2$ from **R** to **R**) is a member of a particular set (the set of all functions). ∎

EXAMPLE 2 Which of these are propositions?

a) $1 > 2$ **b)** $98.61 < 12.1$ **c)** $'B' > 'C'$

d) $\{1, 2\} \cup \{3, 4\} = \{0, 1, 2, 3\}$

SOLUTION   Again, all of them are propositions. They all happen to be false. ∎

EXAMPLE 3 Which of the following are propositions?

a) $1 + 2$ **b)** $'BAC'$ **c)** $a + b = b + a$

SOLUTION   None of these are propositions. The first involves operations on numbers, but there is no relationship that can be tested for truth or falsity. Part (c) might at first appear to be a proposition, but it is not, since a and b are unrestricted and could be anything. Under some interpretations of a and b, such as "let $a = 1$ and $b = 2$," the statement would be true. Under other interpretations, such as "let a be the word *house,* b be the word *fire,* and $+$ mean *concatenate the given words*," the statement is false. ∎

EXAMPLE 4 Test which of the following are propositions.

"Philadelphia is between New York and Washington."

"John is adventurous."

"I'll have the roast beef on kimmelwick and a glass of milk."

SOLUTION   The last statement is definitely not a proposition. Although it is a grammatically correct sentence, it cannot be assigned a value of true or false. The other two statements raise one of the major difficulties that arise in applying logic to natural language, namely, that words in natural languages do not have precise meanings. For instance, in the first sentence, if "between" means that a straight line drawn through the geographic centers of New York and Washington on a map also passes through Philadelphia, then it is false. On the other hand, if it means that most highway routes from New York to Washington pass through or close to Philadelphia, it is true. Similarly, whether John is adventurous depends greatly on what we mean by adventurous. It also depends on our perception of John, which might differ from that of someone else who means the same thing by *adventurous.*

Thus neither the first nor the second sentence is a proposition unless the words used are defined much more precisely. In spite of these difficulties, however, it is often possible to treat statements like the first two as propositions by saying, in essence, "let's suppose that we can agree on an interpretation of the words." Treating sentences that are potentially ambiguous as propositions has the great advantage of allowing us to focus on the form of an argument rather than on its content. That is, if we assume that we can agree on the interpretation of words like *between* and *adventurous,* we can examine the reasoning involved in a discussion apart from the particulars. The danger here is that particulars are often slipped in in subtle and unexpected ways. Also, we often find ourselves saying things like "Well, I agree that your argument is valid, but I still don't accept your premises." ∎

Predicates Let X be some universal set and let x be a variable that assumes values in X. Statements involving x such as $x = 1$, $x > 20$, and $x \in \mathbf{N}$ are very common. Such statements are not propositions because of the presence of x. However, they become propositions when a particular value is substituted for x. For instance, if $x = 2$, the previous three statements become the propositions $2 = 1$, $2 > 20$, and $2 \in \mathbf{N}$. Such statements are dealt with formally by means of the following concept.

Definition Statements that include variables and that become propositions upon substitution of values for those variables are called **predicates** (or **propositional functions** or **open sentences** or **statement forms**). The variables in predicates are called **free variables.**

Predicates may involve more than one free variable, as in $x + y = 2$ or $x^2 + y^2 = z^2$. The notation

$$p(x) = x > 1$$

means "let $p(x)$ denote the predicate $x > 1$." The **truth set** of a predicate is the set of values of its free variable(s) for which it is true.

EXAMPLE 5 Find the truth set of the predicate

$$q(n) = 2^n < 32, \text{ where } n \in \mathbf{Z}^+$$

where $\mathbf{Z}^+$ denotes the set of non-negative integers.

SOLUTION $q(n)$ is true if $n = 0, 1, 2, 3,$ or 4 and false otherwise. Hence the truth set is $\{0, 1, 2, 3, 4\}$. Note that $q(n)$ is a predicate, while $q(0)$ is the proposition $2^0 < 32$, $q(1)$ is the proposition $2^1 < 32$, etc. ∎

EXAMPLE 6 Suppose x is a real number. Find the truth sets of the following.

a) $r(x) = x^2 + x \le 0$ **b)** $s(x) = x^2 + x \ge 0$ **c)** $t(x) = x^2 < 0$

SOLUTION
a) The truth set of $r(x)$ is $\{x \in \mathbf{R} \mid -1 \le x \le 0\}$.
b) The truth set of $s(x)$ is $\{x \in \mathbf{R} \mid x \ge 0\} \cup \{x \in \mathbf{R} \mid x \le -1\}$.
c) The truth set of $t(x)$ is $\varnothing$. ∎

TERMINOLOGY

proposition	propositional function
statement	open sentence
truth value	statement form
simple proposition	free variable
compound proposition	truth set
predicate	

SUMMARY

1. A proposition is a verbal or mathematical statement that is unambiguously true or false.
2. A predicate is a statement involving one or more variables such that it becomes a proposition when values are assigned to its variable(s).
3. The truth set of a predicate is the set of values of its variables that makes it true.

EXERCISES 3.2

Determine which of the following are propositions and give a brief reason for your answer in each case.

1. $10 < 8$ **2.** 'n' $\ge$ 'm'

3. $17.4 + 56.89 = 90.56$

4. $\{-1, 0, 1\} \subset \mathbf{N}$

5. $97^2 - 97 + 1$

6. $\sum_{i=1}^{5} i = 11$

7. e is a vowel.

8. The word *can* has more than one English meaning.

Convert each of the following predicates to two propositions, one true, one false, by making appropriate substitutions for the free variables.

9. $x > 13$

10. $b + 1 \leq 2b + 1$

11. $x + y = 12$

12. Animals of this kind are mammals.

13. That man is an ex-President of the United States.

Find the truth sets of the following predicates. Let the universal set for x and y be $\mathbf{R}$.

14. $x^2 - 2x + 1 \geq 0$

15. $x + y \leq 2$

16. $S \subset \{0, 1, 2\}$

17. This English word has no vowels.

3.3
CONNECTIVES

In algebra there are operations $(+, -, *, /,$ exponentiation$)$ that are used to combine variables and constants into more complex entities called *expressions*. In logic, the operations used to combine propositions are called **connectives** and the resulting expressions are called **compound propositions** or **Boolean expressions.** There are many connectives, but the principal ones we will discuss here are AND, OR, EITHER . . . OR, NOT, the conditional, and the biconditional.

Conjunction

Conjunction refers to the word "AND" and is symbolized mathematically by $\wedge$. In some contexts, it can also refer to other English words such as *since, but,* and *while.* A valid compound proposition using conjunction is of the form $p \wedge q$ or p AND q where p and q are themselves propositions, possibly compound.[†]

[†] This is our first example of a *recursive definition*—something that is defined in terms of itself. It is not the same thing as a circular definition, although it appears that way at first. This is because it is defined in terms of simpler versions of itself. For instance, for p AND q to be a proposition, both p and q must be propositions. But p itself can be a compound expression made up of other symbols that must be checked to determine whether they are propositions. This process does not go on forever, as eventually our expression will be broken down into simple propositions. The concept of a simple proposition is defined without reference to the word *proposition,* and it is this fact—that eventually we get back to something not defined in terms of this word—that makes our definition not circular.

EXAMPLE 1 Which of the following are propositions?

a) $(1 > 2)$ AND $(4 < 5)$ **b)** $0 \in \{0, 1, 2, 3\} \wedge 1 \in \{0, 1, 2, 3\}$

SOLUTION   Both are propositions since each is made up of simple propositions, some true, some false. Note that 1a includes parentheses and 1b does not. Mathematicians omit them if the meaning is clear without them. Computer languages may require that they be left in. ∎

We also use connectives with predicates. The pattern is the same: $p(x)$ AND $q(x)$ or $p(x) \wedge q(x)$. It is possible for one of p AND q to be a predicate and one to be a proposition. This is similar to being able to write $x + 1$ in algebra.

EXAMPLE 2   The following are valid compound predicates.

a) $(x > 1)$ AND $(x < 2)$ **b)** $(y > (2 + 3)) \wedge (y^2 < 1)$
c) $A \subset B$ AND $C \neq \emptyset$ ∎

EXAMPLE 3   Neither of the following are valid compound propositions:

a) $10 > 2$ AND 3   **b)** $1 + 2 \wedge 3 + 4$ ∎

Perhaps the single most common error in writing compound propositions or compound predicates occurs in Example 3a. What is usually meant is

$$10 > 2 \text{ AND } 10 > 3.$$

The rules for using connectives require that both what is to the left of the connective and what is to the right be propositions. "$10 > 2$" is a proposition but "3" is not. Writing "$10 > 2$ AND 3" is a carryover from ordinary language, where such a statement is legitimate. This is yet another example of the ways in which a more formal language requires tighter rules than ordinary language. "$1 + 2 \wedge 3 + 4$" is also invalid since neither what is to the left of the AND symbol nor what is to the right is a proposition. Comments similar to these apply to compound predicates.

Truth Tables

To evaluate the truth value of a compound proposition, we use a tool called a **truth table.** This summarizes, in tabular form, the truth values of a compound proposition based on the truth values of the simple propositions on which it is built. The truth table for the conjunction is

p	q	$p \wedge q$
T	T	T
T	F	F
F	T	F
F	F	F

Thus "p AND q" is true only when both are individually true. Note that the column headed p and the column headed q are set up to list all possible pairs of values for p and q. The easiest way to be sure that you have listed all possibilities is to start at the farthest to the right headed by a variable (in this case, the one headed by q) and list downward, alternating T and F. Then move left one column and alternate two T's and two F's. If there are additional columns, move left again and alternate four T's, four F's, etc.

One way to visualize this is with the **tree diagram** shown in Figure 3.1. Starting at the point on the left (the **root**), p can be either T or F. Treat each of these possibilities as a **branch.** Next, q has two possibilities; hence there should be two branches off of each existing branch. Since there are only p and q, the list of all possibilities is complete. If you follow each branch from left to right, you will see that it corresponds to one line in the truth table. The tree is completed by placing the value of p AND q that corresponds to the possibilities specified by each branch at the end of the branch.

Figure 3.1 Tree diagram showing the two possible values for p followed by the two possible values for q.

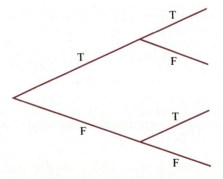

EXAMPLE 4 Find the truth value of the following compound propositions.

a) $(1 > 2)$ AND $(21.06 < 34.9)$ **b)** $(12 > 7)$ AND $(12 < 21)$

SOLUTION

a) $(1 > 2)$ AND $(21.06 < 34.9) =$ F AND T $=$ F

b) $(12 > 7)$ AND $(12 < 21) =$ T AND T $=$ T ∎

The word *but* is logically equivalent to AND; however, in ordinary language it adds an element of contrast. For instance, "Albany's in New York, but Montpelier's in Vermont." is logically the same as "Albany's in New York and Montpelier's in Vermont." The *but* emphasizes the contrast between New York and Vermont. The word *while* can also mean AND, although not always. We can only tell from the context. In the sentence "Albany's in New York, while Montpelier's in Vermont," *while* has the same element of contrast as *but,* only it is not as strong. In "I finished my coffee while I was typing the term paper," *while* has a different meaning—that of simultaneity.

EXAMPLE 5   Symbolize the following sentences using AND, if possible.

The Iroquois lived in New York, but the Navahos lived in the Southwest.

Beethoven was born in 1770, while Bach died in 1750.

Beethoven composed great music while he was deaf.

SOLUTION
Let

$p =$ "the Iroquois lived in New York"

$q =$ "the Navahos lived in the Southwest"

$r =$ "Bach died in 1750"

$s =$ "Beethoven was born in 1770."

The first two can then be symbolized

$$p \text{ AND } q$$
$$r \text{ AND } s$$

The last cannot be symbolized using AND, as this use of *while* means simultaneity. ∎

Negation

Negation refers to the use of the word *not*. It is occasionally called a connective even though it is associated with only one proposition. Mathematically, it is denoted by ¬. The truth table for NOT is

p	$\neg p$
T	F
F	T

EXAMPLE 6 Determine the truth value of each of the following.

a) NOT $(1 > 2)$

b) $2^2 + 6 + 2 \neq 0$ AND $13.5 < 26.7$

SOLUTION

a) NOT $(1 > 2) = \neg(1 > 2) = \neg F = T$

b) $2^2 + 6 + 2 \neq 0$ AND $13.5 < 26.7 = \neg(2^2 + 6 + 2 = 0)$ AND $(13.5 < 26.7) = \neg F$ AND $T = T$ AND $T = T$ ∎

Lewis Carroll, the author of *Alice in Wonderland,* was also a logician. His book *Symbolic Logic* is an excellent source of interesting illustrations of the use of logic in ordinary speech. Sentences in the following example and a number of subsequent examples are from his book.

EXAMPLE 7 Write the following using the NOT explicity with a sentence in propositional form.

a) Wasps are unfriendly.

b) No old misers are cheerful.

SOLUTION

a) NOT (wasps are friendly)

b) NOT (old misers are cheerful)

To make a grammatically correct sentence out of these terms, the word NOT could be replaced by the phrase "IT IS NOT TRUE THAT." ∎

Disjunction

Disjunction refers to the use of the word *or.* It is symbolized mathematically by ∨. The English word *or* is ambiguous in that sometimes it refers to an **exclusive-or** and sometimes to an **inclusive-or.** By an exclusive-or we mean sentences like "I am planning to play golf Thursday or Friday." Typically this means that I plan to play either on Thursday or on Friday, but not both. With an inclusive-or, we allow the possibility of both being true. The exclusive-or is the more common meaning of *or* in ordinary speech. However, in mathematics, OR means the inclusive-or. If they intend the exclusive-or, mathematicians and computer scientists explicitly say so.

The truth tables for OR and XOR (short for exclusive-or) are

p	*q*	*p* OR *q*	*p*	*q*	*p* XOR *q*
T	T	T	T	T	F
T	F	T	T	F	T
F	T	T	F	T	T
F	F	F	F	F	F

EXAMPLE 8 Evaluate the following compound expressions.

a) $(1 > 2)$ OR $(4 > 3)$ **b)** $(22 > 7)$ XOR $(22 < 0)$

SOLUTION

a) $(1 > 2)$ OR $(4 > 3)$ = F OR T = T

b) $(22 > 7)$ XOR $(22 < 0)$ = T XOR F = T ∎

EXAMPLE 9 Symbolize the following.

a) Either he's lazy or very clever.

b) 1 is in the set S or in the set T.

c) She didn't want to come or she forgot the time.

SOLUTION

Let

p = "he's lazy," q = "he's very clever"

s = "1 is in S," u = "1 is in T"

c = "she wanted to come," w = "she forgot the time"

These can then be symbolized as

a) p XOR q **b)** s OR u **c)** $\neg c$ OR w. ∎

Conditional and Implication

Both the **conditional** and the **implication** refer to the relationship "if p then q." But there is a critical difference between them. The conditional refers to hypothesized relationships, i.e., ones that may or may not be true. Implication refers to established relationships, in which it has been proven that q can be deduced from p by valid rules of inference. In other words, implication refers to theorems, the conditional to statements that may or may not be theorems—in fact, that may be demonstrably false. The notation for the conditional is $\boldsymbol{p \to q}$, where both p and q must be propositions. The notation for implication is $\boldsymbol{p \Rightarrow q}$. $p \Rightarrow q$ is not considered a connective since it does not have a truth table; it denotes a statement known to be true. p is called the **premise,** the **hypothesis,** or the **antecedent** and q is called the **conclusion** or the **consequent.**

EXAMPLE 10   Write the following in symbolic form.

a) If $2 > 1$ then $0 > 1$.

b) If $1 \in \mathbf{N}$ then $1 \in \mathbf{Z}$.

c) If ducks waddle, then ducks are graceful.

SOLUTION   All of them can be written in the form $p \to q$. Note that it doesn't matter whether they are true or not, only that p and q are propositions.   ∎

The truth table for the conditional is:

p	q	$p \to q$
T	T	T
T	F	F
F	T	T
F	F	T

The first line of this table is the principal tool used in mathematical arguments. From the table we can see that if $p \to q$ is known to be true and p is true, then it necessarily follows that q is true. This pattern of reasoning was called **modus ponens** by the Greeks and is now occasionally called the **law of detachment.** We will discuss this pattern more fully in later sections. Note that the only time the conditional is false is when p is true and q is false. Intuitively, this says that an argument that starts with true premises and ends with false conclusions must itself be in error. The last two lines are somewhat less obvious. They say that with false premises and any conclusion, whether true or false, the statement "if p then q" is acceptable. With false premises, even if the reasoning is flawless, the conclusions are totally undependable; they could be true or false. Hence a seemingly ridiculous conditional like

If $1 = 2$ then watermelons are strawberries

is considered to be true because the premise is false. After we have discussed quantifiers in the next section, we will understand the principal reason for this situation. However, for now, it is enough to say that it points up the seriousness of having false premises and the critical importance of taking time to verify one's premises before beginning an argument. We may end up with a beautiful argument that establishes nothing if our premises are false.

There is an anecdote about the philosopher Bertrand Russell, who did much of the major work in symbolic logic at the beginning of this century. I have not been able to verify the truth of the anecdote, but it illustrates this point nicely. It seems that someone questioned Russell about the idea that, starting with false premises, any conditional is true, and challenged him to prove that if 1 equaled 2, he was the pope. Russell supposedly thought for a few moments and then said, "Suppose 1 equals 2. The pope and I are two, therefore the pope and I are one." Note that this is the same pattern used previously with the seemingly ridiculous conclusion about watermelons and strawberries. This shows that if $1 = 2$, that conclusion is not so ridiculous; it can actually be *proven* true.

Conditionals are often stated with predicates as well as with propositions. By the substitution of specific values, they can be changed to propositions and tested for truth or falsity for these values. Then their truth values for those particular values of the variables can be found by the truth table.

EXAMPLE 11   Test the following conditionals with the specified values.

a) If $x - 1 = 0$ then $x = 1$. Test with $x = 1$ and $x = 2$.

b) If $a, b,$ and c are the sides of a right triangle, then $a^2 + b^2 = c^2$.

Test with $a = 3, b = 4, c = 5$ and with $a = 1, b = 2, c = 3$.

c) If $x^2 - 1 = 0$, then $x = -1$. Test with $x = -1$ and $x = 1$.

SOLUTION

a) First, let $x = 1$. The conditional simplifies as follows:

$$0 = 0 \rightarrow 1 = 1$$
$$T \rightarrow T$$
$$T.$$

Now let $x = 2$. It becomes

$$1 = 0 \rightarrow 2 = 1$$
$$F \rightarrow F$$
$$T.$$

In both cases the conditional "If $x - 1 = 0$ then $x = 1$" is true.

b) 3, 4, and 5 are the sides of a right triangle and $3^2 + 4^2 = 5^2$. Thus this is of the form

$$T \rightarrow T,$$

and hence the conditional is true in this case. 1, 2, and 3 are not the sides of a right triangle and $1^2 + 2^2 \neq 3^2$. So the conditional is now of the form

$$F \rightarrow F,$$

which is also true.

c) If $x = -1$, the left side is $0 = 0$ and the right is $-1 = -1$. Hence we have the form $T \rightarrow T$, and hence the conditional is true in this case. If $x = 1$, the left side is $0 = 0$ and the right side is $1 = -1$. This time the conditional is of the form $T \rightarrow F$, and hence is false. ∎

The Biconditional

The **biconditional** is symbolized $p \leftrightarrow q$ and means the same thing as $p \rightarrow q$ AND $q \rightarrow p$. It is read as "p if and only if q" and is sometimes written as "p iff q." The truth table for the biconditional is

p	q	$p \leftrightarrow q$
T	T	T
T	F	F
F	T	F
F	F	T

EXAMPLE 12 Determine the truth value of the following.

a) $1 = 2$ if and only if watermelons are strawberries.

b) $x^2 = 1$ if and only if $x = 1$ or $x = -1$. Evaluate for $x = 1$ and 0.

c) R is an isosceles triangle iff R is an equilateral triangle.

SOLUTION

a) This is of the form $F \leftrightarrow F$ and hence is true.

b) When $x = 1$, this becomes $1 = 1 \leftrightarrow (1 = 1$ or $1 = -1)$; that is,

$$T \leftrightarrow (T \text{ or } F)$$
$$T \leftrightarrow T$$
$$T.$$

When $x = 0$, this becomes $F \leftrightarrow F$ and hence is again true.

c) If R is a triangle with sides 2, 2, and 2, this statement is of the form $T \leftrightarrow T$ and hence is true. However, if R is a triangle with sides 2, 2, and 3, the statement is of the form $T \leftrightarrow F$ and hence is false. ∎

Suppose the expression "R is an isosceles triangle iff R is an equilateral triangle" is symbolized as $p \leftrightarrow q$. Then $q \rightarrow p$ is always true but $p \rightarrow q$ is only sometimes true. Care must be exercised in splitting the biconditional "p if and only if q" into its component conditionals. "p if q" can be reworded "if q then p" and hence is symbolized by $q \rightarrow p$. "p only if q" means "if p then q" and is symbolized $p \rightarrow q$.

TERMINOLOGY

connective	negation	premise
compound proposition	disjunction	hypothesis
boolean expression	exclusive-or	antecedent
conjunction	inclusive-or	conclusion
truth table	conditional	consequent
tree diagram	implication	modus ponens
root	$p \rightarrow q$	law of detachment
branch	$p \Rightarrow q$	biconditional

SUMMARY

1. The principal connectives are AND, OR, NOT, the conditional ($\rightarrow$), and the biconditional ($\leftrightarrow$).

2. The truth tables for these connectives are

p	q	p AND q	p	q	p OR q	p	NOT p	p	q	$p \rightarrow q$	p	q	$p \leftrightarrow q$
T	T	T	T	T	T	T	F	T	T	T	T	T	T
T	F	F	T	F	T	F	T	T	F	F	T	F	F
F	T	F	F	T	T			F	T	T	F	T	F
F	F	F	F	F	F			F	F	T	F	F	T

3. Both the conditional and implication are stated using the words "IF ... THEN." The conditional is a hypothesized relationship; it may or may not be true. The implication refers to an IF ... THEN relationship that has been established to be true; it can be proven that assuming the truth of p, the truth of q necessarily follows. Since it is always true, implication does not have a truth table.

EXERCISES 3.3

Write each of the following sentences symbolically, using a letter such as p or q to denote each simple proposition and a connective, or explain why it is not possible.

1. Diane likes golf and John likes tennis.
2. Diane likes golf or John likes tennis.
3. If Diane likes golf then John likes tennis.
4. John likes tennis if and only if Diane does not like golf.
5. John likes tennis, but Diane likes golf.
6. John likes to play tennis while Diane plays golf.

Let p = "misers like money" and q = "my cousins are not greedy." Write the following as ordinary English sentences in as natural a style as possible.

7. p AND q 8. $\neg q$
9. $p \rightarrow \neg q$ 10. $\neg p$ OR q
11. $\neg p$ AND $\neg q$ 12. $p \leftrightarrow q$

Let $p(x) = x^2 > 4$, $q(x) = x \leq 2$, and $r(x) = x \geq 1$. Evaluate each of the following expressions for the given value of x.

13. $p(x)$ AND $q(x)$; $x = 1$ 14. $p(x)$ OR $q(x)$; $x = 1$
15. $p(x) \rightarrow q(x)$; $x = 1.5$ 16. $q(x)$ OR $\neg r(x)$; $x = 0$
17. $\neg p(x)$ OR $q(x)$; $x = 1$ 18. $p(x) \leftrightarrow q(x)$; $x = 2$

19. The connective NAND is defined by the rule that p NAND q is false when both p and q are true; otherwise it is true. Construct a truth table for NAND.
20. Suppose a propositional expression involves three variables, p, q, and r. Draw a tree diagram listing all of the possible combinations of values of p, q, and r that can occur.

3.4

EVALUATING MORE COMPLEX EXPRESSIONS

Simple propositions can be combined into compound propositions, using the connectives of the previous section. These can be combined into still more complex compound expressions. Such expressions arise frequently in mathematical proofs, in large computer programs, and in logical circuit design. Thus it is important to know how to evaluate and how to simplify them.

EXAMPLE 1   Evaluate $(p \text{ AND } q) \rightarrow r$ when p and q are true but r is false.

SOLUTION   We start with the innermost subexpression, substitute the truth values, and evaluate, using the truth tables for AND and $\rightarrow$. In this case

$$(p \text{ AND } q) \rightarrow r$$
$$(\text{T AND T}) \rightarrow \text{F}$$
$$\text{T} \rightarrow \text{F}$$
$$\text{F. } \blacksquare$$

Compound expressions involving predicates can also be evaluated by substitution for their free variables.

EXAMPLE 2   Evaluate $(x > 10) \text{ AND } (x < 20) \rightarrow (x \neq 0)$ when $x = 15$ and $x = 0$.

SOLUTION   When $x = 15$, the expression becomes

$$(15 > 10) \text{ AND } (15 < 20) \rightarrow (15 \neq 0)$$
$$\text{T AND T} \rightarrow \text{T}$$
$$\text{T} \rightarrow \text{T}$$
$$\text{T.}$$

When $x = 0$ the expression becomes

$$(0 > 10) \text{ AND } (0 < 20) \rightarrow (0 \neq 0)$$
$$\text{F AND T} \rightarrow \text{F}$$
$$\text{F} \rightarrow \text{F}$$
$$\text{T. } \blacksquare$$

Sometimes it is necessary to consider not just one possible set of values for the variables in a compound expression but all possible values. In this case, we set up truth tables. That is, we list all of the possible values for the variables and then work from the innermost expression out until the whole expression is evaluated.

EXAMPLE 3 Find the truth table for $(p$ AND $q) \rightarrow r$.

SOLUTION

p	q	r	$p \wedge q$	$(p \wedge q) \rightarrow r$
T	T	T	T	T
T	T	F	T	F
T	F	T	F	T
T	F	F	F	T
F	T	T	F	T
F	T	F	F	T
F	F	T	F	T
F	F	F	F	T

Note that the r column alternates T and F, the q column alternates two T's and two F's, and the p column alternates four T's and four F's. If there were four variables, we would need another column with eight T's followed by eight F's, etc. Once again, the lines in the table could be represented by a tree diagram, as we did in the previous section. See Figure 3.2.

Figure 3.2 Tree diagram indicating all possible values of three propositions.

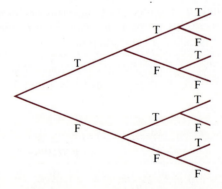

EXAMPLE 4 Find the truth table for $((p$ AND $q)$ OR $r) \rightarrow (p$ OR $\neg r)$.

SOLUTION

p	q	r	$p \wedge q$	$(p \wedge q) \vee r$	$\neg r$	$p \vee \neg r$	$((p \wedge q) \vee r) \rightarrow (p \vee \neg r)$
T	T	T	T	T	F	T	T
T	T	F	T	T	T	T	T
T	F	T	F	T	F	T	T
T	F	F	F	F	T	T	T
F	T	T	F	T	F	F	F
F	T	F	F	F	T	T	T
F	F	T	F	T	F	F	F
F	F	F	F	F	T	T	T

EXAMPLE 5   Find the truth table for $(p \rightarrow q)$ OR $(p \rightarrow \neg q)$.

SOLUTION

p	q	$p \rightarrow q$	$\neg q$	$p \rightarrow \neg q$	$(p \rightarrow q) \vee (p \rightarrow \neg q)$
T	T	T	F	F	T
T	F	F	T	T	T
F	T	T	F	T	T
F	F	T	T	T	T

Note that $(p \rightarrow q)$ OR $(p \rightarrow \neg q)$ is true for all values of p and q. Expressions that have this property are called **tautologies.** An expression that is always false is called a **contradiction.**

EXAMPLE 6   Verify that p XOR $\neg p$ is a tautology.

SOLUTION

p	$\neg p$	p XOR $\neg p$
T	F	T
F	T	T

That p XOR $\neg p$ is a tautology is the law of the excluded middle—that for any proposition p, exactly one of p and $\neg p$ must be true. The words *must be* in the ordinary English suggest the fact that the expression is a tautology.

Definition Two propositional expressions are **equivalent** if they have the same truth value for any values of their propositions. We denote the equivalence of p and q by $p \equiv q$.

EXAMPLE 7 Show that $\neg(p$ AND $q)$ is logically equivalent $(\neg p$ OR $\neg q)$.

SOLUTION

p	q	$p \wedge q$	$\neg p$	$\neg q$	$\neg(p \wedge q)$	$\neg p \vee \neg q$
T	T	T	F	F	F	F
T	F	F	F	T	T	T
F	T	F	T	F	T	T
F	F	F	T	T	T	T

The statements

$$\neg(p \wedge q) \equiv \neg p \vee \neg q$$

and

$$\neg(p \vee q) \equiv \neg p \wedge \neg q$$

are called **DeMorgan's Laws.** We proved the first of these statements in Example 7. The proof of the other is left for the Exercises.

EXAMPLE 8 Show that $p \rightarrow q \equiv \neg p$ OR q.

SOLUTION

p	q	$\neg p$	$p \rightarrow q$	$\neg p$ OR q
T	T	F	T	T
T	F	F	F	F
F	T	T	T	T
F	F	T	T	T

That $p \rightarrow q$ is equivalent to $\neg p$ OR q is especially helpful in deductive proof. We will say more about this in the next two sections.

The Algebra of Propositions

A number of basic properties of connectives can be summarized in the laws of the algebra of propositions.

Laws of the Algebra of Propositions

Idempotent Laws

1a. p AND $p \equiv p$ 1b. p OR $p \equiv p$

Associative Laws

2a. $(p$ AND $q)$ AND $r \equiv$ 2b. $(p$ OR $q)$ OR $r \equiv$
     p AND $(q$ AND $r)$ p OR $(q$ OR $r)$

Commutative Laws

3a. p AND $q \equiv q$ AND p 3b. p OR $q \equiv q$ OR p

Distributive Laws

4a. p OR $(q$ AND $r) \equiv$ 4b. p AND $(q$ OR $r) \equiv$
     $(p$ OR $q)$ AND $(p$ OR $r)$ $(p$ AND $q)$ OR $(p$ AND $r)$

Identity Laws

5a. p AND F $\equiv$ F   5b. p OR F $\equiv p$

6a. p AND T $\equiv p$ 6b. p OR T $\equiv$ T

Complement Laws

7a. p AND $\neg p \equiv$ F   7b. p OR $\neg p \equiv$ T

8a. $\neg\neg p \equiv p$ 8b. $\neg$T $\equiv$ F; $\neg$F $\equiv$ T

DeMorgan's Laws

9a. $\neg(p$ AND $q) \equiv \neg p$ OR $\neg q$ 9b. $\neg(p$ OR $q) \equiv \neg p$ AND $\neg q$

From Lipschutz, *Set Theory and Related Topics*, p. 195, © 1964. Reprinted by permission of McGraw-Hill Book Company, New York.

These laws can be proved by truth tables (Exercises 9 through 16 of this section and 19 through 26 of the supplementary exercises at the end of this chapter). They can also be used to help us simplify a number of expressions.

EXAMPLE 9 Simplify the following.

   a) $\neg(p$ AND $\neg q)$ **b)** $(p \rightarrow q)$ OR $(p \rightarrow \neg q)$

SOLUTION

   a) $\neg(p$ AND $\neg q) \equiv$
         $\neg p$ OR $\neg\neg q \equiv$ (by 9a)
         $\neg p$ OR q (by 8a)

b) $p \rightarrow q$) OR ($p \rightarrow \neg q$) $\equiv$
($\neg p$ OR q) OR ($\neg p$ OR $\neg q$) $\equiv$ (by Ex. 8)
($\neg p$ OR q) OR ($\neg q$ OR $\neg p$) $\equiv$ (by 3b)
$\neg p$ OR (q OR ($\neg q$ OR $\neg p$)) $\equiv$ (by 2b)
$\neg p$ OR ((q OR $\neg q$) OR $\neg p$) $\equiv$ (by 2b)
$\neg p$ OR (T OR $\neg p$) $\equiv$ (by 7b)
$\neg p$ OR T $\equiv$ (by 6b)
T   (by 6b) ∎

Note that Example 9b simplifies to T. This is consistent with Example 5, in which we showed that ($p \rightarrow q$) OR ($p \rightarrow \neg q$) is a tautology.

EXAMPLE 10   A computer program contains the line

IF NOT ((ALPHA $\geq$ 'Z') AND (COUNT = LIMIT)) THEN . . .

where LIMIT is some constant. Simplify the IF portion of this conditional.

SOLUTION   This line is actually made up of two predicates:

$$p(\text{ALPHA}) = (\text{ALPHA} \geq \text{'Z'})$$

and

$$q(\text{COUNT}) = (\text{COUNT} = \text{LIMIT}).$$

The portion between IF and THEN can be written as

$$\neg(p(\text{ALPHA}) \text{ AND } q(\text{COUNT})),$$

which is equivalent to $\neg p(\text{ALPHA})$ OR $\neg q(\text{COUNT})$. Thus the line could be rewritten as

IF (ALPHA $<$ 'Z') OR (COUNT $\neq$ LIMIT) THEN . . .

This is usually regarded as simpler, since it isn't preceded by NOT. ∎

Another way to simplify complicated boolean expressions in computer programs is to introduce new names for predicates. For instance, let

ALPHAOK denote the expression (ALPHA $<$ 'Z')

and

COUNTOK denote (COUNT $\neq$ LIMIT).

Then the conditional could be replaced by the following three statements (the symbol := is read "becomes" and will be explained in Section 4.1):

ALPHAOK := (ALPHA $<$ 'Z');
COUNTOK := (COUNT $\neq$ LIMIT);
IF ALPHAOK OR COUNTOK THEN . . .

TERMINOLOGY

truth table   equivalent expressions

tautology   contradiction

DeMorgan's laws

SUMMARY

1. Truth tables can be used to tabulate the truth values of arbitarily complex compound propositions.
2. The laws of the algebra of propositions summarize several of the main properties of propositions and connectives.

EXERCISES 3.4

Construct truth tables for the following.

1. $(p \text{ OR } q) \rightarrow r$
2. $p \text{ AND } (q \rightarrow r)$
3. $p \text{ AND } \neg q$
4. $r \rightarrow (\neg r \text{ OR } s)$

Evaluate the following predicates for the given values of their free variables.

5. $(x > 1) \rightarrow (x \neq 0); x = 2$
6. $((y > 10) \text{ OR } (y < -10)) \rightarrow (y < 100); y = 0$
7. $(x < 100) \rightarrow (y \neq 0); x = 0, y = 0$
8. $((x > -1) \text{ AND } (y < 1)) \rightarrow (xy < 0); x = 0, y = 0$

Use truth tables to verify the following laws for the algebra of propositions.

9. 1a
10. 2a
11. 3a
12. 4a
13. 5a
14. 6a
15. 7a
16. 8a

Prove that the following are tautologies.

17. $((p \rightarrow q) \text{ AND } p) \rightarrow q$
18. $((p \rightarrow q) \text{ AND } (q \rightarrow r)) \rightarrow (p \rightarrow r)$
19. $((p \rightarrow q) \text{ AND } \neg q) \rightarrow \neg p$
20. $((p \rightarrow (q \text{ AND } \neg q)) \rightarrow p)$

Simplify the following expressions:

21. $\neg(\neg p \text{ AND } \neg q)$
22. $\neg(p \rightarrow \neg q)$

23. $\neg(\neg p \text{ OR } (p \to \neg q))$

24. $(p \to q) \text{ AND } (p \to \neg q)$

25. $p \wedge (p \vee \neg p)$

26. $p \vee (p \wedge q)$

Symbolize the condition in each of the following situations and write it as simply as possible.

27. An employee is to be paid overtime for a particular hour of work if she works more than 40 hours in a week, more than 8 hours in a day, or on Saturday or Sunday.

28. If the number of users is less than 16 and either the memory requested is less than 1 megabyte or there are at least 10 small buffers free, then the system is not overloaded.

3.5
PREDICATE LOGIC

So far in this chapter we have focused primarily on propositional logic—that is, the concept of the proposition, the major connectives, and the evaluation and simplification of compound propositional expressions. We have also seen that predicates become propositions when values are substituted for their free variables and that they can then be treated by the methods of propositional logic.

There are, however, a number of ways in which predicates are quite different from propositions. For instance $x^2 + 5x + 6 = 0$ and $x^2 + 1 \neq 0$ are predicates, each with the free variable x. We often want to say things like "There is an integer x such that $x^2 + 5x + 6 = 0$" and "For every integer $x, x^2 + 1 \neq 0$." It is primarily the use of *there is* and *for every* with predicates that distinguishes them from propositions. These two phrases are called **quantifiers** and are the principal subject of this section. Quantifiers are essential for mathematics, as most theorems are stated in terms of them. Also, they considerably expand our capacity to represent natural language statements symbolically.

Representing Connectives by Venn Diagrams

We will start by analyzing how we can use Venn diagrams to illustrate the use of the major connectives with predicates. Suppose we have some universal set X and a predicate $p(x)$. Let T_p denote the truth set of $p(x)$. That is,

$$T_p = \{x \in X \mid p(x) \text{ is true}\}.$$

Thus T_p can be represented by the following Venn diagram (Figure 3.3):

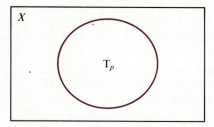

Figure 3.3 Venn diagram for the truth set of a predicate $p(x)$.

Now let $T_q = \{x \in X \mid q(x) \text{ is true}\}$ for some other predicate q. Consider the statement $p(x)$ AND $q(x)$. It is true if and only if both $p(x)$ and $q(x)$ are true for the same x. In other words, its truth set is the intersection of the truth sets of $p(x)$ and $q(x)$. Applying the same kind of reasoning to the other connectives, we have the following correspondences between compound predicates and compound sets:

Expression	Truth Set
$p(x)$ AND $q(x)$	$T_p \cap T_q$
$p(x)$ OR $q(x)$	$T_p \cup T_q$
$\neg p(x)$	T_p'
$p(x) \rightarrow q(x)$	$T_p' \cup T_q$

The truth set for $p(x) \rightarrow q(x)$ is based on the fact that $p \rightarrow q$ is equivalent to $\neg p \lor q$, which was shown earlier. Thus we have the following Venn diagrams for the truth sets of each of the preceding compound predicates (Figure 3.4).

The Venn diagram for the implication is Figure 3.5. $p(x) \Rightarrow q(x)$ is always true. That is, for any x for which $p(x)$ is true, $q(x)$ must also be true. Thus the Venn diagram for $p(x) \Rightarrow q(x)$ is $T_p \subset T_q$.

Figure 3.4 Venn diagrams for compound predicates.

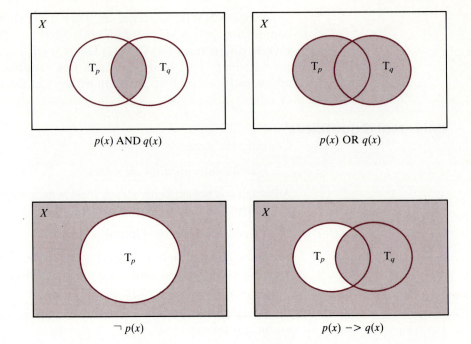

Figure 3.5 Venn diagram for $p(x) \Rightarrow q(x)$.

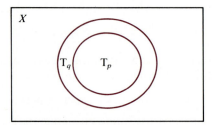

Quantifiers

The two principal quantifiers are *for all* and *there exists. For all* is called the **universal quantifier** and can also be stated as *for every, for each, in every case,* and in other ways as well. It is symbolized by ∀ an upside down uppercase *A. There exists* is called the **existential quantifier** and is often expressed as *there is* or by the use of *some.* It is symbolized by ∃, a backward uppercase E.

Suppose **N** is the universal set. Then $\forall x(p(x))$ is essentially an infinite conjunction: $p(1)$ AND $p(2)$ AND $p(3)$ AND .... Similarly $\exists x(p(x))$ is an infinite disjunction: $p(1)$ OR $p(2)$ OR $p(3)$ OR .... Recall that we called the variable x in a predicate $p(x)$ a **free variable.** When a quantifier is added, as in $\forall x(p(x))$ or $\exists x(p(x))$, we say that the variable is **bound.**

Universal Quantifiers[†]

Consider the sentence

Planets move in elliptical orbits.

Although it is not explicitly stated, there is an implicit "all" in this sentence and others like it. It has the same meaning as

All planets move in elliptical orbits.

Since our aim here is to develop a symbolic means of representing logic, we need to move toward a symbolic representation of sentences like this one. As a first step, we rephrase this sentence as

For all heavenly bodies, if the body is a planet then it moves in an elliptical orbit.

If we replace *heavenly body* by a symbol, we get

For all x, if x is a planet, then x moves in an elliptical orbit.

Finally, if we let $p(x)$ denote *"x is a planet,"* $q(x)$ denote *"x moves in an elliptical orbit,"* and X denote the set of all heavenly bodies, we get

$$\forall x \in X \ (p(x) \rightarrow q(x)).$$

If we can assume that the universal set is something anyone reading what we are writing will know, we can simply write

$$\forall x \ (p(x) \rightarrow q(x)).$$

Note that there is no assumption here that the statement $\forall x \ (p(x) \rightarrow q(x))$ is necessarily true. Similarly there is nothing implicit in the sentence "Planets move in elliptical orbits" that makes it true. We know it to be a fact, but that knowledge requires information outside of the sentence itself. Knowing that an expression of the form $\forall x(p(x) \rightarrow q(x))$ is true is the same as knowing that $p(x) \Rightarrow q(x)$.

There is another way to look at this situation. Suppose that instead of choosing the universal set, X, to be {heavenly bodies}, we had chosen it to

---

[†] The introduction to the universal quantifier given here was inspired by a similar introduction in Hermes's text.

be {planets}. Then the preceding representations would become

For all $x \in X$, x moves in an elliptical orbit.

Symbolically we represent this by

$$\forall x \in X(q(x)).$$

Thus, in summary, *for all* statements can often be written as a conditional with a universal quantifier. If the universal set is chosen appropriately, it is also often possible to represent them more simply as a universal quantifier followed by a single predicate.

EXAMPLE 1 Write the following relationships symbolically, using the universal quantifier.

a) All eggs can be cracked.
b) If $|S| = n$ then $|P(S)| = 2^n$.
c) All triangles have 180°.

SOLUTION

a) First, reword the sentence as follows:

For all edible objects x, if x is an egg, then x can be cracked.

Now let $X = \{$edible objects$\}$, $e(x)$ denote "x is an egg," and $b(x)$ denote "x can be cracked." Then it can be written as

$$\forall x(e(x) \rightarrow b(x)).$$

Alternatively, we could reword the sentence as

For all eggs x, x can be cracked.

Letting E stand for the set of all eggs, this would be represented as

$$\forall x \in E(b(x))$$

b) Theorems like this one typically have a *for all* implicit in them. In this case, we could say,

For all sets S, If $|S| = n$ then $|P(S)| = 2^n$.

If T denotes the set of all sets, $c(S)$ denotes "$|S| = n$," and $p(S)$ denotes "$|P(S)| 2^n$," this theorem can be written as

$$\forall S \in T(c(S) \rightarrow p(S)).$$

c) Let $R = \{$all triangles$\}$ and $D(t) = $ "t has 180°." Then this can be written most conveniently as

$$\forall t \in R(D(t)). \quad \blacksquare$$

Existential Quantifiers

The existential quantifier symbolizes the words *there is* or *for some.* In ordinary English we make a distinction between these two phrases that we do not make in formal logic. That is, the sentence "Some dogs are collies" carries with it a suggestion that there is more than one collie. If, in fact, there were only one collie in the world, most people would regard the statement as false. Symbolic logic, however, ignores this distinction because having only two quantifiers greatly simplifies the process of handling sentences containing negatives, as we shall see shortly. Thus, in symbolic logic, the phrases *there is one, there are some, for some,* and other such statements are all treated as equivalent to each other; they all assert the existence of something. How many exist is ignored.

EXAMPLE 2   Represent the following, using the existential quantifier.

a) Some merchants are rich.

b) There is at least one winner in a tournament.

c) There is an integer solution to the equation $x^2 + 5x + 6 = 0$.

SOLUTION

a) Let $X = \{$merchants$\}$ and $r(x) = $ "x is rich." We first reword the sentence as

$$\text{For some merchants } x, x \text{ is rich.}$$

This can be then be written as

$$\exists \, x \in X \, (r(x)).$$

Compare this result to the sentence "All merchants are rich." This sentence could be reworded

$$\text{For all merchants } x, x \text{ is rich.}$$

This could then be written

$$\forall x \in X \, (r(x))$$

which exactly parallels our expression for "Some merchants are rich." However, if we choose the universal set to be $P = \{$people$\}$ and let $m(x) = $ "x is a merchant," "All merchants are rich" becomes

$$\forall X \in P \, (m(x) \rightarrow r(x)),$$

while "Some merchants are rich" becomes

$$\exists x \in P \, (m(x) \text{ AND } r(x)).$$

The expressions are not parallel. This difference arises because if we allow the universal set to be {people}, we have to take into account the truth value of the statements even when x is *not* a merchant. If we symbolized "Some merchants are rich" by $\exists x \in P\,(m(x) \rightarrow r(x))$, the statement would be true whenever x is not a merchant! Thus the statement could be true even if there were no rich merchants.

When dealing with existential quantifiers, the best way to handle this difficulty is to choose the universal set to be specific to the situation described, if possible.

b) Let $T = $ {participants in the tournament} and $w(t) = $ "t is a winner." This could be written as

$$\exists t \in T\,(w(t)).$$

c) Let $R = $ {triangles} and $i(r) = $ "r is isosceles." One way to write this is

$$\exists r \in R\,(i(r)). \quad \blacksquare$$

Negations Involving Quantifiers

Sentences such as "There is no integer solution of $x^2 + 1 = 0$" and "Not all triangles are equilateral" arise frequently. Such sentences involve both quantifiers and negation. For instance, the first sentence in the preceding statement could also be written as "For all x, $x^2 + 1 \neq 0$" and the second as "There is a triangle that is not equilateral." The relationship between the quantifiers and negation can be symbolized as follows for any predicates $p(x)$ and $q(x)$:

$$\neg \forall x\,(p(x)) \equiv \exists x\,(\neg p(x))$$
$$\neg \exists x\,(q(x)) \equiv \forall x\,(\neg q(x)).$$

These relationships can be easily verified by examining cases in which the predicates are true and in which they are false.

Note that these statements are generalizations of DeMorgan's laws; i.e., if $x \in \mathbf{N}$,

$$\neg \forall x\,(p(x)) = \neg(p(1) \wedge p(2) \wedge p(3) \wedge \cdots)$$

and

$$\exists x\,(\neg p(x)) = \neg p(1) \vee \neg p(2) \vee \neg p(3) \vee \cdots.$$

Similar statements can be made about $\neg \exists x\,(q(x))$ and $\forall x\,(\neg q(x))$.

EXAMPLE 3 Symbolize the following, using both the existential and the universal quantifier.

a) No misers are generous.

b) Some equations do not have solutions.

c) Railways are never ill-managed.

SOLUTION

a) Let $X = \{$misers$\}$ and $g(x) = $ "x is generous." We first reword the statement as

> There does not exist a miser who is generous.

This could be written as

$$\neg \exists x \in X(g(x)).$$

According to the preceding rules, this is equivalent to

$$\forall x \in X(\neg g(x)),$$

which can be reworded as "For all misers x, x is not generous," or more simply, as "All misers are not generous," or even "All misers are stingy."

b) Let $E = \{$equations$\}$ and $s(e) = $ "e has a solution." Then this is

$$\exists e \in E\ (\neg s(e)) \equiv \neg \forall e \in E\ (s(e)).$$

That is, "Not all equations have solutions."

c) Let $L = \{$railways$\}$ and $i(x) = $ "x is ill-managed." Then we can write

$$\forall x(\neg i(x)) \equiv \neg \exists x(i(x))$$

or "There are no railways that are ill-managed." ∎

Multiple Quantifiers

The fundamental theorem of algebra is: "For every polynomial P, there is a complex number x such that $P(x) = 0$." Even though we haven't discussed complex numbers, we can still examine the logical structure of this statement. If we let Y denote the set of all polynomials and $\mathbf{C}$ denote the complex numbers, this can be symbolized as

$$\forall P \in Y\ \exists x \in \mathbf{C}\ (P(x) = 0).$$

Suppose we wanted to negate this statement. Since we know the fundamental theorem of algebra to be true, this negation is certainly false. But we often want to think hypothetically by forming statements like "Suppose this theorem were false. What would that mean?" The negation would be stated as follows:

$$\neg(\forall P \in Y\ \exists x \in \mathbf{C}\ (P(x) = 0)) \equiv$$
$$\exists P \in Y\ (\neg \exists x \in \mathbf{C}\ (P(x) = 0)) \equiv$$
$$\exists P \in Y\ \forall x \in \mathbf{C}\ \neg(P(x) = 0) \equiv$$
$$\exists P \in Y\ \forall x \in \mathbf{C}\ (P(x) \neq 0).$$

That is, the negation of "For every polynomial P, there is a complex number x such that $P(x) = 0$" is "There is at least one polynomial for which $P(x)$ is always not equal to 0" or, more simply, "There is at least one polynomial for which $P(x)$ is never 0."

The process of negating an expression that involves one or more quantifiers in front of a predicate can be summarized with this rule:

Reverse the quantifiers, then negate the predicate.

For instance, consider a four-variable predicate $P(a, b, c, d)$ and the following expression:

$$\neg \forall a\ \exists b\ \forall c\ \exists d(P(a, b, c, d)).$$

This is equivalent to

$$\exists a\ \forall b\ \exists c\ \forall d(\neg P(a, b, c, d)).$$

Proof and Disproof

How would we prove a statement of the form $p(x) \rightarrow q(x)$? T_p is the truth set of $p(x)$, T_q is the truth set of $q(x)$, and so the implication is equivalent to the statement that $\mathrm{T}_p \subset \mathrm{T}_q$. Thus we would prove this statement in the same way we would prove that one set is a subset of another. We take a representative element of the first set and show that it must be an element of the second set. In other words, we take a representative x for which we know that $p(x)$ is true and reason deductively to the conclusion that $q(x)$ must also be true. The actual process of carrying out this reasoning is the subject of the next section.

Disproving a statement of the form $\forall x \in X\ (p(x) \rightarrow q(x))$ is often much easier than proving it. To disprove it, we need to show that its negation is true; i.e., we need to prove that

$$\neg \forall x \in X\ (p(x) \rightarrow q(x)).$$

But this is equivalent to

$$\exists x \in X\ (\neg(p(x) \rightarrow q(x))).$$

That is, we need find only one x for which $p(x) \rightarrow q(x)$ is false. But

$$\neg(p(x) \rightarrow q(x)) \equiv p(x)\ \text{AND}\ \neg q(x).$$

Thus, to disprove $\forall x \in X(p(x) \rightarrow q(x))$, we need only find one example of an x for which $p(x)$ is true and $q(x)$ is false. An example that demonstrates that a universally quantified statement is false is called a **counterexample.** For instance, to disprove the statement "All misers are not generous," we need only find one example of a miser who is generous. To disprove the statement "All triangles are equilateral," we need only find one triangle that is not equilateral.

EXAMPLE 4   Disprove the statement "Every polynomial has at least one root in the real numbers."

SOLUTION   For every real number x, the polynomial $x^2 + 1$ is greater than or equal to 1. Hence it is never 0 and thus has no real roots.   ∎

Note the contrast between this and the fundamental theorem of algebra. There x is allowed to be a complex number. In that case, every polynomial does have a root.

Proving Statements with Existential Quantifiers

Proving statements with existential quantifiers is similar to disproving statements with universal quantifiers. For instance, to prove a statement of the form $\exists x \in X(p(x) \rightarrow q(x))$ or a statement of the form $\exists x \in X(p(x))$, it is sufficient to find an x for which the predicate is true.

EXAMPLE 5   Prove that $x^2 + 5x + 6$ has an integer root.

SOLUTION   Let $x = -2$. Then $(-2)^2 + 5(-2) + 6 = 4 - 10 + 6 = 0$.   ∎

Disproving statements with existential quantifiers will be left for the Exercises.

False Antecedents

Recall that in the previous section we discussed the truth table for $p \rightarrow q$ and pointed out that when p is false, the conditional is true regardless of whether q is true or false. We are now in a position to understand why that is the case.

Earlier we said that having the statement $\forall x \in X(p(x) \rightarrow q(x))$ be true is the same as $T_p \subset T_q$. Thus it is possible that there will be x's in T_q that are not in T_p, and it is also possible that there will be x's in neither. In the first case, substituting x into $p(x) \rightarrow q(x)$ will give $F \rightarrow T$ and in the second case will give $F \rightarrow F$. But the fact that there are such x's in no way violates the truth of the statement $\forall x \in X(p(x) \rightarrow q(x))$. Thus statements of the form $F \rightarrow F$ and $F \rightarrow T$ must be true.

For instance, suppose $\forall x \in X(p(x) \rightarrow q(x))$ is the theorem "If R is a rectangle, then the sum of the angles of R is 360°." If P is a pentagon, substituting P for R in the theorem yields a statement of the form $F \rightarrow F$. Similarly, if Z is a trapezoid, substituting Z for R yields a statement of the form $F \rightarrow T$.

Note also that if $p(x)$ is always false, T_p is the empty set, $\emptyset$. Then $\emptyset \subset T_q$ for any T_q since $\emptyset$ is a subset of every set. Hence if $p(x)$ is always false, $T_p \subset T_q$ and the statement $p(x) \Rightarrow q(x)$ is true. An important consequence of this idea is the concept known as **vacuous proof.** For instance, suppose I make the statement

All the people in that room have red hair.

Suppose you go to that room and come back, saying, "There's no one in that room." In that case, my statement was true. To understand this, let's analyze the statement. Let $P = \{people\}$, $r(x) = $ "x is in that room," and $h(x) = $ "x has red hair." The statement can then be symbolized as

$$\forall x \in P \; (r(x) \rightarrow h(x)).$$

We are dealing with a case in which $r(x)$ is false for all x, i.e., $T_r = \emptyset$. Since $T_r \subset T_h$, the statement is true and we can conclude that $r(x) \Rightarrow h(x)$.

Philosophers tell me that Aristotle would probably not have accepted this statement as true but would have treated it as meaningless. But mathematically, the empty set gives us a way to deal with such situations. To be consistent with the notion that the empty set is a subset of every set, we must treat such statements as true. By the way, after you returned, I could add, "You know, all the people in that room have black hair," and that would also be true.

We will see a number of mathematical examples of vacuous proof later.

TERMINOLOGY

quantifier	bound variable
universal quantifier	counter-example
existential quantifier	vacuous proof
free variable	

SUMMARY

1. The universal quantifier, symbolized as $\forall$, is used with predicates to represent the phrases *for all* and *for every.* The existential quantifier, symbolized by $\exists$, is used with predicates to represent *there is, there exists,* and *for some.* No distinction is made between *there is one* and *there are some.*

2. If X is a universal set and $p(x)$ and $q(x)$ are predicates,

$$\neg \forall x(p(x)) \equiv \exists x(\neg p(x))$$
$$\neg \exists x(q(x)) \equiv \forall x(\neg q(x))$$

3. Suppose $p(x)$ and $q(x)$ are predicates and

$$T_p = \{x \in X \mid p(x) \text{ is true}\}$$

and

$$T_q = \{x \in X \mid q(x) \text{ is true}\}.$$

Then the statement $\forall x(p(x) \rightarrow q(x))$ being true is equivalent to $p(x) \Rightarrow q(x)$, and both are equivalent to $T_p \subset T_q$.

4. To disprove a statement of the form $\forall x(p(x))$, it is only necessary to find one x for which it is false; to prove a statement of the form $\exists x(p(x))$, it is only necessary to find one x for which it is true.

5. The statement $p(x) \Rightarrow q(x)$ is regarded as an implication if it can be shown that $p(x)$ is always false. Such a demonstration is known as a vacuous proof.

EXERCISES 3.5

Let the universal set be **N**, let $p(x) = x^2 < 1$, $q(x) = x - 10 > 0$, and $s(x) = x > 0$. Evaluate the truth value of the following.

1. $\forall x \; p(x)$ 2. $\exists x \; p(x)$
3. $\exists x \; \neg p(x)$ 4. $\forall x \; q(x)$
5. $\forall x \; \neg q(x)$ 6. $\exists x \; \neg q(x)$
7. $\exists x \; (p(x) \text{ AND } s(x))$ 8. $\forall x \; (s(x) \rightarrow q(x))$

Again, let the universal set be **Z**, let $p(x, y) = x + y = 0$, and $q(x, y) = x^2 + y^2 = 1$. Evaluate the truth value of each of the following.

9. $\forall x \exists y \; p(x, y)$ 10. $\exists x \exists y \; p(x, y)$
11. $\forall x \forall y \; q(x, y)$ 12. $\exists x \exists y \; q(x, y)$

Write the following symbolically. In each case, specify clearly what your universal set is and what each symbol stands for.

13. Some skates are fish.

14. All medicine is nasty.

15. Rainy days are tiresome.

16. No judges are unjust.

17. No riddles that can be solved interest me.

18. All green-eyed cats are bad-tempered.

19. A song that lasts for an hour is tedious.

Let the universal set again be the natural numbers. Disprove each of the statements in Exercises 20 through 22 by finding a counter-example.

20. $\forall x \, (x^2 - 1 > 0)$

21. $\forall x \, (x$ divisible by 2 $\Rightarrow x/2$ is even)

22. $\forall x \, (x$ is not an even prime)

23. Find an equivalent expression to $\forall x \, (p(x))$ that uses only $\exists$ and $\neg$.

The statements in Exercises 24–27 can all be written $\exists x \, p(x))$. Prove each by finding x_0 for which $p(x_0)$ is true.

24. Some isosceles triangles are equilateral.

25. There are even numbers divisible by 3.

26. There is an integer solution to the equation $x^2 - 12x + 35 = 0$.

27. There is a prime number between 9^2 and 10^2.

28. In calculus, the following statement is quite important:

$$\forall \epsilon > 0 \; \exists \delta > 0 \, (\,|x - a| \, < \delta \Rightarrow \, |f(x) - f(a)| \, < \epsilon).$$

Write out its negation as simply as possible.

3.6

RULES OF INFERENCE

Why Is Proof Important?

For a mathematician, the ability to read, understand, and do original proofs is essential. The mathematical modeling life cycle depends upon the mathematician's capacity to deduce conclusions and predictions from models. For the computer scientist, social scientist, or user of discrete math from some other discipline, the claim that a certain result is true still needs to be proven. Thus, unless one can read and understand proofs, one is cut off from literature in which new results in mathematics are published. Also, the experience of proving a theorem about some aspect of mathematics is often the key to understanding it. Lastly, for the computer scientist in particular, there are two additional reasons why proof is important. First, computer scientists typically invest much time in the design of algorithms. A proof is very similar to an algorithm. Thus the experience of writing proofs is excellent preparation for the design of algorithms. Second, as computer science matures, there is a growing emphasis on its rigorous aspects. For instance, one important topic in computer science is algorithm verification, or, as some people express it, "proof of programs." This requires a thorough grounding in the process of mathematical proof.

What Is Proof?

Given the importance of proof in mathematics, let's review what a proof is. In formal axiomatics, we start with axioms and undefined terms. Definitions are then added. Theorems are deduced from the axioms, definitions, and previously proven theorems by a finite number of steps that use the allowable rules of inference. Thus proofs are similar to algorithms. Once a theorem has been proven, it can then be used, along with axioms and definitions, in the proof of further theorems. Most theorems are of one of the forms $\forall x(p(x))$, $\forall x(p(x) \rightarrow q(x))$, or $\exists x(p(x))$. To prove each of these theorems, we proceed as follows:

> To prove $\forall x(p(x))$, we must take a representative x in the universal set and show that $p(x)$ is true for that x.
>
> To prove $\forall x(p(x) \rightarrow q(x))$, we must take a representative x, assume $p(x)$, and deduce $q(x)$ by valid rules of inference.
>
> To prove $\exists x(p(x))$, we need to find any one x_0 for which $p(x_0)$ is true.

Thus, to complete our explanation of proof, we need to study the valid rules of inference. In the examples that follow, we will draw most of our examples from set theory, since that is all we have to work with at this point. We will be using a more informal approach than formal axiomatics requires. That is, we will be using the naive approach discussed in Section 3.1. However, we must still be careful to avoid using anything that has not been previously defined or proven and to use valid rules of inference throughout.

Modus Ponens

The first and by far the most important **rule of inference** that we will examine was called **modus ponens** by the Greeks. It is also sometimes called the **law of detachment.** If we let p and q be propositions, it can be stated as follows:

$$p \Rightarrow q$$
$$\underline{p}$$
$$q$$

That is, if $p \Rightarrow q$ and p is true, we can conclude that q must be true. Thus, for instance, suppose $p \Rightarrow q$ is the statement "If R is a rectangle then the angles of R sum to 360°." Suppose also that R is a rectangle. The conclusion that the angles of R sum to 360° is valid, by modus ponens. There are several ways to understand why modus ponens is valid rule of inference.

Consider first the truth table for $p \rightarrow q$. If $p \Rightarrow q$ then $p \rightarrow q$ is true, and we can eliminate the second line as a possibility. Also, if we know that p is true, we have eliminated the third and fourth lines. So the only line left is the first, and we can see from that line that q must be true.

p	q	$p \rightarrow q$
T	T	T
T	F	F
F	T	T
F	F	T

Another way to see this is by the fact that

$$((p \rightarrow q) \text{ AND } p) \rightarrow q$$

is a tautology. This fact tells us that it is impossible to have the left side of the $\rightarrow$ be true and the right side be false. Thus whenever $p \rightarrow q$ and p are true, no matter what propositions p and q might be, we can always conclude that q is true.

Modus ponens can perhaps be better stated in terms of predicates. Here is a classical example:

> All men are mortal.
>
> Socrates is a man.
>
> Therefore, Socrates is mortal.

The first line is of the form $\forall x \in X \, (p(x) \rightarrow q(x))$, or equivalently, $p(x) \Rightarrow q(x)$ for predicates $p(x)$ and $q(x)$ where $p(x) =$ "x is a man" and $q(x) =$ "x is mortal." The second line says that $p(\text{Socrates})$ is true. Thus the argument follows this pattern, using modus ponens:

$$p(x) \Rightarrow q(x)$$
$$\underline{p(\text{Socrates})}$$
$$q(\text{Socrates})$$

Alternatively, we could use the following Venn diagram (Figure 3.6) and conclude $q(\text{Socrates})$ since $T_p \subset T_q$ and Socrates is a member of T_p. In other words, modus ponens, even though it is defined in terms of propositions and implication, is an application of the idea that if $A \subset B$, membership in A also guarantees membership in B.

Figure 3.6 Representation of modus ponens: $p(x_0)$ is true and $p(x) \Rightarrow q(x)$. Hence $q(x_0)$ is true also.

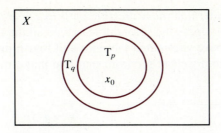

EXAMPLE 1 Prove the following.

a) $A \subset B, A \subset C \Rightarrow A \subset B \cap C$

b) $A \subset B \Rightarrow A \cap B = A$

SOLUTION

a) Since this is our first proof, we will write it in complete detail first. Then we will examine the more condensed form in which such proofs are usually written.

> Assume $A \subset B$ and $A \subset C$.
> $x \in A \Rightarrow x \in B$ by the definition of $A \subset B$.
> $x \in A \Rightarrow x \in C$ by the definition of $A \subset C$.
> Let $x_0 \in A$.
> Hence $x_0 \in B$ by modus ponens.
> $x_0 \in C$, also by modus ponens.
> Hence $x_0 \in B \cap C$ by the definition of intersection.
> Thus $x_0 \in A \Rightarrow x_0 \in B \cap C$.
> Hence $A \subset B \cap C$.

The briefer version is as follows:

> Assume that $A \subset B$ and $A \subset C$.
> Let $x \in A$.
> Then $x \in B$ and $x \in C$ by the definition of a subset.
> Hence $x \in B \cap C$ by the definition of intersection.
> Thus $A \subset B \cap C$.

Note that several differences exist. In the brief version, the distinction between x as a variable that can be any element of A and x_0 as a specific but representative element is omitted. The explicit statement of the definition of $A \subset B$ and $A \subset C$ (for instance, $x \in A \Rightarrow x \in B$) is also not included. And modus ponens is used but not mentioned by name.

Which way is better? Consider the following two arguments from algebra:

$$5x + 1 = 3x + 2$$
$$2x = 1$$
$$x = 1/2$$

$$5x + 1 = 3x + 2$$
$$5x + 1 - 1 = 3x + 2 - 1$$
$$5x = 3x + 1$$
$$5x - 3x = 3x + 1 - 3x$$
$$(5 - 3)x = 3x - 3x + 1$$
$$2x = (3 - 3)x + 1$$
$$2x = 0x + 1$$
$$2x = 1$$
$$(1/2)\,2x = (1/2)\,1$$
$$((1/2) \cdot 2)x = 1/2$$
$$1 \cdot x = 1/2$$
$$x = 1/2$$

Both arguments solve the same problem in the same way, but the one on the right presents the process in considerable detail. The amount of detail that is needed is a subjective judgment made by the author of the proof based on what the author feels can be assumed about the sophistication of the audience. Proofs written in journals for professional mathematics tend to skip many steps and assume that the reader can fill them in. Proofs written for less experienced readers include more steps. However, it's not always true that more steps mean more clarity. For example, in the preceding algebra problem, the amount of detail included on the right is more tedious than helpful. We will include a fair amount of detail in the proofs in this section, but we will reduce it as we go along.

Note that this theorem can be neatly illustrated by Figure 3.7. Drawing the picture, though, is not enough to prove the theorem. It's too easy to build assumptions into a diagram that will make it appear that something is true in general that in fact is true only when those assumptions hold. Attempting to do this is sometimes called *proof by*

Figure 3.7 Venn diagram illustrating that $A \subset B$ and $A \subset C \Rightarrow A \subset B \cap C$.

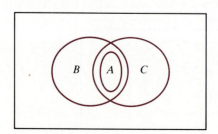

picture and is an invalid form of proof. Diagrams are often very helpful and can give valuable insights, but they are not proofs.

b) Here we want to show that two sets are equal. This time, we will examine only the briefer version of the proof. The usual strategy for showing that two sets, say X and Y, are equal is to show that $X \subset Y$ and $Y \subset X$. We will apply this strategy here. First, we show that $A \cap B \subset A$:

Let $x \in A \cap B$.

Then $x \in A$ and $x \in B$.

Hence $x \in A$.

Thus $x \in A \cap B \Rightarrow x \in A$.

Hence $A \cap B \subset A$.

So we need only show that $A \subset A \cap B$.

Let $x \in A$.

Since $A \subset B$, $x \in B$ also by modus ponens.

Thus $x \in A \cap B$ and thus $A \subset A \cap B$. ∎

Law of Syllogism

The **law of syllogism** is symbolized as follows:

$$p \Rightarrow q$$
$$\underline{q \Rightarrow r}$$
$$p \Rightarrow r$$

This is best justified by the fact that

$$((p \to q) \text{ AND } (q \to r)) \to (p \to r)$$

is a tautology. That is, $p \to r$ cannot be false if both $p \to q$ and $q \to r$ are true. Written in terms of predicates, the law of syllogism becomes

$$p(x) \Rightarrow q(x)$$
$$\underline{q(x) \Rightarrow r(x)}$$
$$p(x) \Rightarrow r(x).$$

This also has a related interpretation in Venn diagrams (see Figure 3.8):

Figure 3.8 The law of syllogism: if $p(x) \Rightarrow q(x)$ and $q(x) \Rightarrow r(x)$, then $p(x) \Rightarrow r(x)$.

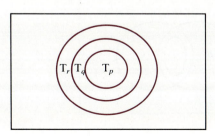

EXAMPLE 2 Proof that $A \subset B$ and $B \subset C$ implies that $A \subset C$.

SOLUTION   Assume $A \subset B$ and $B \subset C$.

> Then $x \in A \Rightarrow x \in B$ and $x \in B \Rightarrow x \in C$ by definition of $\subset$.
> Hence $x \in A \Rightarrow x \in C$ by the law of syllogism.
> Hence $A \subset C$. ∎

Note that in the law of syllogism, the propositions *p, q,* or *r* need not be true. That is, in asserting that

$$x \in A \Rightarrow x \in B \text{ and } x \in B \Rightarrow x \in C,$$

$x \in A$ may be false, although the assertions are true.

EXAMPLE 3 Draw a valid conclusion from each of the following.

a) Dictionaries are useful.   **b)** Collies are dogs.
    Useful books are valuable.   Dogs are mammals.
                                       Mammals are vertebrates.

SOLUTION   Each of these statements is an application of the law of syllogism. From the first, we can conclude.

> Dictionaries are valuable.

From the second,

> Collies are vertebrates. ∎

Indirect Proof

The conditional $p \rightarrow q$ is equivalent to $\neg q \rightarrow \neg p$, the **contrapositive** of $p \rightarrow q$, as can be seen from the truth tables of each. Often it is easier to prove that $\neg q \Rightarrow \neg p$ than to prove directly that $p \Rightarrow q$. Such a proof is called **proof by contrapositive** or **indirect proof.**

EXAMPLE 4 Prove the following.

a) $A \cup B = \varnothing \Rightarrow A = \varnothing$ AND $B = \varnothing$

b) $B = A' \Rightarrow A \cap B = \varnothing$ AND $A \cup B = U$

c) $A \cap B \neq A \Rightarrow A$ not a subset of B

SOLUTION

a) Assume that it is not true that $A = \varnothing$ AND $B = \varnothing$.
    Then $A \neq \varnothing$ OR $B \neq \varnothing$.
    Hence $A \cup B \neq \varnothing$.

b) Suppose it is not true that $A \cap B = \varnothing$ AND $A \cup B = U$.
    Then $A \cap B \neq \varnothing$ OR $A \cup B \neq U$.

Suppose $A \cap B \neq \emptyset$. Then there is an x in both B and A. Hence x is in B and x is not in A'.

Hence $B \neq A'$ in this case.

Suppose $A \cup B \neq U$. Then there is an x in neither A nor B. Hence x is in A' and x is not in B.

Hence $B \neq A'$ in this case also.

Thus, in either case, we can conclude that $B \neq A'$.

c) This is the contrapositive of the theorem proven in Example 1b. Since $A \subset B \Rightarrow A \cap B = B$ has been proven, no further proof is needed. ∎

Modus Tollens

Another rule of inference the Greeks used is called **modus tollens**. Its pattern is:

$$p \Rightarrow q$$
$$\underline{\neg q}$$
$$\neg p.$$

A typical example of the use of modus tollens is a sequence of statements like "If this jewel is really a diamond, then it will scratch glass. This jewel does not scratch glass. Therefore this jewel is not a diamond."

This rule can also be justified from the truth table for $p \rightarrow q$:

p	q	$p \rightarrow q$
T	T	T
T	F	F
F	T	T
F	F	T

Asserting the truth of $p \rightarrow q$ eliminates line 2 as a possibility. Asserting the truth of $\neg q$ leaves us only with line 4. Thus p must be false. This can also be seen by the tautology

$$((p \rightarrow q) \text{ AND } \neg q) \rightarrow \neg p.$$

In case $p(x) \Rightarrow q(x)$, we can illustrate with a Venn diagram that $\neg q(x_0)$ implies that $\neg p(x_0)$. All of the shaded area in Figure 3.9 represents T_p' and the heavily shaded area represents T_q'. Note that $T_q' \subset T_p'$, and hence $\neg q(x_0)$ implies that $\neg p(x_0)$.

Figure 3.9 Modus tollens: if $p(x) \Rightarrow q(x)$ is true and $q(x_0)$ is false, then $p(x_0)$ is false.

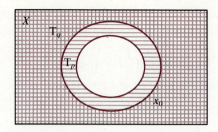

EXAMPLE 5 Prove the following.

a) If $A \subset B$ then $B' \subset A'$.

b) If $A \cap B \neq A$ then A is not a subset of B.

SOLUTION

a) Suppose $A \subset B$.
Then $x \in A \Rightarrow x \in B$.
Let $x_0 \in B'$.
Then $x_0 \in B$ is false.
Hence x_0 is not in A by modus tollens and hence $x_0 \in A'$.

b) Suppose $A \cap B \neq A$.
We have already shown that $A \subset B \Rightarrow A \cap B = A$.
By modus tollens, we can conclude that A is not a subset of B. ∎

EXAMPLE 6 Draw valid conclusions from the following:

Caterpillars are not eloquent.
Jones is eloquent.

For right triangles, $a^2 + b^2 = c^2$.
This triangle has sides 2, 3, and 4.

SOLUTION

Jones is not a caterpillar.
This triangle is not a right triangle. ∎

Proof by Contradiction

Proof by contradiction is a widely used rule of inference, although some logicians argue that it should be avoided. Its pattern is

$$p \Rightarrow (q \text{ AND } \neg q)$$
$$\neg p.$$

The idea is this: If, starting from p, we can deduce two contradictory propositions (q and $\neg q$), we can conclude $\neg p$. This is because q AND $\neg q$ has the truth value false for all propositions q, i.e., it is a contradiction. Thus if $p \Rightarrow F$, p must be false. Hence we can assert that $\neg p$ is true.

EXAMPLE 7 Prove the following:

a) $U' = \emptyset$ **b)** $B \setminus A \subset A'$

SOLUTION

a) Assume $U' \neq \emptyset$.

Then there is an x_0 in U', and thus x_0 is not in U.
But U is the universal set. Hence $\forall x(x \in U)$.
Hence x_0 is in U.
This is a contradiction.
Hence $U' = \emptyset$.

This completes the proof. Note that the pattern is as follows:

> Assume p.
>
> Derive a contradiction.
>
> Conclude $\neg p$.

b) When attempting to prove something of the form $p \Rightarrow q$ by contradiction, a common method is to first assume p and then to assume $\neg q$. If from this we can derive a logical contradiction, we conclude that the second assumption, $\neg q$, was invalid. This amounts to proving $p \Rightarrow (\neg q \Rightarrow F)$. Since $(\neg q \Rightarrow F)$ is equivalent to q, $p \Rightarrow (\neg q \Rightarrow F)$ is equivalent to $p \Rightarrow q$. We will follow this pattern here.

> Let $x \in B \setminus A$.
>
> Suppose x is not in A'.
>
> Then $x \in A$.
>
> But since $x \in B \setminus A$, x is not in A.
>
> This is a contradiction. Hence we conclude that the second assumption, x is not in A', is false and we can conclude $x \in A'$. ∎

It's quite reasonable to ask: Why, after deriving the contradiction, do we negate $\neg q$? Couldn't the contradiction just as well be p's fault? Let's use the same method, negating p instead of $\neg q$:

> Assume p.
>
> Assume $\neg q$.
>
> Derive a contradiction.
>
> Conclude $\neg p$.
>
> Conclude that $\neg q \Rightarrow \neg p$.

We end with the contrapositive of $p \Rightarrow q$, which, of course, is equivalent to it. So, in fact, it makes no difference which assumption we negate after getting a contradiction.

Proofs that do not use contradiction are called **constructive proofs.** Proofs that do are called **nonconstructive proofs.** Constructive proofs are preferred by some mathematicians and logicians because they provide a sequence of steps that lead from the premises to the conclusion. Proof by contradiction will be used in this book. However, if there are two proofs, one constructive, one not, which are equally readable, the constructive one will be chosen.

Use of Connectives

Some proofs are based on the laws of the algebra of propositions. Use of these laws is not a rule of inference, but it does come up often enough to require our attention. Many of the laws of the algebra of sets are proven this way, and occasionally other theorems are as well.

EXAMPLE 8   Prove the following laws of the algebra of sets.

a) $A \cup B = B \cup A$

b) $A \cup (B \cap C) = (A \cup B) \cap (A \cup C)$

SOLUTION

a) Let $x \in A \cup B$. Then $x \in A$ or $x \in B$. Since p OR $q \equiv q$ OR p, we can conclude $x \in B$ or $x \in A$. Hence $x \in B \cup A$. Thus $A \cup B \subset B \cup A$. Similarly, let $x \in B \cup A$. By an analogous argument, $B \cup A \subset A \cup B$.

b) Let $x \in A \cup (B \cap C)$. Then $x \in A$ or $X \in B \cap C$. That is, $x \in A$ or ($x \in B$ and $x \in C$). We know that p OR (q AND r) $\equiv (p$ OR $q)$ AND (p OR r). Thus $x \in A \cup B$ and $x \in A \cup C$. Hence $x \in (A \cup B) \cap (A \cup C)$. Let $x \in (A \cup B) \cap (A \cup C)$. Then $x \in A \cup B$ and $x \in A \cup C$. That is, ($x \in A$ or $x \in B$) and ($x \in A$ or $x \in C$). By the same rule as before, $x \in A$ or ($x \in B$ and $x \in C$). Thus we can conclude that $x \in A \cup (B \cap C)$. ∎

Logical Errors

There are two common logical errors that can be easily described symbolically. One is

$$p \Rightarrow q$$
$$\frac{\neg p}{\neg q.}$$

This is called the **fallacy of the inverse** or the fallacy of **denial of the antece-**

dent. Typically, it occurs in statements like these:

If my spouse likes this dessert, then I know I will like it too.

My spouse does not like this dessert.

I know I won't like it either.

The second fallacy is the **fallacy of the converse,** also known as the fallacy of **affirmation of the consequent.** Its pattern is

$$p \Rightarrow q$$
$$\underline{q}$$
$$p.$$

Typically, it occurs with statements like these:

If you are industrious, then you will get rich.

Joe is rich.

Joe is industrious.

EXAMPLE 9   Identify the fallacies in the following statements.

a) No medicine is nice.
This drink is not nice.
This drink is medicine.

b) All wasps are unfriendly.
This insect is not a wasp.
This insect can be counted on to be friendly.

SOLUTION

a) Let $p(x) =$ "x is a medicine" and $q(x) =$ "x is not nice." Then the first sentence can be symbolized as $p(x) \Rightarrow q(x)$. The second sentence is of the form q ("this drink"). Thus this is an example of the fallacy of the converse.

b) Let $w(x) =$ "x is a wasp" and $u(x) =$ "x is unfriendly." The first sentence is $w(x) \Rightarrow u(x)$. The second sentence is of the form $\neg w$ ("this insect"). Thus this is the fallacy of the inverse.   ■

TERMINOLOGY

rule of inference	law of syllogism
modus ponens	contrapositive
law of detachment	proof by contrapositive

indirect proof   fallacy of the inverse
modus tollens   fallacy of the converse
proof by contradiction   denial of the antecedent
constructive proof   affirmation of the consequent
nonconstructive proof

SUMMARY

Rules of Inference

Rule	Pattern	Tautology
Modus ponens	$p \Rightarrow q$ $\underline{p}$ q	$((p \rightarrow q) \text{ AND } p) \rightarrow q$
Syllogism	$p \Rightarrow q$ $\underline{q \Rightarrow r}$ $p \Rightarrow r$	$((p \rightarrow q) \text{ AND } (q \rightarrow r)) \rightarrow (p \rightarrow r)$
Contrapositive	$\underline{p \Rightarrow q}$ $\neg q \Rightarrow \neg p$	$(\neg q \rightarrow \neg p) \leftrightarrow (p \rightarrow q)$
Modus tollens	$p \Rightarrow q$ $\underline{\neg q}$ $\neg p$	$((p \rightarrow q) \text{ AND } \neg q) \rightarrow \neg p$
Contradiction	$\underline{p \Rightarrow (q \text{ AND } \neg q)}$ $\neg p$	$((p \rightarrow (q \text{ AND } \neg q)) \rightarrow \neg p$

Logical Errors

Fallacy of the inverse	$p \Rightarrow q$ $\underline{\neg p}$ $\neg q$	
Fallacy of the converse	$p \Rightarrow q$ $\underline{q}$ p	

EXERCISES 3.6

Draw valid conclusions from each of the following and indicate which rule of inference you used.

1. All bankers are rich.
   Alphonse is a banker.

2. No doctors are enthusiastic.
   You are enthusiastic.

3. Sugar is sweet.
   Salt is not sweet.

4. Equilateral triangles have three 60° angles.
   T is an equilateral triangle.

5. Every eagle can fly.
   Some pigs cannot fly.

6. No misers are unselfish.
   Only misers save egg shells.

Symbolize each of the following and draw a valid conclusion. Express your conclusion both symbolically and in ordinary English.

7. My saucepans are the only things I have that are made of tin.
   I find all of your presents very useful.
   None of my saucepans are of the slightest use.

8. No one takes in the *Times* unless he is well educated.
   No hedgehogs can read.
   Those who cannot read are not well educated.

9. All of the Eton men in this college play cricket.
   None but the scholars dine at the higher table.
   None of the cricketers row.
   My friends in this college all come from Eton.
   All of the scholars are rowing men.

10. All writers who understand human nature are clever.
    No one is a true poet unless he can stir the hearts of men.
    Shakespeare wrote *Hamlet*.
    No writer who does not understand human nature can stir the hearts of men.
    No one but a true poet could have written *Hamlet*.

In the following proofs, give a reason for each step.

11. Prove that if $A \subset B$ then $B' \subset A'$.

    Proof:

    $$\text{Assume } A \subset B.$$
    $$x \in A \Rightarrow x \in B.$$
    $$x \text{ not in } B \Rightarrow x \text{ not in } A.$$
    $$x \in B' \Rightarrow x \in A'.$$
    $$B' \subset A'.$$

12. Prove that $A \times (B \cup C) \subset (A \times B) \cup (A \times C)$.

Proof:

> Let $(x, y) \in A \times (B \cup C)$.
>
> Then $x \in A$ and $y \in (B \cup C)$.
>
> Hence $y \in B$ or $y \in C$.
>
> Hence $(x \in A$ and $y \in B)$ or $(x \in A$ and $y \in C)$.
>
> Thus $(x, y) \in (A \times B) \cup (A \times C)$.

13. Prove that $(A \setminus B) \cap B = \emptyset$.

Proof:

> Assume $(A \setminus B) \cap B \neq \emptyset$.
>
> Then $\exists x$ such that $x \in A \setminus B$ and $x \in B$.
>
> Thus $x \in B$ and x not in B.
>
> Hence $(A \setminus B) \cap B = \emptyset$.

14. Prove that if $A \subset B$, then $A \cup B = B$.

Proof:

> Suppose $A \subset B$.
>
> Let $x \in A \cup B$.
>
> Then $x \in A$ or $x \in B$.
>
> Suppose x not in A. Then $x \in B$.
>
> Suppose $x \in A$. Then $x \in B$.
>
> Hence $x \in B$.
>
> Thus $x \in A \cup B \Rightarrow x \in B$.
>
> Now suppose $x \in B$.
>
> Then $x \in A \cup B$.
>
> Thus $x \in B \Rightarrow x \in A \cup B$.
>
> Thus $A \cup B = B$.

Provide proofs of each of the following.

15. $B \setminus A \subset A'$

16. $A \cap B = B \cap A$

17. $A \subset B$ and $C \subset D \Rightarrow A \times C \subset B \times D$

18. $A \cap (B \cup C) = (A \cap B) \cup (A \cap C)$

19. $A \times (B \cap C) = (A \times B) \cap (A \times C)$

20. $(A \cap B)' = (A' \cup B')$

21. $A \times (B \cup C) = (A \times B) \cup (A \times C)$

State the converse, inverse, and contrapositive of each of the following.

22. If $f(x)$ is a function, then $f(x)$ will pass the vertical line test.

23. If the throat culture comes back positive, then you have strep throat.

24. If Socrates is a man, then Socrates is mortal.

25. If $A \subset B$ then $A \cap C \subset B \cap C$.

26. Show that $p \rightarrow q$ and its contrapositive are logically equivalent by means of truth tables.

27. Show that the inverse and converse of $p \rightarrow q$ are logically equivalent to each other but are not logically equivalent to $p \rightarrow q$.

Symbolize each of the following arguments and identify the logical error in each.

28. No one who is interested in the theater is ever boring.
Adam is never boring.
Adam must be interested in the theater.

29. All people who fit in well with my political party believe in social justice.
You believe in social justice.
You will fit in well with my political party.

30. People with bad breath are unpopular.
Anyone who uses our product will not have bad breath.
Anyone who uses our product will be popular.

31. Every little girl's doll is loved.
No little girl owns this doll.
This doll is not loved.

Convert each of the following tautologies into a new rule of inference.

32. $(p \rightarrow \neg p) \rightarrow \neg p$

33. $((p \text{ OR } q) \text{ AND } \neg p) \rightarrow q$

3.7

MATHEMATICAL INDUCTION

So far, whenever we have wanted to prove a theorem of the form $\forall x(p(x))$, we have selected a representative x_0 in the universal set and then proven that $p(x_0)$ must be true. This often required the use of one or more of the rules of inference, which we examined in Section 3.6. There is another rule of inference, though, that can be used when the universal set is the natural numbers, the natural numbers with 0 included, or certain subsets of the natural numbers. This rule is called **mathematical induction.** It is of particular importance in discrete mathematics and in computer science because the underlying universal set for both is so often the natural numbers.

Suppose we wanted to give a formal, axiomatic definition of the natural numbers. One way would be to start with 1 and the operation of addition as primitives. Successive integers would then be defined by adding 1 to itself. Thus $2 = 1 + 1, 3 = 2 + 1, 4 = 3 + 1$, etc. If one of the new integers defined in this way is equal to one already defined, the resulting set turns out to be one of the sets $\mathbf{Z}_n$ discussed in Section 2.1. On the other hand, if all of the new integers are distinct, the resulting set is the natural numbers. Thus the idea of a step-by-step process in which each number is formed from the previous one is basic to the natural numbers. We will discuss this topic in more detail in subsequent chapters. It is this property that serves as the basis of mathematical induction. The **first principle of mathematical induction** is a rule of inference based on the step-by-step definition of **N**. If we can establish that a predicate, $p(x)$, is true when $x = 1$ and that assuming the predicate to be true for $x = n$ guarantees its truth for $x = n + 1$, we can conclude that $p(x)$ is true for all x in **N**. Symbolically, this is stated as follows:

1. $p(1)$
2. $\forall n \in \mathbf{N}\ (p(n) \Rightarrow p(n + 1))$
3. $\forall n \in \mathbf{N}\ (p(n))$

The idea is this: Suppose someone wants to walk upstairs. Two things are needed—a means of getting to the first stair and a means of going from each stair to the next. With these two methods, one can climb all of the stairs without being concerned with how to get to any one particular stair. In mathematical induction, proving $p(1)$ is the means of getting to the first stair. This is called the *basis step*. Proving that $\forall n \in \mathbf{N}\ (p(n) \Rightarrow p(n + 1))$ is true is the means of going from one stair to the next and is known as the *induction step*. The statement $\forall n \in \mathbf{N}\ (p(n))$ is the conclusion that $p(n)$ is true for every n and is analogous to being able to climb all of the stairs.[†]

In summary, induction is used when one wants to prove that something is true for every natural number. The process involves two steps—proving that the something is true for 1 and proving that if it is true for n it is also true for $n + 1$. We can conclude that it is true for every n.

Unfortunately, we cannot refer to a truth table to justify this rule of inference. However, we can justify it by modus ponens and an appeal to what we know about the natural numbers. That is, if we know that $p(1)$ is true and that $\forall n \in \mathbf{N}\ (p(n) \Rightarrow p(n + 1))$, we can write the following statements by substituting different integers for n:

$$p(1)$$
$$p(1) \Rightarrow p(2)$$
$$p(2) \Rightarrow p(3)$$
$$p(3) \Rightarrow p(4)$$
$$\text{etc.}$$

[†] Alternative approaches to induction using axioms for **N** and the well-ordering principle will be discussed in Section 5.1.

Combining the first two by modus ponens, we can conclude that $p(2)$ is true. Combining the truth of $p(2)$ with the third one, we can conclude that $p(3)$ is true. Continuing in this fashion, we can conclude that $p(4)$, $p(5)$, $p(6)$, etc. are all true and hence $\forall n \in \mathbf{N}\ (p(n))$ is true.

EXAMPLE 1 Prove that

$$\sum_{i=1}^{n} i = \frac{(n)(n+1)}{2}$$

SOLUTION For simplicity, let LHS denote the left-hand side of this equation and RHS denote the right-hand side. First, let $n = 1$. Then,

$$\text{LHS} = \sum_{i=1}^{1} i = 1$$

and

$$\text{RHS} = (1)(1+1)/2 = 2/2 = 1.$$

In terms of induction, we have now completed the basis step, i.e., we have proven $p(1)$. For the induction step, we must prove that

$$\forall n \in \mathbf{N}\ (p(n) \Rightarrow p(n+1)),$$

i.e, we must let n be an arbitrary natural number and show that $p(n) \Rightarrow p(n+1)$. To do this, we assume that $p(n)$ is true and reason deductively to $p(n+1)$. So in this problem, we assume that

$$\sum_{i=1}^{n} i = \frac{(n)(n+1)}{2}. \tag{3.1}$$

We need to prove that

$$\sum_{i=1}^{n+1} i = \frac{(n+1)(n+2)}{2}. \tag{3.2}$$

The typical strategy in attacking proofs of this sort is to look at Eq. (3.2) and see how we can use what we assumed in Eq. (3.1) to prove Eq. (3.2). In this case, that means rewriting the sum on the LHS of Eq. (3.2) and substituting from Eq. (3.1):

$$\sum_{i=1}^{n+1} i = \sum_{i=1}^{n} i + (n+1) = \frac{(n)(n+1)}{2} + (n+1) = \frac{n^2 + n + 2n + 2}{2}$$

$$= \frac{n^2 + 3n + 2}{2} = \frac{(n+2)(n+1)}{2}.$$

We do get Eq. (3.2). Hence we can conclude that the theorem is true. ∎

At first, it may appear that mathematical induction is circular reasoning. It seems as if we are assuming what we are trying to prove when we assume $p(n)$ and try to deduce $p(n + 1)$. In fact, mathematical induction is not circular reasoning at all. What is actually happening is hypothetical thinking. We are saying, "Suppose $p(n)$ is true. Can we get from there to $p(n + 1)$?" This does not prove $p(n)$ but only that $p(n)$ implies $p(n + 1)$. Consider the analogy of walking up stairs. We are saying, in essence, "Suppose I am on the nth step. Can I get from there to the next one?" This does not say that we can get to the nth step, only that we can go from there to the $(n + 1)$th. It is the combination of the ability to go from each step to the next one and the ability to get to the first one (i.e., prove $p(1)$) that makes it possible to get to the nth.

The rule of inference for mathematical induction was stated previously with the assumption that the universal set was the natural numbers. Hence it required that we first prove $p(1)$ since that is the smallest natural number. But mathematical induction can be used just as effectively on other sets as well. For instance, suppose the universal set is $\mathbf{N} \cup \{0\}$, i.e., $\{0, 1, 2, 3, \ldots\}$. The rule of inference must be modified only slightly so that it looks like this:

$$p(0)$$
$$\frac{\forall n \geq 0 \ (p(n) \Rightarrow p(n + 1))}{\forall n \geq 0 \ (p(n))}$$

Similarly, suppose the universal set were $\mathbf{N}/\{1\}$, i.e., $\{2, 3, 4, \ldots\}$. Then the rule would become

$$p(2)$$
$$\frac{\forall n \in \mathbf{N} \ (p(n) \Rightarrow p(n + 1))}{\forall n \geq 2 \ (p(n))}$$

The notation $\forall n \geq 2 \ (p(n))$ means that $p(n)$ is true for all n greater than or equal to 2, but is not necessarily true when $n \leq 1$.

EXAMPLE 2   Prove that if $r \neq 0$ and $r \neq 1$, then

$$\sum_{i=0}^{n} r^i = \frac{1 - r^{n+1}}{1 - r}$$

SOLUTION   We examined one argument justifying this formula in Section 2.3. This time we will prove it by induction. First, let $n = 0$. The preceding equation becomes

$$\text{LHS} = \sum_{i=0}^{0} r^i = r^0 = 1 \text{ if } r \neq 0$$

$$\text{RHS} = \frac{1 - r}{1 - r} = 1 \text{ if } r \neq 1.$$

Now assume $p(n)$, that is, assume that

$$\sum_{i=0}^{n} r^i = \frac{1 - r^{n+1}}{1 - r}. \tag{3.3}$$

We will try to prove that

$$\sum_{i=0}^{n+1} r^i = \frac{1 - r^{n+2}}{1 - r} \tag{3.4}$$

Adding r^{n+1} to both sides of Eq. (3.3), we get

$$\text{LHS} = \sum_{i=0}^{n} r^i + r^{n+1} = \sum_{i=0}^{n+1} r^i$$

$$\text{RHS} = \frac{1 - r^{n+1}}{1 - r} + r^{n+1} = \frac{1 - r^{n+1} + r^{n+1} - r^{n+2}}{1 - r}$$

$$= \frac{1 - r^{n+2}}{1 - r}. \quad \blacksquare$$

Although this argument is longer than the one given in Section 2.6, it is generally regarded as better. If you recall, that argument involved multiplying $(1 + r + \cdots + r^n)$ by r. The weakness in that argument is in the "$\cdots$". We know intuitively what multiplying every term in that sum by r means and assume that the result is $r + r^2 + \cdots + r^{n+1}$. With induction, nothing is left to intuition or unjustified assumptions; every step in the reasoning is spelled out. That cannot be done in the argument in Section 2.6 since it is impossible to write out $(1 + r + \cdots + r^n)$ explicitly, as n can assume any value in the natural numbers.

$|P(S)| = 2^{|S|}$

EXAMPLE 3   Let S be any finite set and suppose $|S| = n$. Show that

$$|P(S)| = 2^n.$$

SOLUTION   Suppose $n = 0$. Then $S = \varnothing$. Hence $P(S) = \{\varnothing\}$, and hence

$$|P(S)| = 1 = 2^0.$$

Now suppose that $|P(S)| = 2^n$ for all sets S with n members, $n \geq 0$. Let T be an arbitrary set with $n + 1$ members. Denote T by

$$T = \{t_1, t_2, \ldots, t_n, t_{n+1}\}.$$

The subsets of T can be split into two distinct groups—those that include t_{n+1} and those that do not. Every subset of $\{t_1, t_2, \ldots, t_n\}$ is also a subset of T.

Hence there are, by the induction hypothesis, 2^n subsets of T that do not include t_{n+1}. Each set that includes t_{n+1} can be looked upon as the union of a subset of $\{t_1, t_2, \ldots, t_n\}$ and $\{t_{n+1}\}$. Hence there are also 2^n sets that include t_{n+1}. Thus T has $2^n + 2^n = 2 \cdot 2^n = 2^{n+1}$ subsets. ∎

One way to visualize this proof is to list the subsets as follows:

not including t_{n+1}	including t_{n+1}
$\varnothing$	$\{t_{n+1}\}$
$\{t_1\}$	$\{t_1, t_{n+1}\}$
$\{t_2\}$	$\{t_2, t_{n+1}\}$
$\{t_1, t_2\}$	$\{t_1, t_2, t_{n+1}\}$
.	.
.	.
.	.
$\{t_1, t_2, \ldots, t_n\}$	$\{t_1, t_2, \ldots, t_n, t_{n+1}\}$

The left-hand list is all of the subsets of $\{t_1, t_2, \ldots, t_n\}$. The right-hand list is set up so that each set is the union of $\{t_{n+1}\}$ with the corresponding set from the left-hand list. Thus there are 2^n subsets in each list.

Inequalities

Mathematical induction can also be used to prove relationships that involve inequalities as well as equations. Consider the following example.

EXAMPLE 4   Prove that if $n \geq 1$, $n! \geq 2^{n-1}$.

SOLUTION   Let $n = 1$. Then LHS $= 1! = 1$ and RHS $= 2^0 = 1$. Now assume that

$$n! \geq 2^{n-1}. \tag{3.5}$$

Note that since $n \geq 1$, $n + 1 \geq 2$. Multiply both sides of Eq. (3.5) by $n + 1$. We get

$$\text{LHS} = (n + 1)!$$
$$\text{RHS} = (n + 1) \cdot 2^{n-1} \geq 2 \cdot 2^{n-1} = 2^n.$$

Hence

$$n! \geq 2^{n-1} \Rightarrow (n + 1)! \geq 2^n. \quad ∎$$

Mathematical Induction versus Inductive Reasoning

Note that in spite of its name, mathematical induction is actually a form of deductive reasoning, not a form of inductive reasoning. Inductive reasoning is the process typical of science—inferring general principles from specific cases. Thus Kepler used inductive reasoning in inferring his laws for planetary motion from the voluminous data at his disposal. Mathematical induction, however, still follows the typical deductive process—working from axioms, previously proven theorems, and definitions to new theorems by means of rules of inference.

The Second Principle of Induction

There is another form of mathematical induction, called the **second principle of mathematical induction,** which is sometimes used. Its pattern is as follows:

1. $p(1)$
2. $\dfrac{\forall n \in \mathbf{N} \; (p(1), p(2), \ldots p(n)) \rightarrow p(n+1)}{}$
3. $\forall n \in \mathbf{N} \; (p(n))$

The main difference is in the second line: Instead of assuming $p(n)$ and deriving $p(n+1)$, we assume $p(1)$, $p(2)$, etc., up to and including $p(n)$. From this stronger set of assumptions, we then derive $p(n+1)$. Examples of situations that require the use of the second principle of induction will occur in Section 6.6 and in other sections.

An Abuse of Induction

Following is a classical example of the misuse of mathematical induction. See if you can find the error in it before reading the explanation at the end.

EXAMPLE 5   Prove that all horses are the same color.

SOLUTION   Let $p(n)$ be the statement "In any set of n horses, all of them are the same color." First, we must prove $p(1)$—that all horses in a set with exactly one horse are of the same color. But this is obvious—with only one horse, all of the horses are clearly of the same color.

Now we assume $p(n)$—that all horses in a set of n horses are of the same color. Let H denote a set of $n+1$ horses. Remove one of those horses, say, h_1. The remaining n horses are then all of the same color by the induction hypothesis. Now put h_1 back in H and remove a different horse, say, h_2. This time the remaining horses are again all of the same color. Furthermore h_2 has already been shown to be the same color as the n horses in $H \backslash \{h_2\}$. Hence all of the $n+1$ horses are the same color and we have established that $p(n) \Rightarrow p(n+1)$. Thus we have established both the basis and the induction steps and have shown that in any set of horses, no matter how large, all horses are the same color. ∎

Analysis of the Error

Let's take a closer look at this purported proof. Consider the following diagram:

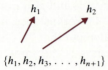

$\{h_1, h_2, h_3, \ldots, h_{n+1}\}$

The idea behind the "proof" is that when h_1 is removed, the remaining horses $h_2, h_3, \ldots h_n, h_{n+1}$ are all of the same color by the induction hypothesis. Similarly, when h_2 is removed, $h_1, h_3, \ldots, h_n, h_{n+1}$ are all of the same color for the same reason. Hence both h_1 and h_2 are the same color as h_3, $\ldots, h_n, h_{n+1}$ and thus all are of the same color. If we consider what happens to this inductive argument when $n = 2$, we will see where the error is. The preceding diagram now looks like this:

$\{h_1, h_2\}$

This time when h_1 and h_2 are removed, there is nothing left with which to compare them. That is, we cannot say that they are the same color since the horses they were supposedly identical to, $h_3, \ldots, h_n, h_{n+1}$, do not exist. In other words, the statement that "h_2 has already been shown to be the same color as the n horses in $H \backslash \{h_2\}$" is false when $H \backslash \{h_2\}$ is only $\{h_1\}$.

In summary, then, mathematical induction involves proving the following statements:

$p(1)$

$\left.\begin{array}{l} p(1) \Rightarrow p(2) \\ p(2) \Rightarrow p(3) \\ p(3) \Rightarrow p(4) \end{array}\right\}$ all included in the single statement $p(n) \Rightarrow p(n+1)$

etc.

Our proof gave the illusion of proving all of these statements but in fact did not prove that $p(1) \Rightarrow p(2)$ (although it did prove all of the rest). Thus the whole argument was invalid.

One possible undesirable result of this example is that it may reduce your confidence in mathematical induction. This should not happen. Mathematical induction is a well-established rule of inference and is of critical importance for discrete mathematics. Rather, it should serve as a warning against jumping too quickly to the induction step. When using induction, it is wise to write out carefully what $p(1)$, $p(2)$, $p(3)$ and perhaps one or two more of the statements mean in the particular situation and see if they make sense before attempting to prove anything. It is also a good idea to try to prove $p(2)$ from $p(1)$ and $p(3)$ from $p(2)$ before attempting to prove that $p(n)$ implies $p(n + 1)$. For instance, in our example regarding the horses, $p(1) \Rightarrow p(2)$ is the statement that "If one horse is the same color as itself, then every pair of horses is the same color," which is obviously nonsense.

We will see many examples of induction in very different settings throughout this book.

TERMINOLOGY

Mathematical induction

First principle of mathematical induction

Second principle of mathematical induction

SUMMARY

1. The principle of mathematical induction is a rule of inference that is often used to prove that something is true for every natural number. As such, it is of special importance in discrete mathematics.

2. The first principle of mathematical induction follows the pattern

    1. $p(1)$

    2. $\forall n \in N \, (p(n) \Rightarrow p(n + 1))$

    3. $\forall \in N \, p(n)$.

3. The second principle of mathematical induction follows the pattern

    1. $p(1)$

    2. $\forall n \in N \, (p(1), p(2), \ldots p(n)) \Rightarrow p(n + 1)$

    3. $\forall n \in N \, p(n)$.

EXERCISES 3.7

Verify the following relationships by mathematical induction:

1. $\displaystyle\sum_{i=1}^{n} i(i + 1) = \frac{n(n + 1)(n + 2)}{3}$

2. $\displaystyle\sum_{i=1}^{n} i^2 = \frac{n(n+1)(2n+1)}{6}$

3. $\displaystyle\sum_{i=1}^{n} i^3 = \frac{n^2(n+1)^2}{4}$

4. $\displaystyle\sum_{i=1}^{n} \frac{n+4}{n(n+1)(n+2)} = \frac{n(3n+7)}{2(n+1)(n+2)}$

5. If $n \geq 4$, $n! \geq 2^n$

6. $(1+x)^n \geq 1 + nx$

7. $(2n+1) \leq 2^n$ for $n = 3, 4, 5, \ldots$

8. $2^n \geq n^2$, $n \geq 4$

In Section 2.5 we proved several properties of exponentials where the exponents were positive integers. Each of these proofs used ellipses. Each can be proven more effectively with mathematical induction. Use induction to prove the following. (*Hint:* let m be an arbitrary but fixed positive integer and use induction on n. Use the following definition of the exponential function: $a^1 = a$; $a^{n+1} = a \cdot a^n$.)

9. $a^{m+n} = a^m a^n$

10. $a^{m-n} = a^m / a^n$

11. $(a^m)^n = a^{mn}$

Prove the following generalizations of DeMorgan's laws:

12. $\left(\displaystyle\bigcap_{i=1}^{n} A_i\right)' = \displaystyle\bigcup_{i=1}^{n} A_i'$ (Note: $\displaystyle\bigcap_{i=1}^{n} A_i$ means $A_1 \cap A_2 \cap A_3 \cap \cdots \cap A_n$. A similar pattern applies to $\displaystyle\bigcup_{i=1}^{n} A_i$.)

13. $\left(\displaystyle\bigcup_{i=1}^{n} A_i\right)' = \displaystyle\bigcap_{i=1}^{n} A_i'$

14. Suppose n straight lines divide a plane in such a way that no two lines are parallel and no three have a common point. Prove that they divide the plane into $(n^2 + n + 2)/2$ regions.

15. Show that the sum of the first n odd numbers is n^2.

16. Suppose a function $f(n)$ is defined by the rule that

$$f(n+1) = n \cdot f(n) \text{ and } f(0) = 1.$$

Show by induction that $f(n) = n!$.

17. Suppose that a function $f(n)$ is defined by the rule that

$$f(n+1) = 2 \cdot f(n) \text{ and } f(0) = 1.$$

Show by induction that $f(n) = 2^n$.

18. Show that when the universal set is the natural numbers, the two principles of induction are logically equivalent.

19. Following the example in the text, create a "pseudo-proof" that all flavors of ice cream are equally tasty.

3.8

COMBINATIONAL CIRCUITS

Combinational circuits or **logic circuits** are critical building blocks of computers and many other digital devices. They are distinguished from other kinds of circuits in that their operation may be completely described by means of truth tables. There is much more that could be learned about them than we have space for here. However, we will see how combinational circuits are an application of propositional logic. This will provide you with adequate background to understand how combinational circuits work.

Gates

Combinational circuits are built of components called **gates.** The three simplest gates are the AND ($\wedge$) OR ($\vee$), and NOT ($\neg$) gates and are based on the corresponding connectives for propositions. These are symbolized as shown in Figure 3.10.

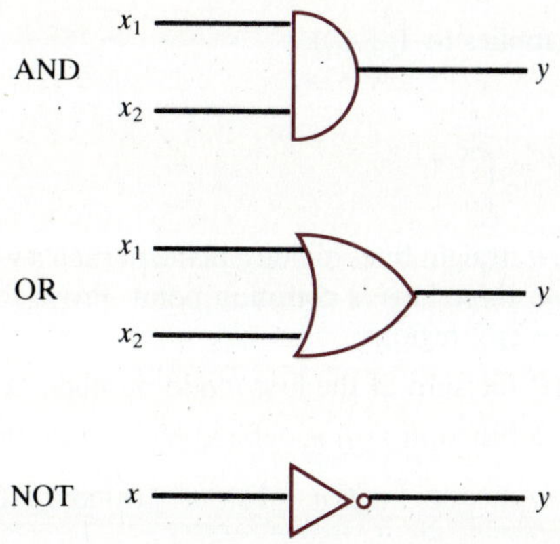

Figure 3.10   AND, OR, and NOT gates.

The operation of these gates follows the same patterns as the truth tables of the previous section. Inputs (coming from the left) are electronic signals, which can have only two possible values. These are labeled 0 and 1. Outputs (on the right) are what the truth table for that connective specifies. Thus we get the following:

AND

x_1	x_2	y
1	1	1
1	0	0
0	1	0
0	0	0

OR

x_1	x_2	y
1	1	1
1	0	1
0	1	1
0	0	0

NOT

x	y
1	0
0	1

Combinational Circuits

Given any truth table, a combinational circuit can be designed whose outputs match or *synthesize* it. For instance,

EXAMPLE 1   Design a circuit to synthesize the expression

$$s = (p \text{ AND } q) \text{ OR } (q \text{ OR } \neg r).$$

SOLUTION   See Figure 3.11.

Figure 3.11   Circuit for $(p \text{ AND } q)$ OR $(q \text{ OR } \neg r)$.

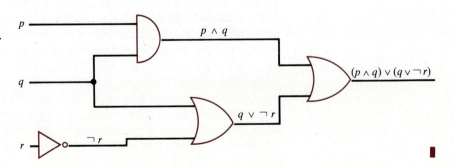

∎

EXAMPLE 2 Consider the relationship shown by the following truth table.

x_1	x_2	y
1	1	0
1	0	1
0	1	1
0	0	0

Find a circuit that will synthesize this relationship.

SOLUTION   This is the truth table for the EXCLUSIVE-OR. It can be interpreted as meaning x_1 OR x_2 but not both. Thus it can be written as

$$(x_1 \text{ OR } x_2) \text{ AND NOT } (x_1 \text{ AND } x_2).$$

A circuit for the EXCLUSIVE-OR based on the latter expression is shown in Figure 3.12.

Figure 3.12 An EXCLUSIVE-OR circuit.

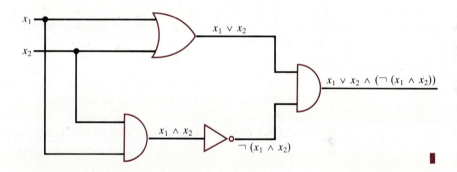

Minterm Representation

Both of the preceding examples depended upon knowing a combinational expression and then designing a circuit to match that expression. However, sometimes we have a table of values that we want the output to follow but we do not have an expression to work from. In such a case, we must construct an expression using AND, OR, and NOT that can then be used to design the circuit. There are several ways of doing this. We will examine one of them: representation of the table by what are called **minterms.** This procedure does not usually give us the simplest possible circuit, but it does work. Simplification of circuits can be achieved by the method of Karnaugh maps, a topic that must be left for a course in logical circuit design.

For instance, consider a truth table that includes the following two lines:

p	q	r	Output
1	0	1	1
0	1	1	1

For the first line, when p is true, q false, and r true, we want the output to be true. If we set

$$\text{output} = p \wedge \neg q \wedge r,$$

we would get that result. Similarly, if we set

$$\text{output} = \neg p \wedge q \wedge r,$$

we would get an output that corresponds to the second line in the table. Thus setting

$$\text{output} = (p \wedge \neg q \wedge r) \vee (\neg p \wedge q \wedge r)$$

would give us an output of 1 in either case. Continuing in this fashion, we can develop an expression to represent an entire truth table. That is, we take each line of the table in which the output should be a 1 and write a maxterm for it. We do this by taking each symbol that is a 1, taking the negation of each symbol that is a 0, and combining the result with AND. The maxterms are then connected by OR and the resulting combinational expression can be used to design the desired circuit.

EXAMPLE 3   Find a minterm representation for the XOR table presented in Example 2.

SOLUTION   The two lines with 1 in the output are

x_1	x_2	x_1 XOR x_2
1	0	1
0	1	1

Thus the desired expression is $(x_1$ AND $\neg x_2)$ OR $(\neg x_1$ AND $x_2)$. Verification of the validity of the expression and the design of the circuit will be left for the Exercises. Note that this is a different expression than the one used earlier.   ∎

EXAMPLE 4 Find a minterm representation for the following truth table.

p	q	r	Output
1	1	1	1
1	1	0	0
1	0	1	0
1	0	0	1
0	1	1	0
0	1	0	0
0	0	1	1
0	0	0	0

SOLUTION   The desired expression is $(p \land q \land r) \lor (p \land \neg q \land \neg r) \lor (\neg p \land \neg q \land r)$. Again, verification of the expression and design of the circuit will be left for the Exercises. ∎

Building Gates from Other Gates

Because of DeMorgan's laws, it is possible to synthesize an OR gate using AND and NOT gates. It is also possible to synthesize an AND gate using NOT and OR gates. This is not merely an intellectual challenge; it also has some valuable practical applications, as we shall see.

EXAMPLE 5 Design an OR circuit using AND and NOT gates.

SOLUTION   By DeMorgan's laws,

$$\neg(x_1 \text{ OR } x_2) = (\neg x_1 \text{ AND } \neg x_2)$$

That is,

$$(x_1 \text{ OR } x_2) = \neg(\neg x_1 \text{ AND } \neg x_2)$$

Thus we can build the following circuit for x_1 OR x_2 (Figure 3.13).

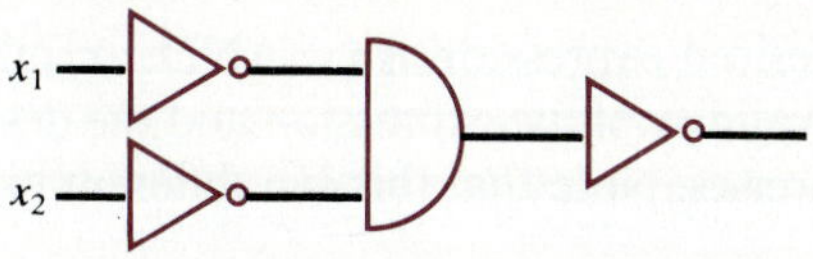

Figure 3.13 An circuit built from AND and NOT gates.

The NAND Gate

Another useful gate is the NAND gate, an abbreviation for NOT AND. Its truth table and symbol (Figure 3.14) are as follows.

x_1	x_2	y
1	1	0
1	0	1
0	1	1
0	0	1

Figure 3.14 NAND gate.

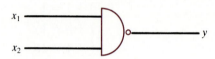

All three gates—AND, OR, and NOT—can be synthesized using just the NAND gate. Thus it can be used as a single building block from which any logical circuit can be constructed.

EXAMPLE 6 Synthesize a NOT gate using only NAND gates.

SOLUTION See Figure 3.15.

Figure 3.15 A NOT gate built from a NAND gate.

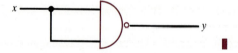

EXAMPLE 7 Synthesize an AND gate using NAND gates.

SOLUTION See Figure 3.16.

Figure 3.16 An AND gate built from NAND gates.

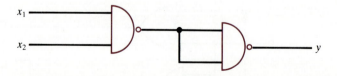

The strategy in developing this circuit is first to use the NAND gate and then to negate its output using the circuit of Example 4 to get an AND gate. ∎

Note that we now have circuits for NOT and AND built from NAND. Using these as building blocks, we could substitute them for the AND and NOT circuits used earlier and build an OR from NAND gates alone. Thus any combinational circuit can be built exclusively from NAND gates.

Arithmetic

EXAMPLE 8 Design a combinational circuit that will do addition.

SOLUTION Suppose one is doing binary arithmetic. The addition table for adding 1-bit numbers ("bit" is short for binary digit) is as follows:

$$0 + 0 = 0$$
$$0 + 1 = 1$$
$$1 + 0 = 1$$
$$1 + 1 = 10$$

If we let s stand for the rightmost or **sum bit** of this table and let c stand for the leftmost or **carry bit,** the following table summarizes the preceding additions in a format that looks more like the truth tables we have been seeing:

x_1	x_2	s	c
1	1	0	1
1	0	1	0
0	1	1	0
0	0	0	0

We can design a circuit to synthesize this table by noting that the column under the sum bit is an EXCLUSIVE-OR and the column under the C bit is an AND. This circuit (Figure 3.17) is called a **half adder.**

Figure 3.17 Half adder circuit for adding two 1-bit numbers.

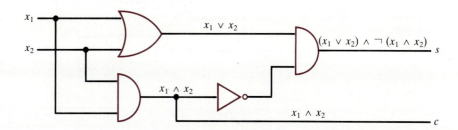

EXAMPLE 9 Verify that the half adder works in the case where $x_1 = 0$ and $x_2 = 1$.

SOLUTION See Figure 3.18. By placing the initial values of the inputs on the circuit diagram, the outputs can be determined by tracing through each gate. The resulting values agree with the desired truth table.

Figure 3.18 Half adder with sample values assigned to the inputs.

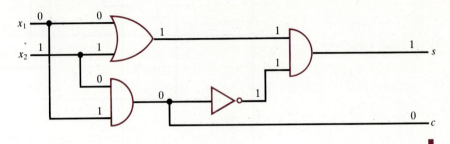

The Full Adder

The half adder is fine if we only want to add 1-bit numbers. For larger numbers, we need a larger circuit, the **full adder.** For instance, suppose we want to compute the following binary sum:

$$\begin{array}{r} 101 \\ 11 \\ \hline 1000 \end{array}$$

The half adder could compute the sum of the rightmost 2 bits, but couldn't go any further since successive sums require adding 3 bits—the 2 in the original number and the carry bit.

EXAMPLE 10 Synthesize a circuit that can add 3 bits.

SOLUTION The full adder allows addition of 3 bits and produces a sum and a carry. Its table is as follows:

x_1	x_2	x_3	s	c
1	1	1	1	1
1	1	0	0	1
1	0	1	0	1
1	0	0	1	0
0	1	1	0	1
0	1	0	1	0
0	0	1	1	0
0	0	0	0	0

Figure 3.19

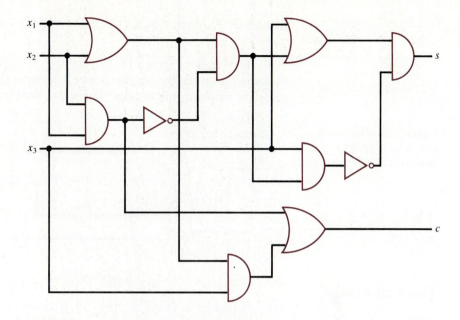

We could construct a maxterm expression for both s and c. These expressions could then be used to design the circuit. The circuit shown in Figure 3.19 accomplishes the same task somewhat more simply. For more on this topic, see the Exercises. ∎

Combining Adders

Note that the preceding full adder combines only three 1-bit numbers. Suppose that we wanted to add two 3-bit binary numbers:

$$x_1 \ x_2 \ x_3$$
$$\underline{y_1 \ y_2 \ y_3}$$
$$s_1 \ s_2 \ s_3$$

A circuit to do it can be built using the half adder and full adder as building blocks (Figure 3.20). This circuit is called a *ripple adder*. The c bit at the very end detects an "overflow" condition, i.e., a sum larger than the 3-bit number $s_1 s_2 s_3$ can hold. As we can see, this circuit can be extended indefinitely to add binary numbers of any length and thus could be used in the

Figure 3.20 Ripple adder for adding 3-bit numbers.

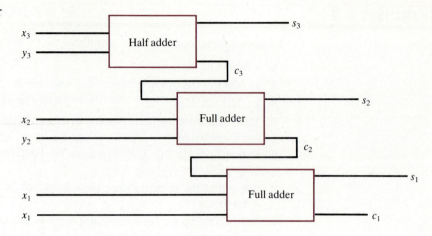

arithmetic-logic unit of a computer. Its disadvantage is that it is relatively slow; the carries have to "ripple through" the entire adder.

TERMINOLOGY

combinational circuits	sum bit
logic circuits	carry bit
gate	half adder
AND, OR, NOT, NAND gates	full adder
maxterm	

SUMMARY

1. Combinational circuits are electronic circuits whose operation can be described in terms of truth tables. Such circuits can be built from gates; the simplest gates are the AND, OR, and NOT gates.

2. A combinational circuit can be designed to synthesize any boolean expression; using minterm representation, a circuit can be synthesized for any truth table.

3. All combinational circuits can be built using just the NAND gate.

4. Circuits that do binary arithmetic are combinational circuits; two of the most basic are the half adder and the full adder.

EXERCISES 3.8

Design combinational circuits to synthesize the following expressions.

1. x AND (y OR z)
2. $\neg x$ OR $\neg y$
3. (x AND y) OR ($\neg w$ OR $\neg z$)
4. $x \rightarrow y$
5. x AND y AND z
6. x OR y OR z OR w

7. Verify that the minterm expression developed in Example 3 does represent the XOR connective.

8. Design a circuit for XOR based on the minterm expression developed in Example 3.

9. Verify that the minterm expression developed in Example 4 represents the given table.

10. Design a circuit for the output table specified in Example 4 based on the minterm expression developed there.

Construct minterm expressions for the following truth tables.

11.

p	q	r	Output
1	1	1	0
1	1	0	1
1	0	1	0
1	0	0	1
0	1	1	1
0	1	0	0
0	0	1	0
0	0	0	1

12.

p	q	r	Output
1	1	1	1
1	1	0	0
1	0	1	0
1	0	0	0
0	1	1	1
0	1	0	0
0	0	1	0
0	0	0	1

13. Design an OR circuit using only NAND gates.

14. Design an AND circuit using only OR and NOT gates.

15. Verify that the half adder circuit gives the correct outputs for $x_1x_2 = 11, 10,$ and 00.

16. Verify that the full adder gives the correct outputs for the cases where $x_1x_2x_3 = 110, 010,$ and 001.

17. Design a full adder circuit by means of minterm representations for s and c.

18. An important component of many electronic devices is a light-emitting diode (LED) display.

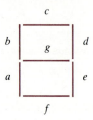

Of the seven diodes, different ones are illuminated to display different numerals or characters. For simplicity, consider an LED used to display any of the digits 0 through 7. The digits are coded according to their binary equivalents:

Digit $x_1x_2x_3$	Binary Equivalent
0	000
1	001
2	010
3	011
4	100
5	101
6	110
7	111

$x_1x_2x_3$ are the three inputs to the LED circuitry.

Design combinational circuits for diodes a, b, and c. Verify that the circuits correctly illuminate the diodes for 0 and 6. (*Hint:* Write the digits 0 through 7 as they would appear in block form on the display. Set up a truth table for each diode based on when you would want to have it illuminated where 1 means illuminated and 0 means not illuminated.)

REVIEW EXERCISES—CHAPTER 3

Write each of the following sentences symbolically, using a letter such as p or q to denote each simple proposition and a connective, or explain why it is not possible.

1. Diane likes golf and John does not like tennis.

2. Diane likes golf, while John likes tennis.

3. John likes to play tennis while Diane plays golf.

Let p = "misers like money" and q = "my cousins are not greedy." Write the following as ordinary English sentences in as natural a style as possible.

4. $q \rightarrow \neg p$ **5.** p XOR q **6.** $\neg(p$ OR $q)$ **7.** $p \leftrightarrow \neg q$

Let $p(x)$ = "$x^2 > 4$," $q(x)$ = "$x \leq 2$," and $r(x)$ = "$x \geq 1$." Evaluate each of the following expressions for the given value of x.

8. $p(x) \rightarrow q(x)$; $x = 2$ **9.** $r(x) \rightarrow \neg p(x)$; $x = 1$

10. $(p(x)$ OR $q(x))$ OR $r(x)$; $x = -3$ **11.** $p(x) \leftrightarrow q(x)$; $x = 0$

12. The connective NOR is defined by the rule that p NOR q is true when both p and q are false; otherwise it is false. Construct a truth table for NOR.

Construct truth tables for the following.

13. $(p$ OR $q) \rightarrow (r$ OR $s)$ **14.** $(p$ OR $\neg q) \rightarrow q$

Evaluate the following predicates for the given values of their free variables.

15. $(x > 1) \rightarrow (x \neq 0)$; $x = 0$

16. $((y > 10)$ OR $(y < -10)) \rightarrow (y < 100)$; $y = 20$

17. $(x < 100) \rightarrow (y \neq 0)$; $x = 90$, $y = 1$

18. $((x > -1)$ AND $(y < 1)) \rightarrow (xy < 0)$; $x = 1$, $y = -1$

Use truth tables to verify the following laws for the algebra of propositions.

19. 1b   **20.** 2b   **21.** 3b   **22.** 4b

23. 5b   **24.** 6b   **25.** 7b   **26.** 9b

Prove that the following are tautologies.

27. $(\neg q \rightarrow \neg p) \leftrightarrow (p \rightarrow q)$

28. $((p \rightarrow q)$ AND $(p \rightarrow r)) \rightarrow (p \rightarrow (q$ AND $r))$

Let the universal set be **N**, let $p(x) = x^2 < 1$, $q(x) = x - 10 > 0$, and $s(x) = x > 0$. Evaluate the truth value of the following.

29. $\forall x \, \neg p(x)$ **30.** $\exists x \, q(x)$

31. $\forall x \, (p(x)$ OR $s(x))$ **32.** $\exists x \, (p(x) \rightarrow s(x))$

Again, let the universal set be **N**, let $p(x, y)$ = "$x + y = 0$," and let $q(x, y)$ = "$x^2 + y^2 = 1$." Evaluate the truth value of each of the following.

33. $\exists x \forall y \, p(x, y)$ **34.** $\forall x \forall y \, p(x, y)$

35. $\forall x \exists y \, q(x, y)$ **36.** $\exists x \forall y \, q(x, y)$

Write the following symbolically. In each case, specify clearly what your universal set is and what each symbol stands for.

37. There are no fish that cannot swim.

38. No skeletons are fat.

39. Some youths are not studious.

40. Some long lectures are hard to endure.

41. Wise musicians play a variety of pieces.

42. Let the universal set be the set of all college professors, $p(x) = $ "x has a beard," $q(x) = $ "x smokes a pipe," and $s(x) = $ "x has a puppy dog." Write each of the expressions in Exercises 1 through 12 of Section 3.5 in ordinary English, being as conversational as possible.

43. Find an expression equivalent to $\exists x \, (p(x))$ that uses only $\forall$ and $\neg$.

44. Negate the following statement, again expressing the negation in the simplest possible form:

$$\forall \epsilon > 0 \; \exists N > 0 \; \forall n > N \, (\, |x_n| \, < \epsilon)$$

Draw valid conclusions from each of the following and indicate which rule of inference you used.

45. If x is a prime number, then $\sqrt{x}$ is not an integer.
29 is a prime number.

46. "I saw it in a newspaper."
"All newspapers tell nothing but lies."

47. If $1 > 2$ then all integers are positive.
$1 > 2$.

48. Gold is heavy.
Nothing but gold will satisfy him.

Symbolize each of the following and draw a valid conclusion. Express your conclusion both symbolically and in ordinary English.

49. No experienced person is incompetent.
Jenkins is always blundering.
No competent person is always blundering.

50. All hummingbirds are richly colored.
No large birds live on honey.
Birds that do not live on honey are dull in color.

51. No shark ever doubts that it is well fitted out.
A fish that cannot dance a minuet is contemptible.
No fish is quite certain that it is well fitted out unless it has three rows of teeth.
All fishes except sharks are kind to children.
No heavy fish can dance a minuet.
A fish with three rows of teeth is not to be despised.

Provide proofs of the following. Justify each step.

52. $A \subset B \Rightarrow A \cup (B \setminus A) = B$

53. If $A \cap B = \emptyset$ then $A \subset B'$

54. $A \cap B = \emptyset \Rightarrow B \cap A' = B$

55. $\emptyset' = U$

Symbolize each of the following arguments and identify the logical error in each.

56. Anyone who drinks Slushy likes good taste.
You like good taste.
You should drink Slushy.

57. Susan only likes red outfits.
This outfit is red.
Susan will like this outfit.

Convert each of the following tautologies into a new rule of inference.

58. $((p \rightarrow r) \text{ OR } (q \rightarrow r)) \rightarrow ((p \text{ OR } q) \rightarrow r))$

59. $(((p \rightarrow q) \text{ AND } r \rightarrow s)) \text{ AND } (p \text{ OR } r)) \rightarrow (q \text{ OR } s)$

60. Design a NOT circuit using only NOR gates. (See Exercise 12.)

61. Design an OR circuit using only NOR gates.

62. Design an AND circuit using only NOR gates.

Combinational circuits can also be designed using maxterms instead of minterms. For example, if a row in a truth table has a final value of 0 and if $p = 0$, $q = 0$, and $r = 1$, the maxterm corresponding to that row is $p \vee q \vee \neg r$. Maxterms are then joined by AND.

63. Find a maxterm expression representing the following truth table and verify its correctness.

p	q	r	
1	1	1	0
1	1	0	1
1	0	1	1
1	0	0	0
0	1	1	1
0	1	0	1
0	0	1	0
0	0	0	1

64. Explain why the minterm and maxterm representations of a truth table do in fact have a truth table that agrees with the original.

REFERENCES

Carroll, Lewis, *Symbolic Logic.* Originally pub. 1897. New York: Dover Press, republished 1958.

Eves, Howard, and Carroll V. Newsom, *The Foundations and Fundamental Concepts of Mathematics,* rev. ed. New York: Holt, Rinehart and Winston, 1965.

Greenberg, Marvin Jay, *Euclidean and Non-Euclidean Geometries.* San Francisco, Calif.: W. H. Freeman, 1974.

Hermes, H., *Introduction to Mathematical Logic.* New York: Springer-Verlag, 1973.

Rubinstein, Moshe, *Patterns of Problem-Solving,* Englewood Cliffs, N.J.: Prentice-Hall, 1975.

Stanat, Donald, and David McAllister, *Discrete Mathematics in Computer Science.* Englewood Cliffs, N.J.: Prentice-Hall, 1977.

Algorithms

4

One thread unifying discrete mathematics is the fact that all of the mathematics it includes is ultimately based on the natural numbers. Another unifying factor, though, is that almost all of discrete mathematics uses computational procedures. Such procedures are known as **algorithms;** they form the subject of this chapter. Algorithms, along with logic, provide the basic tools for our subsequent study of various discrete models. If you have already had an introductory programming course in Pascal, you can probably read Section 4.1 very quickly. Be sure, though, that you become familiar with the particular form of pseudo-code used here.

4.1

ALGORITHMS

A Brief History

Although the study of algorithms does not have as long a history as the study of logic, the subject is still ancient. One of the most famous algorithms is the Euclidean algorithm for finding the greatest common divisor of two positive integers. It was part of Euclid's *Elements,* written around 300 B.C. The Babylonians and Egyptians developed many algorithms, as was pointed out in Section 3.1 The word *algorithm* itself derives from the name of a famous Persian textbook author, Adu Ja'far Mohammed ibn Musa al-Khowarizmi, who lived around 825 A.D. The word was originally written as *algorism* and referred to processes for carrying out arithmetic computations using Arabic numerals (in contrast to doing arithmetic with Roman numerals, which were the principal tool in use at the time). It eventually became *algorithm* by hybridizing with the word *arithmetic.* It has become widely known in recent years in the *algorithm* form because of its adoption by computer science.

The Concept of an Algorithm

In brief, an algorithm is a systematic, step-by-step process for solving a particular type of problem. For instance, suppose one wanted to find the largest of three integers, *a, b,* and *c.* Algorithm 1 would solve this problem.

We will discuss the form in which this is written later. For now, note that the symbol := is read "becomes" or "gets" and means that the variable on the left-hand side acquires the value of the variable or expression on the right.

ALGORITHM 1   Finding the Largest of Three Integers

Procedure Find_the_largest(a,b,c; var largest);

{a, b, c, largest are integers.}

begin

    largest := a;

    if $b >$ largest then largest := b;

    if $c >$ largest then largest := c

end;

The key point to notice is that the algorithm operates on a particular set of three integers. Such a set is an instance of the more general concept *set of three integers.* We will use the term **data structures** to refer to abstract concepts in which the underlying set is discrete (i.e., N or a subset of N). We can now define an algorithm more precisely. An algorithm consists of an **input** (a data structure), an **output** (also a data structure, which may be very elementary; in Algorithm 1 it is only a single integer), and a finite sequence of steps that transform the input into the output. We will soon discuss the rules governing what is allowed in those steps. Note the similarity between the idea of an algorithm and the idea of a proof. Proofs begin with hypotheses (inputs) and terminate with conclusions (outputs). In between there is a finite sequence of steps that transform the hypotheses into the conclusions. The steps in a proof must conform to certain rules, just as an algorithm allows only steps of a certain limited type.

Algorithmic Language

The language we shall use to express algorithms is commonly called **pseudo-code.** It resembles a compute language but is not as formal. It also varies from one writer to another in some of the notational conventions used and in the degree of informality allowed. However, most versions of pseudo-code in popular use are very similar to the Pascal language. For our purposes, the language will consist of the following:

The words *begin* and *end*—these mark the boundaries of the algorithm or major sections of it. The portion between the first *begin* and the last *end* is called the **body of the algorithm.**

Imperative verbs—like *get, put, read,* and *write.*

Constants, variables, and data structures—the nouns that the verbs act on.

Assignment statements—statements of the form

variable := constant or variable or expression

where expressions can be algebraic or boolean.

Comments—verbal statements intended to explain to a reader what the algorithm is doing at various points. Comments are enclosed in braces: {and}. Comments are also used at the beginning of the algorithm to explain the rules governing the appropriate inputs and outputs.

Logical structures—procedures, conditionals, and loops.

The last item requires more explanation.

Procedures and Procedure Calls

Every algorithm written here will begin with the word **procedure** followed by the name of the algorithm. Also, any variables that the algorithm needs to carry out its task will be listed in parentheses after the name. For instance:

Procedure Find_largest(a,b,c;var largest);

Find_largest is the name of the procedure, and the integer variables *a, b, c,* and *largest* are the variables that *Find_largest* needs. Variables listed in parentheses after the name of the procedure are called **parameters.** Variables needed for input will be listed first, followed by output variables. Following the Pascal convention, a semicolon and the word *var* will be used to separate the input and output variables. If a variable is used for both input and output, it will be included with the output variables.

Often one wants to isolate a particular task or group of tasks from a procedure and regard it as a procedure in its own right. This is frequently done for clarity or simplicity. In this case, that operation will be also considered a procedure. Whenever a larger procedure uses such a smaller procedure, we say that it **calls the subprocedure.** For instance:

EXAMPLE 1

Procedure Solve_a_problem;
{*a, b,* and *c* are integers}
begin
  read(*a,b,c*);
  Find_the_largest(*a,b,c*,largest);
   .
   .
   .

The procedure *Solve_a_problem* calls *Find_the_largest* as a subprocedure. The variables *a, b, c,* and *largest* are parameters. Note that the word *var* is not used in the procedure call, although it is used in the procedure declaration. ∎

The point of this example is to illustrate the patterns we will follow in declaring procedures and writing procedure calls. We shall soon look at some realistic procedures that use these conventions.

Conditionals

We will use two forms of the **conditional:**

if condition then action

and

if condition then action1 else action2.

The **condition** here is a simple or compound predicate of the form discussed in previous sections. **Actions** are one or more imperative verbs followed by appropriate nouns. In **if...then,** if the condition is true, the action is performed. If it is not true, nothing is done. In **if...then...else,** if the condition is true, action1 is performed. If it is not true, action2 is performed.

Note that this is different from the *if...then*'s ($\rightarrow$ and $\Rightarrow$) of the previous chapter. Here what follows the word *then* is an action or actions to be performed, whereas $\rightarrow$ and $\Rightarrow$ connect propositions or predicates to be checked for truth or falsity. Nevertheless, the *if condition then action* statement is more like the implication ($\Rightarrow$) than the conditional ($\rightarrow$). That is, if the condition is true, the action must be performed, just as, in $p \Rightarrow q$, if p is true, so must q be true.

Also, note that *if condition then action* is really the same as *if condition then action else do nothing;* i.e., *if...then* is really a special case of *if...then...else.*

EXAMPLE 2   Write conditionals that will do the following.

a) Suppose a test included extra credit questions. Reduce the score of any student earning more than 100 points to 100.

b) If an individual works on Saturday or more than eight hours, pay time and a half for overtime; otherwise pay at the regular rate.

c) If the first of two numbers is larger than the second, reverse their order. Otherwise, leave them as they are.

SOLUTION

a) if $x > 100$ then $x := 100$;

b) if day='saturday' or hours_worked > 8

then pay_rate := base_pay * 1.50

else pay_rate := base_pay;

c) if $x1 > x2$ then

begin

temp := $x1$;

$x1$:= $x2$;

$x2$:= temp

end;   ∎

Loops

Loops are used whenever we want to repeat a particular group of statements several times. We will use three loops: *repeat, while,* and *for.* Their patterns are as follows. The word *condition* again refers to a simple or compound predicate.

repeat

list of actions

until condition;

while condition do

begin

list of actions

end;

for counter := 1 to limit do

begin

list of actions

end;

EXAMPLE 3   Write a loop that will count the number of times a real number x has to be divided by 2 until it is less than some other positive, real number $\in$, assuming $x > \in$ to start.

SOLUTION

count := 0;

repeat

x := $x/2$;

count := count + 1

until $x < \in$;   ∎

We stated earlier that x was greater than $\in$. Often we would like to write a loop that allows the possibility that the loop will never be executed. To do this, we need the while loop.

EXAMPLE 4   Solve the problem of Example 3 using the while loop.

SOLUTION

> count := 0;
> while $x > \in$ do
>   begin
> $x := x/2$;
>   count := count + 1
>   end;

In this case, if x is initially less than or equal to $\in$, count is simply assigned the value zero. ∎

Figure 4.1 Flow charts for the repeat and while loops.

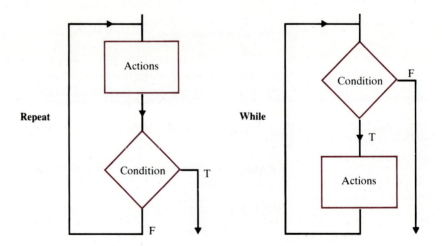

Figure 4.1 provides a way to visualize the differences between the repeat and while loops. The main difference is that with repeat, the actions are performed first; then the condition is tested. With while, the situation is reversed. Thus the actions in repeat are always performed at least once, whereas with while, the actions may not be performed at all if the condition is never true.

The third kind of loop, the *for* loop, is used when one wants to repeat an action or a group of actions a specific number of times.

EXAMPLE 5   Write a loop to sum 10 integers.

SOLUTION

> total := 0;
> for count := 1 to 10 do begin
>   read(x);
>   total := total + x
> end; ∎

Notice that the for loop is really a variation of the while loop; i.e.,

> for count := 1 to limit do begin
>   actions
> end;

has exactly the same effect as

> count := 1;
> while count < limit do begin
>   actions
>   count := count + 1
> end;

In both cases, the sections are repeated exactly limit times, and if limit is less than 1, they are never performed at all.

Formality versus Informality

EXAMPLE 6   Write a loop to describe the process of reading text one character at a time.

SOLUTION

> get a character;
> repeat
>   write the character;
>   get the next character
> until all characters are written; ∎

This is a standard example of what computer scientists call a *read-write loop*. Note that it is written in a less formal style than the previous examples. *Write* and *get* are not followed by variable names, and there is no indication of precisely how to tell when all of the characters are written. Example 3 and 4 are actual Pascal code; Example 6 is not. We will use the more informal style of writing algorithms whenever the technical details needed to implement the algorithm in a computer language or a particular computer system would distract from the main point of the algorithm.

Some Complete Algorithms

EXAMPLE 7 Write an algorithm for finding the largest element in a nonempty unsorted list.

SOLUTION

ALGORITHM 2   Finding the Largest Element in a Nonempty Unsorted List

Procedure Find_max(n, list; var max);

{list is an unordered set of integers; n is its length; max is an integer. We know n is at least 1.}

begin

1. max := first element of list;   {initialize max}
2. current := 2;   {set current to 2nd item in list}
3. while current $\leq n$ do
        begin
4.   if current item > max   {compare item to max}
5.   then max := current item;   {if bigger, replace max}
6.   current := current + 1   {count items}
    end {while}
end;{find_max} ■

The numbers on each executable line are not essential; they are included for later reference. Using arrays (Appendix A), lists can be handled more precisely. For now we will think of lists as ordered sets, i.e., sets in which it is possible to distinguish the first, second, third, and subsequent elements.

Testing Algorithms

One way to test an algorithm is by **tracing.** This means choosing a sample data set and carrying out the algorithm on that sample to see if it performs correctly. This is an effective means of detecting errors. It is very much like attempting to disprove a theorem by trying to find a counter-example. Unfortunately, an algorithm may work in every case traced, but there may be others in which it does not work. We will take up the subject of proving algorithms correct in Section 4.3.

EXAMPLE 8   Trace Algorithm 2 for the list {11, 12, 17, 15}.

SOLUTION   A **detailed trace** consists of a statement of any changes made in any variable each time a line is executed. It also indicates any decisions that are made.

1. max := 11
2. current := 2
3. is 2 ≤ 4? yes
4. is 12 > 11? yes
5. max := 12
6. current := 3
3. is 3 ≤ 4? yes
4. is 17 > 12? yes
5. max := 17
6. current := 4
3. is 4 ≤ 4? yes
4. is 15 > 17? no
6. current := 5
3. is 5 ≤ 4? no

Note that line 4 is not executed the last time.

A **brief trace** consists of just a list of the variables and the successive values they take on, concluding with the final values. For instance:

max 11 12 17

current 2 3 4 5

A detailed trace is valuable the first few times one traces a particular algorithm; after that a brief trace usually suffices.   ∎

Expressing Familiar Algorithms in Pseudo-Code

EXAMPLE 9   Using pseudo-code, write the standard algorithm taught in elementary school for multiplying integers two or more digits in length.

SOLUTION   This algorithm assumes that the user knows the multiplication tables from 0 to 9 and can multiply a number of any length by a one-digit number. It then provides a mean of multiplying integers when both numbers have two or more digits. We will work through an example before generalizing:

$$
\begin{array}{r}
69 \\
\times 34 \\
\hline
276 \\
207 \\
\hline
2346
\end{array}
$$

This involves the following steps:

Selecting the rightmost digit of 34, namely, 4

Multiplying 69 by this digit

Recording the result

Selecting the next digit of 34, namely, 3

Multiplying 69 by that digit

Recording this result shifted one place to the left

Adding the two results.

The algorithm to carry out these operations is a generalization of each of the preceding steps. However, an algorithm is usually not totally independent of the technology used to implement it. For instance, the preceding sequence of steps is set up assuming multiplication by pencil and paper. Hence intermediate results must be written down. Written in a low-level computer language, there would be analogous steps such as "multiply the numbers" and "store the result." In our algorithmic language, we assume that the variables themselves hold the intermediate results and that no special step is needed to record them. Instead, the algorithm must keep a running total of the numbers to be added, rather than recording all of them and then adding. It also shifts the 69 to the left and then multiplies, rather than multiplying and then shifting the answer. Thus we do not have to worry about how many places to shift the product. A brief note on notation: **mod** refers to the remainder of ordinary integer division and **div** to the quotient. Thus 69 mod 10 equals 9, while 69 div 10 equals 6.

ALGORITHM 3   Multiplication

Procedure Multiply(a,b; var product);

{a, b, product, d are integers.}

begin

   product := 0;   {initialize running total}

```
repeat
    d := b mod 10;              {d := rightmost digit of b}
    product := product + a * d  {keep running total}
    a := a * 10                 {shift a left}
    b := b div 10               {shift b right and drop
                                 rightmost digit}

    until b = 0
end;  ∎
```

Characteristics of Algorithms

Recall that at the beginning of this section we compared an algorithm to a proof in that both involved input, output, and a sequence of steps. Having seen a few algorithms, we can now state some restrictions that are placed on the sequence of steps. For a more extensive discussion of these ideas, see *Fundamental Algorithms* by Knuth, Chapter 1.1 (see References).

An algorithm must have the following characteristics:

1. Finiteness
2. Definiteness
3. Effectiveness

Finiteness does not refer to the fact that an algorithm is a finite list of operations. It means that an algorithm must always terminate in a finite number of steps, including repetitions. For instance, the following collection of steps do not constitute an algorithm:

```
Repeat
    x := 1;
    x := x + 1;
until x = 10;
```

This is an example of an **infinite loop** since x never gets to 10. Some algorithms may take billions of steps to complete. Such algorithms may not be practical, but they do satisfy the requirement of finiteness.

Definiteness means that every step of the algorithm must be precisely and unambiguously defined. This is a totally inappropriate requirement for poetry, but it is essential for algorithms.

Effectiveness means that every step can be carried out, and in a finite amount of time. For instance, a statement like

Get all data necessary to complete this problem

is finite and definite, but, if the data are not available, it is ineffective.

At this point, we have completed our coverage of algorithmic language. We have no way to evaluate the quality of algorithms, however. That will be the subject of the next three sections.

TERMINOLOGY

algorithm	actions
data structures	loops
input	tracing
output	detailed trace
pseudo-code	brief trace
body of the algorithm	mod
procedure	div
parameter	finiteness
procedure call	infinite loop
conditional	definiteness
conditions	effectiveness

SYNTAX

if . . . then	while . . . do
if . . . then . . . else	for . . . do
repeat . . . until	

SUMMARY

1. An algorithm consists of an input (a data structure), an output (also a data structure), and a finite sequence of steps that transforms the input to the output.

2. Algorithms are the basic tool for carrying out computations involving data structures. These computations may be logical as well as arithmetic.

3. We will express algorithms in pseudo-code, an informal language similar to Pascal and consisting of the words *begin* and *end,* imperative verbs, constants and variables, assignment statements, comments, and logical structures.

4. The logical structures include procedures, procedure calls, conditionals—*if . . . then* and *if . . . then . . . else*—and loops—*repeat, while,* and *for.*

5. One way to test algorithms is by tracing. Although this is a good way to find errors, it cannot prove that an algorithm is correct.

6. An algorithm must be finite (terminate in a finite number of steps), definite (unambiguous), and effective (capable of being carried out in a finite amount of time).

EXERCISES 4.1

1. Modify Algorithm 1 to find the smallest of $a, b,$ and $c.$

2. Modify Algorithm 2 to find the minimum element in the list.

Trace the following pieces of algorithms and indicate what will be written in each case.

3. Trace with $x = 1, -12, 10$:
      if $x * x < 100$

      then write('x is small')

      else write('x is large');

4. Trace with kind $= 1,$ quantity $= 10,$
                 kind $= 2,$ quantity $= 10,$
                 kind $= 1,$ quantity $= 100,$
                 kind $= 2,$ quantity $= 100$:

      if kind $= 1$

         then

            if quantity > 100

            then write('rate $= .05$')

            else write('rate $= .08$')

         else

         write('rate $= .10$');

5. This algorithm finds the location of the largest element in a list, rather than finding the element itself. Trace first for list $= [13, 1, 17]$ and then for list $= [5, 4, 3, 2, 1]$.

      $i := 0;$

      largest $:= -\infty;$

      repeat

         $i := i + 1;$

```
        get ith from list;
        if number > largest
        then begin
            location := i;
            largest := number
        end
    until whole list is examined;
    write (location);
```

6. Trace for list = [1, 17, 15] and [19, 6, 3, 19, 21].

```
    count := 0;
    largest := —∞;
    while there are unexamined numbers in the list do begin
        count := count + 1;
        get next number;
        if number > largest
        then begin
            location := count;
            largest := number;
        end
    end;
    write(location);
```

7. Trace for list = [1, 17, 2, 16, 32, 14, 18]:

```
    largest := —∞;
    for i := 1 to 7 do begin
        get ith number
        if number > largest
        then begin
            location := i;
            largest := number
        end
    end;
    write(location);
```

Write each of the following as parts of algorithms, using pseudo-code.

8. If the number of participants exceeds 22, reduce that number to 22.

9. A list of names is to be counted. Read each name on the list and, at the end, print the count.

10. Read through a list of 10 names and see if anyone is called Smith. If so, set the variable *Present* to true; if not, set *Present* to false. (*Hint:* initialize *Present* to false in the first line of your algorithm.)

Write complete algorithms for the following.

11. "If your sales for the day were over $1000, your commission is 10%. Otherwise it is 9% if you sold one of the items on special and 8% if you did not. List every contact you made today, the amount sold to each, and keep a running total of your sales. At the end, calculate the amount of your commission and write it at the bottom."

12. Read a sentence one character at a time, counting the number of commas in the sentence.

13. Write the algorithm for adding two numbers more than one digit in length in pseudo-code. You may assume that the user has an addition table for the digits 0 through 9. Be careful to handle carries correctly.

14. Write the algorithm for subtracting two numbers more than one digit in length in pseudo-code. Be careful to handle borrows correctly.

Why are the following not algorithms?

15. procedure count_the_integers (var count);

```
    begin
        count := 0;
        repeat
            count := count + 1
        until count < 1;
        write(count)
    end;
```

16. procedure get_rich_quick(var big_bucks);

```
    begin
        for every possible lottery ticket number
            buy one ticket;
        big_bucks := amount of winnings
    end;
```

17. procedure tom_sawyer;

```
    begin
        while there are tasks undone do begin
            identify next task to be done;
            find someone else to do it
        end; {while}
    end;
```

4.2

ASYMPTOTIC DOMINATION

Typically, the larger the set of data an algorithm must act upon, the more time and space it needs to complete its task. For instance, in an algorithm to find the largest member of a set, the larger the set, the more time and space the algorithm will require. Thus the **efficiency** of an algorithm is usually measured as a function, $f(n)$, where n is the size of the data set that the algorithm receives as input. Efficiency can refer to either the amount of time or the amount of space the algorithm requires, although we will think of it here primarily in terms of time. If we are to compare the efficiency of algorithms, then, we will need some mathematical tools to enable us to compare functions. Under most circumstances, small sets of data take little time, regardless of what algorithm is used, and so comparisons are not very important. But with large sets of data, they become critical. (Recall the 38,573 years it would take the brute force algorithm to solve the traveling salesman problem for 20 cities as opposed to the 3.8×10^{-3} seconds the nearest neighbor algorithm requires.) The concept of **asymptotic domination** is used to compare the behavior of functions for large values of n.

The Basic Definition

Definition Let $g, f: N \rightarrow R$. g **asymptotically dominates** f if there are positive integers m and N such that for every n greater than N, $|f(n)| \leq m|g(n)|$.

With quantifier notation, this definition can be written

$$\exists N \, \exists m \text{ such that } \forall n > N \, (|f(n)| \leq m|g(n)|).$$

Examples

Consider Figures 4.2 and 4.3. In both of these figures, the g function asymptotically dominates the f function. In Figure 4.2, g is at first smaller than f, but eventually becomes larger and stays that way. In Figure 4.3, though, g never becomes permanently larger than f. However, a constant multiple of g will always exceed f; hence we say that in this case also, g asymptotically dominates f.

Figure 4.2 g asymptotically dominates f since g exceeds f for large values of x.

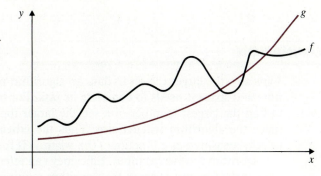

Figure 4.3 g asymptotically dominates f since a constant times g exceeds f for large values of x.

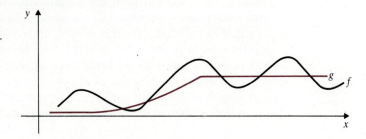

EXAMPLE 1 In the following table, each function in the column headed $f(n)$ is asymptotically dominated by the corresponding function in the $g(n)$ column. Show this to be true.

$f(n)$	$g(n)$
a) n	n^2
b) n^2	n^3
c) $\log_2 n$	n
d) $2n$	$3n$
e) $3n$	$2n$

SOLUTION   All of these relationships follow a similar pattern. The definition of asymptotic domination includes two $\exists$s followed by one $\forall$. Thus a proof that $g(n)$ dominates $f(n)$ must include a presentation of the suitable values for N and m, followed by a proof that for any representative $n \geq N$, $|f(n)|$ is indeed less than or equal to $m|g(n)|$.

a) When showing that one nonnegative function asymptotically dominates another, the best way to start is to sketch a graph of each to determine if one ever permanently exceeds the other, as shown in Figure 4.4. Since $g(n)$ exceeds $f(n)$ for all n values greater than or equal to 1, we let $N = 1$. We also set $m = 1$ since g is itself bigger than f. It doesn't need to be multiplied by a constant to exceed f. For many purposes, this graphic demonstration is sufficient and we can stop here. However, a proof that g dominates f would require us to show algebraically that the rest of the definition is satisfied. That is, we need to show that for all $n > 1$, $n^2 \geq n$. In this case, this is very easy to do. We know that

$$n > 1.$$

Multiplying both sides by n, we get

$$n^2 > n.$$

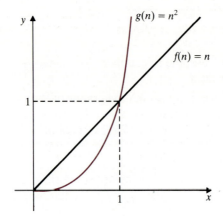

Figure 4.4 g dominates f since g exceeds f for $n \geq 1$.

b) We will omit the graph on this one. However, we must again let $N = 1$ and $m = 1$. Then

$$n > 1.$$

Multiplying both sides by n^2, we get $n^3 > n^2$.

c) We can see immediately from the graphs of these two functions (Figure 4.5) that g asymptotically dominates f. Proving it algebraically requires the use of mathematical induction. We can again let $N = 1$ and $m = 1$. Then $\log_2 1 = 0$, which is less than 1, so $f(1) \leq g(1)$. Now let $n \geq 1$ and assume that $\log_2 n \leq n$. The latter inequality implies that $n \leq 2^n$, and thus, multiplying both sides by 2, $2n \leq 2^{n+1}$. Since $n \geq 1$, adding n to both sides, $2n \geq n + 1$. Hence

$$n + 1 \leq 2^{n+1}, \text{ and hence } \log_2(n + 1) \leq (n + 1).$$

Thus $\log_2 n \leq n$ for all $n > 1$.

Figure 4.5 $g(n) = n$ dominates $f(n) = \log_2 n$.

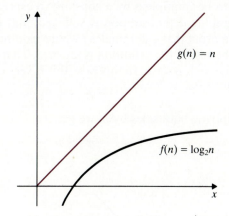

d) In this case, the selection of $N = 1$ and $m = 1$ are obvious since $2n \leq 3n$ for all n.

e) Once again, we sketch the graphs of the two functions. See Figure 4.6. This time we can take $N = 1$, but we must take $m \geq 1.5$ in order to get $m \, | \, g(n) \, | \, \geq \, | \, f(n) \, |$. We can take $m = 2$ and get

$$m \, | \, g(n) \, | \, = mg(n) = 2 \cdot 2n = 4n \geq 3n = f(n) = \, | \, f(n) \, |.$$

This example is particularly important because it illustrates the case in which g dominates f even though its graph never crosses the graph of f.

Figure 4.6 $g(n) = 2n$ asymptotically dominates $f(n) = 3n$ since any constant of 1.5 or larger multiplied by g exceeds f.

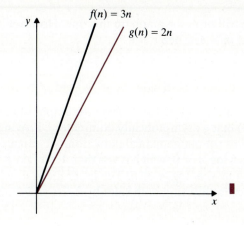

EXAMPLE 2 Show that each function in the right column asymptotically dominates the corresponding function in the left column. k is any positive integer; c is a real number greater than 1.

a) n 2^n

b) n^k c^n

c) 2^n 3^n

d) c^n $n!$

SOLUTION

a) Let $N = 1$ and $m = 1$. This follows immediately from Example 1c.

b) This is also an especially important relationship. To justify it intuitively, we can use the fact that the inequality

$$n^k \leq c^n$$

is equivalent to

$$n \leq (c^{1/k})^n.$$

Since $c^{1/k}$ is a constant greater than 1, $(c^{1/k})^n$ will exceed n for large n. A formal proof requires calculus so we will not prove it here.

c) Again, let $N = 1$ and $m = 1$. Clearly, $2 < 3$. Hence $2^n < 3^n$ for all $n \geq 1$. (This can be proven by induction. See Exercise 20.)

d) This one is also hard to illustrate graphically, so we will proceed algebraically. Let $N = \lfloor c \rfloor$; that is, if c is an integer, $N = c$; otherwise, $N + 1$ is the smallest integer greater than c. Let $m = c^N/N!$. Then if $n > N$,

$$\frac{c^n}{n!} = \frac{c^{n-N}c^N}{n(n-1)\cdots N(N-1)\cdots 1} = \frac{c}{n}\frac{c}{n-1}\cdots\frac{c}{N+1}\frac{c^N}{N!} \leq m.$$

Hence $c^n \leq m \cdot n!$ for $n \geq N$.

In this case, the choice of N and m was motivated by the observation that if $c < n$, c/n will be less than 1. Thus the $n!$ in the expression $c^n/n!$ must be split into two parts, those terms greater than c and those less than or equal to c. The latter terms, together with N c's, are incorporated into m. ∎

In summary, then, when we want to find out if one nonnegative function asymptotically dominates another, we usually start by graphing both functions. If one exceeds the other for all values beyond some point, that one dominates. This can be verified algebraically by taking $m = 1$ and $N =$ any integer greater than or equal to the point at which the one function crosses the other before permanently exceeding it. If, however, the two functions begin to either parallel each other or increase at a similar rate, the smaller function may still asymptotically dominate the larger. This occurs if a constant multiple of the smaller function actually exceeds the larger one beyond some point. The graphs help us choose the point and the constant multiple. The fact that the smaller function dominates must then be checked algebraically.

Definition   The set of all functions asymptotically dominated by a particular function g is denoted $O(g)$ (usually read *big-oh of g*). Thus, if f is asymptotically dominated by g, we write $f \in O(g)$.

EXAMPLE 3   Applying Examples 1 and 2, we have:

$n \in O(n^2)$ $n^2 \in O(n^3)$

$\log_2 n \in O(n)$ $2n \in O(3n)$

$3n \in O(2n)$ $n \in O(2^n)$

$n^k \in O(2^n)$ $2^n \in O(3^n)$

$c^n \in O(n!)$ ▍

Note that $2n \in O(3n)$ and $3n \in O(2n)$.

Definition   Two functions are **asymptotically equivalent** if each asymptotically dominates the other. Thus $2n$ and $3n$ are asymptotically equivalent. We shall see in the next theorem that this means that $O(2n) = O(3n)$.

Big-oh notation is the principal tool used in computer science for describing the efficiency of algorithms. The following theorem and corollary are used frequently when computing the efficiency of a particular algorithm and when comparing the efficiency of different algorithms.

THEOREM 4.1   Let f, g, and h be functions on the positive integers and let c be any real constant. Then

a) $f \in O(f)$.

b) $f \in O(g) \Rightarrow cf \in O(g)$.

c) $f \in O(g) \Rightarrow O(f) \subset O(g)$.

d) $f, h \in O(g) \Rightarrow f + h \in O(g)$.

e) If $\exists N$ is such that $n \geq N \Rightarrow f(n) \geq 1$, then $g \in O(fg)$.

PROOF   For simplicity, we shall assume that f and g are positive functions and $c > 0$. The generalization of the proof to cases where f and g are not always positive is left for the Exercises. Since $|f(n)| = f(n)$ for all n

and $|g(n)| = g(n)$ for all n, we will drop the absolute value signs throughout this proof.

a) Let $N = 1$ and $m = 1$. Since $f(n) \leq f(n)$ for all n, f is itself a member of the set of functions asymptotically dominated by f.

b) Suppose $f \in O(g)$. Then $\exists N$ and m such that $n > N \Rightarrow f(n) \leq mg(n)$. Hence $cf(n) \leq cmg(n)$. Using the same value of N and replacing m by cm, we can conclude that $cf \in O(g)$.

c) Suppose $f \in O(g)$ and $h \in O(f)$. Then $\exists N$ and m such that $f(n) \leq mg(n)$ and $\exists N'$ and m' such that $h(n) \leq m' f(n)$. Let N'' be the larger of N and N' and let $m'' = mm'$. Then

$$h(n) \leq m' f(n) \leq m'mg(n) = m''g(n)$$

whenever $n > N''$. Thus $h \in O(g)$ and hence $O(f) \subset O(g)$.

d) Suppose both f and h are members of $O(g)$. Then $\exists N$ and m such that $n > N \Rightarrow f(n) \leq mg(n)$. Also, $\exists N'$ and m' such that $n > N' \Rightarrow h(n) \leq m'g(n)$. Let N'' be the larger of N and N' and let $m'' = m + m'$. Then $f(n) + h(n) \leq mg(n) + m'g(n) = (m + m')g(n) = m''g(n)$. Thus

$$f + h \in O(g).$$

e) Suppose $\exists N$ such that $n \geq N \Rightarrow f(n) \geq 1$. Then if $n \geq N$, $g(n) \leq f(n)g(n)$. Taking $m = 1$, we conclude that $g \in O(fg)$. ☐

This theorem tells us a number of valuable things. For instance, if two functions are both in $O(g)$, so is their sum. Similarly, if a function is in $O(g)$, so is any constant multiple of it and of any function it asymptotically dominates. It also has a number of valuable corollaries.

COROLLARY 4.2 Suppose $k > 2$ is an integer and $c > 1$ is a real number.

a) $f \in O(g) \Rightarrow f + g \in O(g)$.

b) $f \in O(g), g \in O(f) \Rightarrow O(f) = O(g)$.

c) $O(1) \subset O(\log n) \subset O(n) \subset O(n \log n) \subset O(n^2) \subset \cdots \subset O(n^k) \subset \cdots \subset O(c^n) \subset O(n!)$.

PROOF

a) This follows directly from Theorem 4.1, parts a and d.

b) This follows from Theorem 4.1, part c.

c) These follow from Theorem 4.1 and Examples 1 and 2. ☐

EXAMPLE 4 Prove the following:

a) $n^2 + 3n + 2 \in O(n^2)$.

b) If f is a polynomial of degree k, then $f \in O(n^k)$.

c) If $c \geq 1$ and $d \geq 1$ and $c < d$, then $O(c^n) \subset O(d^n)$.

SOLUTION

a) By Corollary 4.2 and Theorem 4.1, both $3n$ and 2 are members of $O(n^2)$. Thus, by Theorem 4.1, part d, $n^2 + 3n + 2 \in O(n^2)$.

b) Let $f(n)$ be denoted $a_k{}^k + a_{k-1}{}^{k-1} + \cdots + a_0$. The remainder of the proof is a generalization of part a. (See the Exercises.)

c) This is a generalizaton of Example 2, part c. (See the Exercises.)   ▐

The result we will use the most often from this section is Corollary 4.2, part c:

$$O(1) \subset O(\log n) \subset O(n) \subset O(n \log n) \subset O(n^2) \subset \cdots \subset O(n^k) \subset \cdots$$
$$\subset O(c^n) \subset O(n!).$$

In each case, the set on the left is not just a subset of the set on the right, but is a proper subset.

EXAMPLE 5   Show that

a) $O(n)$ is a proper subset of $O(n^2)$.

b) $O(1)$ is a proper subset of $O(\log_2 n)$.

c) $O(\log_2 n)$ is a proper subset of $O(n)$.

SOLUTION

a) To show that $O(n)$ is a proper subset of $O(n^2)$, we must show that $O(n^2)$ is not a subset of $O(n)$. That is, we must find a function $f \in O(n^2)$ that is not in $O(n)$. The easiest way to do this is to show that n^2 itself is not in $O(n)$. To show that any function f is not asymptotically dominated by g, we must show that it is not true that

$$\exists N \, \exists m \text{ such that } \forall n > N \mid f(n) \mid \, \leq m \mid g(n) \mid .$$

That is, we must show that for arbitrary integers N and m, there is an $n > N$ such that

$$\mid f(n) \mid \, \geq m \mid g(n) \mid .$$

Let $f(n) = n^2$ and $g(n) = n$. Let m and N be any fixed positive integers. Let n be any integer greater than both N and m. Then

$$n > m$$
$$\Rightarrow \quad n^2 > mn.$$

Hence we have shown that there is an $n > N$ such that $\mid f(n) \mid \, \geq m \mid g(n) \mid$.

b) Following the same pattern as part (a), we shall show that $\log_2(n)$ is not an element of $O(1)$. Let $f(n) = \log_2 n$ and let $g(n) = 1$. Let N and m be fixed positive integers. Let n be greater than N and 2^m. Then

$$n > 2^m$$
$$\Rightarrow \quad \log_2 n > m$$
$$\Rightarrow \quad \log_2 n > m \cdot 1.$$

Hence we have shown that there is a $n > N$ such that $|f(n)| \geq m|g(n)|$.

c) We shall show that n is not an element of $O(\log_2 n)$. Let $f(n) = \log_2 n$ and $g(n) = n$. Let N and m be fixed positive integers. We know that there is an N' such that $n > N'$ implies that

$$n^m \leq 2^n$$
$$\Rightarrow \quad m \log_2 n \leq n.$$

Hence, if n is greater than both N and N',

$$n \geq m \log n.$$

Hence we have shown that there is an $n > N$ such that $|f(n)| \geq m|g(n)|$. ∎

We will examine some of the applications of these results in the next two sections. For a more extensive discussion of the ideas included in this section, see the book by Stanat and McAlister or the book by Liu (see References).

TERMINOLOGY

efficiency

asymptotic domination

$O(g)$

asymptotically dominates

asymptotically equivalent

SUMMARY

1. We say that a function g asymptotically dominates another function f if there are positive numbers N and M such that for every $n \geq N$, $|f(n)| \leq m|g(n)|$.

2. One way to find out whether one function, g, asymptotically dominates another function, f, is to sketch the graphs of each. If g or a constant multiple of g exceeds f for all n after a certain point, g dominates f.

3. The set of all functions asymptotically dominated by g is called $O(g)$.

4. If f, g, and h are functions on the positive integers and c is any real constant:

   a) $f \in O(f)$.

   b) $f \in O(g) \Rightarrow cf \in O(g)$.

   c) $f \in O(g) \Rightarrow O(f) \subset O(g)$.

   d) $f, h \in O(g) \Rightarrow f + h \in O(g)$.

   e) If $\exists N$ such that $n \geq N \Rightarrow f(n) \geq 1$, then $g \in O(fg)$.

   f) $f \in O(g) \Rightarrow f + g \in O(g)$.

   g) $f \in O(g)$, $g \in O(f) \Rightarrow O(f) = O(g)$.

   h) $O(1) \subset O(\log n) \subset O(n) \subset O(n \log n) \subset O(n^2) \subset \cdots \subset O(n^k) \cdots \subset O(c^n)$
      $\subset O(n!)$.

5. Each set in 4h is a proper subset of the one to its right.

EXERCISES 4.2

Sketch graphs of the following pairs of functions and illustrate that in each case $g(n)$ asymptotically dominates $f(n)$.

1. $f(n) = 1$, $g(n) = \log_2 n$ 2. $f(n) = n + 6$, $g(n) = n^2$

3. $f(n) = n^2 + 10$, $g(n) = 2^n$ 4. $f(n) = 3n^2 + 10$, $g(n) = n^3$

5. Let $f(n) = 16n + 1$ and $g(n) = 0.1n^2 + n + 1$. For what values of n is $f(n) > g(n)$? For what values is $g(n) > f(n)$?

6. Separate the following list of functions into collections that are asymptotically equivalent to each other:

$\log n$	$3n + 2$
$n + \log n$	$5n^2 + 2 \log n$
37	$10^{-5}n^2$
$2^{\log_2 n}$	$13n + \log n + 6$
$n^2 + 16n + 1$	$(3n + 1)(2 + \log n)$
$n + 18/n$	$16n + 5n \log n$

For each of the following pairs of functions, show directly from the definition (as in Example 1 in the text) that f asymptotically dominates g.

7. $f(n) = n^4$, $g(n) = n^3$ 8. $f(n) = \log n$, $g(n) = 1$

9. $f(n) = n$, $g(n) = n/2$

10. Find an example of a function in $O(1)$ that is not a constant function.

11. Find examples of functions f and g such that f is not in $O(g)$ and g is not in $O(f)$.

Fill in the details in the proofs of the following parts of Corollary 4.2:

12. Part a   **13.** Part b

14. $O(1) \subset O(\log n)$ **15.** $O(n^k) \subset O(2^n)$

16. Suppose f and g are asymptotically equivalent. Show that cf and g are asymptotically equivalent.

Prove the following:

17. $f, h \in O(g) \Rightarrow f - h \in O(g)$.

18. $f \in O(g) \Rightarrow fh \in O(gh)$.

19. If f is a polynomial of degree k, then $f \in O(n^k)$. (Example 4b)

20. Prove by induction that $\forall n \geq 1, 2^n < 3^n$.

21. Write out in symbolic notation the negation of

$$\exists N \; \exists m \text{ such that } \forall n > N, |f(n)| \leq m|g(n)|.$$

Show each of the following relationships to be false.

22. $O(n) \supset O(n \log n)$

23. $O(n^2) \subset O(n \log n)$

Generalize the following parts of the proof to Theorem 4.1 to handle the cases when f and g are not restricted to be positive functions.

24. Part b   **25.** Part c

26. Part d   **27.** Part e

4.3

ANALYSIS AND VERIFICATION OF ALGORITHMS

As we saw earlier, the Babylonians and Egyptians discovered many mathematical truths and developed many computational procedures. But they lacked a means of proving their theorems or verifying their procedures. Hence some things they thought were correct were not. The Greeks, on the other hand, developed methods to prove theorems, and we use these same methods today. We can also use them to verify the correctness of our algorithms, and we will begin to see how to do that in this section. But in the study of algorithms, their quality, as well as their correctness, must be considered. For instance, the time required for the brute force algorithm to solve the traveling salesman problem for 20 cities makes it totally unacceptable even though the algorithm is correct. When computer scientists or applied mathematicians talk about the quality of an algorithm, there are four criteria that are often evaluated: **time efficiency, space efficiency, elegance,** and the presence of **side effects.**

Efficiency and Elegance

With time efficiency, faster is regarded as better. With space efficiency, smaller is better. Elegance, however, refers to more subtle qualities. The principal qualities of an elegant algorithm are clarity and simplicity. Other descriptive terms, such as *clean, neat, insightful, brilliant, clear, wow!,* and a low whistle are often used. Thus elegance is harder to measure than efficiency. There are specific mathematical tools for measuring efficiency, and these tools are critical parts of discrete mathematics. There are, however, no such tools for measuring elegance. Hence little space will be devoted to it even though it is a valued quality in algorithms.

Use of Test Data

There are two approaches that can be used to measure the efficiency of an algorithm. One is to code the algorithm as a computer program and run it for several different sets of test data. The principal advantage of this approach is that it shows how the algorithm performs with data that are representative of the realistic situations with which the algorithm is to be used. But it has several significant disadvantages. First, if the algorithm is hopelessly inefficient, it may not be worth coding. It would be nice to know this before going to the effort of coding it. Second, using test data suffers from the same limitation as any experimental procedure: The algorithm may perform well on the specific test data chosen, but this is no guarantee that it works well in general. Third, the results are clouded by the fact that performance depends not only on the algorithm and the test data but also on the programming language, the operating system of the computer, and the hardware used.

Similarly, test data are frequently used to verify the correctness of algorithms. As mentioned earlier, this is very much like trying out a conjecture on several examples to see if one proves to be a counter-example. This is an extremely valuable way to find errors; however, it does not guarantee correctness.

Asymptotic Complexity

Another approach to use in measuring efficiency is to express the algorithm in pseudo-code and determine the number of computations the algorithm would require as a function of the size of the input. This can usually be done by careful counting. We then use $O(g)$ notation to express the result. Thus we can say that a particular algorithm is an $O(n)$ algorithm and another is an $O(\log n)$ algorithm. An expression of the number of computations an algorithm requires in $O(g)$ notation is called its **asymptotic com-**

plexity. It is used for at least two reasons. First, knowing that an algorithm takes precisely $3n + 1$ steps is usually no more helpful than knowing that it is an $O(n)$ algorithm since the time that it will actually take to run on a computer will be only approximately proportional to n. This is because computations vary greatly in their execution time when they finally become machine language. Stating the efficiency as $O(n)$ properly emphasizes the approximate nature of the estimate. Second, computations of efficiency are done primarily to allow comparisons of the behavior of algorithms for large values of n. Stating the efficiency as $O(n^2)$, for instance, calls more attention to the factor that has the greatest influence on efficiency than does a formula like $n^2 + 3n + \log n + 2$.

In practice, the asymptotic complexity is often computed. Algorithms that are asymptotically equivalent are then often compared by means of test data.

EXAMPLES

EXAMPLE 1 Compute the asymptotic efficiency of Algorithm 2 from Section 4.1.

SOLUTION First, let's repeat the algorithm with the line numbers.

	begin	
1.	max := first element of list;	{initialize max}
2.	current :=2;	{set current to second item in list}
3.	while current ≤ n do	
	begin	
4.	if current item > max	{compare item to max}
5.	then max := current item;	{if bigger, replace max}
6.	current := current + 1	{count items}
	end {while}	
	end; {find_max}	

Now assume there are n items in the list and compute the number of computations in each step. Each assignment, comparison, and individual arithmetic operation $(+, -, *, /)$ is counted as one computation.

Step Number		Number of Computations
1		1
2		1
3		n
4		$n-1$
5	x	$0 \le x \le n-1$
6		$2(n-1)$

The x indicates that line 5 may never be performed (if the first member of the list is already the largest) or may be performed as many as $n-1$ times (if the list is sorted from smallest to largest). Then n occurs for line 3 because the algorithm must check if current $\le n$ for each of the remaining $n-1$ members of the set after the first and must do one extra check when current becomes $n+1$. The total is

$$f(n) = 4n - 1 + x.$$

We distinguish a **best case,** an **average case,** and a **worst case,** depending on the value of x. This time the best case occurs when $x = 0$ (giving $f(n) = 4n - 1$) and the worst case occurs when $x = n - 1$ (giving $f(n) = 5n - 2$). Finding the average case is often very difficult and requires some knowledge of probability, as one must consider all of the possible ways the data could be arranged. We generally compute the worst case, although at times we state or compute the average case as well. In this algorithm, we can say that both the best and worst cases are $O(n)$. Since the average must be somewhere between them, we can conclude that the average case is $O(n)$ also. ∎

Verification of an Algorithm

A formal approach to verifying algorithms typically involves the use of **preconditions** and **postconditions.** A precondition is a precise statement of what conditions hold before the algorithm is carried out; a postcondition is a precise statement of what conditions are expected to hold when the algorithm is completed. The verification is a careful demonstration that the algorithm transforms the precondition into the postcondition. In this book, we will use an informal approach to verification; pre- and postconditions will be described informally rather than rigorously, and informal arguments will be used to verify that the algorithm transforms the inputs into the desired outputs.

The main difficulty in verifying algorithms is to show that the loops work correctly, since conditionals can be checked by tracing each of their branches with representative data elements. Loops are normally verified by mathematical induction. A careful use of induction with loops also requires a precise formulation of a statement as to exactly what is supposed to be true at each iteration of a loop. Such a statement is called a **loop invariant.** Here, too, we will use an informal approach and will not attempt to formulate our invariants rigorously.

We will now demonstrate this informal approach to verification by verifying Algorithm 2. Initially, list is simply a collection of n integers. This statement is the precondition. At the end, we want max to equal the largest element of the list. This is the postcondition. The loop invariant is this: After i passes through the loop, max equals the largest of items 1 through $i + 1$. To verify by induction that the invariant holds at each pass through the loop, we begin with the basic step and set $i = 0$. After zero passes, the algorithm is at the beginning of the While loop; however, the algorithm has set max to the first element in the list. That is, max equals the largest of items 1 through 1, which is correct after 0 passes through the loop. Now for the induction step, we suppose that after the first i passes, max equals the largest of the first $i + 1$ items. On the $i + 1$th pass, the $i + 2$th item is compared to max; if it is greater, line 4 will detect this and line 5 will replace max by that $i + 2$th item. If the $i + 2$th item is not greater than max, then max will not be changed. Thus, in either case, the algorithm correctly sets max equal to the largest of the $i + 2$ items and hence the loop invariant has been verified. As for the algorithm as a whole, after $n - 1$ passes through the loop, max will be the largest of the n elements. Also, after $n - 1$ passes, the algorithm halts and thus the postcondition has also been verified.

Another Algorithm for Finding the Maximum of n Items

Consider the following algorithm for finding the max of n items.

ALGORITHM 4   Find the Maximum of a List of n Items

Procedure Bubble_down(list; var max);

{list is an unordered set of integers; max, i, and temp are integers.}

begin

1.   for $i := 1$ to $n - 1$ do

2.   if ith item $> i + 1$th item then

            begin

3.   temp := ith item;

4. ith item := $i + 1$th item;

5. $i + 1$th item := temp

        end;

6.   max := nth item in the list

  end;

The strategy behind this algorithm is to compare the first item in the list to the second and, if the first is greater, to switch them. Then the second is compared to the third and a similar switch is made if necessary. This continues until, at the end of the loop, the largest element is at the bottom. Max is then set equal to that bottom element.

EXAMPLE 2   Compute the complexity of Algorithm 4.

SOLUTION

Step Number	Number of Computations
1	$2n$
2	$n - 1$
3	x $0 \leq x \leq n - 1$
4	x
5	x
6	1

Step 1 requires $2n$ computations, since each execution of it involves adding 1 to i and comparing i to $n - 1$. Hence $f(n) = 3n + 3x$ and we have a best case of $f(n) = 3n$ and a worst case of $f(n) = 6n - 3$. Thus the best, worst, and average cases are all $O(n)$. ∎

EXAMPLE 3   Verify Algorithm 4.

SOLUTION   Our pre- and postconditions are the same as for Algorithm 2. Our loop invariant is that after i passes though the loop, the $i + 1$th item in the list is the largest of items 1 through $i + 1$. Once again, we proceed by induction and first assume that $i = 0$. The for loop has not yet been entered and item 1 is indeed the largest of items 1 through $i + 1$, since $i + 1 = 1$. Now assume that after i passes, the largest item in the list is in the $i + 1$th place in the list and consider what happens for the $i + 1$th pass. If the $i + 2$th item is greater than or equal to the $i + 1$th, the condition in line 2 will be false and lines 3 through 5 will be skipped. If the $i + 2$th item is less than the $i + 1$th, lines 3 through 5 will be performed and the $i + 1$th and

$i + 2$th items will be switched. Thus, in either case, the resulting $i + 2$th will be the largest of the $i + 2$ elements. Hence the loop is verified. Therefore, after $n - 1$ passes, the largest item will be in position n and thus max receives the correct value in line 6. ∎

Comparison of the Algorithms

If we compare these two algorithms in terms of time efficiency, space efficiency, elegance, and side effects, we first note that their time efficiencies are asymptotically equivalent, but not identical. That is, for the first, its worst case is $4n - 1$; for the second, its worst case is $5n - 3$. The second algorithm uses one more variable than the first and so is slightly less space efficient, but that difference is negligible. In terms of elegance, neither algorithm is significantly more or less elegant than the other. The second one does, however, have a significant side effect, namely, that the list has been altered when it completes its task. This does not happen with the first algorithm. Hence, for this problem, the first algorithm is preferable, since it fares at least as well as or slightly better on time and space efficiency and on elegance. It is much better, though, in its absence of side effects. Sometimes algorithms do better on one criterion and worse on another. In this case, value judgments have to be made as to which criterion is more important in the particular situation.

An $O(n^2)$ Algorithm

The following is an example of an $O(n^2)$ algorithm. It sorts a list of integers from smallest to largest.

ALGORITHM 5   Sorting a List

Procedure Bubble_sort(n; var list);

{list is an ordered set of integers; n, temp, i, j are integers.}

begin

1.   for $i := 1$ to $n - 1$ do

2.   for $j := 1$ to $n - i$ do

3.   if jth element $> j + 1$th then

              begin

4.   temp := jth element;

5. jth element := $j + 1$th element

6. $j + 1$th := temp

            end

   end;

Note that using Algorithm 4, *bubble_down,* this could be written as follows:

ALGORITHM 5a   Sorting an Ordered List
Procedure Bubble_sort(n; var list);
{list is an ordered set of integers; n is an integer.}
begin
1.   for $i := 1$ to n do
2.   bubble_down($n - i + 1$, list)
    end;

The strategy behind Algorithm 5 is first to shift the largest element in the list to the bottom—the nth position. Then the algorithm repeats the same process with a sublist—the first $n - 1$ elements. This places the second largest element in the $n - 1$th position, and so on. The strategy is the same with Algorithm 5a. The one potentially confusing point is the use of $n - i + 1$ in line 2. Consider the following table of values for $n - i + 1$:

i	$n - i + 1$
1	n
2	$n - 1$
3	$n - 2$
.	.
.	.
.	.
$n - 1$	2
n	1

Thus, when i is 1, *bubble_down* moves the largest element of the list of n items to the bottom. When i is 2, it moves the largest element of the sublist of the remaining $n - 1$ elements to the bottom and so on.

Verification of Algorithm 5

Since we have already verified Algorithm 4, it is easier to verify Algorithm 5a than Algorithm 5. The process we are using is very similar to the process of proving mathematical theorems: Once a theorem has been proven, it can be used without proof in subsequent theorems. Similarly, once an algorithm has been verified, it can be used as a procedure within other algorithms without further verification.

The precondition is simply that the list consists of n integers. The postcondition is that the list is sorted from smallest to largest. The loop invariant is that after i passes through the loop,

item $n \geq$ item $n - 1 \geq \cdots \geq$ item $n - i + 1 \geq$ items 1 through $n - i$.

First, let $i = 1$. Then after one pass, item $n \geq$ items 1 through $n - 1$, since we have already verified algorithm *bubble_down*. We now let $i > 1$ and assume that the invariant holds. Then, on pass $i + 1$, *bubble_down* moves the largest of items 1 through $n - i$ to position $n - i$. Items in positions greater than $n - i$ are untouched. Thus, after this pass,

$$\text{item } n \geq \text{item } n - 1 \geq \cdots \geq \text{item } n - i + 1 \geq \text{item } n - i$$
$$\geq \text{items 1 through } n - i - 1$$

and the loop invariant is verified. Thus, setting $i = n$, after completion of the loop

$$\text{item } n \geq \text{item } n - 1 \geq \cdots \geq \text{item } 1$$

and the algorithm is verified.

Complexity of Algorithm 5

To determine the asymptotic efficiency of Algorithm 5, we again count the number of computations required by each step.

Step Number	Number of Computations
1	$2n$
2	$2(n + (n - 1) + (n - 2) + \cdots + 2)$
3	$(n - 1) + (n - 2) + \cdots + 2 + 1$
4	x
5	x
6	x

where $0 \leq x \leq (n - 1) + (n - 2) + \cdots + 2 + 1$. An explanation follows.

Step 1: The loop is executed $n - 1$ times. One more time is added because the algorithm must come back the last time and check that i has actually exceeded the upper limit, $n - 1$. The 2 is included since each execution of this line involves one addition and one comparison.

Step 2: When step 2 is encountered the first time, i is 1. Thus line 2 becomes "for $j := 1$ to $n - 1$." As in step 1, this is executed n times. When these n executions are completed, the algorithm returns to step 1 and i becomes 2. This time, when line 2 is executed, it becomes "for $j := 1$ to $n - 2$." Hence it is executed $n - 1$ times. Continuing in this pattern, the number of times it executes decreases by 1 until finally $n - i$ becomes 1. In that case, line 2 is executed twice.

Step 3: The explanation here is identical to that of step 2 except that the extra execution caused by going back to check that the limit is exceeded does not occur. Thus every term of this sum is one less than the previous sum.

Steps 4–6: These are executed only when the test in line 3 is true. Hence x can be anywhere between 0 and $(n - 1) + (n - 2) + \cdots + 2 + 1$.

At this point, we use the following formula from Section 2.6, which was proven by induction in Section 3.7:

$$\sum_{i=1}^{n} i = \frac{n(n + 1)}{2}$$

Hence

$$(n - 1) + (n - 2) + \cdots + 2 + 1 = \frac{(n - 1)n}{2}$$

and

$$n + (n - 1) + (n - 2) + \cdots + 2 = \frac{n(n + 1)}{2} - 1 = \frac{n^2 + n - 2}{2}.$$

Thus $f(n) = \dfrac{3n^2}{2} + \dfrac{5n}{2} - 2 + 3x$. In the best case

$$f(n) = \frac{3n^2}{2} + \frac{5n}{2} - 2$$

and in the worst case

$$f(n) = 3n^2 - n - 2.$$

Hence we can conclude that the best, average, and worst cases are all $O(n^2)$.

Comparison of the Algorithms

Both algorithms used to find the maximum of a set were $O(n)$; the sorting algorithm is $O(n^2)$. The difference is in the loops. Both of the $O(n)$ algorithms involved a single loop whose lines were executed n times. All other lines in the algorithms were executed only once. Thus the entire algorithm depended directly on n. The sorting algorithm involved *nested for loops*—

one for loop inside another. Both the outside and the inside loop depended on n, although the number of executions of the inner one dropped by one with each execution of the outer loop. This double dependence on n produces an efficiency that depends on $n \times n$ rather than just on n. Similarly, triply nested for loops, that is, a loop structure of the form

> for $i := 1$ to n do
>> for $j := 1$ to some function of i do
>>> for $k := 1$ to some function of i and j do

typically produce an efficiency of $O(n^3)$.

The idea that loops nested k deep will produce an efficiency of $O(n^k)$ is a helpful starting point but is not always true, especially if other types of loops are used. In the next section, we will see an algorithm with doubly nested loops that has an efficiency of $O(2^n)$, which is much worse than $O(n^k)$ no matter how large k is.

The Effect of Different Efficiency Levels

The difference between the times that two different algorithms take to solve the same problem can be enormous (see Table 4.1).

EXAMPLE 4 Assume that each computation of an algorithm takes one microsecond (10^{-6} sec). Use Table 4.1 to determine how long it would take algorithms requiring n, n^2, and 2^n steps to complete their task if $n = 10, 50,$ or 1000.

TABLE 4.1 Time in Seconds to Execute an Algorithm with Complexity $f(n)$, Assuming That Each Computation Takes One Microsecond

				n			
$f(n)$	5	10	20	50	100	1000	10^6
1	10^{-6}	10^{-6}	10^{-6}	10^{-6}	10^{-6}	10^{-6}	10^{-6}
$\log_2 n$	$2 \cdot 10^{-6}$	$3 \cdot 10^{-6}$	$4 \cdot 10^{-6}$	$6 \cdot 10^{-6}$	$7 \cdot 10^{-6}$	10^{-5}	$2 \cdot 10^{-5}$
n	$5 \cdot 10^{-6}$	10^{-5}	$2 \cdot 10^{-5}$	$5 \cdot 10^{-5}$	10^{-4}	10^{-3}	1
$n \log_2 n$	10^{-5}	$3 \cdot 10^{-5}$	$4 \cdot 10^{-5}$	$3 \cdot 10^{-4}$	$7 \cdot 10^{-4}$	10^{-2}	20
n^2	$3 \cdot 10^{-6}$	10^{-4}	$4 \cdot 10^{-4}$	$3 \cdot 10^{-3}$	10^{-2}	1	10^6 (= 12 days)
n^3	10^{-4}	10^{-3}	$8 \cdot 10^{-3}$	0.1	1	16.7 min	31,710 yr
2^n	$3 \cdot 10^{-5}$	10^{-3}	1	36 yr	$4 \cdot 10^{16}$ yr	$3 \cdot 10^{287}$ yr	$3 \cdot 10^{30,1016}$ yr
$n!$	$1 \cdot 10^{-4}$	3.6	77,413 yr	$9.7 \cdot 10^{50}$ yr	.	.	.

SOLUTION

Number of Steps	n	Time
n	10	10×10^{-6} sec
n	50	50×10^{-6} sec
n	1000	1×10^{-3} sec
n^2	10	1×10^{-4} sec
n^2	50	3×10^{-3} sec
n^2	1000	1 sec
2^n	10	1×10^{-3} sec
2^n	50	36 yr
2^n	1000	3×10^{287} yr ∎

The time that a particular algorithm that is $O(n)$, $O(n^2)$, or $O(2^n)$ takes will almost certainly not be precisely what is given in the table. However, the table does give us crude estimates of the time. These estimates are best used to compare algorithms rather than as a basis for predicting the actual time a particular algorithm will take. Generally speaking, algorithms that are $O(\log n)$ can be used with extraordinarily large sets of data; algorithms that are $O(n)$ can be used with quite large collections of data (n in the billions). $O(n \log n)$ algorithms can be used with reasonably large sets of data (n in the millions). $O(n^2)$ or $O(n^3)$ algorithms can be used with moderate-sized data sets (n in the thousands). $O(2^n)$ or $O(n!)$ algorithms are useful only with very small data sets (up to 25–30 for $O(2^n)$ and 10–12 items for $O(n!)$). These numbers are rough estimates but are included to give some concrete idea of the effect of different levels of efficiency.

TERMINOLOGY

time efficiency	average case
space efficiency	worst case
elegance	precondition
side effects	postcondition
asymptotic complexity	loop invariant
best case	bubble sort

SUMMARY

1. Four factors are considered in evaluating the quality of an algorithm: time efficiency, space efficiency, elegance, and the presence of side effects.

2. The asymptotic complexity of many algorithms can be approximated by counting the number of steps the algorithm takes when acting on a data set of size n. The complexity is expressed in $O(g)$ notation; a best case, a worst case, and an average case are often distinguished.

3. The correctness of algorithms must be verified; this can often be done by applying mathematical induction to verify the loops in an algorithm.

EXERCISES 4.3

In each of the following, assume that the expression given is the number of steps an algorithm requires. Assume further that each step requires one microsecond. Using the table from this section, determine the time such an algorithm would take for $n = 10$, 50, and 1000.

1. 1

2. $\log n$

3. n

4. $n \log n$

5. n^2

6. 2^n

7. Determine the complexity of the algorithm for multiplication given in Section 4.1 (Algorithm 3). Let n stand for the number of digits in the numbers being multiplied.

8. Determine the complexity of the algorithm for addition written in Exercise 13 of Section 4.1.

9. Determine the complexity of the following algorithm for finding where the largest element appears in an unsorted list.

ALGORITHM 6   Locate the Largest Element in a List

```
Procedure Locate_largest(list; var location);
{list is an ordered set of integers; count and largest are integers.}
begin
    count := 1;
    location := 1;
    largest := first entry in list;
    while there are unexamined members of the list do
    begin
        count := count + 1;
        get next number;
```

```
        if new number > largest
        then begin
            location := count;
            largest := number
        end
    end
end;
```

10. Verify Algorithm 6.

11. Compute the complexity of the following algorithm for calculating the value of the polynomial, $a_0x^n + a_1x^{n-1} + \cdots + a_n$, for a particular x.

ALGORITHM 7   Evaluate a Polynomial

```
Procedure Poly1(x,a_0,a_1, · · · ,a_n; var value);
{x, a_0,a_1, · · · a_n, y, value are real; i is an integer.}
begin
    y := 1;
    value := a_0;                    {start at a_0 this time}
    for i := 1 to n do begin
        y := y * x;                  {compute each power of x}
        value := value + a_{n-1} * y;    {keep running total of terms}
    end;
end;
```

---

12. Trace Algorithm 7 for $p(x) = x^2 + 3x + 2$ and $p(x) = x^3 + 4x^2 + 3x + 1$.

13. Compute the complexity of the following algorithm for calculating the value of the polynomial, $a_0x^n + a_1x^{n-1} + \cdots + a_n$, for a particular x.

ALGORITHM 8   Evaluate a Polynomial

```
Procedure Poly2(x,a_0,a_1, · · · ,a_n; var value);
{x, a_0,a_1, · · · a_n, y, value are real; i is an integer.}
begin
    value := a_0;                    {start at a_0 this time}
    for i := 1 to n do
        value := x * value + a_i
end;
```

[Note: this algorithm is based on the following formula, known as *Horner's method*:

$$a_n + a_{n-1}x + \cdots + a_0 x^n = x(\cdots(x(a_0 x + a_1) + a_2) + \cdots) + a_n \quad]$$

14. Trace Algorithm 8 for the same polynomials used in Exercise 12.

15. Show that procedure Poly2 is more efficient than Poly1, since it requires fewer computations than Poly1. (In fact, this method has been shown to use the minimum possible number of computations.)

16. Verify Poly1.

17. Verify Poly2. (*Hint:* write the loop invariant as the *i*th step in Horner's method.)

4.4
INEFFICIENT ALGORITHMS AND INTRACTABLE PROBLEMS

Asymptotically Optimal Solutions

Suppose we have an algorithm that is $O(g(n))$ for some $g(n)$. We often want to know whether this is the most efficient algorithm for that particular problem. For instance, in the last section we looked at two algorithms for finding the largest member of an ordered set. Both were $O(n)$. Can we improve this result? The answer is no unless the items are sorted in some way. This is because finding the largest element of an unsorted set requires the examination of every member of the set. With an n-element set, this involves a minimum of n steps. Hence any algorithm to solve this problem is at best $O(n)$ and possibly worse. Since we have $O(n)$ solutions, we can conclude that an $O(n)$ algorithm is the best possible solution for the problem. We call such algorithms **asymptotically optimal.**

Complexity of a Problem versus Complexity of Its Solution

Our bubble-sort algorithm was $O(n^2)$. There are, in fact, $O(n \log n)$ sorting algorithms; one of them is discussed in Section 6.6. Thus we distinguish between the **complexity of an algorithm** and the **complexity of a problem.** The complexity of an algorithm is its asymptotic efficiency. The complexity of a problem is the asymptotic efficiency of an asymptotically optimal solution. Thus, for instance, the complexity of the problem of finding the largest element of a unsorted set is $O(n)$. It can be shown that the complexity of the sorting problem is $O(n \log n)$, and hence $O(n \log n)$ solutions are asymptotically optimal. Usually, however, determining the complexity of a problem is extremely difficult.

Consider the traveling salesman problem again. Recall that we had a set of, say, $n + 1$ cities and wanted to start from one of them, travel to each of the others, and return to the starting city, minimizing the total distance of travel. Informally stated, the brute force algorithm for doing this operation is

1. List all of the possible routes.
2. Find the length of each route.
3. Select the shortest.

Since there are $n!$ different orders in which to list the n cities other than the first, the first line will take a minimum of $n!$ steps. Hence this algorithm is at best $O(n!)$. Since listing each route will take roughly the same time, the first step will require approximately a constant multiple of $n!$ units of time. Similarly, finding the length of each route will take approximately the same time, so that step will also take approximately a constant multiple of $n!$ time. The last step is an application of a slightly modified version of the *find_max* algorithm of Section 4.3 and hence is also $O(n!)$, since there are $n!$ members of the set it must examine. Thus we conclude that this algorithm is precisely $O(n!)$. With more than about 10 to 12 cities, this algorithm will probably take an unreasonable amount of time to execute.

The nearest neighbor algorithm is as follows:

1. For each city
2. Find the nearest city not yet visited and visit it
3. Return home

If there are n cities, the first step involves a for loop, the body of which must be executed $n - 1$ times. Finding the nearest city in the second line involves examining each of the other $n - 1$ cities to see if it has been visited yet and to see if it is closest. In other words, it involves another for loop, which will be passed through $n - 1$ times. It is also a modification of *find_max* and hence is itself $O(n)$. Thus the nearest neighbor algorithm involves for loops nested two deep and is $O(n^2)$. It can reasonably be used with 1000 or more cities.

These two algorithms provide a dramatic demonstration of the difference between an $O(n!)$ and an $O(n^2)$ algorithm. Recall that the nearest neighbor algorithm is an example of a greedy algorithm. It involves making what seems to be the best choice at each step without looking more than one step ahead. Suppose we decided to look two steps ahead. Then there would be n^2 routes to consider in the inner loop previously discussed, and hence the algorithm would become $O(n^3)$. Similarly, looking three steps ahead would alter the efficiency of the algorithm to $O(n^4)$. Thus, if we know how large an n we want to work with, we can decide how much efficiency we need and compute how many steps ahead we can reasonably look.

Algorithms that have a complexity of $O(n^k)$ or better are called **efficient** and are said to **run in polynomial time.** Those that have a complexity of $O(2^n)$ or worse are called **inefficient** and are said to **run in exponential time.** Similarly, problems that have a complexity of $O(n^k)$ or better are called **tractable,** and those that have a complexity of $O(2^n)$ or worse are called **intractable.** We know that the brute force algorithm for the traveling salesman problem is inefficient, but at this time no one knows whether the problem itself is intractable. This is one of the major unsolved problems of theoretical computer science. In fact, there is an entire class of problems, of which several hundred have been identified, called **NP-complete problems.** (NP stands for nondeterministic polynomial.) The traveling salesman problem belongs in this category. It has been shown that all of these problems are either tractable or all are intractable. For more information, see Aho, et al., *Design and Analysis of Computer Algorithms,* or Garey and Johnson, *Computers and Intractability, A Guide to the Theory of NP-Completeness* (see References).

The Knapsack Problem

Besides the traveling salesman problem, there is another NP-complete problem that is of special interest because it arises so often. This is the **knapsack problem.** There are many variations of this problem, but they all follow a pattern similar to this one:

Suppose you are going on a camping trip and can carry only a certain number of pounds in your knapsack. The items you would like to take have a total weight that exceeds your maximum. Find the most satisfactory combination of items you can take without exceeding your maximum weight.

Note that the problem does not say "Select as many items as you can take without exceeding your maximum." That problem would be too easy to solve: Eliminate the heaviest item, then the second heaviest, and so on, until you dropped below your allowable weight.

Knapsack problems all involve selecting the subset of a given set that best satisfies some condition. Thus one algorithm for solving a knapsack problem is this:

1. List all subsets of the given set.

2. Select the one that best satisfies the given condition.

Since a set of n elements has 2^n subsets, this is clearly an inefficient algorithm. But, like the traveling salesman problem, it is not known whether the problem itself is intractable. As a concrete example of an $O(2^n)$ algorithm, here is an algorithm for carrying out the first step in the preceding solution of the knapsack problem.

ALGORITHM 9   List all Subsets of $\{1, 2, \ldots, n\}$

Procedure Subsets(n; var list);

$\{n, i, j$, size_of_list are integers; S is a set of integers; list is a set of subsets of S.$\}$

begin

1.   put the empty set in list;   {initialize list}

2.   size_of_list := 1;

3.   for $i := 1$ to n do   {loop through the set}

       begin

4. $j := 1$;   {go to start of list}

         repeat

5.   let $S := j$th element from   {get next set from
             the list;   list}

6.   add $S \cup \{i\}$ to end of list   {add i to it and
             of subsets   place in list}

7. $j := j + 1$ {move to next
                                                      member of list}

8.   until $j >$ size_of_list

9.   size_of_list := 2 $*$ size_of_list;

       end {for}

end;

The strategy behind this algorithm is based on the strategy behind the inductive proof in Section 3.7 that if $|S| = n$, then $|P(n)| = 2^n$. It proceeds as follows: We first place the empty set in the list. Then we append the union of $\{1\}$ with each set in the list (which at this point only contains $\varnothing$) to the end of the list. The list is now $\varnothing$, $\{1\}$. Next, we combine $\{2\}$ with the two sets we have, thereby adding $\{2\}$, and $\{1, 2\}$ and giving a list that consists of $\varnothing$, $\{1\}$, $\{2\}$, and $\{1, 2\}$. Then the union of each of these with $\{3\}$ is added. Continuing in this fashion, we generate all the subsets of $\{1, 2, \ldots, n\}$. Algorithm 9 provides yet another illustration of the close link between mathematical proof and algorithm development.

Verification

The precondition is that n is an integer and list contains nothing. The postcondition is that list contains all of the subsets of $\{1, \ldots, n\}$ and size_of_list $= 2^n$. The loop invariant for the for loop is that after i passes, list contains the subsets of $\{1, \ldots, i\}$ and $size\_of\_list = 2^i$. Verification then proceeds by induction. First, let $i = 0$; that is, we are examining the status of list and $size\_of\_list$ immediately before the first pass through the for loop. The empty set goes into the list in line 1 and $size\_of\_list$ is set to 1 in line 2. Hence the invariant is verified in this case.

Now suppose that after i passes, list contains all subsets of $\{1, 2, \ldots i\}$ and $size\_of\_list$ is 2^i. We examine the $i + 1$th pass through the for loop. j is initialized to 1 and the repeat loop involves getting each of the subsets of $\{1, 2, \ldots i\}$, taking the union of $\{i + 1\}$ with each and adding that new set to the end of the list. The repeat loop ends when the 2^i subsets of $\{1, \ldots, i\}$ have all been retrieved and $\{i + 1\}$ has been combined with each. At this point, there are 2^{i+1} subsets in the list—the original 2^i subsets of $\{1, \ldots, i\}$ and 2^i additional subsets that include $i + 1$. These are precisely the subsets of $\{1, \ldots, i + 1\}$, and hence the algorithm generated the correct list. In line 10, $size\_of\_list$ is doubled and becomes 2^{i+1}; thus the loop is verified. Hence, after n passes, list contains the subsets of $\{1, \ldots, n\}$ and $size\_of\_list = 2^n$.

Complexity

We now compute the complexity of Algorithm 9.

Step	Number of Times
1	1
2	1
3	$2(n + 1)$
4	n
5	$1 + 2 + 4 + \cdots + 2^{n-1}$
6	$1 + 2 + 4 + \cdots + 2^{n-1}$
7	$1 + 2 + 4 + \cdots + 2^{n-1}$
8	$1 + 2 + 4 + \cdots + 2^{n-1}$
9	n

Steps 1 and 2: These are not part of any loop and so are executed only once.

Step 3: This is the first line in a for loop.

Step 4: This is in the body of the for loop but not the inner loop.

Steps 5–8: The number of passes through this loop is regulated by the value of $size\_of\_list$. It is initially 1 but doubles each time until it becomes 2^{i-1} at the start of the ith pass through the loop.

Step 9: Like step 4, this is in the body of the for loop but not in the inner loop.

From Section 2.6, we know that

$$\sum_{i=0}^{n-1} 2^i = 2^n - 1.$$

Thus $f(n) = 4n + 4 + 4(2^n - 1)$. Hence $f(n)$ is in $O(2^n)$. Note that there are no best, worst, or average cases since there are no conditionals.

TERMINOLOGY

asymptotically optimal algorithm

complexity of an algorithm

complexity of a problem

efficient algorithm

algorithm that runs in polynomial time

inefficient algorithm

algorithm that runs in exponential time

tractable problem

intractable problem

NP-complete problem

knapsack problem

SUMMARY

1. An asymptotically optimal algorithm is one whose efficiency, $f(n)$, is such that any other algorithm for solving the same problem has an efficiency no better than $O(f(n))$.

2. We distinguish between the comlexity of an algorithm and the complexity of a problem. By the complexity of an algorithm, we mean its asymptotic efficiency. By the complexity of a problem, we mean the asymptotic efficiency of an asymptotically optimal solution.

3. Algorithms that have a complexity of $O(n^k)$ or better are called efficient and are said to run in polynomial time. Algorithms that have a complexity of $O(2^n)$ or worse are called inefficient and are said to run in exponential time. Similarly, problems that have a complexity of $O(n^k)$ or better are called tractable and those that have a complexity of $O(2^n)$ or worse are called intractable.

4. We know that the brute force algorithm for the traveling salesman problem is inefficient, but at this time we do not know whether the problem itself is intractable.

5. The knapsack problem is another problem that has no known efficient algorithm at this time.

EXERCISES 4.4

1. Trace the algorithm for listing all subsets of $\{1, 2, \ldots, n\}$ for $n = 2, 3,$ and 4.

2. Suppose a computer executes one instruction in a microsecond and an algorithm is known to have a complexity of $O(2^n)$. If a maximum of 24 hours of computer time can be given to this algorithm, determine the largest value of n for which the algorithm can be used.

3. In the problem of Exercise 2, suppose that technological advances multiply the speed of the computer by 10 so that now it only takes 0.1 microsecond to execute an instruction. Now determine the largest value of n for which the algorithm can be used.

4. Rewrite the nearest neighbor algorithm to allow a two-city look ahead rather than just a one-city look ahead.

5. Determine the complexity of your algorithm for Exercise 4.

Solve the following knapsack problems.

6. Consider the digits 2, 3, 5, 7, and 11. Find all combinations of these digits that sum to a multiple of 3.

7. In starting on a camping trip, Maria has five items she particularly wants to take. Each has a weight and a value to her given by the following table:

Item	Weight	Value
1	22	33
2	13	20
3	27	34
4	15	15
5	11	14

Her maximum load is 40 pounds. Which selection of items will give the maximum value without exceeding her maximum weight?

8. Sam likes to buy used motorcycles, fix them up, and resell them for a profit. He goes to a reclaimed vehicle sale with $2000 to spend. The cycles available sell for the following amounts, with the profit Sam estimates he can make from each.

Cycle	Sale Price	Profit
1	1200	350
2	950	200
3	1800	575
4	800	175

How should he spend his $2000?

4.5

SEARCHING ALGORITHMS

Sets of data are often organized into arrays. (See Appendix A for an introduction to one-dimensional arrays and Appendix B for an introduction to two-dimensional arrays.) With large data arrays, two types of problems often arise—finding a particular item in an array (or determining if it is there at all) and sorting the data according to some specified order. Both of these problems have been extensively studied. In this section, we will consider some standard algorithms that address the first of these problems, searching. As for sorting, we examined one algorithm, bubble sort, in the previous section. Although easy to understand, it is not very efficient. We will treat a more efficient algorithm, quicksort, in Section 6.6.

Linear Search

One way to find out if a particular element is in an array is to start at the beginning of the array and examine every element. This method is not very efficient, but is effective. It is analogous to finding a name in the telephone book by starting at the beginning and examining every name until the desired one is found. This is called **linear search,** and an algorithm for it follows. We will use the notation $A[1 \ldots n]$ to stand for an entire array, with subscripts that run from 1 to n, and the notation $A[1]$, $A[2]$, etc., to stand for individual elements of the array. For simplicity, we will assume that all arrays are made up of integers, although the algorithms given here could be used with any data type. We will also use the symbol X to stand for the item being sought.

ALGORITHM 10 Locating an Element in an Array

Procedure Linear_search(n, A, X; var location);

{A is an array of n integers, $n \geq 0$; all other variables are integers; location = 0 means the item has not been found.}

begin

1.	location := 0;	{assume item not found until located}
2.	counter := 0;	{set counter to start of the array}
3.	while (counter $< n$) and (location = 0) do begin	{loop through array}

> **4.**   counter := counter + 1;
> **5.**   if A[counter] = X
> **6.**   then location := counter   {mark location if found}
>   end {while}
>   end; {linear_search}

    The important thing to note here is the use of the variable *location*. This algorithm is based on an "innocent until proven guilty" strategy. It assumes that X is not in the array (by setting *location* at zero) until it is located. Note that if X appears in the array more than once, only the first appearance is reported.

Variations on Linear Search

    Following is a variation of Algorithm 10 that tests whether an item was found but does not tell where it was found. In the Exercises, you will be asked to modify Algorithm 10 to count the number of occurrences of X. In the following algorithm, *found* is a Boolean variable that takes on the value true if X has been found and is false otherwise.

ALGORITHM 11   **Test for the Presence of a Particular Item**
    Procedure Is_it_there(n, A, X; var found);
    {A is an array of n integers; X is an integer; found is Boolean.}
    begin
1.   found := false;
2.   counter := 0;
3.   while (counter < n) and (not found) do begin
4.   counter := counter + 1;
5.   if A[counter] = X
6.   then found := true
     end {while}
   end; {is_it_there}

Complexity of Linear Search

The complexity of Algorithm 10 is computed as follows:

Step	Number of Times	
1	1	
2	1	
3	$x + 1$	where $0 \leq x \leq n$
4	x	
5	x	
6	1	
Total $= 3x + 4$		

In the best case, X is found on the first step. In the worst case, X is not found at all. In the average case, X is found in the middle of the array at $(n + 1)/2$. Thus we get the following:

	Total
Best case	4
Average case	$\dfrac{3n + 11}{2}$
Worst case	$3n + 4$

Both the worst and average case are $O(n)$. Hence, as a rule of thumb, it is not unreasonable to think of using linear search by computer on arrays up to perhaps 1 million entries. There is a much more efficient way, however, when the array is sorted.

Binary Search

When we look up a name in a telephone directory, say, John Stapleton, we guess about where in the book the Stapletons are listed and open to that place. We look at the page to see if we are too far forward or too far back. We then choose a second location and check that page. We continue in this fashion until we get close to the name Stapleton and then we leaf through page by page until we find the right one. (I have just tried this; it took me seven pages.) Once on the right page, we do the same thing: a couple of guesses to get close, followed by scanning name by name until we find the right one.

This process uses two types of search. The first uses guesses to get close to the right location and the second is a linear search to find the right name or page. **Binary search** is an algorithm similar to the first type of search and can be used not just to get close to the item being sought, but actually to find it. It is an extremely efficient algorithm, as we shall see. It is also an example of a class of algorithms known as **divide and conquer algorithms.** All of these algorithms involve dividing a set of data into subsets and seeing which subset has the desired property. The algorithm is then applied again to the subset. This process continues as long as necessary to find an item or items or to perform some operation. The reason these algorithms are so efficient is that large problems are quickly reduced to small problems.

Binary search involves an attempt to find a particular item in an array and requires that the array be sorted. It proceeds by first looking at the *middle* of the array. If the middle item is not the desired one, a decision is made as to which half of the array contains it. The algorithm is applied again to the appropriate half of the array. Thus, with each step, we are dividing the size of the array by successive powers of 2. Binary search is so efficient because the exponential function grows so rapidly. Here is the algorithm. The operation *div* is an integer division in which the remainder is dropped.

ALGORITHM 12   Binary Search of a Sorted Array

Procedure Binary_search(n, A, X; var location, found);

{found is boolean; low, high, X, and location are integers; A is an array of n integers.}

begin

1.   found := false;

2.   low := 1;

3.   high := n;

4.   while (not found) and (high $\geq$ low) do begin

5.   location := (low + high) div 2;   {find the middle}

6.   if A[location] = X {have we found X?}

7.   then found := true
       else

8.   if A[location] > X {No?—decide which half X is in}

9.   then high := location − 1   {X in lower half—change upper limit}

> **10.**   else low := location + 1   {X in upper half—
> change lower limit}
>
>   end {while}
>   end; {binary_search}

Tracing Binary Search

In order to illustrate how binary search works, we will trace it for two different values of X and the following array:

$$A = \begin{bmatrix} 1 \\ 2 \\ 8 \\ 13 \\ 19 \\ 27 \\ 31 \\ 91 \end{bmatrix}$$

Suppose first that $X = 8$. Here is a detailed trace of *binary_search* in this case:

Line	Action
1	found is set to false
2	low := 1
3	high := 8
4	not found = true and ($8 \geq 1$) so continue
5	location := $(8 + 1)$ div $2 = 4$
6	does $A[4] = 8$? no—skip to line 8
8	is $A[4] > 8$? yes—do line 9, skip line 10
9	high := 3
4	not found = true and ($3 \geq 1$) so continue
5	location := $(3 + 1)$ div $2 = 2$
6	does $A[2] = 8$? no—skip to line 8
8	is $A[2] > 8$? no—skip to line 10
10	low := 3
4	not found = true and ($3 \geq 3$) so continue
5	location := $(3 + 3)$ div $2 = 3$

6	does $A[3] = 8$? yes—continue to line 7
7	found := true
4	not found = false so end the loop

Note that the final value of location is 3, so *binary_search* correctly reports where the value of X we were searching for is located.

Suppose X is given a value that is not in A, for instance, $X = 87$. Following is a trace in this case.

Line	Action
1	found := false
2	low := 1
3	high := 8
4	not found = true and $(8 \geq 1)$ so continue
5	location := 4
6	does $A[4] = 87$?—no—skip to line 8
8	is $A[4] > 87$? no—skip to line 10
10	low := 5
4	not found = true and $(8 \geq 5)$ so continue
5	location := 6
6	is $A[6] = 87$? no—skip to line 8
8	is $A[6] > 87$? no—skip to line 10
10	low := 7
4	not found = true and $(8 \geq 7)$ so continue
5	location := 7
6	does $A[7] = 87$? no—skip to line 8
8	is $A[7] > 87$? no—skip to line 10
10	low := 8
4	not found = true and $(8 \geq 8)$ so continue
5	location := 8
6	is $A[8] = 87$? no—skip to line 8
8	is $A[8] > 87$? yes—continue to line 9
9	high := 7
4	$(7 \geq 8)$ is false so halt

Note that found = false and location = 8, which is where 87 would be inserted if it were added to the array.

Complexity of Binary Search

For now, we will only do a worst case analysis of *binary_search*; as we shall see shortly, even the worst case is extremely fast. For the average case, see the Exercises of Section 7.7. Let us assume that we have an array of size N and an element X that is not in the array. At the first iteration of the loop, the middle of the array is found. If N is odd, the exact middle is found (e.g., if $N = 9$, location $= 5$). If N is even, the number a half-step below the middle is chosen (e.g., if $N = 8$, $(N + 1)/2 = 4.5$, which is truncated to 4). If *location* does not contain the desired element, the process is repeated on either the upper or lower half-array. If N is odd, both the upper and lower half-arrays have a fraction less than $N/2$ elements; if N is even, the upper half has exactly $N/2$ elements, while the lower half has a fraction less. Thus the array that's left to search at the next step has, at most, $N/2$ elements. This process continues until there is only one element left. Thus, if we tabulate the number of the iteration about to be performed and the size of the array about to be examined at each iteration, we get the following:

Iteration	Maximum Size of Array to Be Examined
1	N
2	$N/2$
3	$N/4$
4	$N/8$
.	.
.	.
.	.
k	$N/2^{k-1}$

If $N/2^{k-1} \leq 1$ and $N/2^{k-2} > 1$, then the kth pass through the loop will be the last and the maximum number of iterations needed in the worst case is no more than k. We can solve this inequality for k.

If

$$N/2^{k-1} \leq 1,$$

then

$$N \leq 2^{k-1}$$
$$\log_2 N \leq k - 1$$
$$k \geq 1 + \log_2 N.$$

Hence, if we let $k = 1 + \lceil \log_2 N \rceil$, we know that the number of iterations required is at most k. We can now complete our worst case analysis of the algorithm:

Step	Number of Computations
1	1
2	1
3	1
4	$3(k + 1)$
5	$2k$
6	k
7	0
8	k
9 ⎱ 10 ⎰	k

Thus Total $= 6 + 8k = 14 + 8 \lceil \log_2 N \rceil = O(\log_2 N)$.

Comparison to Linear Search

Suppose, for instance, that an array has 30,000 entries. In the worst case, linear search takes $3n + 4$ steps, in this instance over 90,000. The average case takes over 45,000. Binary search, however, takes at most 126 steps since $\log_2(30,000)$ is between 14 and 15. This comparison is one of the best examples of how analyzing the complexity of algorithms can help us establish that one algorithm is preferable to another.

TERMINOLOGY

linear search   divide and conquer algorithm
binary search

SUMMARY

1. Two important searching algorithms are linear search and binary search.
2. The average and worst cases of linear search are $O(n)$; the worst case for binary search is $O(\log_2 n)$.

3. Binary search is an example of a divide and conquer algorithm, a problem-solving strategy that involves breaking large problems down into successively smaller problems of the same form until the resulting problem can be done quickly.

EXERCISES 4.5

Trace Algorithm 10, linear search, for the following arrays.

1. Item searched for $= 63$

   [0, 89, 63, 13, 5]

2. Item searched for $= 13$

   [0, 89, 63, 13, 5]

3. Item searched for $= 0$

   [0, 19, 18, 23, 4]

4. Item searched for $= 7$

   [19, 8, 1, 7]

5. Item searched for $= 3$

   [8, 1, 7, 6, 4]

6. Item searched for $= 16$

   [11, 16, 13, 14, 1ˆ, 19]

Trace Algorithm 11 for the following arrays.

7. See Exercise 1.

8. See Exercise 2.

9. See Exercise 3.

10. See Exercise 4.

11. See Exercise 5.

12. See Exercise 6.

13. Modify Algorithm 10 so that it counts the number of occurrences of the item being searched for.

14. Modify Algorithm 10 so that it counts the frequencies of the positive and negative elements of the array.

15. Compute the complexity of your answer to Exercise 13.

16. Compute the complexity of your answer to Exercise 14.

17. Write a precondition, a postcondition, and a loop invariant; then verify Algorithm 10.

Let A be the array:

$$\begin{bmatrix} 1 \\ 18 \\ 72 \\ 93 \\ 102 \\ 105 \\ 113 \\ 121 \end{bmatrix}$$

In Exercises 18–22, trace Algorithm 12, binary search, for the array A and the given value of X.

18. $X = 93$

19. $X = 72$

20. $X = 105$ **21.** $X = 1$

22. $X = 95$

23. Modify Algorithm 12 so that it counts the number of occurrences of the item being searched for. (*Hint:* Your algorithm should use binary search to find the item and then use linear search in both directions to count the frequencies.)

The game "Guess My Number" is played like this: The first player selects a number between 1 and 100. The second player makes a guess. The first player responds by telling whether the guess is too high, too low, or correct. The second player makes another guess. The object of the game is to find the correct number using as few guesses as possible.

24. Write an algorithm that the second player could use to play Guess My Number based on binary search.

25. What is the maximum number of guesses your algorithm should require?

26. Find a number that the first player could use that would cause the second player to use the maximum number of guesses.

27. Suppose the range of allowable numbers was 1 to N, for some N, rather than 1 to 100. Now what is the maximum number of guesses your algorithm would require?

Recall that the average case for linear search required $(3n + 11)/2$ steps and that the worst case for binary search required $14 + 8 \lceil \log_2 n \rceil$ steps.

28. Graph these two functions on the same set of axes.

29. Estimate the range of values of n such that binary search is the more efficient algorithm and the range of values such that linear search may be preferable.

Using Table 4.1 in Section 4.3, for the following values of n, estimate the average case time for linear search, the worst case time for linear search, and the worst case time for binary search.

30. $n = 9$ **31.** $n = 50$

32. $n = 1000$ **33.** $n = 10^6$

REVIEW EXERCISES—CHAPTER 4

Trace the following pieces of algorithms and indicate what will be written in each case.

1. Trace with $N = 3, 0, 6$.

    if N mod $2 = 1$

    then write('odd number')

    else write('even number')

2. $x := 2;$

    count := 1;

    repeat

        $x := x * x;$

        count := count + 1;

        write(count,x)

    until count > 5

3. $x := 2$;
   count $:= 1$;
      while count < 4 do begin
         $x := x * x + 1$;
         write(count,x)
   end;

4. $x := 2$;
      for count $:= 1$ to 5 do begin
         $x := x * x - 1$;
         write(count,x);
      end;

Write each of the following as parts of algorithms, using pseudo-code.

5. For employees with a code of 0, write that they are hourly employees. For others, write that they are salaried.

6. A list of names includes at least one Smith. Read through the list until the first Smith is found, counting the number of names before him. Print out this count.

7. Rewrite the multiply algorithm (Algorithm 3) for use by someone doing multiplication with pencil and paper rather than by computer.

8. Trace the multiply algorithm (Algorithm 3) with $a = 476$, $b = 123$.

9. Separate the following list of functions into collections that are asymptotically equivalent to each other.

$\log 2n$	$n + 1$
$n^2 - 3n + 2$	$2^{\log 2n}$
$1 - \log_2 n$	2^{n+1}
$(n - 1)!$	$(2n + 1) \log_2 n$
$n^2 + 2^n$	$\log_2 n + n^2$

Show directly from the definition that $f(n)$ dominates $g(n)$:

10. $f(n) = 4^n$, $g(n) = 3^n$

11. $f(n) = n/2$, $g(n) = n$

Prove:

12. $O(n \log n) \subset O(n^2)$

13. $O(2^n) \subset O(n!)$

14. $f \in O(g)$, $g \in O(h) \Rightarrow f \in O(h)$.

15. If $\exists N$ is such that $n > N \Rightarrow 0 \le f(n) \le 1$, then $fg \in O(g)$.

16. If $c \ge 1$ and $d \ge 1$ and $c < d$, then $O(c^n) \subset O(d^n)$. (Example 4c)

17. $O(n^3) \not\subset O(n^2)$.

18. Write an algorithm that will determine both the largest and smallest elements in an ordered set. Determine its complexity and verify it.

19. The following algorithm is called *interchange sort*.

ALGORITHM 13   Sort a List of Integers

    Procedure Interchange_sort(n; var list);
    {n, i, j, temp are integers; list is a set of integers.}

```
begin
    for i := 1 to n − 1 do
        for j := i + 1 to n do
            if ith element > jth element
            then begin
                temp := ith element;
                ith element := jth element;
                jth element := temp
            end
end;
```

Determine the best and worst case complexity of *interchange_sort*.

20. Verify *interchange_sort*.

21. Suppose we have a collection of colored marbles. Here is an algorithm for listing the different colors that are represented in the collection. Initially, *marbles* is a list of colors that may include duplicates; at the completion of the algorithm, *color_list* will contain all of the colors in *marbles* but without duplicates.

ALGORITHM 14   List Colors

Procedure List_colors(marbles; var color_list);

{both marbles and color_list are sets of colors. n is the length of the list of marbles; initially, no color in marbles is "blank".}

```
begin
    for i := 1 to n do begin
        if ith marble ≠ "blank"
        then begin
            add color of ith marble to color_list;
            for j := i + 1 to n do
                if color of ith equals color of jth
                then replace color of jth by "blank"
        end
    end
end;
```

Compute the best and worst case complexity of *list_colors*.

22. Verify *list_colors*.

23. Repeat Exercise 2 of Section 4.4 with an algorithm that is $O(n!)$.

24. Repeat Exercise 3 of Section 4.4 with an algorithm that is $O(n!)$.

Trace Algorithm 10, linear search, for the following arrays.

25. Item searched for $= 12$

[12, 13, 14, 18]

26. Item searched for $= 19$

[18, 1, 17, 13, 9]

27. Item searched for $= 12$

[9, 1, 7, 6, 13]

28. Item searched for $= 24$

[24, 12, 111, 32, 24, 24]

Trace Algorithm 11 for the following arrays.

29. See Exercise 25.

30. See Exercise 26.

31. See Exercise 27.

32. See Exercise 28.

In Exercises 33–35, trace Algorithm 12, binary search, for the Array A and the given value of X.

33. $X = 18$

34. $X = 121$

35. $X = 127$

Another important variation of binary search involves the problem of finding roots of an equation. For instance, suppose $x^2 + 3x - 2 = 0$. If $x = 0$, the left-hand side of this equation will be -2, and if $x = 1$, the left-hand side will be $+2$. Hence the equation has a root between 0 and 1. Choose the number midway between 0 and 1, namely, 0.5, and evaluate the left-hand side, getting -0.25. Since this value is negative, there must be a root between 0.5 and 1. Now take a value midway between 0.5 and 1 and evaluate the left-hand side. Continue in this fashion until the difference between the numbers is arbitrarily close to zero (for instance, less than 10^{-6}).

36. Suppose you are given the equation $g(x) = 0$. Write an algorithm for finding a root of this equation, using the preceding process. You may assume that x_1 such that $g(x_1) < 0$ and x_2 such that $g(x_2) > 0$ are known and may be treated as inputs to your algorithm.

37. In the example of $x^2 + 3x - 2 = 0$, $x_1 = 0$, and $x_2 = 1$. How many steps are required until the difference between x_1 and x_2 is less than 10^{-6}?

Using the Table 4.1 in Section 4.3, for the following values of n estimate the average case time for linear search, the worst case time for linear search, and the worst case time for binary search.

38. $n = 12$

39. $n = 100$

40. $n = 10^5$

REFERENCES

Aho, Alfred V., John E. Hopcroft, and Jeffrey D. Ullman, *Design and Analysis of Computer Algorithms*. Reading, Mass.: Addison-Wesley, 1974.

Anderson, Robert B., *Proving Programs Correct.* New York: Wiley, 1979.

Garey, Michael R., and David S. Johnson, *Computers and Intractability, A Guide to the Theory of NP-Completeness.* San Francisco, Calif.: W. H. Freeman, 1979.

Knuth, Donald L., *The Art of Computer Programming,* Vol. 1, *Fundamental Algorithms.* Reading, Mass.: Addison-Wesley, 1973.

Liu, C. L., *Elements of Discrete Mathematics.* New York: McGraw-Hill, 1977.

Stanat, Donald, and David McAllister, *Discrete Mathematics in Computer Science.* Englewood Cliffs, N.J.: Prentice-Hall, 1977.

Elementary Number Theory

Theory

5

Discrete models and algorithms are unifying themes that tie discrete mathematics together. However, mathematical models and algorithms are concepts usually associated with applied mathematics. Thus, it might seem strange that at this point we take an excursion into number theory—a branch of mathematics that has traditionally been part of pure mathematics. But there is a good reason for it. Discrete mathematics actually encompasses all of mathematics, pure and applied, in which the underlying sets are finite or are subsets of the set of integers; number theory is principally a study of the properties of the integers. If we are to use discrete models effectively and even develop our own discrete models, we must have a thorough understanding of the properties of the integers. Furthermore there are a number of concepts in number theory that have direct application to questions in computer science, most notably data encryption.

5.1

INTRODUCTION

The goal of this first section is to introduce the structure of the natural numbers by looking at them axiomatically. Subsequent sections will develop further properties of the natural numbers and the integers.

The Natural Numbers

What do we mean by the set **N**? What exactly is a natural number?* One approach is this: Take a specific number, say, 187, and ask, "What exactly do we mean by 187?" We could say that 187 means one 100, eight 10s, and seven 1s. This answer assumes that we know what 1, 7, 8, 10, and 100 mean. It also assumes that we understand addition and multiplication, and implicitly it assumes that we understand exponentiation as well (1 being 10^0, 10 being 10^1, 100 being 10^2, etc.). These are not unreasonable assumptions of people who have completed the study of arithmetic, and we will pursue this approach in Section 5.4. But there is a simpler method that can give us a better understanding of the natural numbers. Instead, we could say, "187 means one more than 186." At first glance, this appears to be a response of the "Ask a silly question, get a silly answer" type. But there is more to it. We could then say, "186 means one more than 185." The process could be continued until we ask, "What do we mean by 2?" and the

---

* Some years ago, the comedian Shelly Berman did a routine in which he imitated a rather pompous philosopher, saying things like "This is a glass of water. But is it a glass of water? How do we know it is a glass of water?" When I ask questions like this one, I suspect I sound like that philosopher. But I'll risk it.

answer would be "one more than 1." Thus we could answer our question about 187 by using only two concepts, 1 and $+$, as long as we are willing to apply these concepts enough times. An axiomatic approach to the natural numbers does just this. As we saw in Chapter 3, the use of the axiomatic method requires the use of primitive terms and axioms. The following set of axioms for the natural numbers was developed by the Italian mathematician Guiseppi Peano (1858–1932). As primitive terms, Peano used *one,* **successor,** and *natural number.* The phrase "the successor of x" is abbreviated "Succ(x)" and usually means $x + 1$. Thus Peano's approach allows us to define the natural numbers using just 1 and $+$.

Axioms for the Natural Numbers

1. 1 is a natural number.

2. Each natural number has one and only one successor, Succ(x).

3. 1 is not the successor of any natural number.

4. If Succ(x) = Succ(y), then $x = y$.

5. If **M** is a subset of the set of natural numbers such that
   (i) $1 \in$ **M,**
   (ii) $x \in$ **M** $\Rightarrow$ Succ(x) $\in$ **M,**
   then **M** is the set of all natural numbers.

Discussion of the Axioms

The axioms consist of two parts—the primitive terms and the statements about the primitive terms. Axiom 1 gives us our first natural number, namely, 1. Axiom 2 tells us that Succ is a function. Axiom 3 tells us that in a sense not yet defined, 1 is the smallest natural number. Axiom 4 tells us that Succ is one-to-one. Axiom 5 is called the **finite induction postulate.** Although it was not explicitly stated, we used it in our justification of mathematical induction in Section 3.7. Our argument was as follows: Suppose $p(1)$ is true and $p(n) \Rightarrow p(n + 1)$ for all n. Then we have the following statements, all of which are true:

$p(1)$

$p(1) \Rightarrow p(2)$

$p(2) \Rightarrow p(3)$

etc.

Combining the first two statements by modus ponens, we concluded that $p(2)$ is true. Combining the truth of $p(2)$ with the third statement, we concluded that $p(3)$ is true. Continuing in this fashion, we concluded that $p(n)$ is true for all n. The key phrase in this argument is "Continuing in this fashion." Asserting that it is even possible to continue in this fashion implicitly assumes Axiom 5, since this is the only axiom that lets us say something about all natural numbers.

The picture we normally associate with the natural numbers is Figure 5.1:

Figure 5.1 A visual representation of the natural numbers.

1   2   3   4   5   6   7   8   9   10   11   12   13   14   15

But there are several other ways that the natural numbers could conceivably be represented. (Figure 5.2) We can better understand the axioms if we see how they exclude each of these possibilities.

Figure 5.2 Other conceivable structures for the natural numbers.

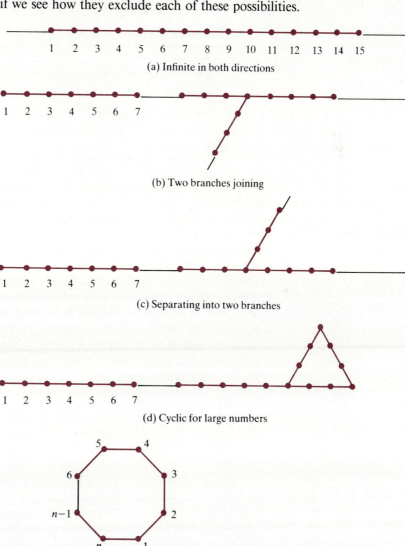

1   2   3   4   5   6   7   8   9   10   11   12   13   14   15

(a) Infinite in both directions

1   2   3   4   5   6   7

(b) Two branches joining

1   2   3   4   5   6   7

(c) Separating into two branches

1   2   3   4   5   6   7

(d) Cyclic for large numbers

(e) Cyclic for all numbers

EXAMPLE 1 Show from Peano's axioms that Figures 5.2(a) and (d) cannot be the structure of the set defined by those axioms.

SOLUTION Axiom 3 states that 1 is not the successor of any natural number. But in Figure 5.2(a), 1 is the successor of another number. Thus Figure 5.2(a) is excluded. Axiom 4 states that if two natural numbers have the same successor, they are equal. But in Figure 5.2(d), two distinct numbers have the same successor. Thus the figure is also excluded. ∎

The demonstration that the rest of the figures are inconsistent with the axioms will be left for the Exercises.

Properties of the Natural Numbers

Peano's axioms have several attractive features. They are simple and intuitively appealing. They define the entire set **N** in just five succinct statements. And they reveal very clearly how integral mathematical induction is to the natural numbers. Furthermore the common operations of arithmetic—addition and multiplication—can be defined in terms of the successor operation. Once this is done, it is then possible to verify that all of the familiar properties of arithmetic hold for these operations. We will not attempt to go through that verification here. What we will do is to state the definitions of *addition, multiplication, exponentiation,* and *less than* and state a number of results that can be proven from the axioms. For an excellent discussion of these ideas and several of the proofs, see Section 7.3 of *An Introduction to the Foundations and Fundamental Concepts of Mathematics,* revised edition, by Eves and Newsom.

Definitions Let x and n be in **N**. We define:

$$\text{Succ}^1(x) = \text{Succ}(x)$$
$$\text{Succ}^{n+1}(x) = \text{Succ}(\text{Succ}^n(x)), \; n \geq 1$$
$$x + n = \text{Succ}^n(x)$$
$$x \cdot 1 = x$$
$$x \cdot (n + 1) = x + x \cdot n, \; n \geq 1$$
$$x^1 = x$$
$$x^{n+1} = x \cdot x^n, \; n \geq 1$$

$x < n$ if there exists $a \in$ **N** such that $x + a = n$

Pred is the inverse of the successor function.

Pred(x) is usually denoted $x - 1$.

Note that the definition of $\text{Succ}^n(x)$ consists of two parts—the statement $\text{Succ}^1(x) = \text{Succ}(x)$ and the statement $\text{Succ}^{n+1}(x) = \text{Succ}(\text{Succ}^n(x))$,

$n \geq 1$. This is an important example of a **recursive definition**—a form of definition in which an entity is defined in terms of simpler versions of itself, starting from an elementary case. In this case, $\text{Succ}^n(x)$ is defined first by defining $\text{Succ}^1(x)$ (the elementary case) and then by defining $\text{Succ}^{n+1}(x)$ as the successor of $\text{Succ}^n(x)$ (the definition in terms of a simpler version of itself). Note the similarity between this definition and mathematical induction; the definition of Succ^1 is like the basis step of an inductive proof, and the definition of Succ^{n+1} is like the induction step. Notice that addition and multiplication are also recursively defined. Recursive definitions are extremely common in discrete mathematics, and we will examine them throughout this text. In the next chapter, we will be studying recursively defined functions and recursive algorithms as well. We will see a very close tie between all of these and mathematical induction.

Using these definitions and the axioms, all of the common properties of the natural numbers can be derived. Here is a partial list.

Properties of the Natural Numbers

For all natural numbers a, b, and c:

1.	$a + b = b + a$.	Commutativity of addition
2.	$a \cdot b = b \cdot a$.	Commutativity of multiplication
3.	$a + (b + c) = (a + b) + c$.	Associativity of addition
4.	$a \cdot (b \cdot c) = (a \cdot b) \cdot c$.	Associativity of multiplication
5.	$a \cdot (b + c) = a \cdot b + a \cdot c$.	Distributive property
6.	If $c + a = c + b$, then $a = b$.	Cancellation for addition
7.	If $c \neq 0$ and $c \cdot a = c \cdot b$, then $a = b$.	Cancellation for multiplication
8.	One and only one of the following holds: $a = b$, $a < b$, $a > b$.	Trichotomy
9.	If $ae = a$, then $e = 1$.	Uniqueness of the multiplicative identity
10.	$a \cdot 0 = 0 \cdot a = 0$	Existence of zero
11.	If $a < b$ and $b < c$, then $a < c$.	Transitivity of less than
12.	If $a < b$ and $c < d$, then $a + c < b + d$.	Additivity for inequalities
13.	If $a < b$ then $ac < bc$.	Multiplicativity for inequalities
14.	$1 \leq a$.	1 is the least natural number
15.	There is no natural number n such that $a < n < a + 1$.	
16.	If $h < k + 1$, then $h \leq k$.	
17.	$1^a = 1$.	
18.	$(ab)^c = a^c b^c$.	
19.	$a^b a^c = a^{b+c}$.	

The Well-Ordering Principle

There is one additional property that will be particularly important in the next two sections.

THEOREM 5.1 (Well-ordering principle) Every nonempty subset of the natural numbers has a smallest element.

PROOF Suppose there is a nonempty subset of **N** that has no smallest element; call it M. Let $T \subset \mathbf{N}$ consist of those natural numbers that are less than or equal to m for every m in M. Now $M \cap T = \emptyset$: Any element common to both would be a member of M that is less than or equal to every member of M; hence it would be a least element of M. Also, $1 \in T$ since, by property 14 of the natural numbers, 1 is less than or equal to all elements of **N**, and since $M \subset \mathbf{N}$, $1 \leq m$ for all m in M. Now let n be an element of T and suppose that $k \leq n$. Since $n \leq m$ for all $m \in M$, $k \leq m$ for all $m \in M$ as well. Hence k is in T for all $k \leq n$. Thus, if $n + 1 \in M$, $n + 1$ would be the least element of M. Hence $n + 1$ is not in M either. Thus all elements of M are greater than $n + 1$ as well as greater than n, and hence $n + 1 \in T$ as well. Hence $n \in T$ implies that $n + 1 \in T$. Thus, by Axiom 5, T contains all of the natural numbers, and hence M is empty. This is a contradiction, and hence every nonempty set of the natural numbers has a least element. □

Some writers prefer to assume the well-ordering principle and derive mathematical induction from it, perhaps feeling that the well-ordering principle seems more acceptable intuitively than mathematical induction. Such an approach requires a different set of axioms for **N**.

Note that so far, we have only discussed the natural numbers, not the integers. We say that the natural numbers are **closed under addition and multiplication,** meaning that the sum or product of two natural numbers is again a natural number. The natural numbers are not closed under subtraction and division, however. For instance, $6 - 7$ is not a natural number nor is 6/7. If we extend the natural numbers to include 0 and the negative numbers, we get **Z**, the set of integers. **Z** is closed under addition, multiplication, and subtraction. Furthermore, all of the above properties of the natural numbers except 13 and 14 are true of the integers also. Division is more complex, however. The divisibility properties of the integers are the subject of the next two sections.

TERMINOLOGY

successor
finite induction postulate
predecessor

recursive definition
well-ordering principle
closure under an operation

SUMMARY

1. The natural numbers can be defined by three primitive terms—1, successor, and natural number—and the following axioms:

---

1. 1 is a natural number.
2. Each natural number has one and only one successor, Succ(x).
3. 1 is not the successor of any natural number.
4. If Succ(x) = Succ(y), then $x = y$.
5. If M is a subset of the natural numbers such that
   (i) $1 \in M$,
   (ii) $x \in M \Rightarrow$ Succ(x) $\in M$,

   then M is the set of all natural numbers.

Axiom 5 is called the *finite induction postulate.*

2. Recursive definitions—definitions in which an entity is defined in terms of simpler version of itself, starting from an elementary case—occur throughout the study of discrete mathematics.

3. The well-ordering principle states that every nonempty set of natural numbers has a least element. Given a suitable formulation of the axioms for **N**, this principle is equivalent to the finite induction postulate.

4. **N** is closed under addition and multiplication but not under subtraction and division. **Z** is closed under addition, multiplication, and subtraction.

EXERCISES 5.1

Write each of the following expressions, statements, and equations, using successor and predecessor notation.

1. $a + 1$
2. $a + 2$
3. $a + 3$
4. $a - 1$
5. $a - 2$
6. For all natural numbers a, $(a + 1) - 1 = a$
7. If $a \neq 1$, then $(a - 1) + 1 = a$
8. a is never equal to its successor
9. Every natural number has a successor
10. 1 does not have a predecessor

Answer each of the following questions in a few sentences.

11. The phrase "A rose is a rose is a rose" appears at first to be a recursive definition in the sense that we are using the word *recursive* here. But it is not. What is the difference?

12. Explain why the statement "Every natural number has a predecessor" is false.

13. In the programming language Pascal, the statement "1 has a predecessor" is true. Explain why.

14. If Pred is the inverse of Succ, what are its domain and codomain?

Show that each of the following figures is inconsistent with Peano's axioms for the natural numbers.

15. 5.2(b)   **16.** 5.2(c)

17. 5.2(e)

18. Suppose $a,b \in$ **N,** $a > b$. $a - b$ could be defined as $\text{Pred}^b(a)$. Give a recursive definition for $\text{Pred}^b(a)$.

19. A string is a list of characters. Formulate a recursive definition for the word *string*.

20. Derive the principle of mathematical induction from the axiom of induction.

5.2

PRIME NUMBERS

As we discussed at the end of the last section, the natural numbers are closed under multiplication and addition, and the integers are closed under subtraction. However, neither are closed under division. That is, sometimes division of one integer by another (such as 72/9) yields an integer and sometimes (such as 73/9) it doesn't. Questions about divisibility have received a great deal of attention in the history of mathematics. In this section, we will look especially at prime numbers—numbers that can be divided only by themselves and 1—and some of their properties. The study of prime numbers is ancient, dating at least to the Greeks. Even today there are a number of significant unsolved problems. To begin, we must develop some basic concepts of divisibility of integers.

Definition Suppose $a,b \in$ **Z,** $a \neq 0$. We say that a **divides** b if there is another integer c such that $b = a \cdot c$. The phrase "a divides b" is denoted by **a | b.** In this case, we say that a is a **divisor** of b and b is a **multiple** of a.

The symbol "$|$" is also used when denoting sets—for example, in $\{x \in \mathbf{N} \mid x > 10\}$. There is no connection between the two uses.

EXAMPLE 1 $3 \mid 6$ since $6 = 3 \cdot 2$. $9 \mid 54$ since $54 = 9 \cdot 6$. ∎

Elementary Divisibility Properties

There are a number of elementary properties of divisibility. These are collected together in the following theorem. Note that whenever we write $a \mid b$, we will assume that $a \neq 0$.

THEOREM 5.2   For all $a, b, c \in \mathbf{Z}$,

a) $1 \mid a$, $a \mid a$, and $a \mid 0$.

b) If $a \mid b$ and $b \mid a$, then $a = \pm b$.

c) If $a \mid b$, then $a \mid bc$.

d) Suppose that x, y, and $z \in \mathbf{Z}$ and $x = y \pm z$. If a divides x and y, then a also divides z.

e) If $a \mid b$ and $a \mid c$, then $a \mid bx + cy$ for all $x, y \in \mathbf{Z}$.

PROOF

a) Note that $a = 1 \cdot a$. Hence $1 \mid a$ and $a \mid a$. Also, $0 = 0 \cdot a$. Hence $a \mid 0$. (We are assuming here the elementary properties of the natural numbers stated in the previous section and the corresponding rules for the integers.)

b) If $a \mid b$, then $b = ma$ for some integer m. If $b \mid a$, then $a = m' b$. Hence $b = m(m'b) = (mm')b$ and hence $mm' = 1$. Since m and m' are integers, $m = m' = \pm 1$. Thus $b = \pm a$.

c) (Left for the Exercises.)

d) Suppose $x = y - z$ and $a \mid x$ and $a \mid y$. Then $x = am$ and $y = an$ for some integers m and n. Again, using elementary properties of the integers, we have

$$am = an - z$$
$$am - an = -z$$
$$an - am = z$$
$$a(n - m) = z$$
$$a \mid z.$$

The other case is similar; some variations of this case are left for the Exercises.

e) If $a \mid b$ and $a \mid c$, then $b = am$ and $c = an$ for some integers m and n. Therefore $bx + cy = (am)x + (an)y = a(mx) + a(ny) = a(mx + ny)$.

Since **Z** is closed under addition and multiplication, the latter expression is an integer. Thus, by part (c), $a \mid (bx + cy)$. □

EXAMPLE 2 Show that

a) $2 \mid n(n + 1)(2n + 1)$ and

b) $3 \mid n(n + 1)(2n + 1)$ for any n.

SOLUTION

a) If n is even, then $2 \mid n$. If n is odd, $n + 1$ is even and thus $2 \mid (n + 1)$. Hence, in either case, $2 \mid n(n + 1)(2n + 1)$.

b) If $3 \mid n$, we are done. If not, then if $3 \mid n + 1$, we are done. If neither is true, then n is not of the form $3k$ or $3k - 1$ for any k. Hence it must be of the form $3k - 2$ (since $3k - 3$ would be the same form as $3k$, $3k - 4$ would be the same form as $3k - 1$, etc.). Thus

$$2n + 1 = 2(3k - 2) + 1 = 6k - 4 + 1 = 6k - 3 = 3(k - 2)$$

which is divisible by 3 by Theorem 5.2(c). ∎

Definition A number $p \in \mathbf{N}$, $p > 1$, is called **prime** if its only divisors in **N** are 1 and p itself. A number that is not prime is **composite.**

EXAMPLE 3 The following numbers are prime: 2, 3, 5, 7, 11, 13, and 17. The other integers between 2 and 17 are composite:

$4 = 2 \cdot 2$ $6 = 3 \cdot 2$

$8 = 4 \cdot 2$ $9 = 3 \cdot 3$

$10 = 5 \cdot 2$ $12 = 6 \cdot 2$

$14 = 7 \cdot 2$ $15 = 5 \cdot 3$

$16 = 8 \cdot 2$ or $4 \cdot 4$ ∎

How Can We Decide If a Number Is Prime?

One of the earliest questions asked about prime numbers was this: How can one tell if a given number is prime? For small numbers, a trial-and-error approach is sufficient. For instance, to decide if 119 is prime, we could try dividing it first by 2, then by 3, etc. If it has no divisors less than 119, we know it is prime. For large numbers this is not very practical, however. Try computing how long it would take to decide if $2^{42} - 1$ is prime ($2^{42} > 4$ trillion) even for a computer that could do one division every 10^{-6} seconds. One improvement that can be made in this approach is to note that if a number r can be written as $n \cdot m$, one of m and n must be no larger than $\sqrt{r}$ and one must be no smaller. Thus, to find out if a number is prime, we need only examine integers up to its square root as possible divisors. In fact, we only need to examine *prime numbers* up to its square root. A computer program to decide if $2^{42} - 1$ is prime would take an unreasonable amount of time dividing by every natural number up to $2^{42} - 1$. Using the symbol $\cong$ to mean "is approximately equal to," $\sqrt{(2^{42} - 1} \cong 2^{21} \cong 2$ million. Eliminating the even numbers as possible divisors, since they are not prime, brings it down even further, to around 1 million. Thus a computer program to decide if $2^{42} - 1$ is prime by trial and error is quite reasonable. However, for larger numbers, it too will eventually be too slow. At the present time, there is no satisfactory algorithm for taking an arbitrary integer and deciding if it is prime. The largest prime number known at present is $2^{216,091} - 1$. This number contains 65,050 digits.

Listing Primes

Another related problem is listing all of the primes up to a certain number. One ancient approach used to solve this problem was developed by the Greek mathematician Eratosthenes and is commonly called the **sieve of Eratosthenes.** It is a favorite programming exercise for introductory computer science courses. It works as follows: First, write down all of the integers from 2 up to the number that is to be the upper limit. Then take the first prime (i.e., 2) and cross out all of its multiples except itself. Take the next number not crossed out (i.e., 3). It will be prime. Cross out all of its multiples except itself. Continue in this manner until you reach the square root of your upper limit. All of the numbers that remain in the list and haven't been crossed out will be prime. For instance, Figure 5.3 shows the primes less than 100.

2 3 ~~4~~ 5 ~~6~~ 7 ~~8~~ ~~9~~ ~~10~~ 11 ~~12~~ 13 ~~14~~ ~~15~~ ~~16~~
17 ~~18~~ 19 ~~20~~ ~~21~~ ~~22~~ 23 ~~24~~ ~~25~~ ~~26~~ ~~27~~ ~~28~~ 29 ~~30~~ 31
~~32~~ ~~33~~ ~~34~~ ~~35~~ ~~36~~ 37 ~~38~~ ~~39~~ ~~40~~ 41 ~~42~~ 43 ~~44~~ ~~45~~ ~~46~~
47 ~~48~~ ~~49~~ ~~50~~ ~~51~~ ~~52~~ 53 ~~54~~ ~~55~~ ~~56~~ ~~57~~ ~~58~~ 59 ~~60~~ 61
~~62~~ ~~63~~ ~~64~~ ~~65~~ ~~66~~ 67 ~~68~~ ~~69~~ ~~70~~ 71 ~~72~~ 73 ~~74~~ ~~75~~ ~~76~~
~~77~~ ~~78~~ ~~79~~ ~~80~~ ~~81~~ ~~82~~ 83 ~~84~~ ~~85~~ ~~86~~ ~~87~~ ~~88~~ 89 ~~90~~ ~~91~~
~~92~~ ~~93~~ ~~94~~ ~~95~~ ~~96~~ 97 ~~98~~ ~~99~~ ~~100~~

Figure 5.3 The sieve of Eratosthenes.

One obvious limitation of this approach is that it takes so long to write down all of the integers up to the limit unless the limit is relatively small. That time limitation can be avoided if the sieve is implemented as a computer program; however, a space limitation then arises. There was an Austrian astronomer named Kulik who lived in the nineteenth century. He constructed an enormous sieve and computed all of the primes up to 100,000,000. It took him 20 years, off and on. (See Dudley, *Elementary Number Theory*, Section 2; see References.)

The Existence of Infinitely Many Primes

Another ancient question is, "How many primes are there?" The answer is "infinitely many." This question was answered very early; the following theorem appears in Euclid's *Elements* (c. 300 B.C.). Although we ordinarily associate the *Elements* with geometry, of the original 13 books into which it was divided, 3 deal primarily with number theory. Before we can prove that there are infinitely many primes, though, we need the following lemma. (A *lemma* is a small theorem, i.e., a result that we need to prove but that is seen more as an intermediate step on the way to an important theorem than as an end in itself.)

LEMMA 5.3   Every integer $n > 1$ is divisible by at least one prime.

PROOF   Suppose the statement is false. Then there is a nonempty collection, S, of integers that are not divisible by any prime. By the well-ordering principle, S contains a smallest integer, m. If m is prime, it is divisible by a prime, namely, itself. This is a contradiction. If m is not a prime, it can be written as $m = m_1 m_2$, where $1 < m_1 < m$ and $1 < m_2 < m$. Since m is the smallest integer not divisible by a prime, m_1 is divisible by a prime, p. By Theorem 5.2c, m is also divisible by p. This is also a contradiction. Hence S is empty. Thus every integer $n > 1$ is divisible by at least one prime.   ☐

THEOREM 5.4   There exist infinitely many primes.

PROOF   Suppose there are only finitely many primes. Label them $p_1, p_2, \ldots, p_n$. Then let

$$p = p_1 \cdot p_2 \ldots p_n + 1.$$

Then p is larger than each of $p_1, \ldots, p_n$, and since $p_1, \ldots, p_n$ are the only primes, p is not prime. Hence, by Lemma 5.3, p is divisible by at least one of $p_1, \ldots, p_n$. Call this prime p_i. Clearly, $p_1 \cdots p_n$ is also divisible by p_i. Thus, by Theorem 5.2(d), 1 is divisible by p_i. But this is a contradiction since all primes are greater than 1. Hence there must be infinitely many primes.   ☐

The Distribution of Primes

Another question that has been frequently asked is, "How are the primes distributed among the other integers?" One possible answer to this question would be a formula $p(n)$ that would give the nth prime when a particular integer is substituted for n. No such formula has been found. Figure 5.4 shows the location of the primes from 0 to 400.

Figure 5.4 Distribution of primes between 0 and 400.

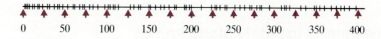

Primes seem to be distributed very irregularly among the rest of the integers. However, some significant results dealing with the distribution of primes have been discovered. For instance, we can see from Figure 5.4 that primes seem to become more scarce as numbers get larger. The most famous such theorem is Theorem 5.5. Although no subsequent material in this text depends on it, it is helpful in developing our understanding of the structure of the natural numbers. Its proof, however, is beyond the scope of this text. Before stating the theorem, suppose we let $\pi(x)$ stand for the number of primes between 1 and x. We also let $\ln(x)$ stand for so-called natural logarithm, i.e., the log base e of x where $e = 2.71828 \ldots$. If we tabulate the values of $\pi(x)$ and $x/\ln(x)$ for a few values of x, we get Figure 5.5.

Figure 5.5 Table relating $\pi(x)$ and $x/\ln(x)$.

x	100	500	1000	10^4	10^5	10^6	10^7	10^8
$\ln(x)$	4.61	6.21	6.91	9.21	11.51	13.82	16.12	18.42
$x/\ln(x)$	21.71	80.46	144.76	1085.7	8685.9	72,382.4	620,420.7	5,428,681.0
$\pi(x)$	25	95	168	1229	9592	78,498	664,579	5,761,455
$\dfrac{\pi(x)}{(x/\ln(x))}$	1.15	1.18	1.16	1.13	1.10	1.08	1.07	1.06

THEOREM 5.5 (Prime number theorem) Let $\pi(x)$ denote the number of primes between 1 and x.
Then

$$\pi(x) \cong x/\ln(x).$$

More precisely, as x approaches ∞, the ratio $\pi(x)$ divided by $x/\ln(x)$ approaches 1. The word *approaches* can be defined even more precisely using the notion of limit; however, this is a calculus topic and we will not go into it here. ⬚

EXAMPLE 4 Estimate the proportion of integers between 1 and 1000 and between 1 and 100,000 that are prime.

SOLUTION If $x = 1000$, $\ln(x) = 6.91$. Thus there are roughly 1000/7 primes between 1 and 1000, i.e., about 1/7 of the integers between 1 and 1000 are prime. $\ln(100,000) \cong 11.5$. Thus between 1/11 and 1/12 of the integers between 1 and 100,000 are prime. ■

TERMINOLOGY

divides   composite

divisor   sieve of Eratosthenes

multiple   Prime number theorem

prime number

SUMMARY

1. Integers greater than 1 can be classified as either prime—divisible only by themselves and 1—or composite.

2. There is no altogether satisfactory algorithm for deciding if an arbitrary integer n is prime. One way is to try dividing by all primes up to $\sqrt{n}$. Another is the sieve of Eratosthenes, which lists all primes up to and including n.

3. There are infinitely many primes. This was proven by the ancient Greeks and was included in Euclid's *Elements*.

4. If we let $\pi(x)$ denote the number of primes between 1 and x, $\pi(x) \cong x/\ln(x)$ where $\ln(x)$ denotes the natural logarithm of x.

EXERCISES 5.2

1. Find all prime numbers between 2 and 200 by constructing a sieve of Eratosthenes.

Decide if each of the following numbers is prime or composite.

2. 511   3. 241

4. 377   5. 671

6. 575   7. 823

8. 409   9. 667

Prove each of the following.

10. $2 \mid n(n + 1)$ 11. $3 \mid n(n + 1)(n + 2)$

12. $4 \mid n^2(n + 1)^2$ 13. $2 \mid n(3n + 7)$

14. $6 \mid n(n^2 - 1)$

If each of the following statements is true, prove it. If it is false, disprove it.

15. $a \mid b$ and $c \mid d \Rightarrow ac \mid bd$ 16. $d \mid a$ and $d \mid b \Rightarrow d^2 \mid ab$

17. $a \mid bc \Rightarrow a \mid b$ or $a \mid c$ 18. $a \mid b$ and $b \mid c \Rightarrow a \mid c$

19. $a \mid b \Rightarrow a \mid bc$ 20. $a > b \Rightarrow a$ does not divide b

Suppose x, y, and z are integers and $x = y \pm z$. Show that

21. If a divides y and z, then a divides x.

22. If a divides x and z, then a divides y.

Write algorithms for each of the following.

23. To decide if an integer N is prime.

24. To print all positive divisors of N.

Calculate $\pi(x)/(x/\ln(x))$ for the following values of x.

25. 50

26. 100

27. 150

28. 200

Estimate the proportion of primes between the following.

29. 1 and 500

30. 1 and 10^5

31. 1 and 10^7

32. 1 and 10^8

5.3

DIVISIBILITY PROPERTIES

Our first concern in this section is with the situation in which we have two integers, neither of which divides the other. We will study several results that apply in this case. We will also prove the fundamental theorem of arithmetic—that every integer has a unique factorization as a product of primes.

The Division Algorithm

The earliest kind of division problem we encountered in elementary school was something like this:

PROBLEM   Divide 30 by 7.

SOLUTION

$$
\begin{array}{r}
4 \text{ Rem } 2 \\
7 \overline{)30} \\
28 \\
\hline
2
\end{array}
$$

At that level, we took it for granted that it is always possible to find an answer to such a problem and that there is only one answer that is correct. The theorem that this is in fact the case is called the **division algorithm** and is another result discovered by ancient mathematicians. It is probably so familiar to you that it hardly seems necessary to prove it. However, like any mathematical result, it does require proof. Furthermore, the ideas involved in the proof are important and come up again in various ways. It is not an algorithm in the sense that we customarily think of it. It does involve inputs and unique outputs for those inputs, but it does not specify the process by which the transformation takes place. ∎

THEOREM 5.6 **(Division algorithm)** Suppose a and b are integers, $b > 0$. Then there exist unique integers q and r such that $a = bq + r$, $0 \leq r < b$.

PROOF This proof consists of two parts—first, a proof that the specified integers exist and, second, a proof that they are unique.
First, we establish the existence of q and r. Let t be any integer. Consider the set

$$S = \{a - bt \mid t \in \mathbf{Z}\}.$$

Then S always has nonnegative elements:

Suppose $a > 0$. Letting $t = 0$, we get $a - bt > 0$.

Suppose $a \leq 0$. If we set $t = a$, then, since $b \geq 1$ and since t is negative, $a - bt \geq 0$.

If S contains 0, let $r = 0$. If S does not contain 0, by the well-ordering principle, S has a least nonnegative element, which we will denote r. Let q denote the value of t such that

$$r = a - bq.$$

Then $a = bq + r$. We will show that $0 \leq r < b$. Since $r \in S$, $r \geq 0$. If $r \geq b$, there is an $r' \geq 0$ such that $r = r' + b$. Hence

$$a = bq + r' + b = b(q + 1) + r'$$

and thus $a - b(q + 1) \geq 0$. Thus $r' \in S$, $r' \geq 0$ and $r' < r$. This contradicts the fact that r was the least nonnegative element in S. Hence $r < b$.

Now we show that q and r are unique. Suppose there are q_1, r_1, q_2, and r_2 such that

$$0 \leq r_1 < b, \qquad 0 \leq r_2 < b$$

and

$$a = bq_1 + r_1 = bq_2 + r_2.$$

Then $b(q_1 - q_2) = r_2 - r_1$. Hence $b \mid r_2 - r_1$. But since $0 \leq r_2 < b$,

$$-r_1 \leq r_2 - r_1 < b - r_1,$$

and hence $-b < r_2 - r_1 < b$, since $0 \leq r_1 < b$. Thus $r_2 - r_1 = 0$ since b also divides $r_2 - r_1$. Hence $b(q_1 - q_2) = 0$ and thus $q_1 - q_2 = 0$. Thus $r_1 = r_2$ and $q_1 = q_2$, and thus q and r are unique. ∎

EXAMPLE 1   Let $a = 127$, $b = 23$. Find q and r.

SOLUTION $127 = 5 \cdot 23 + 12$, so $q = 5$, $r = 12$. Note that q is a symbol for what is commonly called a DIV b and r is a symbol for what is called a MOD b. ∎

The Euclidean Algorithm

The algorithm we will examine next, the Euclidean algorithm, is one of the earliest examples of what we call an algorithm today. For this reason, it has special importance. But it is also the foundation of much of number theory and is a valuable addition to our understanding of the structure of the integers. It is also perhaps the earliest example of an algorithm that was proven correct. As we saw earlier, rigorous verification of the correctness of algorithms is becoming increasingly important in contemporary computer science. A few concepts must be reviewed before we examine the algorithm.

Definitions   Suppose a and b are integers, d is a nonzero integer, and $d \mid a$ and $d \mid b$. Then d is called a **common divisor** of a and b. Suppose d' is a common divisor of a and b, with the property that any other common divisor, d, of a and b is less than d'. Then d' is called the **greatest common divisor (GCD)** of a and b. We denote the GCD of a and b by (a,b).

Note that the notation (a,b) for the GCD is the same as the notation for the ordered pair (a,b); this won't prove to be a problem, as it will be clear from the context which one is meant.

How can we find the GCD of two integers? One way is to use a fact that we know from our experience with arithmetic even though we haven't proven it yet—that every integer has a unique factorization as a product of primes. We can then identify the primes common to both expressions, and their product will be the GCD. For instance, to find the GCD of 12 and 30, we can write $12 = 2 \cdot 2 \cdot 3$ and $30 = 2 \cdot 3 \cdot 5$. Thus $(12,30) = 2 \cdot 3 = 6$. This is a reasonable approach for small numbers, but it can be very time-consuming for large numbers whose prime factorization is not obvious. (For instance, try to find the GCD of 20,064 and 34,580.) An alternative method is the following method, called the **Euclidean algorithm.**

ALGORITHM 15   Find the GCD of Two Integers
    Procedure Find_GCD(a,b; var GCD);
    {all variables are integers.}
    begin

```
1.    while b ≠ 0 do begin    {if b = 0, then GCD = a}
2.        q := a DIV b;
3.        r := a MOD b;        {find q and r such that a =
                               bq + r, 0 ≤ r < b}
4.        a := b;
5.        b := r;              {replace a by b, b by r, and try
                               again}
      end; {while}
6.    GCD := a
   end;
```

Although the pseudo-code form of writing this algorithm is helpful in clari-
fying the loop structure, it has traditionally been written in a different way,
one that makes it easier to prove that it works. In essence, the traditional
form is a trace of the algorithm, showing the successive applications of the
division algorithm. This is listed in Figure 5.6, using subscripts to label the
successive values of q and r.

Figure 5.6 A trace of the
Euclidean algorithm for
arbitrary a and b.

$$a = bq_1 + r_1$$
$$b = r_1q_2 + r_2$$
$$r_1 = r_2q_3 + r_3$$

.

.

.

$$r_{k-1} = r_kq_{k+1} + r_{k+1}$$
$$r_k = r_{k+1}q_{k+2}$$

In this formulation, the GCD is r_{k+1}. The process can be summarized
as follows: If we think of each line as being of the form $a = bq + r$, we can
think of a, b, and r as three positions. As we pass from each step to the next,
a new quotient, q, is found, b and r are each shifted one place to the left,
and a new remainder, r, is brought in. Note that the unwritten r_{k+2} in the
last step is zero, i.e., we continue until the remainder after division of a by b
is zero.

Here is a proof that the algorithm works.

PROOF   From the last step in Figure 5.6, it is clear that r_{k+1} divides r_k. In
the next to last step, then, since r_{k+1} divides itself and r_k, it also divides
r_{k-1}. Continuing in this fashion, it is clear that r_{k+1} divides both a and b,
and hence is a common divisor. Let d be any other common divisor. Since

d divides a and b, it divides r_1. Since it divides b and r_1, it divides r_2. Continuing in this way, we conclude that d divides r_{k+1} and is hence less than r_{k+1}. Thus r_{k+1} is the GCD.   ◻

EXAMPLE 2   Find the GCD of 800 and 640.

SOLUTION   Following the pseudo-code form of the algorithm, we get this trace:

1. $640 \neq 0$, so proceed to step 2
2. $q := 800 \text{ DIV } 640 = 1$
3. $r := 800 \text{ MOD } 640 = 160$
4. $a := 640$
5. $b := 160$
1. $160 \neq 0$, so proceed to step 2
2. $q := 640 \text{ DIV } 160 = 4$
3. $r := 640 \text{ MOD } 160 = 0$
4. $a := 160$
5. $b := 0$
1. $0 = 0$ so continue to step 6
6. $\text{GCD} := 160$

Written in traditional form, the sequence of steps is as follows:

$$800 = 1 \cdot 640 + 160$$
$$640 = 4 \cdot 160.$$

Hence $\text{GCD} = 160$.   ∎

EXAMPLE 3   Find the GCD of 20064 and 34580.

SOLUTION   Let a denote 34580 since it is larger, and let b denote 20064. Then

$$34580 = 1 \cdot 20064 + 14516$$
$$20064 = 1 \cdot 14516 + 5548$$
$$14516 = 2 \cdot 5548 + 3420$$
$$5548 = 1 \cdot 3420 + 2128$$
$$3420 = 1 \cdot 2128 + 1292$$
$$2128 = 1 \cdot 1292 + 836$$
$$1292 = 1 \cdot 836 + 456$$

$$836 = 1 \cdot 456 + 380$$
$$456 = 1 \cdot 380 + 76$$
$$380 = 5 \cdot 76.$$

Thus $(20064, 34580) = 76$. ∎

EXAMPLE 4 Show that the GCD of two distinct prime numbers is 1.

SOLUTION Let the numbers be denoted by p_1 and p_2. Since p_1 is prime, it is divisible only by p_1 and 1. Similarly, p_2 is divisible only by p_2 and 1. Thus the only common divisor is 1, and hence 1 is the GCD. ∎

Definition Two integers are **relatively prime** if their GCD is 1. Note that neither of the numbers needs to be prime.

EXAMPLE 5 Show that 84 and 55 are relatively prime.

SOLUTION
$$84 = 1 \cdot 55 + 29$$
$$55 = 1 \cdot 29 + 26$$
$$29 = 1 \cdot 26 + 3$$
$$26 = 8 \cdot 3 + 2$$
$$3 = 1 \cdot 2 + 1$$
$$2 = 2 \cdot 1$$

Hence $(84, 55) = 1$, and hence they are relatively prime. ∎

Two Additional Results

The next two theorems are used in many ways in number theory. However, the only use we will make of them here is in proving that every integer has a unique prime factorization.

THEOREM 5.7 Suppose $(a, b) = d$. Then there exist integers x and y such that $ax + by = d$.

PROOF The idea is to go through the Euclidean algorithm backward, repeatedly substituting an expression involving r_{k-1} and r_k for r_{k+1} as k becomes successively smaller. For instance, we showed previously that $(84, 55) = 1$. Starting from the next to last line, we can write

$$1 = 3 - 1 \cdot 2 = 3 - 2.$$

The second to last line can be used to write 2 in terms of 26 and 3:

$$1 = 3 - (26 - 8 \cdot 3) = 3 - 26 + 8 \cdot 3 = 9 \cdot 3 - 26.$$

Thus the third last line can be used to write 3 in terms of 26 and 29:

$$1 = 9(29 - 26) - 26 = 9 \cdot 29 - 10 \cdot 26.$$

Continuing this pattern

$$1 = 9 \cdot 29 - 10(55 - 29)$$
$$= 9 \cdot 29 - 10 \cdot 55 + 10 \cdot 29 = 19 \cdot 29 - 10 \cdot 55$$

and

$$1 = 19(84 - 55) - 10 \cdot 55$$
$$= 19 \cdot 84 - 19 \cdot 55 - 10 \cdot 55 = 19 \cdot 84 - 29 \cdot 55.$$

This completes the process, since we have written $1 = x \cdot 84 + y \cdot 55$, where $x = 19$ and $y = -29$. Generalizing this process into a proof of the theorem is left as an exercise. $\square$

THEOREM 5.8   If $d \mid ab$ and $(d, a) = 1$, then $d \mid b$.

PROOF   Since $(d, a) = 1$, there are x and y such that $dx + ay = 1$. Hence $bdx + bay = b$. Since $d \mid bdx$ and $d \mid bay$, we can conclude that $d \mid b$ as well. $\square$

The Fundamental Theorem of Arithmetic

We are finally ready to prove that every integer greater than 1 has a unique factorization in terms of primes. This is called the **fundamental theorem of arithmetic** and is also included in Euclid's *Elements*. When we say "unique," we do not distinguish factorizations that include the same primes written in different orders. For instance, 100 can be factored as $5 \cdot 5 \cdot 2 \cdot 2$ and as $2 \cdot 5 \cdot 5 \cdot 2$. We regard these as the same factorization. In the following proof, we will use the convention that factorizations are written like this: $100 = 2^2 \cdot 5^2$, i.e., in order from the smallest to the largest factor, with duplicates written in exponential form. As we have seen earlier, the proof will be divided into two parts, an existence part and a uniqueness part. The existence part will be very similar to Lemma 5.3 of the previous section. Note that 100 is a product of four primes—two 2's and two 5's. We also treat a prime number such as 7 as a product of primes by regarding it as a "product" of one prime—namely, itself.

THEOREM 5.9   Any integer $n > 1$ can be uniquely written as a product of primes (up to the order written).

PROOF   First, we establish the existence of a prime factorization. If n is prime, it can be written as a product of primes, namely, itself. Thus we can assume that n is not prime. Suppose that n cannot be written as a product of primes. Then the set, S, consisting of those natural numbers that cannot be written as a product of primes is nonempty. By the well-ordering principle, there is a least element, m, of S. Since m is not prime, it can be written

$m = m_1 m_2$, where $1 < m_1 < m$ and $1 < m_2 < m$. Since m was the least integer that cannot be written as a product of primes, both m_1 and m_2 can be so written. Substituting their prime factorizations into the equation $m = m_1 m_2$, we conclude that m can also be written as a product of primes. This is a contradiction, and thus S is empty. Hence any $n > 1$ can be written as a product of primes.

We establish uniqueness by mathematical induction. Note that we use the form of induction we called *strong induction* in Chapter 3. As the basis step, we note that 2 has a unique prime factorization, namely, itself. Suppose, then that $2, 3, \ldots, n - 1$ all have unique prime factorizations and let

$$n = p_1^{r_1} \cdot p_2^{r_2} \cdots p_k^{r_k} = q_1^{s_1} \cdot q_2^{s_2} \cdots q_j^{s_j}$$

where $p_1, \ldots, p_k, q_1, \ldots, q_j$ are all primes and $r_1, \ldots, r_k, s_1, \ldots, s_j$ are integers greater than or equal to 1. Also, suppose that the factors of n are arranged so that $p_1 < p_2 < \cdots < p_k$ and $q_1 < q_2 < \cdots < q_j$. Since $p_1 \mid n$, $p_1 \mid q_1^{s_1} \cdot q_2^{s_2} \cdots q_j^{s_j}$ also. By repeated applications of Theorem 5.8 and Example 4, $p_1 = q_i$ for some i. Thus $p_1^{r_1 - 1} \cdot p_2^{r_2} \cdots p_k^{r_k} = q_1^{s_1} \cdots q_i^{s_i - 1} \cdots q_j^{s_j}$. But since this expression denotes an integer less than n, it has a unique prime factorization. Thus these two expressions are identical. Multiplying by $p_1 = q_i$, we conclude that $p_1^{r_1} \cdot p_2^{r_2} \cdots p_k^{r_k}$ and $q_1^{s_1} \cdot q_2^{s_2} \cdots q_j^{s_j}$ are also identical, and hence $j = k$, $p_i = q_i$ for all i and $r_i = s_i$ for all i. Thus every integer greater than 1 has a unique prime factorization. ☐

EXAMPLE 6 Write 294060 as a product of prime factors in standard form.

SOLUTION We break the number into its prime factors by successive division then arrange them in standard order:

$$294060$$
$$= 2 \cdot 147030$$
$$= 2 \cdot 2 \cdot 73515$$
$$= 2 \cdot 2 \cdot 3 \cdot 24505$$
$$= 2 \cdot 2 \cdot 3 \cdot 5 \cdot 4901$$
$$= 2 \cdot 2 \cdot 3 \cdot 5 \cdot 13 \cdot 377$$
$$= 2 \cdot 2 \cdot 3 \cdot 5 \cdot 13 \cdot 13 \cdot 29$$
$$= 2^2 \cdot 3^1 \cdot 5^1 \cdot 13^2 \cdot 29^1 \quad \blacksquare$$

TERMINOLOGY

division algorithm	Euclidean algorithm
common divisor	relatively prime numbers
greatest common divisor (GCD)	fundamental theorem of arithmetic

SUMMARY

1. Given any two integers a and b, there are unique integers q and r, $0 \le r < b$ such that $a = bq + r$. q is a DIV b and r is a MOD b.
2. The GCD of any two integers can be found by the Euclidean algorithm.
3. If $(a,b) = d$, then there exist integers x and y such that $ax + by = d$.
4. The fundamental theorem of arithmetic states that every natural number can be written as a product of primes and that this product is unique up to the order in which the product is written.

EXERCISES 5.3

Using the division algorithm, find q and r for each of the following pairs of numbers.

1. 611, 39
2. 1025, 125
3. 6331, 87
4. 601, 600
5. 600, 601
6. 739, 1

Find the GCD of each of the following pairs of integers by using the Euclidean algorithm.

7. 382, 24
8. 1024, 236
9. 1922, 2018
10. 2310, 2730

Which of the following pairs are relatively prime?

11. 793, 1037
12. 124, 111
13. 1024, 729
14. 770, 1547

In each of the following equations, you are given a, b, and their GCD, d. Find x and y such that $ax + by = d$.

15. $(312, 384) = 24$
16. $(72, 65) = 1$
17. $(82, 26) = 2$
18. $(126, 294) = 42$

Find the prime factors of each of the following.

19. 210
20. 900
21. 360
22. 1617
23. 6256
24. 4199

Let n be any integer. Find:

25. $(n, 0)$
26. $(n, 1)$

27. $(n, -n)$ **28.** (a, na), $a \in N$

29. (p_1, p_2) where p_1 and p_2 are prime   **30.** all primes of the form $n^3 - 1$

Prove each of the following. All symbols denote integers.

31. If $a = bq + r$, then $(a, b) = (b, r)$.

32. If $a \mid b$ and $a > 0$, then $(a, b) = a$.

33. $(n, n + 1) = 1$ for all integers n.

34. If $(a, b) = 1$ and $c \mid a$, then $(c, b) = 1$.

35. If $(a, b) = 1$ then $(a + b, a - b) = 1$ or 2.

36. $ax + by = m \Rightarrow (a, b) \mid m$

37. $(na, nb) = n(a, b)$

38. If $p \mid ab$ and p is prime, then $p \mid a$ or $p \mid b$.

39. Show that the worst-case complexity of the Euclidean algorithm is no worse than $O(\log n)$ where n is the larger of the two numbers.

5.4
POSITIONAL NOTATION

At the beginning of Section 5.1, we talked about two ways to explain what 187 means: one 100, eight 10s and seven 1s or one more than 186. Working from the latter idea, we developed an understanding of the natural numbers based on three primitive terms—natural number, 1, and +. The key idea was that the natural numbers are defined recursively, i.e., starting at 1, each natural number is defined as one more than the previous one. We also observed that not only are the natural numbers themselves defined recursively, but so are the basic operations of addition, multiplication, and exponentiation. We then proceeded in Sections 5.2 and 5.3 to develop the basic divisibility properties of the integers, including the concept of prime number and the fundamental theorem of arithmetic.

It is important to note that in none of the results or proofs presented in those sections did it matter what system we used to denote the integers. That is, all of the results in Sections 5.1 through 5.3 are just as true if we represent integers using Roman numerals (or any other system) instead of the decimal system. The properties presented in those sections are properties of the integers themselves, not of how we represent them. In this section, we turn our attention to representation.

Definition   Suppose b and n are natural numbers. A **positional representation** of n with base b is a sequence of symbols $d_k d_{k-1} \cdots d_1 d_0$, where $0 \le d_i < b$ for each i. We think of the sequence of symbols as a shorthand notation for

$$d_k b^k + d_{k-1} b^{k-1} + \cdots + d_1 b + d_0.$$

Such an approach to representing the natural numbers is called a **positional numbering system**, and b is called its **base.**

EXAMPLE 1   The decimal system is a positional system with base 10. For instance,

$$1234 = 1 \cdot 10^3 + 2 \cdot 10^2 + 3 \cdot 10^1 + 4 \cdot 10^0.$$

Roman numerals are not a positional system in this sense. For instance, I means 1 but II means two; it does not mean $1 \cdot b + 1$ for any b. The system is not totally independent of position, though; for instance, IV does not represent the same number as VI. ∎

The decimal system and other positional systems have several advantages over other numbering systems such as roman numerals. They are much more effective in handling large numbers; they give more compact representations of numbers unless b is very small (compare 388 to CCCLXXXVIII); similar-sized numbers have similar-sized representations (compare 88 and 90 to LXXXVIII and XC); and algorithms for addition and multiplication are simple and easy to learn. But the positional idea is a very sophisticated concept. It uses a single distinct symbol for each of the digits 0 through $b - 1$ and it combines an understanding of addition, multiplication, and exponentiation. So it is not hard to understand that its development took a long time. In fact, it did not find its way into European usage until after 1100 A.D. in spite of the fact that the results for integers that we saw in the previous two sections were known by 300 B.C. and earlier.

There is no reason that the base of a positional system must be 10. The fact that our system is based on 10 has been called "an accident of physiology." Other bases, notably 2, 8, and 16, have had particular importance in computer science. In this section, we will first prove a theorem that is the foundation of positional systems—the every natural number has a unique positional representation for a given base. Then we will examine counting and arithmetic in various bases and conversion between bases, all with the purpose of developing such ease in working with positional systems that we can work in any base and can choose whatever base is most convenient in a given situation.

The Basic Theorem

THEOREM 5.10 Let n and b be any natural numbers, $b \geq 2$. Then n has a unique positional representation with base b.

PROOF Like many theorems in number theory, this one is divided into an existence part and a uniqueness part.

First, the existence part. This is primarily a repeated application of the division algorithm:

$$n = bq_1 + r_1, \ 0 \leq r_1 < b.$$

q_1 can then be written

$$q_1 = bq_2 + r_2.$$

Continuing in this way, we can write

$$q_k = bq_{k+1} + r_{k+1}.$$

Thus we get a sequence of quotients $q_1 > q_2 > \cdots > q_k$, and eventually, for some K, q_{K+1} will be zero simply because the q_k values are a decreasing sequence of positive integers. When that happens, from the last preceding equation, $q_K = r_{K+1}$. Hence we know from the division algorithm that q_K and all of the r_i values will be greater than or equal to zero and less than b. Thus we can write

$$n = bq_1 + r_1 = b(bq_2 + r_2) + r_1 = b^2 q_2 + br_2 + r_1$$
$$= b^2 (bq_3 + r_3) + br_2 + r_1 = b^3 q_3 + b^2 r_3 + br_2 + r_1$$

$$\cdot$$
$$\cdot$$
$$\cdot$$

$$= b^K q_K + b^{K-1} r_K + \cdots + br_2 + r_1, \ 0 \leq q_K, r_K, \cdots, r_1 < b$$

and this latter expression is a positional representation for n.

As for uniqueness, suppose we have two representations for n,

$$n = b^k q_k + b^{k-1} r_k + \cdots + br_2 + r_1 = b^j t_j + b^{j-1} s_j + \cdots + bs_2 + s_1.$$

We can assume that $k = j$ since if either representation is shorter than the other, we can simply add zero terms to the shorter one to make them the same length. Thus we can write

$$b^k q_k + b^{k-1} r_k + \cdots + br_2 + r_1 = b^k t_k + b^{k-1} s_k + \cdots + bs_2 + s_1.$$

Hence

$$b^k(q_k - t_k) + b^{k-1}(r_k - s_k) + \cdots + b(r_2 - s_2) + (r_1 - s_1) = 0.$$

Since $b \mid (b^k(q_k - t_k) + b^{k-1} (r_k - s_k) + \cdots + b(r_2 - s_2))$, $b \mid (r_1 - s_1)$. But since $-s_1 < r_1 - s_1 < b - s_1$ and $0 < s_1 < b$, $r_1 - s_1 = 0$. We will wait for the Exercises to show that it follows from this that $(r_2 - s_2) = 0, \cdots,$ $(r_k - s_k) = 0$, and $(q_k - t_k) = 0$, and hence that the representation is unique. $\square$

Counting in Bases Other Than 10

To represent numbers in a positional system for any base b, the first thing to note is that it takes b distinct symbols to do so—one for each number from 0 up to but not including b. Thus in base 2 our usual symbols are 0 and 1, although we could have used F and T, a and b, or any other pair of distinct symbols. Similarly, in base 5 we use 0, 1, 2, 3, and 4, and in base 8 we use 0, 1, 2, 3, 4, 5, 6, and 7. In base 16, we use 0, 1, 2, 3, 4, 5, 6, 7, 8, and 9 and add the letters A, B, C, D, E, and F to represent 10, 11, 12, 13, 14, and 15, respectively. The A, B, C, D, E, and F are needed for the following reason. Since we want to represent numbers in a positional system, 10 (in base 16) means $1 \cdot 16 + 0 \cdot 1$, which is certainly not the number that follows 9. Thus 10 (and similarly, 11 through 15) must be replaced by a symbol that uses only one place. Thus in Figure 5.7 we can see a tabulation of the numbers from 0 through 99 in different bases.

We use subscripts to indicate the base in which a number is written. Thus, for instance,

$$(31)_{16} = (49)_{10}$$

and

$$(340)_5 = (137)_8.$$

EXAMPLE 2 Write $(100)_{10}$ in bases 2, 5, 8, and 16.

SOLUTION Continuing the counting process used in Figure 5.7, we get

$$(100)_{10} = (1100100)_2 = (400)_5 = (144)_8 = (64)_{16}. \quad \blacksquare$$

Arithmetic in Other Bases

Now that we have seen representations of numbers in other bases, we are ready to examine arithmetic in other bases. In base 10, this involves two things—tables for the sums and products of the digits from 0 through 9 (which are memorized, usually in third grade or so) and some algorithms for carrying out arithmetic on numbers of two or more digits. The same procedure works for other bases as well. For instance, in base 10, we have the familiar tables found in Figure 5.8.

To add two numbers in base 10, such as $(73)_{10}$ and $(58)_{10}$, we use an algorithm. We write

$$\begin{array}{r} 73 \\ +59. \end{array}$$

10	2	5	8	16		10	2	5	8	16		10	2	5	8	16
		Base						Base						Base		
0	0	0	0	0		34	100010	114	42	22		67	1000011	232	103	43
1	1	1	1	1		35	100011	120	43	23		68	1000100	233	104	44
2	10	2	2	2		36	100100	121	44	24		69	1000101	234	105	45
3	11	3	3	3		37	100101	122	45	25		70	1000110	240	106	46
4	100	4	4	4		38	100110	123	46	26		71	1000111	241	107	47
5	101	10	5	5		39	100111	124	47	27		72	1001000	242	110	48
6	110	11	6	6		40	101000	130	50	28		73	1001001	243	111	49
7	111	12	7	7		41	101001	131	51	29		74	1001010	244	112	4A
8	1000	13	10	8		42	101010	132	52	2A		75	1001011	300	113	4B
9	1001	14	11	9		43	101011	133	53	2B		76	1001100	301	114	4C
10	1010	20	12	A		44	101100	134	54	2C		77	1001101	302	115	4D
11	1011	21	13	B		45	101101	140	55	2D		78	1001110	303	116	4E
12	1100	22	14	C		46	101110	141	56	2E		79	1001111	304	117	4F
13	1101	23	15	D		47	101111	142	57	2F		80	1010000	310	120	50
14	1110	24	16	E		48	110000	143	60	30		81	1010001	311	121	51
15	1111	30	17	F		49	110001	144	61	31		82	1010010	312	122	52
16	10000	31	20	10		50	110010	200	62	32		83	1010011	313	123	53
17	10001	32	21	11		51	110011	201	63	33		84	1010100	314	124	54
18	10010	33	22	12		52	110100	202	64	34		85	1010101	320	125	55
19	10011	34	23	13		53	110101	203	65	35		86	1010110	321	126	56
20	10100	40	24	14		54	110110	204	66	36		87	1010111	322	127	57
21	10101	41	25	15		55	110111	210	67	37		88	1011000	323	130	58
22	10110	42	26	16		56	111000	211	70	38		89	1011001	324	131	59
23	10111	43	27	17		57	111001	212	71	39		90	1011010	330	132	5A
24	11000	44	30	18		58	111010	213	72	3A		91	1011011	331	133	5B
25	11001	100	31	19		59	111011	214	73	3B		92	1011100	332	134	5C
26	11010	101	32	1A		60	111100	220	74	3C		93	1011101	333	135	5D
27	11011	102	33	1B		61	111101	221	75	3D		94	1011110	334	136	5E
28	11100	103	34	1C		62	111110	222	76	3E		95	1011111	340	137	5F
29	11101	104	35	1D		63	111111	223	77	3F		96	1100000	341	140	60
30	11110	110	36	1E		64	1000000	224	100	40		97	1100001	342	141	61
31	11111	111	37	1F		65	1000001	230	101	41		98	1100010	343	142	62
32	100000	112	40	20		66	1000010	231	102	42		99	1100011	344	143	63
33	100001	113	41	21												

Figure 5.7 Numbers from 0 through 99 in bases 10, 2, 5, 8, and 16.

Figure 5.8 Base 10 addition and multiplication tables.

+	0	1	2	3	4	5	6	7	8	9
0	0	1	2	3	4	5	6	7	8	9
1	1	2	3	4	5	6	7	8	9	10
2	2	3	4	5	6	7	8	9	10	11
3	3	4	5	6	7	8	9	10	11	12
4	4	5	6	7	8	9	10	11	12	13
5	5	6	7	8	9	10	11	12	13	14
6	6	7	8	9	10	11	12	13	14	15
7	7	8	9	10	11	12	13	14	15	16
8	8	9	10	11	12	13	14	15	16	17
9	9	10	11	12	13	14	15	16	17	18

*	0	1	2	3	4	5	6	7	8	9
0	0	0	0	0	0	0	0	0	0	0
1	0	1	2	3	4	5	6	7	8	9
2	0	2	4	6	8	10	12	14	16	18
3	0	3	6	9	12	15	18	21	24	27
4	0	4	8	12	16	20	24	28	32	36
5	0	5	10	15	20	25	30	35	40	45
6	0	6	12	18	24	30	36	42	48	54
7	0	7	14	21	28	35	42	49	56	63
8	0	8	16	24	32	40	48	56	64	72
9	0	9	18	27	36	45	54	63	72	81

We add the 3 and the 9, using the addition table, getting 12. We write the 2 under the 9 and treat the 1 as a carry. We then add the 7 and the 5, getting 12, and then add the carry to that sum for a total of 13. Putting this number under the 5, we get a total of $(132)_{10}$:

$$\begin{array}{r} 73 \\ +59 \\ \hline 132 \end{array}$$

Multiplication follows a similar algorithm. We use the table to get the product of the one-digit numbers and use a system of carries to handle products that are more than one digit in length.

The process of adding and multiplying in other bases is exactly the same. For instance, suppose we want to do arithmetic in base 8. The base 8 addition and multiplication tables can be easily constructed by simply counting in base 8 or by converting the corresponding numbers in the base 10 tables. Using these tables, the computations can be done in exactly the same way as with base 10 numbers (Figure 5.9).

+	0	1	2	3	4	5	6	7		*	0	1	2	3	4	5	6	7
0	0	1	2	3	4	5	6	7		0	0	0	0	0	0	0	0	0
1	1	2	3	4	5	6	7	10		1	0	1	2	3	4	5	6	7
2	2	3	4	5	6	7	10	11		2	0	2	4	6	10	12	14	16
3	3	4	5	6	7	10	11	12		3	0	3	6	11	14	17	22	25
4	4	5	6	7	10	11	12	13		4	0	4	10	14	20	24	30	34
5	5	6	7	10	11	12	13	14		5	0	5	12	17	24	31	36	43
6	6	7	10	11	12	13	14	15		6	0	6	14	22	30	36	44	52
7	7	10	11	12	13	14	15	16		7	0	7	16	25	34	43	52	61

Figure 5.9 Base 8 addition and multiplication tables.

EXAMPLE 3 Compute (a) $(27)_8 + (44)_8$ and (b) $(27)_8 * (44)_8$.

SOLUTION   Using the tables we have

a)
$$\begin{array}{r} 27 \\ +44 \\ \hline 73_8 \end{array}$$

b)
$$\begin{array}{r} 27 \\ *44 \\ \hline 134 \\ 134 \\ \hline 1474_8 \end{array}$$ ∎

Of course, these computations could be done by changing to base 10, doing the arithmetic there, and changing back to base 8. But doing so would provide less insight into this aspect of arithmetic.

Base 2 Arithmetic

There is one base other than 10 for which it is particularly easy to remember the tables, and that is base 2. Thus arithmetic in this base is especially simple. The tables are

+	0	1
0	0	1
1	1	10

*	0	1
0	0	0
1	0	1

EXAMPLE 4   Add and multiply $(1000101)_2$ and $(1101011)_2$.

SOLUTION

$$
\begin{array}{r}
1000101 \\
+1101011 \\
\hline
10110000
\end{array}
$$

$$
\begin{array}{r}
1000101 \\
*1101011 \\
\hline
1000101 \\
1000101 \\
1000101 \\
1000101 \\
1000101 \\
\hline
1110011010111
\end{array}
$$

Note that multiplication in base 2 involves only two operations—shifting and adding; a multiplication table isn't necessary. It is this observation that is the basis for the algorithms that digital computers use to carry out multiplication. Such algorithms may be programmed or may be built into the hardware.

Writing Numbers in Various Bases

The process of changing bases is not hard to understand if we keep in mind what positional notation means. That is, when we write a number as a sequence of symbols $d_k d_{k-1} \cdots d_1 d_0$, we mean

$$n = d_k b^k + d_{k-1} b^{k-1} + \cdots + d_1 b + d_0.$$

EXAMPLE 5   To change 432_8 to base 10, we use the expanded notation:

$$
\begin{aligned}
432_8 &= 4 \cdot 8^2 + 3 \cdot 8^1 + 2 \cdot 8^0 \\
&= 4(64) + 24 + 2 \\
&= 256 + 26 \\
&= 282_{10}.
\end{aligned}
$$

EXAMPLE 6   Change $3E0_{16}$ to base 10.

SOLUTION   We expand 3E0 and substitute base 10 symbols:

$$3E0_{16} = 3 \cdot 16^2 + E \cdot 16^1 + 0 \cdot 16^0$$
$$= 3(256) + 14(16)$$
$$= 768 + 224$$
$$= 992_{10}. \blacksquare$$

The process followed in Examples 5 and 6 is the one generally used in changing from any nondecimal base to base 10. But this is only because base 10 arithmetic is so familiar to us. It could (at least in principle) be used to change from base 10 to other bases. For instance, suppose we wanted to change 242_{10} to base 8. Following the preceding pattern, we would write

$$242_{10} = (1 \cdot 10^2 + 4 \cdot 10^1 + 2 \cdot 10^0)_{10}$$
$$= (2 \cdot 10^2 + 4 \cdot 10^1 + 2 \cdot 10^0)_{10}$$
$$= (2 \cdot 12^2 + 4 \cdot 12 + 2)_8 \qquad \text{(Using that } 10_{10} = 12_8.)$$
$$= (2 \cdot 144 + 50 + 2)_8$$
$$= (310 + 50 + 2)_8$$
$$= 362_8.$$

Unfortunately, a table for base 8 multiplication is not usually available, and without this table, we can't easily carry out this process. Hence it is normally used only for changing from nondecimal bases to base 10. But it is important to realize that this is a property of human limitations, not of decimal or nondecimal number bases. For a machine, in the appropriate situation, this might be exactly the process to use even when not changing to base 10.

There is another way to change bases. This method is based on the division algorithm. That is, if we want to write a number n in some base b, we can find its representation in that base by successive divisions by b. The remainders of that division will form the representation.

EXAMPLE 7   Write $(410)_{10}$ in base 8.

SOLUTION $410 = 51 \cdot 8 + 2$
$ 51 = 6 \cdot 8 + 3$
$ 6 = 0 \cdot 8 + 6.$

Hence $(410)_{10} = (632)_8$.

An abbreviated version of this process is often written as in Figure 5.10:

Figure 5.10 $410_{10} = 632_8$.

$$
\begin{array}{r|rl}
8 & 410 & \\
8 & 51 & 2 \\
8 & 6 & 3 \\
8 & 0 & 6 \quad \blacksquare
\end{array}
$$

EXAMPLE 8   Write $(213)_{10}$ in base 2.

SOLUTION   Using the abbreviated form of the preceding process, we get

Figure 5.11 $213_{10} =$ 11010101_2.

$$
\begin{array}{r|rl}
2 & 213 & \\
2 & 106 & 1 \\
2 & 53 & 0 \\
2 & 26 & 1 \\
2 & 13 & 0 \\
2 & 6 & 1 \\
2 & 3 & 0 \\
2 & 1 & 1 \\
2 & 0 & 1 \quad \blacksquare
\end{array}
$$

This particular procedure is used primarily for changing from base 10 to nondecimal bases. But just as with the first process, this is a matter of human limitation rather than of any property of the number bases. The first approach allows us to change from nondecimal bases to base 10 using only decimal arithmetic, and this second process allows us to change from base 10 to nondecimal bases using only decimal arithmetic.

Changing Between Bases 2 and 2^n

There is a nice trick that is commonly used and can save considerable time in changing between base 2 and any base of the form 2^n. In fact, it can be used between any base b and another base b^n. The method is as follows: Suppose we have a number in base 2 such as

$$10001011101$$

and want to change it to base 8. Group the bits into threes from right to left. Then write each of those groups in base 8;

$$10\ 001\ 011\ 101 = 2\ 1\ 3\ 5.$$

That is, $10001011101_2 = 2135_8$.

To go from base 8 to base 2, we simply reverse the process:

$$17635_8 = 001\ 111\ 110\ 011\ 101_2.$$

With base 16, the procedure is exactly the same, except that we group by four bits.

EXAMPLE 9   Write $(10001100101011101)_2$ in base 16 and $(C3D0)_{16}$ in base 2.

SOLUTION

$$10001100101011101 = 1\ 0001\ 1001\ 0101\ 1101 = 1195D_{16}$$
$$C3D0 = 1100\ 0011\ 1101\ 0000 = 1100001111010000_2 \quad \blacksquare$$

Why does this trick work? One of the best examples of counting in a positional number system is the odometer in a car. Suppose an odometer starts at zero. There are 10 symbols—0 through 9—and the odometer goes sequentially through all of them, returning to zero again after 9. But in returning to zero, it advances the 10s digit to a 1. It then proceeds through all of the symbols, again eventually advancing the 10s digit. Now consider a base 8 odometer. It works the same way, except that it has only eight symbols—0 through 7. When it has advanced through all eight of them, it returns to zero, advancing the number in the 8s place. Whether the symbols are denoted 0, 1, 2, 3, 4, 5, 6, 7, 10 or 000, 001, 010, 011, 100, 101, 110, 111, 1000 doesn't matter. There is a one-to-one correspondence between the symbols in base 2 when grouped by threes and the symbols in base 8. Counting in either base involves stepping through these two sets of symbols in a way that maintains the correspondence between them. Thus counting in base 2 with the bits grouped by threes and counting in base 8 are the same.

Similarly, with base 16, grouping the bits in fours, starting from the right, yields a collection of 16 symbols 0000 through 1111. Counting in base 2 cycles through this collection of 16 symbols in precisely the same order that the symbols 0, 1, . . . , E, F are cycled through and thus base 2 grouped by fours can be considered as simply another notation for base 16.

TERMINOLOGY

positional representation of n
positional numbering system
base

SUMMARY

1. The properties of the integers developed in Sections 5.1 through 5.3 are inde-

pendent of the way in which integers are denoted and are thus properties of the integers themselves, not of their representation.

2. Let b be any base. Then every integer has a unique positional representation in base b.

3. Arithmetic in base 10 proceeds by using tables for one-digit sums and products and standard algorithms for sums and products of larger numbers and for subtraction and division. Arithmetic in other bases can be carried out using different tables but the same algorithms.

4. Base 2 multiplication can be done using only the operations of shifting and adding, and hence is easily implemented on a computer.

5. There are two algorithms for changing bases. One uses the positional representation in the original base and substitutes the symbols of all numbers using the new base. This algorithm is commonly used to change from other bases to base 10. A second algorithm uses repeated applications of the division algorithm. It is commonly used to change from base 10 to other bases.

6. Changing between base 2 and base 2^n can be done by grouping bits and replacing the groups by their symbols in base 2^n or, in reverse, by replacing symbols by groups of bits.

EXERCISES 5.4

Use Figure 5.7 to find the base 10 equivalents of the following.

1. $(13)_8$
2. $(24)_5$
3. $(5E)_{16}$
4. $(10011)_2$

Use Figure 5.7 to answer the following questions.

5. Find all numbers in the table that have a representation of 10. What do they have in common?

6. Find all numbers in the table that have a representation of 100. What do they have in common?

7. From your answers to Exercises 5 and 6, infer the base 10 values of 1000_5, 1000_8, and 1000_{16}.

8. Look up $(100 - 1)$ in bases 5 and 8. Infer $(100 - 1)$ in base 16.

9. Under what circumstances are the rightmost digits of the base 5 and base 10 representations of a number the same? Why?

10. Under what circumstances are the rightmost digits of the base 8 and base 16 representations of a number the same? Why?

11. Extend Figure 5.7 to include numbers from 100_{10} to 125_{10}.

12. For what range of values do base 2 numbers have two bits? Three bits? Four bits? n bits? Generalize your answer to other bases.

Perform the following arithmetic operations.

13. $17_8 + 34_8$ **14.** $100011_2 + 111011_2$

15. $342_5 + 221_5$ **16.** $E13A_{16} + 2727_{16}$

17. $17_8 * 34_8$ **18.** $62_8 * 31_8$

19. $100011_2 * 111011_2$ **20.** $342_5 * 221_5$

Construct addition tables in the following bases.

21. 16   **22.** 4

23. 7   **24.** 12

Construct multiplication tables in the following bases.

25. 16   **26.** 4

27. 7   **28.** 12

The standard algorithm for base 10 subtraction can also be used with other bases. Using this algorithm and the tables in the text, compute the following.

29. $743_8 - 372_8$ **30.** $341_8 - 242_8$

31. $3424_5 - 2332_5$ **32.** $101010_2 - 100111_2$

Using the tables in the text and generalizing the standard algorithm for division of integers, compute the following.

33. $674_8 / 72_8$ **34.** $341_8 / 22_8$

35. $3424_5 / 32_5$ **36.** $100111_2 / 101_2$

Change the following numbers from the indicated base, b, to base 10 by substitution for b.

37. 731_8 **38.** 232_5

39. $10E_{16}$ **40.** 10001100_2

Change the following numbers from base 10 to the indicated base by substituting the representation of 10 in the other base and doing the arithmetic in that base.

41. 61; base 8   **42.** 29; base 5

43. 32; base 2   **44.** 63; base 2

Change the following numbers from base 10 to the indicated base by using division.

45. 421; base 8   **46.** 417; base 16

47. 139; base 2   **48.** 600; base 5

Change the following from base 2 to bases 8 and 16.

49. 100100100   **50.** 11001100

51. 111010100   **52.** 11111111

Change the following to base 2.

53. 3764_8 **54.** 1177_8

55. $3E0F_{16}$ **56.** $ABCD_{16}$

5.5

CONGRUENCE

There is one more topic we must discuss in order to complete our study of the basic properties of the natural numbers: congruence. It has provided the tools with which many significant theoretical discoveries in number theory have been made. But it is also of considerable practical importance, notably in computer science. This section consists of a brief introduction to the concept of congruence and some of its basic properties and applications. Its purpose is to enhance your understanding of the structure of the integers and to indicate a number of other directions that can be pursued. For more information, three excellent number theory texts are listed in the references.

Basic Concepts

Definition   Let a and b be integers and let m be a positive integer. a **is congruent to** b **modulo** m, denoted $a \equiv b(\text{mod } m)$, if $m \mid (a - b)$.

EXAMPLE 1   Show that $9 \equiv 1(\text{mod } 8)$, $1 \equiv 9(\text{mod } 8)$, and $16 \equiv 0(\text{mod } 8)$.

SOLUTION   Since $8 \mid (9 - 1)$, $9 \equiv 1(\text{mod } 8)$. And since $8 \mid (1 - 9)$, $1 \equiv 9 \ (\text{mod } 8)$. Lastly, since $8 \mid (16 - 0)$, $16 \equiv 0(\text{mod } 8)$.   ∎

EXAMPLE 2   Show that for any integer k, $k + 8 \equiv k(\text{mod } 8)$.

SOLUTION $(k + 8 - k) = 8$ and hence is divisible by 8. Thus $k + 8 \equiv k(\text{mod } 8)$.   ∎

THEOREM 5.11   Let m be any positive integer.

a) Every integer, z, is congruent mod m to exactly one of $\{0, 1, 2, 3, \ldots, m - 1\}$.

b) $a \equiv b(\text{mod } m)$ if and only if there is an integer k such that $a = b + km$.

c) $a \equiv b(\text{mod } m)$ if and only if both a and b leave the same remainder upon division by m.

PROOF

a) By the division algorithm, there are unique q and r such that $z = qm + r$, $0 \le r < m$. Hence $qm = z - r$ and hence $m \mid z - r$. Thus $z \equiv r(\text{mod } m)$ and $r \in \{0, 1, 2, \ldots, m - 1\}$.

b) $a \equiv b(\mod m) \Leftrightarrow m \mid (a - b) \Leftrightarrow a - b = km$ for an integer $\Leftrightarrow$ $a = b + km$.

c) Suppose first that $a \equiv b(\mod m)$. By the division algorithm, $a = mq + r$, $0 \le r < m$. By part (b), $b + km = mq + r$ and hence $b = m(q - k) + r$. Thus a and b yield the same remainder upon division by m. As for the converse, suppose both leave the same remainder, r, upon division by m. Then $a = mq + r$ and $b = mq' + r$. Hence $(a - b) = m(q - q')$. Thus $m \mid (a - b)$ and thus $a \equiv b(\mod m)$. ☐

When we write $b \mod m$ (as in 8 div 5 = 1, 8 mod 5 = 3) we are referring to the integer z of the first part of this theorem—the unique integer in $\{0, \ldots, m - 1\}$ such that $z \equiv b(\mod m)$. Using this notation, part (c) could be stated as $a \equiv b(\mod m)$ if and only if $a \mod m = b \mod m$. $b \mod m$ is also called the **least residue of b modulo m**.

A Geometric Interpretation of Congruence

The underlying relationship in Theorem 5.11 is shown in Figure 5.12: Imagine a circle of circumference n. If the number line were wrapped around this circle, n would fall on 0, $n + 1$ on 1, $n + 2$ on 2, $2n$ on 0, $2n + 1$ on 1, etc. Numbers that "land on top of each other" are congruent mod n. It is this cyclic quality that is the essence of congruence. Note that, in terms of Theorem 5.11, part (a) states that every integer lands on top of one of the integers $0, \ldots, n - 1$. Part (b) states that two numbers that land on the same integer differ by a multiple of n that can be thought of as the number of times the number line is wrapped around the circle between the two numbers. Part (c) gives us a nice geometric interpretation of the division algorithm: If we write $a = nq + r$, q is the number of times we must

Figure 5.12 Illustration of congruences modulo n.

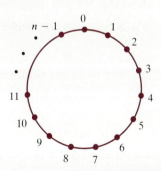

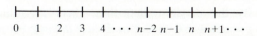

wrap the number line around a circle of circumference n until a is placed on the circle. r is the distance around the circle from 0 at which n is placed. Since n can be any integer, this cyclic quality is an intrinsic and important property of the integers.

EXAMPLE 3 Determine whether $143 \equiv 13(\text{mod } 11)$.

SOLUTION   11 does not divide $(143 - 13)$. Hence 143 is not congruent to $13(\text{mod } 11)$. Alternatively, using Theorem 5.11c, 13 mod $11 = 2$, while 143 mod $11 = 0$. Thus they do not leave the same remainder upon division by 11 and hence are not congruent mod 11. ∎

EXAMPLE 4 Find five distinct integers, each of which is congruent to $13(\text{mod } 11)$.

SOLUTION   Using Theorem 5.11b, we need only select five different values of k for the equation $a = 13 + 11k$. Using $k = 1, 2, 3, -1$, and -2, we get 24, 35, 46, 2, and -9 as the five integers. ∎

Recall that in Chapter 2 we defined the set $\mathbf{Z}_n$ to be $\{0, 1, 2, \ldots, n - 1\}$. We can now see why $\mathbf{Z}_n$ is called the set of **integers mod n.** $\mathbf{Z}_n$ is sometimes denoted $\{\overline{0}, \overline{1}, \overline{2}, \ldots, \overline{n-1}\}$, where $\overline{0}$ denotes the set of all integers congruent to 0 mod n, namely, $\{-2n, -n, 0, n, 2n, 3n, \ldots\}$, $\overline{1}$ denotes the set of all integers congruent to 1 mod n, namely, $\{-2n + 1, -n + 1, 1, n + 1, 2n + 1, \ldots\}$, etc. In this interpretation, each element of $\mathbf{Z}_n$ is thought of as a set of integers congruent to an integer i such that $0 \leq i \leq n - 1$.

It is because of the cyclic property that congruence is so important. Positional number systems exploit this property. Counting from 0 to 10 (base 10) causes one to cycle through the digits from 0 to 9 and return to 0. The odometer on a car is a physical implementation of this property. The digits from 0 through 9 are placed on a circle, and as it rotates, counting takes place. When a circle passes from 9 to 0, the next circle to the left is incremented by 1. Ultimately the odometer returns to zero again (if the car lasts for 100,000 miles). Thus the entire odometer is an implementation of $\mathbf{Z}_{100,000}$, just as each ring of digits that make it up are implementations of $\mathbf{Z}_{10}$. Computer science is also very dependent on this property. For instance, a byte is an eight-bit number ranging from 00000000 to 11111111; adding 1 more to 11111111 brings it back to 00000000 again. This transition is usually recorded as an overflow. But counting in a computer acts on exactly the same principle as it does in the odometer. Also, no matter how large the computer is, it is still a finite machine. Thus every effort to deal with the integers is basically an approximation of the integers by $\mathbf{Z}_n$ for some large n. This fact, combined with the cyclic nature of $\mathbf{Z}_n$, is the basis for the standard algorithms used to generate pseudo-random numbers. Furthermore, periodicity occurs so frequently in nature that congruence appears in many applications of discrete mathematics.

EXAMPLE 5   If we count in base 5 starting from 0, we get the sequence 0, 1, 2, 3, 4, 10, 11, 12, etc. Note that the rightmost digits cycle through $\mathbf{Z}_5$, and the digit to its left indicates the number of times we have cycled through.   ∎

Further Properties of Congruence

The following properties of congruence are often useful:

THEOREM 5.12   Let a, b, c, and m be integers, $m > 0$. If $a \equiv b(\text{mod } m)$, then

a) $(a + c) \equiv (b + c) \ (\text{mod } m)$.

b) $ac \equiv bc(\text{mod } m)$.

PROOF

a) Since $m \mid (a - b)$, $m \mid ((a + c) - (b + c))$.

b) (Left for the Exercises.)   ☐

EXAMPLE 6   Knowing that $6 \equiv 1(\text{mod } 5)$, we can apply part (a) with $c = 3$ and conclude that $9 \equiv 4(\text{mod } 5)$. Applying part (b) with $c = 7$, we have $42 \equiv 7(\text{mod } 5)$.   ∎

THEOREM 5.13   If $ac \equiv bc(\text{mod } m)$ and $(c, m) = d$, then $a \equiv b(\text{mod } m/d)$.

PROOF   If $ac \equiv bc(\text{mod } m)$, then $m \mid (ac - bc)$ and hence $m \mid c(a - b)$. Now $d \mid c$ and $d \mid m$. For some k, $c(a - b) = km$; hence $(c/d)(a - b) = k(m/d)$. Since c/d and m/d are relatively prime, $(m/d) \mid (a - b)$, and thus $a \equiv b(\text{mod } m/d)$.   ☐

EXAMPLE 7 $121 \equiv 66 \ (\text{mod } 55)$ and $(11, 55) = 11$. Hence $11 \equiv 6 \ (\text{mod } 5)$.   ∎

The next theorem summarizes several properties that will be especially important in Chapter 8.

THEOREM 5.14   Let a, b, c, and m be integers, $m > 0$. Then

a) $a \equiv a(\text{mod } m)$.

b) $a \equiv b(\text{mod } m) \Rightarrow b \equiv a(\text{mod } m)$.

c) $a \equiv b(\bmod\ m)$ and $b \equiv c(\bmod\ m) \Rightarrow a \equiv (c\ \bmod\ m)$.

PROOF   (Left for the Exercises.) ☐

Arithmetic in Z_n

Definition   Define $a \oplus b$ to mean $(a + b)\ (\bmod\ n)$.

Note that $\mathbf{Z}_n$ is closed under this definition of addition, since $a \oplus b$ is an integer and every integer is congruent mod n to some element of $\mathbf{Z}_n$. For instance, in $\mathbf{Z}_{12}$, $10 \oplus 3 = 1$. That is, we start at 10 and advance three more places around the circle, ending at 1. The fact that we pass 0 en route means that we subtract 12 from what would be the sum in $\mathbf{Z}$.

EXAMPLE 8   Compute the following sums in $\mathbf{Z}_8$:

$$3 \oplus 5$$
$$6 \oplus 7$$
$$2 \oplus 5$$

SOLUTION

$$3 \oplus 5 = (3 + 5)\ \bmod\ 8 = 8\ \bmod\ 8 = 0$$
$$6 \oplus 7 = (6 + 7)\ \bmod\ 8 = 13\ \bmod\ 8 = 5$$
$$2 \oplus 5 = (2 + 5)\ \bmod\ 8 = 7\ \bmod\ 8 = 7. \quad ∎$$

Further study of the properties of arithmetic in $\mathbf{Z}_n$ leads to the study of the theory of **groups.** The study of groups is a branch of mathematics in itself and one of the most fruitful in consolidating and integrating results from many diverse situations. We will not pursue it further here; for more information, see a textbook on abstract algebra, such as the ones listed in the References.

TERMINOLOGY

a congruent to b modulo m $(a \equiv (b\ \bmod\ m))$

least residue of b modulo m

integers mod m

group

complete residue system modulo n

SUMMARY

1. Two integers are congruent modulo n if, in wrapping a number line around a circle of circumference n, they fall at the same point on the circle. This cyclic property of the integers mod n can be exploited in many ways including answering theoretical questions about the integers, designing mechanical counting devices such as odometers and computers, and writing algorithms.

2. **a)** Every integer, z, is congruent mod m to one of $\{0, 1, 2, 3, \cdots, m - 1\}$.

   b) $a \equiv b(\mod m)$ if and only if there is an integer k such that $a = b + km$.

   c) $a \equiv b(\mod m)$ if and only if both a and b leave the same remainder upon division by m.

   d) If $a \equiv b(\mod m)$, then $(a + c) \equiv (b + c) \ (\mod m)$.

   e) If $a \equiv b(\mod m)$, then $ac \equiv bc(\mod m)$.

   f) If $ac \equiv bc(\mod m)$ and $(c, m) = d$, then $a \equiv b(\mod m/d)$.

   g) $a \equiv a(\mod m)$.

   h) $a \equiv b(\mod m) \Rightarrow b \equiv a(\mod m)$.

   i) $a \equiv b(\mod m)$ and $b \equiv c(\mod m) \Rightarrow a \equiv c(\mod m)$.

EXERCISES 5.5

Find the least residues of:

1. 13 mod 5   **2.** 37 mod 13

3. 42 mod 14   **4.** 104 mod 12

5. n^2 mod n **6.** -2 mod 7

Show whether each of the following is true, or false.

7. $17 \equiv 2(\mod 5)$ **8.** $51 \equiv 0(\mod 17)$

9. $-88 \equiv 8(\mod 8)$ **10.** $an \equiv a(\mod n)$

Each of the following is of the form $a \equiv b(\mod n)$. Find k such that $a = b + nk$.

11. $19 \equiv 5(\mod 7)$ **12.** $81 \equiv 0(\mod 9)$

13. $-3 \equiv 5(\mod 8)$ **14.** $-102 \equiv 8(\mod 11)$

A **complete residue system modulo n** is a set $\{\overline{m}_1, \ldots, \overline{m}_n\}$ such that each of $\{\overline{0}, \overline{1}, \ldots, \overline{n-1}\}$ is included in it. Which of these are complete residue systems modulo the given value of n?

15. $\{0, 3, 6, 9, 12\}$; $n = 5$ **16.** $\{1, 8, -2, -3\}$; $n = 4$

17. $\{1, 8, 13, 26, 27, 37, 54\}$; $n = 8$ **18.** $\{22, 24, 26, 28, 30, 32, 34\}$; $n = 7$

Find values of x such that the following are true:

19. $18 \equiv x(\mod 7)$ **20.** $31 \equiv 3(\mod x)$

21. $-17 \equiv 3(\mod x)$ **22.** $-21 \equiv x(\mod 11)$

Add the following in Z_{12}.

23. $8 \oplus 7$

24. $11 \oplus 1$

25. $3 \oplus 8$

26. $1 \oplus 2 \oplus \cdots \oplus 11$

Prove:

27. $a \equiv b(\bmod\ m) \Rightarrow ac \equiv bc(\bmod\ m)$

28. $ac \equiv bc(\bmod\ m) \nRightarrow a \equiv b(\bmod\ m)$

29. $a \equiv a(\bmod\ m)$

30. $a \equiv b(\bmod\ m) \Rightarrow b \equiv a(\bmod\ m)$

31. $a \equiv b(\bmod\ m)$ and $b \equiv c(\bmod\ m) \Rightarrow a \equiv c(\bmod\ m)$.

REVIEW EXERCISES—Chapter 5

Prove or disprove the following.

1. $8 \mid n^2 - 1$; n any odd integer

2. $64 \mid (n^2 - 1)(m^2 - 1)$; n, m odd integers

3. $x = y + z$ and a divides any 2 of x, y, and $z \Rightarrow a$ divides the third

4. Every integer divides zero.

Find the GCD, d, for the following pairs of numbers, and find x and y such that $ax + by = d$.

5. $a = 18$, $b = 216$

6. $a = 91$, $b = 312$

7. $a = 19$, $b = 489$

8. $a = 720$, $b = 1080$

Prove each of the following.

9. If p and q are prime and $p \mid q$, then $p = q$.

10. If $b \mid a$ and $b \mid a + 2$, then $b = 1$ or $b = 2$.

11. $a \mid c$ and $b \mid c \nRightarrow ab \mid c$. Find a property of a and b that would make this relationship true and prove that it works.

Add the following numbers.

12. $62_8 + 31_8$

13. $1001_2 + 1010_2$

14. $114_5 + 331_5$

15. $3AE_{16} + 432_{16}$

Multiply the following numbers.

16. $1001_2 * 1010_2$

17. $114_5 * 331_5$

Using the standard algorithm for subtraction, compute the following.

18. $2113_5 - 323_5$

19. $110011_2 - 101010_2$

Using the tables in the text and generalizing the standard algorithm for division of integers, compute the following.

20. $2113_8/323_8$

21. $110011_2/101_2$

Change the following numbers from the indicated base, b, to base 10 by substitution for b.

22. 1234_8

23. 4310_5

24. $3F07_{16}$

25. 1100110_2

Change the following numbers from base 10 to the indicated base by using division.

26. 4181; base 8

27. 5081; base 16

28. 65; base 2

29. 99; base 5

Find the least residues of the following.

30. $(n + k)\bmod n$

31. $-99 \bmod 12$

Show whether each of the following congruences is true or false.

32. $3 \equiv 3 \bmod 11$

33. $a^2 \equiv a \bmod n$

Add the following in $\mathbf{Z}_{16}$.

34. $9 \oplus 10$

35. $8 \oplus 5$

36. $11 \oplus 15$

37. $15 \oplus 14$

Prove the following.

38. $p \mid a \Leftrightarrow a \equiv 0 \bmod p$

39. If $p(x)$ is a polynomial with integer coefficients and $a \equiv b \bmod m$, then $p(a) \equiv p(b) \bmod m$.

REFERENCES

Dudley, Underwood, *Elementary Number Theory.* San Francisco, Calif.: W.H. Freeman, 1978.

Eves, Howard, and Carroll V. Newsom, *An Introduction to the Foundations and Fundamental Concepts of Mathematics.* New York: Holt, Rinehart, and Winston, rev. ed., 1965.

Fraileigh, John B., *A First Course in Abstract Algebra.* Reading, Mass.: Addison-Wesley, 1982.

Gallian, Joseph A., *Contemporary Abstract Algebra.* Lexington, Mass.: D.C. Heath, 1986.

Herstein, I. N., *Topics in Algebra.* Waltham, Mass.: Ginn, 1964.

LeVeque, William Judson, *Elements of Number Theory.* Reading, Mass.: Addison-Wesley, 1962.

Rosen, Kenneth H., *Elementary Number Theory and Its Applications.* Reading, Mass.: Addison-Wesley, 1984.

Recursion

6

In Section 5.1, we saw that the natural numbers can be **defined recursively;** that is, 1 was taken as a primitive and the rest of the natural numbers were defined by the use of the successor function. Thus every natural number except 1 was defined in terms of its predecessor. Since the natural numbers are very important in discrete mathematics and are themselves defined recursively, recursive definitions arise very often in discrete mathematics. In fact, whenever we want to define a concept, function, or algorithm in terms of what it does to an arbitrary natural number, n, we are likely to use a recursive definition. Since recursion is so fundamental, this chapter will address it. First, we will look at recursive definitions of functions, and then we will turn our attention to recursively defined algorithms. As for concepts that are defined recursively, we will encounter them throughout the rest of this book and will address them as they arise.

6.1

INTRODUCTION TO RECURRENCE EQUATIONS

Once again, let's start with a problem:

Suppose n *diplomats are at a party, and during the course of the festivities each shakes hands with every other diplomat exactly once. How many handshakes occur?*

Before reading on, see if you can solve this problem yourself.

One way to approach this problem is as follows. (We will see another in the next chapter.) First, suppose that there are only two diplomats. Then there is only one handshake. Now suppose that there are $n - 1$ diplomats and let $h(n - 1)$ denote the number of handshakes that occur. Then, if one new diplomat arrives, $h(n)$ handshakes will occur. But the nth diplomat has to shake hands with all of the $n - 1$ people already present. Also, the same number of handshakes occurred among the other $n - 1$ people as would have occurred if the nth diplomat had not come. Thus $h(n) = (n - 1) + h(n - 1)$. Combining these observations we derive a pair of equations of the form

$$h(2) = 1$$
$$h(n) = (n - 1) + h(n - 1), n > 2.$$

Letting n first be 3, then 4, etc., we get the following:

$$h(3) = 2 + 1$$
$$h(4) = 3 + 2 + 1$$
$$h(5) = 4 + 3 + 2 + 1$$

and we can infer that, in general,

$$h(n) = (n - 1) + (n - 2) + \cdots + 2 + 1.$$

Thus we conclude that the number of handshakes is $n(n-1)/2$ (although we haven't proven it, since we jumped from $h(5)$ to $h(n)$; we will prove it in Example 3 of Section 6.2).

Note that our recursive definition of $h(n)$ in this example consisted of two parts—an equation for $h(n)$ in terms of $h(n-1)$ and a value for $h(2)$. This situation is formalized in the following definition:

Definition Let f be a function with domain $\mathbf{N}$ and suppose that for some n_0 and for all $n > n_0$, $f(n)$ is defined in terms of one or more of $f(n-1)$, $f(n-2)$, ..., $f(1)$. Then f is said to be **recursively defined**. The equation that expresses $f(n)$ in terms of $f(n-1), \ldots, f(1)$ is called a **recurrence relation** or **recurrence equation**. If, in addition, one or more values are given for specific initial n's (such as $f(1)$, $f(2)$, etc.), these are called **initial conditions**.

EXAMPLE 1 In the handshake example,

$$h(n) = (n-1) + h(n-1), \, n > 2, \, h(2) = 1,$$

$h(n) = (n-1) + h(n-1)$, $n > 2$ is the recurrence equation. There is only one initial condition, that $h(2) = 1$. ■

At times, recurrence equations are given for functions defined on $\mathbf{N} \cup \{0\}$ or $\mathbf{N} \setminus \{1\}$ or other such domains. The definition given previously can be easily generalized to handle such cases. Also, recurrence equations frequently occur for functions that are not defined on the integers but rather are defined on sets of the form $\{0, t, 2t, 3t, 4t, \ldots\}$ where t is a real number that may be either quite large or quite small. Such functions can also be treated by the ideas developed here. For instance, suppose f is defined by the equation $f(nt) = g(n, f((n-1)t))$. If we treat t as a constant and let $h(n)$ denote $f(nt)$, we can rewrite this equation as $h(n) = g(n, h(n-1))$. h is then defined on the domain $\mathbf{N} \cup \{0\}$ and is recursively defined.

EXAMPLE 2 Here are several examples of recurrence equations with initial conditions:

$$f(n) = 3 + f(n-1), \, f(1) = 0$$
$$f(n) = 2 f(n-1), \, f(1) = 1$$
$$f(n) = 4 + 2 f(n-1), \, f(3) = 4$$
$$f(n) = 2n + 1 + f(n-1), \, f(2) = 2$$
$$f(n) = n! + n + 1 + f(n-1), \, f(1) = 0$$
$$f(n) = f(n-1) + f(n-2), \, f(1) = 1, \, f(2) = 1$$
$$f(n) = 4 f(n-1) - 4 f(n-2), \, f(0) = 0, \, f(1) = 4 \quad \blacksquare$$

Note that these equations could have been written

$$f(n + 1) = 3 + f(n), f(1) = 0$$
$$f(n + 1) = 2 f(n), f(1) = 1$$

.

.

.

by replacing n by $n + 1$. We will use both forms at different times.

Sources of Recurrence Equations

Recursion is a phenomenon that is so common that we easily miss it. Then when it finally does catch our attention, we begin to see it everywhere. For instance, here are some situations that involve recursion:

The amount of corn that a farmer can grow in a field can be estimated on the basis of what the field was able to produce over the last few years plus other factors such as weather, amount of fertilizer used, etc.

A child's height this year depends on his or her height last year plus other factors such as age, genetics, diet, etc.

The amount of work completed on a construction project as of the end of day n depends on the amount completed at the end of day $n - 1$ plus what was done on the nth day.

The amount a rat has learned after n attempts to run a maze depends on what was learned after $n - 1$ trials plus a factor depending on the rat's experience on the nth trial.

The progress a student has made toward completion of a computer program or English paper after n hours depends on the progress made after $n - 1$ hours plus the progress during the nth hour.

The national income in any quarter depends on the previous quarter's income plus the consumption and investment occurring during that quarter.

Notice that in all of these examples the independent variable is time. While the time interval may vary in length, in each case some quantity is being measured that depends on time, and that quantity is examined periodically. Attempts to model such situations mathematically are a major source of recurrence equations, since much of what is studied, notably in the social sciences and biology, is the result of periodic measurements. But not all recursive definitions of functions depend on time. For instance:

The cost of moving n pieces of heavy equipment depends on the cost of moving $n - 1$ pieces plus an incremental cost for moving the nth piece.

The gasoline efficiency of a truck with n units of weight is measured in terms of gallons consumed per unit of weight. Thus it depends on the efficiency with $n-1$ units minus some factor for the nth unit (i.e., assuming it is more efficient with more weight).

The cost of ordering n items in a bulk shipment depends on the cost of ordering $n-1$ items plus an incremental cost.

The time required for an investment to grow to n dollars in value depends on the time required to grow to $n-1$ dollars plus a factor that depends on n.

The time required for a certain algorithm to sort n items depends on the time it takes to sort $n-1$ items plus the additional time required to put the nth in its proper place.

Note that in the last two examples, time is the dependent variable, whereas the number of dollars or the size of a list is the independent variable.

At this point, we have seen that there are many situations in which quantities that we may want to study can be modeled by recursively defined functions. Following are a number of examples that will show more precisely how the recurrence equation can be found. We will see how to solve these equations in the next three sections.

EXAMPLE 3 Consider the traveling salesman problem again. Find a recurrence equation for the number of routes that must be examined by the brute force algorithm.

SOLUTION Recall that in the brute force algorithm each route that visits every city once and returns to the original starting point must be examined. Let $f(n)$ denote the number of possible routes for an n-city problem. Since with one city there are no routes and with two cities there is one route, $f(1) = 0$ and $f(2) = 1$. In order to see the pattern, consider a four-city problem (Figure 6.1). The number of routes to be examined is $f(4)$. One such route is $ABCDA$. Let us focus on this route for a moment. If a fifth city, E, is added to the problem, we must decide if it is better to go directly from A to E, or go to B first and then E, or go to B and C first and then E,

Figure 6.1 Map of four cities.

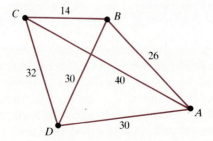

etc. That is, we must check the following routes: *AEBCDA, ABECDA, ABCEDA,* and *ABCDEA.*

To find the total number of routes for five cities, we can repeat the same process for each route in the four-city problem; all routes must be checked with *E* in each of the four different positions in which it could be placed. (One of the exercises asks you to show that no routes are counted twice in this process.) Thus we can say that $f(5) = 4f(4)$. For *n* cities, we can follow the same pattern. Each of the $f(n-1)$ routes associated with $n-1$ cities has to be modified to allow for the *n*th city occurring in any of $n-1$ positions. Hence we get the recurrence equation

$$f(2) = 1$$
$$f(n) = (n-1)f(n-1). \quad \blacksquare$$

EXAMPLE 4   Recurrence equations provide a clear way to illustrate the difference between simple and compound interest. Suppose an amount of money P_0 is invested at interest rate *r*. Let P_n denote the value of the investment after *n* periods, which could be years, months, days, etc. With simple interest, the amount of interest is computed using the original principal, P_0. That is,

$$P_{n+1} = P_n + rP_0$$

or

$$P_{n+1} = c + P_n \qquad \text{where } c \text{ denotes } rP_0.$$

Compound interest is computed using the current principal P_n. That is,

$$P_{n+1} = P_n + rP_n$$
$$P_{n+1} = (1+r)P_n$$

or

$$P_{n+1} = c' P_n \qquad \text{where } c' \text{ denotes } (1+r).$$

Thus, in computing simple interest, a constant amount is added to the principal at each interval; in computing compound interest, the principal is multiplied by a constant at each interval. This process is sometimes described by saying that simple interest follows an arithmetic progression, while compound interest follows a geometric progression. $\blacksquare$

EXAMPLE 5   In amortization of a loan, a fixed payment is made in each repayment period to pay both the interest due and part of the principal. Thus, at the end of each period, the amount of interest due becomes smaller. Hence the amount of the fixed payment applied to the principal becomes larger. The object is to select the fixed payment in such a way that the principal will be completely paid off at the end of a specified number of periods. Let P_n stand for the amount of principal still outstanding after n periods and let P_0 denote the original indebtedness. Let R denote the fixed payment and r the interest rate. Then we get the following recurrence equation:

$$\text{New principal} = \text{old principal} + \text{interest} - \text{payment}$$
$$P_{n+1} = P_n + rP_n - R$$
$$= (1 + r)P_n - R.$$

For instance, if r is 7% and R is $500 per month, the equation becomes

$$P_{n+1} = 1.07P_n - 500.$$

The object is, first, to solve the equation $P_{n+1} = (1 + r)P_n - R$ for P_n in terms of n, r, and R. If the debt is to be paid off in N periods, set $P_N = 0$ and solve for R.

There is an alternative way to look at this problem. Instead of trying to write P_{n+1} in terms of P_n, we let ΔP_n (Δ is read *delta*) denote the change in the principal as we go from the nth to the $(n + 1)$th period. There are two sources of this change: The debt will increase due to the interest and will decrease because of the payment. Hence

$$\Delta P_n = rP_n - R.$$

This can be written

$$P_{n+1} - P_n = rP_n - R$$

or

$$P_{n+1} = P_n + rP_n - R,$$

which is the same equation as before, although it is obtained in a different and perhaps easier way. ∎

Note that in the previous examples, the notation P_n was used, while in the other examples, the notation $f(n)$ was used. Functions whose domain is **N** can be described using either subscripts or function notation. In this text, the function notation will most often be used.

Definition Suppose f is a function with domain **N**. The **first difference of f** is a function whose value at n is given by $f(n + 1) - f(n)$. We denote this function Δf and its value at n by $\Delta f(n)$ or Δf_n. An equation of the form $\Delta f(n) = g(n, f(n), f(n - 1), \ldots)$ is called a **difference equation** or sometimes a **finite difference equation**.

The equation $\Delta P_n = rP_n - R$ from the previous example is an example of a difference equation. Note that any difference equation is equivalent to a recurrence equation, and vice versa. That is,

$$\Delta f(n) = g(n, f(n), f(n-1), \ldots)$$

can be written

$$f(n+1) = f(n) + g(n, f(n), f(n-1), \ldots)$$

which is a recurrence equation. Similarly,

$$f(n) = g(n, f(n-1), f(n-2), \ldots, f(2), f(1))$$

can be written as a difference equation by subtracting $f(n-1)$ from both sides. Older texts on this topic tend to talk primarily about difference equations; newer ones talk about recurrence equations or recurrence relations. This reflects an important change in mathematical thinking since the 1950s; difference equations are primarily seen as an approximation of differential equations, a calculus topic. Recurrence equations are seen as an important topic in their own right. The shift in names suggests the growing recognition of the importance of discrete mathematics.

There are some situations in which the recurrence equation can be derived directly and others that are easier to express in terms of differences (see Example 6).

EXAMPLE 6 Consider the following table:

n	$f(n)$	$\Delta f(n)$
1	0	3
2	3	5
3	8	7
4	15	9
5	24	11
.	.	.
.	.	.
.	.	.

Express f using a difference equation.

SOLUTION   It can be seen immediately that

$$\Delta f_n = 2n + 1$$

and hence

$$f_{n+1} = f_n + 2n + 1.$$

We first looked at this table in Chapter 2, where you were asked to guess the next value. At that time, we inferred that $f(n) = n^2 - 1$. But that equation is only one way to express the pattern observed in the table. It is equally legitimate to say that the differences between successive values of $f(n)$ increase by two with each step. Difference equations provide a mathematical tool to formulate this way of looking at the pattern. ∎

EXAMPLE 7 A population (e.g., of rabbits) reproduces at certain seasons of the year. Model the growth of this population.

SOLUTION As stated, the question is not well specified; this vagueness is typical of many situations we are asked to model. Thus, before going further, we must make some simplifying assumptions. Suppose that the rabbits have a known average litter size, L, which is independent of the season. Suppose that the rabbits mature in one generation and that any that do not survive the first generation are not included in L. Also, suppose that every female has a litter in each generation and that there is an equal number of males and females. Lastly, assume that there is a constant death rate; i.e., from one season to the next, a constant proportion, r, of the total population will not survive. Let P_n denote the population in the nth generation. Then the number of females in the nth generation is $(1/2)P_n$, and thus the number of newborn rabbits is $L(1/2)P_n$. Similarly the number that die is rP_n. Hence

$$\Delta P_n = L(1/2)P_n - rP_n = (L/2 - r)P_n$$

or

$$P_{n+1} = P_n + (L/2 - r)P_n = (L/2 - r + 1)P_n.$$

For instance, if the average litter size is four and the death rate $r = 1/2$, $P_{n+1} = 5/2\ P_n$. ∎

EXAMPLE 8 Here is a simplified model of a deer population using recurrence equations. Suppose that deer reproduce once a year (in the spring) and that every female 1 year old or more gives birth to exactly two fawns. Suppose that males and females occur in equal numbers in the population and that each deer dies after its second year. Also, suppose that only a proportion, k_1, of the newborn survive to the next spring. Suppose a proportion, k_2, of 1-year-olds survive to the next year. Model the deer population using recurrence equations.

SOLUTION One helpful way to look at this problem is to break up the population into **cohorts**; i.e., in year n, we split the population into 0-year-olds, 1-year-olds, and 2-year-olds. We denote the entire population in year

n by P_n and we denote the cohorts by z_n, y_n, and t_n. From the survival ratios given, we get

$$y_n = k_1 z_{n-1} \tag{6.1}$$

and

$$t_n = k_2 y_{n-1} = k_2 k_1 z_{n-2}. \tag{6.2}$$

We also get the following equation for the number of newborn in year n:

$$z_n = 2(1/2)y_n + 2(1/2)t_n = y_n + t_n. \tag{6.3}$$

Thus, since

$$P_n = z_n + y_n + t_n, \tag{6.4}$$

we can combine the last two equations and conclude that

$$z_n = (1/2)P_n. \tag{6.5}$$

Similarly

$$z_{n-1} = (1/2)P_{n-1} \tag{6.6}$$
$$z_{n-2} = (1/2)P_{n-2}. \tag{6.7}$$

Combining Eqs. (6.1), (6.2), and (6.3), we get

$$z_n = k_1 z_{n-1} + k_2 k_1 z_{n-2}. \tag{6.8}$$

Substituting Eqs. (6.5), (6.6), (6.7) into Eq. (6.8), we have

$$(1/2)P_n = k_1(1/2)P_{n-1} + k_2 k_1(1/2)P_{n-2}.$$

Thus

$$P_n = k_1 P_{n-1} + k_2 k_1 P_{n-2}.$$

To compute the population in year n, we must know the population in the previous 2 years. However, once we know that, we can compute the population for all succeeding years. This, then, is an example of a deterministic model. The fact that it is deterministic shows some of its limitations. Unpredictable factors such as the intensity of the winter have a great impact on a deer population. However, it is a reasonable starting point and gives us some insight into deer population dynamics. It is also gives us insight into the modeling process. ∎

EXAMPLE 9   Consider the sequence of integers

$$0, 1, 1, 2, 3, 5, 8, 13, 21, 34, \ldots$$

This is known as the **Fibonacci sequence** and models a number of natural phenomena. For instance, the number of seeds in the successive rings of a sunflower follow this pattern. The sequence satisfies the following recurrence equation and initial conditions:

$$f(n) = f(n-1) + f(n-2), f(0) = 0, f(1) = 1.$$

We will discuss this example more extensively in Sections 6.4 and 6.5. ∎

TERMINOLOGY

recursive definition	first difference of f
recursively defined function	difference equation
recurrence relation	finite difference equation
recurrence equation	cohort
initial conditions	Fibonacci sequence

SUMMARY

1. Since the integers are themselves recursively defined, many concepts, functions, and algorithms that occur in discrete mathematics are also recursively defined.

2. Any situation in which the current state of an entity one wishes to model depends (at least in part) on its previous states is very likely to lead to a recurrence equation.

EXERCISES 6.1

For Exercises 1–5, write a recurrence equation and an initial condition that model the given situation and assumptions.

1. The cost of moving n items of heavy equipment depends on the cost of moving $n-1$ items plus an incremental cost for the nth item. Assume that the incremental cost is constant.

2. This is the same as Exercise 1, but suppose that the incremental cost is proportional to $1/n$.

3. The gasoline efficiency of a truck transporting n units of weight equals its efficiency in transporting $n-1$ units minus an incremental cost for the nth unit. Assume that the incremental cost is inversely proportional to n^2.

4. This is the same as Exercise 3, but suppose that the incremental cost is zero.

5. The time that a certain algorithm requires to sort n items depends on the time it requires to sort $n - 1$ items plus the time needed to put the nth item in its proper place. Suppose that the time required for the nth is proportional to n.

Write a difference equation for the functions whose values are tabulated in Exercises 6–9. Also, rewrite each difference equation as a recurrence equation.

6. n	$f(n)$	7. n	$f(n)$	8. n	$f(n)$	9. n	$f(n)$
1	1	1	6	1	1	1	1
2	4	2	5	2	7	2	1
3	7	3	4	3	17	3	2
4	10	4	3	4	31	4	6
						5	15

10. In Example 3, show that no route is counted twice when the paths for the n-city problem are generated by inserting an nth city in every possible position in the collection of all routes for the $n - 1$ city problem.

11. In Example 7, it was assumed that every female has a litter of size L. Suppose, instead, that litter size decreases with population, i.e., that the litter size is inversely proportional to the population. Find an expression for P_n in this case.

12. Modify the equation of Example 7 in such a way that the death rate, r, is proportional to the size of the population instead of being constant.

Let S_n denote the finite series $\sum_{i=1}^{n} a_i$. Then the series can also be represented by the difference equation $S_n = a_n + S_{n-1}$. Represent each of the following series by such an equation.

13. $\sum_{i=1}^{n} i$

14. $\sum_{i=0}^{n} i$

15. $\sum_{i=0}^{n} i^2$

16. $\sum_{i=2}^{n} \log_2 i$

17. In a particular decision tree, there are two choices at each choice point. Write a recurrence equation for the total number of options available after n choice points. (See Figure 6.2.)

18. Modify your answer to Exercise 17 to allow N_i choices at the ith choice point.

19. A ball is dropped from height h. On any bounce, it rebounds to 60% of the height of its previous bounce. Write a recurrence equation for the height it attains on the nth bounce.

20. In the Towers of Hanoi puzzle, n disks of decreasing size are placed on one of three pegs. The disks can be moved, one at a time, to another peg. The object is to move all n disks from a first peg to a second peg, one at a time, without ever putting a larger disk on top of a smaller one. (See Figure 6.3.) The strategy is recursive. That is, if there is only one disk, move it. If there are $n > 1$ disks,

Figure 6.2 Tree diagram with two options at each choice point.

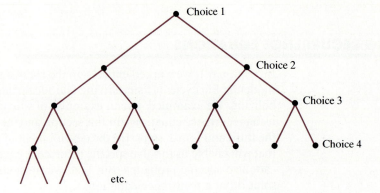

Figure 6.3 Towers of Hanoi puzzle with eight disks.

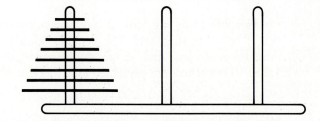

move the first $n - 1$ to the third peg, move the nth to the second peg, and then move the $n - 1$th to the second peg. Write a recurrence equation and an initial condition for $s(n)$, the number of moves required to implement this strategy.

21. n lines are drawn on a plane in such a way that no two are parallel and no three intersect at the same point. Write a recurrence equation and an initial condition for the number of regions into which the plane is divided.

22. In the *sinking fund* method of repaying a debt, it is assumed that the entire principal remains outstanding until the debt matures. Each periodic payment consists of an interest payment and a fixed principal payment that is invested at rate r. When the investment accumulates to the point where its value equals the value of the debt, the debt is repaid. Write two difference equations, one for the amount that has accumulated toward the principal after n periods and one for the total amount that has been spent in repaying the debt. Include initial conditions. (This problem is adapted from Samuel Goldberg, *Introduction to Difference Equations: With Illustrative Examples from Economics, Psychology, and Sociology.* John Wiley & Sons, 1958; Dover Publications, Inc., 1986. By permission.)

23. Modify the formula for amortization of a loan (Example 5) to model a situation in which the amount, R, of repayment increases by a fixed amount at each period.

6.2

SOLVING RECURRENCE EQUATIONS

In Section 6.1, we worked entirely on the fist two steps of the mathematical modeling life cycle: (1) simplifying and idealizing a problem and (2) symbolizing. We examined several examples of situations that can be modeled using recurrence equations. In this section and the next two, we move on to the third and fourth steps: (3) the development of theorems and techniques that will enable us to solve specific recurrence equations and (4) their application to specific problems. But first, we must clarify what we mean by a solution of a recurrence equation.

Definition   A **sequence** (or **infinite sequence**) is a function, f, with domain **N**.

In some cases, we look at sequences defined over $\mathbf{N} \cup \{0\}$ or over $\mathbf{N} \backslash \{1\}$ or other such sets. The generalization of this definition to such domains is obvious, and we will say no more about it. When written, the values of sequences are presented either as lists of the form

$$f(1), f(2), f(3), \ldots$$

(as long as the pattern given by the ellipses is clear) or as expressions $f(n)$. In the latter case, we say that the sequence is presented in **closed form.**

EXAMPLE 1   The following are sequences:

a) $1, 2, 3, 4, \ldots$ This is the sequence $f(n) = n$.

b) $0, 1, 1, 2, 3, 5, 8, \ldots$ Each term of this sequence from the third on is the sum of the previous two terms.

c) $0, 3, 8, 15, 24, 35, \ldots$ In closed form, this sequence can be written $f(n) = n^2 - 1$.   ∎

Suppose we have a recurrence equation and initial condition(s) such as the equation

$$h(n) = (n - 1) + h(n - 1), \; n > 2, \; h(2) = 1,$$

which we saw in the last section. The next definition explains what we mean by a solution of such an equation.

Definition   Suppose we have a recurrence equation and initial condition(s). A **solution** of the recurrence equation is a sequence, f, such that its elements satisfy both the equation and its initial conditions. If no initial conditions are specified, we say that a sequence, f, is a solution of the recurrence equation if f is a solution for some set of initial conditions.

EXAMPLE 2 Consider the recurrence equation

$$f(n + 1) = 1 + f(n), \ n \geq 1, f(1) = 1.$$

Then the sequence 1, 2, 3, ... (which can be expressed as $f(n) = n$) is a solution. To see this, we note that $f(1) = 1$, so the initial condition is satisfied, and substituting for f gives $n + 1 = 1 + n$, which is an identity. ∎

EXAMPLE 3 Show that $h(n) = n(n - 1)/2$ is a solution of the equation from the handshake problem.

SOLUTION To verify this, we first check the initial condition: $h(2) = (2 \cdot 1)/2 = 1$. We then proceed by induction. We have already completed the basis step, so we assume that

$$h(i) = i(i - 1)/2, \ i \leq n - 1$$

and seek to show that $h(n) = n(n - 1)/2$.
Then

$$
\begin{aligned}
h(n) &= (n - 1) + h(n - 1) \\
&= (n - 1) + (n - 1)(n - 2)/2 \\
&= (n - 1)(1 + n/2 - 1) \\
&= n(n - 1)/2.
\end{aligned}
$$

Hence the function $h(n) = n(n - 1)/2$ is a solution of the equation. ∎

Existence and Uniqueness

Before attempting to solve an equation, we would like to know that it actually has a solution. This is the **existence** question. We would also like to know if a solution that we have found is the only one or if we can expect to find more than one. This is the **uniqueness** question. Some recurrence equations have unique solutions (such as the two preceding examples), some have no solutions, and some have infinitely many solutions. Before examining techniques for solving recurrence equations, we must consider briefly the existence and uniqueness of solutions.

EXAMPLE 4 Show that the recurrence equation $nf(n + 1) = f(n)$, $n \geq 0$, $f(0) = 0$ has infinitely many solutions.

SOLUTION Taking $n = 0$ in the recurrence equation, we have $0 \cdot f(1) = 0$, which is true no matter what value $f(1)$ assumes. Suppose then that $f(1) = c$ for any constant c. Then

$$1 \cdot f(2) = c,$$
$$2 \cdot f(3) = f(2),$$
$$3 \cdot f(4) = f(3),$$
$$\cdot$$
$$\cdot$$
$$\cdot$$

Hence $f(2) = c, f(3) = c/2, f(4) = c/3 \cdot 2$, etc. Thus we have a solution for any value of c, and hence the equation has infinitely many solutions. ∎

EXAMPLE 5 Show that the recurrence equations

a) $f(n + 1) = 3 + f(n), n \geq 0, f(0) = 1, f(1) = 2$

and

b) $f(n + 1) = \dfrac{1}{1 - f(n)}, n \geq 0, f(0) = 0$

have no solutions with the given initial conditions.

SOLUTION

a) The recurrence equation and the initial condition $f(0) = 1$ would lead us to conclude that $f(1)$ must equal 4, not 2, as required by the second initial condition. Thus the second initial condition is inconsistent with the equation together with the first initial condition.

b) We get $f(1) = 1$ and $f(2) = 1/0$. Since $1/0$ is undefined, this problem has no solution. ∎

Classification of Recurrence Equations

For one important class of equations, the linear equations, the existence of solutions is guaranteed if we have the appropriate initial conditions. We will restrict our attention to this class of equations. After defining a few preliminary concepts, we will return to the existence and uniqueness questions.

Definition A recurrence equation is said to be **linear** if it can be written in the form

$$a_0(n) f(n) + a_1(n) f(n - 1) + \cdots + a_{k-1}(n) f(n - k + 1) + a_k(n) f(n - k) = h(n),$$

where f is a function on $\mathbf{N}$, $a_0(n), \cdots, a_k(n)$, and $k(n)$ are real valued functions of n.

EXAMPLE 6   Consider the following list of recurrence equations:

$$f(n) = 3 + f(n-1) \tag{6a}$$
$$f(n) = 2f(n-1) \tag{6b}$$
$$f(n) = 4 + 2nf(n-1) \tag{6c}$$
$$f(n) = 2n + 1 + f(n-1) \tag{6d}$$
$$f(n) = n! + n + 1 + f(n-1) \tag{6e}$$
$$f(n) = 2 + (f(n-1))^2 \tag{6f}$$
$$f(n) = f(n-1) + f(n-2) \tag{6g}$$
$$f(n) = 4f(n-1) - 4f(n-2) \tag{6h}$$
$$f(n) = n^2 f(n-1) \tag{6i}$$

Equation (6a) can be written

$$f(n) - f(n-1) = 3,$$

which is the form of a linear equation with $k = 1$, $a_0(n) = 1$, $a_1(n) = -1$, and $h(n) = 3$. Equation (6i) is also linear as can be seen if we write it

$$f(n) - n^2 f(n-1) = 0.$$

In this case, $a_0(n) = 1$, $a_1(n) = -n^2$, $k = 1$, and $h(n) = 0$. In fact, all of the preceding equations are linear except for (6f). ∎

A linear recurrence equation is said to have **order k** if k is the largest integer for which both $a_0(n)$ and $a_k(n)$ are nonzero when the equation is written in the preceding form. Equations (6a) through (6e) have order 1 and Eqs. (6g) and (6h) have order 2. A linear recurrence equation is said to have **constant coefficients** if each of $a_0(n), \ldots, a_k(n)$ are constant. If any coefficient is not constant, the equation does not have constant coefficients. All of the preceding equations except (6c) and (6i) have constant coefficients. If $h(n) = 0$, we say that the equation is **homogeneous.** Equations (6b) and (6g) are homogeneous. Equations that are not homogeneous are called **nonhomogeneous.** If an equation is nonhomogeneous, it has another equation associated with it that is formed by setting $h(n) = 0$. This is called the **reduced equation.** For instance, Eq. (6a), $f(n) = 3 + f(n-1)$, is not homogeneous. The equation $f(n) - f(n-1) = 0$ is its reduced equation.

EXAMPLE 7   Classify the following equations as linear or nonlinear. Also classify them by their order, by their homogeneity, and by whether they have constant coefficients.

a) $f(n) = 3 + 2f(n-1)$ **b)** $f(n) = 2f(n-1) - 2nf(n-2)$
c) $f(n) = 1/(1 - f(n-1))$

SOLUTION

a) This equation is linear, nonhomogeneous, and has constant coefficients. It has order 1.

b) This equation is also linear, is homogeneous, and has order 2. It does not have constant coefficients.

c) This equation is nonlinear. We will attempt no further classification of nonlinear equations. ∎

The Main Theorem

THEOREM 6.1   The linear recurrence equation of order k

$$a_0(n) f(n) + a_1(n) f(n-1) + \cdots$$
$$+ a_{k-1}(n) f(n-k+1) + a_k(n) f(n-k) = h(n), \qquad n > k$$

has a unique solution if $f(1), \ldots, f(k)$ are specified.

PROOF   We must find a sequence that satisfies both the equation and the initial conditions, and we must show that this is the only such sequence. Such a sequence is defined inductively as follows:

$$f(1), \ldots, f(k) \text{ are as given by the initial conditions.}$$
$$f(k+1) = (1/a_0(k+1))(h(k+1) - (a_1(k+1) f(k) + \cdots$$
$$+ a_{k-1}(k+1) f(2) + a_k(k+1) f(1)))$$

This defines a value of $f(k+1)$ that satisfies the equation and is in fact the only possible value consistent with both the equation and its initial conditions.

Now suppose that $f(m), \ldots, f(1), m > k+1$, are uniquely determined by the equation and the initial conditions. Let

$$f(m+1) = (1/a_0(m+1))(h(m+1) - (a_1(m+1) f(m) + \cdots$$
$$+ a_{k-1}(m+1)) f(m-k+2) + a_k(m+1) f(m-k+1)).$$

This also determines a unique value for $f(m+1)$, and thus f is uniquely defined for all $n \geq 1$. ∎

EXAMPLE 8   Which of the following problems have unique solutions?

a) $h(n) = (n-1) + h(n-1), n > 2, f(2) = 1$

b) $f(n) = f(n-1) + f(n-2), n > 2, f(1) = 0, f(2) = 1$

c) $f(n) = 2 + 2 f(n-1) \ n \geq 1$

SOLUTION

a) This equation does; it is of order 1 and has one initial condition.

b) This equation is of order 2 and has initial conditions for $n = 1$ and $n = 2$; hence it also has a unique solution.

c) This equation has no initial conditions and thus does not have a unique solution. In fact, it has infinitely many solutions, one for each possible value of $f(1)$. ∎

EXAMPLE 9 Which of the following problems have unique solutions?

a) $f(n) = 4f(n-1) - 4f(n-2)$, $n > 1$, $f(1) = 2$

b) $f(n) = 3 + f(n-1)$, $n > 1$, $f(1) = 1$, $f(2) = 5$

c) $f(n) = f(n-1) + f(n-2)$, $n > 1$, $f(1) = 0$, $f(3) = 1$

SOLUTION

a) This equation is of order 2 but has only one initial condition. It has infinitely many solutions.

b) In this case, the number of initial conditions exceeds the order of the equation. Taking $n = 2$ in the recurrence equation, we get

$$f(2) = 3 + f(1).$$

Hence $f(2) = 4$, which is inconsistent with the initial condition $f(2) = 5$. Thus there are no solutions that satisfy the equation and both initial conditions.

c) This equation is of order 2 and has two initial conditions, but they are specified for $n = 1$ and $n = 3$. Thus the equation does not satisfy the conditions of the theorem. However, taking $n = 3$ in the equation, we get

$$f(3) = f(2) + f(1).$$

Applying the initial conditions, $1 = f(2) + 0$, and thus $f(2) = 1$. Using this value for $f(2)$, the conditions of the theorem are satisfied and the given value of $f(3)$ is consistent with $f(1)$ and $f(2)$. Thus the problem has a unique solution. ∎

Note that the theorem can be easily modified to apply to the situation in which the initial conditions are given at any k *successive* points, not just at $n = 1$, $n = 2$, etc. If no initial conditions are specified for a linear recurrence equation, the equation has infinitely many solutions.

Solution by Iteration

We are now ready to look at some systematic techniques for solving problems involving recurrence equations. The first of these is called **solution by iteration.** The next two sections will be devoted to further techniques. We will illustrate solution by iteration with an example first and then define and discuss it afterward.

EXAMPLE 10   In Section 6.1, we derived the equation

$$P_{n+1} = (5/2)P_n$$

for the growth of a rabbit population. Assuming an initial population, P_0, of 100 rabbits, solve this problem for P_n.

SOLUTION   We can compute the values recursively as follows:

$$P_1 = (5/2)P_0 = (5/2) \cdot 100$$
$$P_2 = (5/2)P_1 = (5/2)^2 \cdot 100$$
$$P_3 = (5/2)P_2 = (5/2)^3 \cdot 100$$
$$\vdots$$

In this case, it is not hard to infer a pattern in the numbers and to assert that $P_n = (5/2)^n \cdot 100$. However, we have *proven* this to be a solution only in the few cases we have actually computed, namely, $n = 0, 1, 2,$ and 3. To determine that this formula is the solution for any value of n, we proceed by induction. Any of the cases we have already checked will serve as the basis step. We then assume that $P_n = (5/2)^n \cdot 100$. Then

$$P_{n+1} = (5/2)P_n = (5/2)^{n+1} \cdot 100.$$

This verifies that $P_n = (5/2)^n \cdot 100$ is a solution for all n.   ∎

Solution by iteration, then, is the combination of

a) repeated application of the recurrence equation to compute a sequence of values,

b) inferring a pattern to the sequence,

c) the use of induction to check that the inferred pattern is correct for all values of n.

The difficulty with this method occurs in step (b). Inferring the pattern often involves more than a little creative guessing and may be all but impossible. Thus this method is quite limited in its applicability. Still, it is not worthless for the following reasons: (1) There are a number of cases in which we can infer the pattern. (2) In many situations, what is needed is not a closed form expression for the solution but the value of $f(N)$ for some specific N or N's. Then step (a) of this method is all that is needed. In Section 6.5 we will examine an algorithm to carry out this operation. (3) Using this method may be very helpful in developing intuition about a particular situation one is trying to model by seeing the particular numerical values and how they are generated. (4) This method can be applied even if the recurrence equation is not linear and/or for some reason does not fit the criteria needed to apply any other method. In such a case, even if we cannot infer a pattern, we at least have a list of values of the function for as many n's as we are willing to calculate.

We will look at two more examples of the application of solution by iteration and then move on to more powerful methods.

EXAMPLE 11   Find the solution of $f(n + 1) = 3 + f(n)$, $n > 0$, $f(0) = 0$.

SOLUTION   We can generate the following table:

n	$f(n)$
0	0
1	3
2	6
3	9
4	12
.	
.	
.	

We can infer from the table that $f(n) = 3n$. To verify this, we first note that $f(0) = 3 \cdot 0 = 0$. Then, assuming that $f(n) = 3n$, $f(n + 1) = 3 + 3n = 3(n + 1)$ and the correctness of this solution is verified.   ∎

EXAMPLE 12   Let $f(n + 1) = 2 + 2f(n)$, $n > 1$, $f(1) = 1$. Find $f(n)$.

SOLUTION   We get the following table:

n	$f(n)$
1	1
2	4
3	10
4	22
5	46
.	
.	
.	

This time the pattern is far from obvious, and a more powerful method is needed. ∎

Fundamental Properties of Solutions

Two important properties of solutions of recurrence equations will be used in developing methods more powerful than iteration. They are presented in the next two theorems. The first property is called the **principle of superposition**.

THEOREM 6.2   Suppose f_1 and f_2 are both solutions to a given linear, homogeneous recurrence equation and c_1 and c_2 are real numbers. Then $c_1 f_1 + c_2 f_2$ is also a solution.

PROOF   First, we write the equation in standard form:

$$a_0(n) f(n) + a_1(n) f(n-1) + \cdots + a_k(n) f(n-k) = 0.$$

Then, if f_1 and f_2 are both solutions, we can say that

$$a_0(n) f_1(n) + a_1(n) f_1(n-1) + \cdots + a_k(n) f_1(n-k) = 0$$

and

$$a_0(n) f_2(n) + a_1(n) f_2(n-1) + \cdots + a_k(n) f_2(n-k) = 0 \text{ for all } n.$$

Multiplying the first equation by c_1 and the second by c_2, we get:

$$c_1 a_0(n) f_1(n) + c_1 a_1(n) f_1(n-1) + \cdots + c_1 a_k(n) f_1(n-k) = 0$$

and

$$c_2 a_0(n) f_1(n) + c_2 a_1(n) f_1(n-1) + \cdots + c_2 a_k(n) f_1(n-k) = 0 \text{ for all } n.$$

Adding and rearranging terms, we get

$$a_0(n)[c_1 f_1(n) + c_2 f_2(n)] + a_1(n)[c_1 f_1(n-1) + c_2 f_2(n-1)] + \cdots$$
$$+ a_k(n)[c_1 f_1(n-k) + c_2 f_2(n-k)] = 0.$$

That is,

$$a_0(n)(c_1 f_1 + c_2 f_2)(n) + a_1(n)(c_1 f_1 + c_2 f_2)(n-1) + \cdots$$
$$+ a_k(n)(c_1 f_1 + c_2 f_2)(n-k) = 0 \text{ for all } n.$$

Thus $(c_1 f_1 + c_2 f_2)$ is also a solution to the equation.   ☐

THEOREM 6.3   Suppose f is a solution of a nonhomogeneous recurrence equation and g is the solution of its reduced equation. Then $f + g$ is also a solution of the nonhomogeneous equation.

PROOF   Once again, we first write the equation in standard form, assuming that f is a solution:

$$a_0(n)\,f(n) + a_1(n)\,f(n-1) + \cdots + a_k(n)\,f(n-k) = h(n) \text{ for all } n.$$

Then, if g is a solution of the reduced equation,

$$a_0(n)\,g(n) + a_1(n)\,g(n-1) + \cdots + a_k(n)\,g(n-k) = 0 \text{ for all } n.$$

Adding them together, we get

$$a_0(n)(f+g)(n) + a_1(n)(f+g)(n-1) + \cdots$$
$$+ a_k(n)(f+g)(n-k) = h(n) \text{ for all } n.$$

Hence $f + g$ is also a solution of the original equation.   □

EXAMPLE 13   Consider again the recurrence equation

$$f(n) = 4f(n-1) - 4f(n-2).$$

The functions $f_1(n) = 2^n$ and $f_2(n) = n2^n$ are both solutions. (We will leave the verification of this to the Exercises.) Then, by Theorem 6.2, $c_1 2^n + c_2 n 2^n$ is a solution for any values of c_1 and c_2. ■

EXAMPLE 14   Consider again the recurrence equation

$$f(n) = 3 + f(n-1).$$

A solution to this equation is $f(n) = 3n$. The reduced equation is

$$f(n) - f(n-1) = 0,$$

and it can be easily verified that $g(n) = c$ is a solution to the reduced equation for any constant c. Hence, according to Theorem 6.3, $(f+g)(n) = 3n + c$ is a solution to the original equation for any value of c. ■

TERMINOLOGY

sequence	order k
infinite sequence	constant coefficients
closed form	homogeneous equation
solution of a recurrence equation	nonhomogeneous equation
existence of solutions	reduced equation
uniqueness of solutions	solution by iteration
linear recurrence equations	principle of superposition

SUMMARY

1. By a solution to a recurrence equation, we mean a sequence $a_1, a_2, \ldots, a_n, \ldots$ that satisfies the equation for every n, as well as the initial condition(s), if any.

2. Some recurrence equations have no solutions, some have unique solutions, and some have many solutions. One condition that guarantees unique solutions of recurrence equations is the following: The linear recurrence equation of order k

$$a_0(n) f(n) + a_1(n) f(n-1) + \cdots$$
$$+ a_{k-1}(n) f(n-k+1) + a_k(n) f(n-k) = h(n)$$

   has a unique solution if $f(1), \ldots, f(k)$ are specified.

3. Recurrence equations can be classified as linear or nonlinear. Linear equations are further classified as homogeneous or nonhomogeneous, constant or non-constant coefficients, and by order.

4. Linear, homogeneous, recurrence equations without initial conditions satisfy the principle of superposition, i.e., if f_1 and f_2 are solutions, so is $c_1 f_1 + c_2 f_2$ for any real constants c_1 and c_2.

5. If f is a solution of a linear, nonhomogeneous recurrence equation without initial conditions and g is a solution of its reduced equation, $f + g$ is also a solution of the original equation.

6. One method of solving recurrence equations is the method of iteration. It involves

   a) using the initial conditions plus repeated application of the recurrence equation to compute a sequence of values,

   b) inferring a pattern to the sequence,

   c) using mathematical induction to check that the inferred pattern is correct for all n.

EXERCISES 6.2

In Exercises 1–4, check if the given function is a solution of the given recurrence equation.

1. $f(n) = 2 + f(n-1); f(n) = 2n$ 2. $f(n) = 3 + f(n-1); f(n) = 3n + 5$
3. $f(n) = 2f(n-1); f(n) = 21 \cdot 2^n$ 4. $f(n+1) = 1 + f(n)^2; f(n) = n^2$

Show that the following recurrence equations have no solution by attempting to calculate the values of $f(1), f(2)$, etc.

5. $f(n) = \sqrt{(1 - f(n-1))}; f(0) = 2$ 6. $f(n) = 1/(3 - f(n-1)); f(0) = 21/8$

7. $f(n) = \dfrac{1 + f(n-1)}{1 - f(n-1)}; f(0) = 0$ 8. $f(n) = \dfrac{1}{\sqrt{f(n-1)}}; f(0) = 0$

Classify each of the following equations as linear or nonlinear; for the linear ones, identify the order, state whether the coefficients are constant, and state whether the equation is homogeneous.

9. $f(n) = \sqrt{5 - f(n-1)}$

10. $f(n) = \dfrac{1}{2 - f(n-1)^2}$

11. $f(n) + 2f(n-1) = 3n$

12. $f(n) = 3f(n-1) + nf(n-2) - \dfrac{1}{n}$

13. $f(n+2) = 3f(n) + 2$

14. $f(n) + f(n-1) + f(n-2) + f(n-3) = 0$

Apply Theorem 6.1 to decide whether or not each of the following linear recurrence equations has a unique solution.

15. $f(n) = 2 + 3f(n-1); f(1) = 1$

16. $f(n+1) = 2 - f(n); f(0) = 1$

17. $f(n) = f(n-1) + f(n-2); f(0) = 0, f(1) = 1$

18. $f(n) = f(n-1) + 2; f(1) = 1$

19. $f(n+1) = 2 + f(n-1); f(0) = 0, f(1) = 1$

20. $f(n) = f(n-1) + 3f(n-2) + n; f(1) = 2$

21. Prove Theorem 6.2 in the case where the equation is first order; i.e., show that if the equation

$$a_0(n) f(n) + a_1(n) f(n-1) = 0$$

has solutions f_1 and f_2, then $c_1 f_1 + c_2 f_2$ is also a solution for any c_1 and c_2.

22. State and prove Theorem 6.3 in the case where the equation is first order.

Solve each of the following by the iteration method and, in each case, verify your solution by mathematical induction.

23. $f(n) = 2 + f(n-1); f(1) = 3$

24. $f(n) - 3f(n-1); f(0) = 2$

25. $f(n) = nf(n-1); f(0) = 1$

26. $f(n) = (n-1) + f(n-1); f(2) = 1$

27. $f(n) = n^2 + f(n-1); f(0) = 0$

6.3

LINEAR, FIRST-ORDER RECURRENCE EQUATIONS

We concluded the last section by examining the iteration technique for solving recurrence equations. This technique is very general; it can be used with any recurrence equation for which a solution exists. However, in all but the simplest cases, the best it can yield is a sequence of values rather than a closed form solution. Determining closed form solutions for general recurrence equations is often extremely difficult. For linear equations, however, there are very powerful solution techniques. We will focus on these techniques here and in the next section. The methods presented in this section are all referred to as the **method of substitution.**

We will start with the first-order, linear equation. In its most general form, it can be written

$$a_0(n) f(n) + a_1(n) f(n-1) = h(n),$$

where $a_0(n)$ and $a_1(n)$ are both nonzero for all n. This can be written in the form

$$f(n) = A(n) f(n-1) + B(n)$$

by dividing through by $a_0(n)$ and rearranging and renaming the terms. Equations of this form can be classified even further by whether they are homogeneous or nonhomogeneous and whether they have constant coefficients. Thus the linear, first-order equation may appear in any of the following forms:

	Constant Coefficients	**Nonconstant Coefficients**
Homogeneous	$f(n) = Af(n-1)$	$f(n) = A(n) f(n-1)$
Nonhomogeneous	$f(n) = Af(n-1) + B(n)$	$f(n) = A(n) f(n-1) + B(n)$

We will discuss each of these categories separately.

Homogeneous with Constant Coefficients

Homogeneous, linear, first-order, recurrence equations with constant coefficients are the simplest variety. A solution can be easily found for all equations in this class.

Before proving Theorem 6.4, let's see how this solution can be obtained. Let $f(0) = c$. We follow the same pattern used with the iteration method:

$$f(1) = Af(0) = cA$$
$$f(2) = Af(1) = cA^2$$
$$f(3) = Af(2) = cA^3$$
$$f(4) = Af(3) = cA^4$$
$$\vdots$$

We then infer that $f(n) = cA^n$.

THEOREM 6.4   The recurrence equation

$$f(n + 1) = Af(n), f(0) = c$$

has the unique solution $f(n) = cA^n$.

PROOF   We want to verify that $f(n) = cA^n$ for all n. First, let $n = 0$. Then $f(0) = cA^0 = c$ and hence the proposed solution satisfies the initial condition. Now assume that $f(n) = cA^n$. Then

$$f(n + 1) = Af(n) = AcA^n = cA^{n+1}.$$

Hence the solution $f(n) = cA^n$ is correct for all n. We know from Theorem 6.1 that this is the only solution.   □

If the equation treated in the previous theorem had been stated without an initial condition, the function $f(n) = cA^n$ would be a solution for any value of c. Since $f(0)$ must assume some value, and since we know that the solution is unique for every such value, we know that the equation $f(n) = cA^n$ must include all possible solutions of $f(n) = Af(n - 1)$. We call $f(n) = cA^n$ the **general solution** of the equation, since it includes every possible solution as a special case. General solutions typically involve arbitrary constants. One of the most common strategies for solving recurrence equations is to find the general solution first and then use the initial conditions to solve for the arbitrary constants that appear in the general solution.

EXAMPLE 1   There are many situations that lead to equations of the form $f(n + 1) = Af(n)$. The best example is the exponential function itself, for which the following recursive definition was given in Section 5.1:

$$a^0 - 1$$
$$a^{n+1} = a \cdot a^n, n > 0$$

In this case, the recurrence equation is used to define what we mean by a^n. Thus, if we say that $f(n) = a^n$ is the solution of the equation $f(n + 1) = af(n), f(0) = 1$, we are simply applying the definition of a^n.   ∎

EXAMPLE 2   Show that the number of subsets of a set with n elements is 2^n, using recurrence equations.

SOLUTION   The set with no elements has one subset, namely, itself. Adding another element to a set doubles the number of subsets, since every previous subset is still a subset and the union of the new element with every previous subset is also a subset. Thus, if $f(n)$ denotes the number of subsets of a set of size n, we have

$$f(n + 1) = 2f(n), f(0) = 1.$$

The general solution of the recurrence equation is $f(n) = c \cdot 2^n$. Applying the initial condition, we get $f(n) = 2^n$. $\blacksquare$

Nonhomogeneous Equations with Constant Coefficients

The nonhomogeneous, linear, first-order equation with constant coefficients can be written as

$$f(n + 1) = Af(n) + B(n), f(0) = c.$$

First, suppose that B is also constant. Then the equation becomes

$$f(n + 1) = Af(n) + B.$$

Once again, let's see first how the solution of this equation can be obtained. Then we will prove that it is correct. As before, we observe that

$$f(1) = Af(0) + B = cA + B$$
$$f(2) = Af(1) + B = A(cA + B) + B = cA^2 + AB + B$$
$$f(3) = Af(2) + B = A(cA^2 + AB + B) + B = cA^3 + A^2B + AB + B$$
$$f(4) = Af(3) + B = A(cA^3 + A^2B + AB + B) + B =$$
$$cA^4 + A^3B + A^2B + AB + B.$$

In general, then, we can infer that

$$f(n) = cA^n + B(A^{n-1} + A^{n-2} + \cdots + A + 1)$$

$$= cA^n + B \sum_{i=0}^{n-1} A^i$$

$$= \begin{cases} cA^n + \dfrac{B(1 - A^n)}{(1 - A)} & \text{if } A \neq 1 \\ cA^n + Bn & \text{if } A = 1 \end{cases}$$

THEOREM 6.5   The recurrence equation

$$f(n + 1) = Af(n) + B, f(0) = c$$

has the unique solution

$$f(n) = \begin{cases} cA^n + \dfrac{B(1 - A^n)}{(1 - A)} & \text{when } A \neq 1 \\ c + Bn & \text{when } A = 1. \end{cases}$$

PROOF   Once again, we proceed by induction. First, we assume that $A \neq 1$, let $n = 0$, and get

$$f(0) = cA^0 + \frac{B(1 - A^0)}{(1 - A)} = c,$$

which is consistent with the initial condition and serves as the basis step. As the induction hypothesis, we assume that

$$f(n) = cA^n + \frac{B(1 - A^n)}{(1 - A)}.$$

Then

$$
\begin{aligned}
f(n + 1) &= Af(n) + B \\
&= A\left(cA^n + \frac{B(1 - A^n)}{(1 - A)}\right) + B \\
&= cA^{n+1} + \frac{BA - BA^{n+1} + B - BA}{1 - A} \\
&= cA^{n+1} + \frac{B(1 - A^{n+1})}{(1 - A)}
\end{aligned}
$$

This verifies the correctness of the proposed solution when $A \neq 1$. The proof when $A = 1$ is left for the Exercises. ☐

EXAMPLE 3   Solve the recurrence equation

$$f(n + 1) = 3 + f(n), \ n \geq 0, f(0) = 1.$$

SOLUTION   Apply Theorem 6.5 with $A = 1$, $B = 3$, and $c = 1$. We conclude that

$$f(n) = c + Bn = 1 + 3n. \ ∎$$

EXAMPLE 4   In Section 6.1 we derived the equation

$$P_{n+1} = (1 + r)P_n - R$$

for the amortization of a loan. P_0 was the original principal due, r was the interest rate, and R was the fixed payment for each period. The amortization problem is to find R such that P_N will be zero after exactly N periods. Solve this problem.

SOLUTION   The equation is a linear, first-order, recurrence equation with constant coefficients. So we can apply Theorem 6.5 with $A = 1 + r$, $B = -R$, and $c = P_0$. We conclude that

$$
\begin{aligned}
P_n &= cA^n + \frac{B(1 - A^n)}{(1 - A)} \\
&= P_0(1 + r)^n + \frac{R(1 - (1 + r)^n)}{r}.
\end{aligned}
$$

Setting $P_N = 0$ and solving for R, we get

$$R = \frac{P_0 r}{1 - (1 + r)^{-N}}.$$

For instance, for a $40,000 mortgage at a 10% annual rate over 25 years, we would take $N = 300$ periods (one per month for 25 years) and $r = 0.0083333$ (0.10 divided by 12 months). This would yield a monthly payment of $363.48. If the mortgage were over 30 years rather than 25, we would repeat the same process with $N = 360$. This would yield a monthly payment of $351.03. ∎

Now we relax the restriction that B has to be a constant. We get the following:

THEOREM 6.6   The recurrence equation

$$f(n + 1) = Af(n) + B(n), f(0) = c$$

has the unique solution

$$f(n) = cA^n + \sum_{i=0}^{n-1} B(i)A^{n-1-i}, \ n \geq 1.$$

PROOF   This proof is left for the Exercises. ⬜

EXAMPLE 5   Consider the equation we derived for the handshake problem at the beginning of Section 6.1:

$$f(n) = (n - 1) + f(n - 1), \ n \geq 2, f(1) = 0$$

Use Theorem 6.6 to solve this equation.

SOLUTION   The equation can be rewritten to put it in the form required for Theorem 6.6 by replacing each occurrence of n by $n + 1$:

$$f(n + 1) = f(n) + n, \ n \geq 1, f(1) = 0.$$

We can see immediately that $A = 1$ and $B(n) = n$. But we need $f(0)$, not $f(1)$. Even though the rewritten recurrence equation applies only for $n \geq 1$, we can replace the condition "$n \geq 1$" with "$n \geq 0$" by selecting a value of $f(0)$ consistent with the equation and with $f(1) = 0$. That is, using the recurrence equation, we must have the following expression for $f(0)$:

$$f(1) = f(0) + 0.$$

Thus $f(0) = 0$ will give us an equation in the proper form. Hence

$$f(n) = cA^n + \sum_{i=0}^{n-1} B(i)A^{n-1-i}$$

$$= 0 \cdot 1^n + \sum_{i=0}^{n-1} B(i)1^{n-1-i}$$

$$= \sum_{i=0}^{n-1} i$$

$$= \frac{n(n-1)}{2}.$$

This is the same answer we inferred at the beginning of Section 6.1, but now we have a method to use in solving not only this problem but a whole class of similar problems. ∎

Nonconstant Coefficients

Recall the notation

$$\sum_{i=1}^{n} A(i)$$

for the sum $A(1) + A(2) + \cdots + A(n)$. There is a similar notation for the *product* of n values.

Definition The notation

$$\prod_{i=1}^{n} A(i)$$

means $A(1) \cdot A(2) \cdot \cdots \cdot A(n)$.

Using this notation, we have the following theorem:

THEOREM 6.7 The homogeneous recurrence equation

$$f(n+1) = A(n) f(n), f(0) = c$$

has the unique solution

$$f(n) = c \prod_{i=0}^{n-1} A(i), \; n \geq 1.$$

PROOF This proof is left for the Exercises. ☐

EXAMPLE 6 Solve the recurrence equation

$$f(n+1) = (n+1) f(n) \; n \geq 0, f(0) = 1.$$

SOLUTION   By Theorem 6.7, the solution is

$$f(n) = c \prod_{i=0}^{n-1} (i+1)$$

$$= n!. \; ∎$$

Theorems 6.5 through 6.7 can be generalized to handle the nonhomogeneous, nonconstant coefficient cases as well. These generalizations and their proof are also left for the Exercises.

TERMINOLOGY

method of substitution

general solution

$$\prod_{i=1}^{n} A(i)$$

SUMMARY

1. Linear, first-order recurrence equations may be classified as homogeneous or nonhomogeneous and as constant or nonconstant coefficient.

2. The recurrence equation $f(n + 1) = Af(n)$, $f(0) = c$ has the unique solution $f(n) = cA^n$. The expression $f(n) = cA^n$ is also the general solution to the equation $f(n + 1) = Af(n)$ if c is regarded as an arbitrary constant.

3. The recurrence equation

$$f(n + 1) = Af(n) + B, f(0) = c$$

has the unique solution

$$f(n) = \begin{cases} cA^n + \dfrac{B(1 - A^n)}{(1 - A)} & \text{when } A \neq 1 \\ c + Bn & \text{when } A = 1. \end{cases}$$

4. The recurrence equation

$$f(n + 1) = Af(n) + B(n), \quad f(0) = c$$

has the unique solution

$$f(n) = cA^n + \sum_{i=0}^{n-1} B(i)A^{n-1-i}.$$

5. The homogeneous recurrence equation

$$f(n + 1) = A(n)f(n), f(0) = c$$

has the unique solution

$$f(n) = c \prod_{i=0}^{n-1} A(i).$$

EXERCISES 6.3

Solve the following recurrence equations.

1. $f(n+1) = 3f(n); f(0) = 1$

2. $f(n+1) = 4f(n); f(0) = 4$

3. $f(n+1) = \frac{1}{2}f(n); f(0) = 2$

4. $f(n+1) = \frac{1}{a}f(n); f(0) = b$

5. $f(n+1) = kf(n); f(0) = 0$

6. $f(n+1) = \frac{1}{k}f(n); f(0) = 0$

7. $f(n+1) = 3 + f(n); f(0) = 3$

8. $f(n+1) = 1 + 2f(n); f(0) = 1$

9. $f(n+1) = 1 + 3f(n); f(0) = 1$. How does this solution differ from the solution to Exercise 1 and why?

10. $f(n+1) = 4 + 4f(n); f(0) = 4$. How does this solution differ from the solution to Exercise 2 and why?

11. Suppose a binary number has n bits and $f(n)$ denotes the number of possible values for this number. Adding one more bit doubles the number of possible binary numbers that can be formed. Find and solve a recurrence equation for $f(n)$.

12. This is the same as Exercise 11, but for base 3 numbers.

Solve the recurrence equations developed in each of the following exercises from Section 6.1.

13. Exercise 6

14. Exercise 7

15. Exercise 15

16. Exercise 16

17. Exercise 17

18. Exercise 18

19. Exercise 21

20. Exercise 22

21. Prove Theorem 6.5 in the case where $A = 1$.

22. Prove Theorem 6.6.

23. Prove Theorem 6.7.

24. Modify Theorem 6.6 to handle the case where A is not constant but B is, and prove it.

25. Modify Theorem 6.6 to handle the case where neither A nor B is constant and prove it.

Use your answers to Exercises 24 and 25 or Theorem 6.6 to solve the following and verify that your solution is correct.

26. $f(n+1) = (n+1) + f(n); f(0) = 1$

27. $f(n+1) = n \cdot f(n); f(0) = 1$

28. $f(n+1) = n + f(n); f(0) = 1$

29. $f(n+1) = (n+1) + n \cdot f(n);$ $f(0) = 1$

30. Exercise 5, Section 6.1

31. Exercise 11, Section 6.1

6.4

LINEAR, SECOND-ORDER RECURRENCE EQUATIONS

In this section, we will continue the process of showing how to find closed form solutions to linear recurrence equations. The last section dealt with first-order equations; this one will cover second-order equations. At the end of the section, we will briefly discuss higher-order linear equations.

The general form for second-order, linear, recurrence equations is

$$a_0(n) f(n) + a_1(n) f(n-1) + a_2(n) f(n-2) = h(n),$$

where a_0 and a_2 are both nonzero for all n. As in the previous section, we can express this equation in an alternative form by dividing through by $a_0(n)$ and rearranging terms:

$$f(n) + b_1(n) f(n-1) + b_2(n) f(n-2) = h'(n).$$

Note that all of the terms involving f are collected on the left-hand side, unlike the form used for the first-order equation. The last form presented is usually more convenient for the analysis of second-order equations, as we shall see. However, we will occasionally write second-order equations in the form

$$f(n) = d_1(n) f(n-1) + d_2(n) f(n-2) + h'(n).$$

We can also classify equations of this form by whether they have constant coefficients and whether they are homogeneous. We derive the following general forms:

	Constant Coefficients	Nonconstant Coefficients
Homogeneous	$f(n) + b_1 f(n-1) + b_2 f(n-2) = 0$	$f(n) + b_1(n) f(n-1) + b_2(n) f(n-2) = 0$
Nonhomogeneous	$f(n) + b_1 f(n-1) + b_2 f(n-2) = h'(n)$	$f(n) + b_1(n) f(n-1) + b_2(n) f(n-2) = h'(n)$

For second-order equations, we will restrict our attention to constant coefficients simply because the others are more complex than we have space to address in an introductory text.

Homogeneous Equations with Constant Coefficients

We will consider homogeneous equations with constant coefficients first. In the last section, we selected an arbitrary value, c, for the initial value of the solution, generating the successive values of $f(n)$ from the recurrence equation, inferring the pattern $f(n) = cA^n$, and verifying the correctness of the solution. We then used the uniqueness theorems from Section 6.2 to argue that this was the only possible solution. However, because the previous strategy doesn't work in this case, our method here will be different. For instance, consider the equation

$$f(n) = f(n-1) + f(n-2).$$

If we were to proceed as in Section 6.3, we would have to select two arbitrary, initial values for f, since the uniqueness theorem requires two initial values for an equation of order 2. Thus we let $f(0) = c_0$ and $f(1) = c_1$. Computing successive values, we get

$$f(2) = c_1 + c_0$$
$$f(3) = 2c_1 + c_0$$
$$f(4) = 3c_1 + 2c_0$$
$$f(5) = 5c_1 + 3c_0$$
$$\vdots$$

We could continue, but no obvious pattern emerges, so this approach is not helpful.

Instead we use a different strategy. For the first-order equation, we found that the solution was $f(n) = cA^n$, an exponential function. We will try an exponential function as a possible solution for the second-order equation as well. It turns out that this guess does give a solution, although it will take a few pages to work out all of the details. It is possible to verify that this "lucky guess" system (more properly called the **method of characteristic roots**) does, in fact, give us the general solution. (See Goldberg, *Introduction to Difference Equations,* Sections 3.3 and 3.4 for the proof; see the References.) Let's start with an example and then generalize.

EXAMPLE 1 Solve the recurrence equation

$$f(n) = 2f(n-2) - f(n-1).$$

SOLUTION   We assume that there is a solution of the form $f(n) = t^n$ for some real number $t \neq 0$. Substitute this proposed solution into the equation and attempt to evaluate t. That is, if $f(n) = t^n$,

$$t^n = 2t^{n-2} - t^{n-1}$$
$$1 = 2t^{-2} - t^{-1} \text{ (dividing by } t^n)$$
$$t^2 = 2 - t \text{ (multiplying by } t^2)$$
$$t^2 + t - 2 = 0$$
$$(t + 2)(t - 1) = 0$$
$$t = -2 \text{ or } t = 1.$$

Hence both $f_1(n) = (-2)^n$ and $f_2(n) = 1^n = 1$ appear to be solutions. To verify this, we substitute each function into the right-hand side of the recurrence equation and see if it becomes $f(n)$:

$$
\begin{aligned}
\text{RHS} &= 2(-2)^{n-2} - (-2)^{n-1} \\
&= -(-2)^{n-1} - (-2)^{n-1} \\
&= (-2)(-2)^{n-1} \\
&= (-2)^n.
\end{aligned}
$$

Similarly substituting $f(n) = 1$ gives

$$
\text{RHS} = 2 \cdot 1 - 1 = 1.
$$

Thus both functions are solutions. By Theorem 6.2 of Section 6.2, we can conclude that

$$
\begin{aligned}
f(n) &= c_1(-2)^n + c_2(1) \\
&= c_1(-2)^n + c_2
\end{aligned}
$$

is a solution for any real constants c_1 and c_2. ∎

The technique can be generalized to any second-order, linear recurrence equation with constant coefficients. That is, if we have an equation of the form

$$
f(n) + b_1 f(n-1) + b_2 f(n-2) = 0,
$$

we can assume a solution of the form $f(n) = t^n$, substitute, and solve for t. This operation is summarized in the following theorem.

THEOREM 6.8   The second-order recurrence equation

$$
f(n) + b_1 f(n-1) + b_2 f(n-2) = 0
$$

has as a solution the function

$$
f(n) = c_1 t_1^n + c_2 t_2^n,
$$

where t_1 and t_2 are roots of the quadratic equation

$$
t^2 + b_1 t + b_2 = 0.
$$

PROOF   Let $f(n) = t^n$ where t is one or the other of the values given previously. We know that $t \neq 0$ since $b_2 \neq 0$. Substituting into the recurrence equation, we get

$$
\begin{aligned}
t^n + b_1 t^{n-1} + b_2 t^{n-2} &= 0 \\
t^2 + b_1 t + b_2 &= 0 \text{ (dividing by } t^{n-2}) \\
t^2 + b_1 t + b_2 &= 0,
\end{aligned}
$$

which is an identity, since t is one of the roots of this quadratic equation. Applying Theorem 6.2, we conclude that $f(n) = c_1 t_1{}^n + c_2 t_2{}^n$ is in fact a solution. ▯

Definition The equation $t^2 + b_1 t + b_2 = 0$ is called the **characteristic equation** of the recurrence equation

$$f(n) + b_1 f(n-1) + b_2 f(n-2) = 0.$$

A solution, t, is called a **root** of the characteristic equation.

EXAMPLE 2 Consider the recurrence equation

$$f(n) = f(n-1) + f(n-2), f(0) = 0, f(1) = 1,$$

which defines the Fibonacci sequence we looked at in Section 6.1. Solve this equation.

SOLUTION First, we write the equation in the form

$$f(n) - f(n-1) - f(n-2) = 0.$$

From the previous theorem, we can conclude that

$$f(n) = c_1 t_1{}^n + c_2 t_2{}^n,$$

where t_1 and t_2 are solutions of the equation

$$t^2 - t - 1 = 0.$$

Using the quadratic formula, we conclude that

$$t = \frac{1 \pm \sqrt{5}}{2}.$$

Hence $f(n) = c_1 \left(\dfrac{1+\sqrt{5}}{2}\right)^n + c_2 \left(\dfrac{1-\sqrt{5}}{2}\right)^n$

is a solution of the equation $f(n) = f(n-1) + f(n-2)$. However, we also had two initial conditions, $f(0) = 0$ and $f(1) = 1$. Thus we must select the values of the constants, c_1 and c_2, so that $f(n)$ also satisfies the initial conditions. We get

$$f(0) = c_1 \left(\frac{1+\sqrt{5}}{2}\right)^0 + c_2 \left(\frac{1-\sqrt{5}}{2}\right)^0 = 0$$

and

$$f(1) = c_1 \left(\frac{1+\sqrt{5}}{2}\right)^1 + c_2 \left(\frac{1-\sqrt{5}}{2}\right)^1 = 1.$$

That is,

$$c_1 + c_2 = 0$$

and

$$c_1(1 + \sqrt{5}) + c_2(1 - \sqrt{5}) = 2$$
$$c_1 = -c_2$$
$$c_1(1 + \sqrt{5}) - c_1(1 - \sqrt{5}) = 2$$
$$c_1 = \frac{1}{\sqrt{5}} \text{ and } c_2 = -\frac{1}{\sqrt{5}}.$$

Hence

$$f(n) = \frac{1}{\sqrt{5}}\left(\left(\frac{1 + \sqrt{5}}{2}\right)^n - \left(\frac{1 - \sqrt{5}}{2}\right)^n\right)$$

and this function, by Theorem 6.1, is the unique solution of the original equation. ∎

We now have a solution technique for second-order, linear, homogeneous recurrence equations with constant coefficients. It involves solving a quadratic equation and, if there are initial conditions, solving a pair of simultaneous equations in order to match the solution to those initial conditions. However, there are two complications we must take into account —the possibility that the roots of the quadratic equation are equal and the possibility that they are complex numbers. Let's address these concerns one at a time.

Equal Roots of the Characteristic Equation

It is not unusual for the roots of the characteristic equation to be equal. Such roots are called **multiple roots**. For instance, consider the recurrence equation

$$f(n) - 4f(n - 1) + 4f(n - 2) = 0.$$

This has characteristic equation

$$t^2 - 4t + 4 = 0,$$

which has equal roots, both of value $t = 2$. In this case, we know from Theorem 6.8 of this section that

$$c_1 2^n + c_2 2^n$$

will be a solution for any constants c_1 and c_2. But this expression is just

$$(c_1 + c_2)2^n,$$

which reduces to

$$c2^n,$$

since c_1 and c_2 are both arbitrary constants. Thus, in this case, our two solutions reduce to only one solution. There is, however, another solution, and that is $f(n) = n \cdot 2^n$. We can check this in the same way we checked that t^n was a solution in the first place:

$$
\begin{aligned}
\text{RHS} &= 4 f(n - 1) - 4f(n - 2) \\
&= 4(n - 1)2^{n-1} - 4(n - 2)2^{n-2} \text{ (substituting } n2^n \text{ for } f(n)) \\
&= 4((n - 1) \cdot 2 - 4(n - 2)) \cdot 2^{n-2} \\
&= (8n - 8 - 4n + 8) \cdot 2^{n-2} \\
&= 4n \cdot 2^{n-2} \\
&= n \cdot 2^n \\
&= \text{LHS.}
\end{aligned}
$$

Hence $n \cdot 2^n$ is also a solution. Thus, by Theorem 6.2, we can conclude that

$$f(n) = c_1 2^n + c_2 n \cdot 2^n$$

is a solution for any constants c_1 and c_2. This can be generalized to the following theorem:

THEOREM 6.9 Suppose the characteristic equation

$$t^2 + b_1 t + b_2 = 0$$

of the recurrence equation

$$f(n) + b_1 f(n - 1) + b_2 f(n - 2) = 0$$

has equal roots, t. Then the equation has solution

$$f(n) = c_1 t^n + c_2 n \cdot t^n.$$

PROOF The proof is left for the Exercises. It is a generalization of the argument used previously to solve the equation

$$f(n) = 4 f(n - 1) - 4 f(n - 2). \quad \square$$

EXAMPLE 3 Solve the equation

$$f(n) = 6 f(n - 1) - 9 f(n - 2), f(0) = 1, f(1) = 2.$$

SOLUTION The characteristic equation is

$$t^2 - 6t + 9 = 0.$$

Thus, $t = 3$, where 3 is a multiple root. Hence

$$f(n) = c_1 3^n + c_2 n \cdot 3^n.$$

Applying the initial conditions, we get

$$f(0) = c_1 = 1$$
$$f(1) = 3c_1 + 3c_2 = 2$$

Hence

$$3c_2 = -1$$
$$c_2 = -1/3.$$

Thus

$$f(n) = 3^n - (1/3)n \cdot 3^n$$
$$f(n) = 3^n - n \cdot 3^{n-1}. \quad \blacksquare$$

Complex Roots of the Characteristic Equation

Consider the recurrence equation

$$f(n) + f(n-1) + f(n-2) = 0.$$

The characteristic equation is

$$t^2 + t + 1 = 0,$$

which has roots

$$t = \frac{-1 \pm \sqrt{-3}}{2}.$$

These roots can be written

$$t = \frac{-1 \pm i\sqrt{3}}{2},$$

where i denotes $\sqrt{-1}$. From Theorem 6.8, we can conclude that the function

$$f(n) = c_1 \left(\frac{(-1 + i\sqrt{3})}{2} \right)^n + c_2 \left(\frac{(-1 - i\sqrt{3})}{2} \right)^n$$

is a solution of our recurrence equation for any constants, c_1 and c_2. But this poses a problem for us. We started with an equation with all real number coefficients that presumably modeled a realistic situation. But we have ended with a solution that involves what are sometimes called *imaginary numbers*.

There is a way to resolve this paradox. Suppose the characteristic equation has roots $a \pm bi$. It turns out that $f(n)$ can also be written

$$f(n) = r^n(k_1 \cos n\theta + k_2 \sin n\theta)$$

where $r = \sqrt{(a^2 + b^2)}$ and θ is given by $\tan \theta = b/a$. Note that this form of the solution does not use i. See Appendix E for the details.

Summary of the Homogeneous Case

We can summarize all of these results as follows:
The recurrence equation

$$f(n) + af(n - 1) + bf(n - 2) = 0$$

has the characteristic equation

$$t^2 + at + b = 0.$$

Then if the roots t_1 and t_2 of the characteristic equation are

a) *real and unequal,* the equation has the solution $f(n) = c_1 t_1{}^n + c_2 t_2{}^n$;
b) *real and equal,* the equation has the solution $f(n) = c_1 t^n + c_2 n t^n$;
c) *complex with real initial conditions,* the equation has the solution

$$f(n) = r^n(k_1 \cos n\theta + k_2 \sin n\theta)$$

where $r = \sqrt{(a^2 + b^2)}$ and $\tan \theta = b/a$.

Nonhomogeneous Equations

The key to solving nonhomogeneous, linear, recurrence equations is Theorem 6.10. This theorem states that if we can find *any* one solution of a nonhomogeneous equation, we can find any other solution by adding a solution of the reduced equation to the one solution we found.

THEOREM 6.10 Suppose g is a solution of the recurrence equation

$$f(n) + b_1 f(n - 1) + b_2 f(n - 2) = h(n).$$

Then any solution of the equation can be written in the form $f + g$, where f is a solution of the corresponding reduced equation.

PROOF Let s be any solution of the given recurrence equation. Then

$$s(n) + b_1 s(n - 1) + b_2 s(n - 2) = h(n).$$

Since g is also a solution, we have

$$g(n) + b_1 g(n - 1) + b_2 g(n - 2) = h(n).$$

Subtracting, we get

$$(s - g)(n) + b_1(s - g)(n - 1) + b_2(s - g)(n - 2) = 0.$$

Thus, if we let f denote $s - g$, f is a solution of the reduced equation and $s = f + g$. $\quad \square$

This result gives rise to the most popular method for solving nonhomogeneous equations, the **method of undetermined coefficients.** The idea behind it is this: Suppose we have a nonhomogeneous, second-order, linear, recurrence equation with constant coefficients:

$$f(n) + b_1 f(n - 1) + b_2 f(n - 2) = h(n).$$

If $h(n)$ is a constant, a polynomial, or an exponential function, we can usually find *a* solution of the equation, as we shall see. We then solve the reduced equation, add the two solutions together, and derive the general solution. No other solutions are possible. If, in addition, we have two initial conditions, we use them to solve for the arbitrary constants that arise from the solution of the reduced equation. The resulting solution is then the unique solution of the equation by Theorem 6.1. First, we need a lemma.

LEMMA 6.11   Two polynomials are equal if and only if they are of the same degree and each of their coefficients are equal.

PROOF   The proof can be done by induction. It is left for the Exercises.

$\quad \square$

EXAMPLE 4   Solve the equation

$$f(n) - 4f(n - 1) - 5f(n - 2) = 1.$$

SOLUTION   First, find any solution of the nonhomogeneous equation. We do this by noting that the right-hand side is a constant, and hence we try as a possible solution $f(n) = k$ for some constant k. We then substitute this into the equation and see if we can solve for k. We get

$$k - 4k - 5k = 1$$
$$-8k = 1$$
$$k = -1/8.$$

The reduced equation is $f(n) - 4f(n - 1) - 5f(n - 2) = 0$, which has as its characteristic equation $t^2 - 4t - 5 = 0$. The roots of this characteristic equation are $t = 5$ and $t = -1$. Thus, by Theorem 6.8, the solution to the reduced equation is

$$f(n) = c_1 5^n + c_2(-1)^n.$$

Hence the general solution to the original equation is

$$f(n) = c_1 5^n + c_2(-1)^n - 1/8. \quad \blacksquare$$

EXAMPLE 5 Solve the equation

$$f(n) - 4f(n-1) - 5f(n-2) = n^2 + 3n + 1.$$

SOLUTION We start by attempting to find a solution of the nonhomogeneous equation. We assume a solution of the form

$$an^2 + bn + c,$$

i.e., also a polynomial of degree 2, but with unknown coefficients. It is from this assumption that the method gets its name, the method of undetermined coefficients. Substituting, we get

$$an^2 + bn + c - 4(a(n-1)^2 + b(n-1) + c) -$$
$$5(a(n-2)^2 + b(n-2) + c) = n^2 + 3n + 1$$
$$an^2 + bn + c - 4an^2 + 8an - 4a - 4bn + 4b - 4c -$$
$$5an^2 + 20an - 20a - 5bn + 10b - 5c = n^2 + 3n + 1$$
$$-8an^2 + (-8b + 28a)n + (-8c - 24a + 14b) = n^2 + 3n + 1.$$

By the previous lemma, we have

$$-8a = 1 \Rightarrow a = -1/8$$
$$-8b + 28a = 3 \Rightarrow b = -13/16$$
$$-8c - 24a + 14b = 1 \Rightarrow c = -75/64.$$

Thus a solution of the equation is

$$f(n) = (-1/8)n^2 - (13/16)n - (75/64),$$

and hence, by Theorem 6.10, the complete solution is

$$f(n) = c_1 5^n + c_2(-1)^n + (-1/8)n^2 - (13/16)n - (75/64). \quad \blacksquare$$

EXAMPLE 6 Solve the equation

$$f(n) - 4f(n-1) - 5f(n-2) = 2^n.$$

SOLUTION Using the method of undetermined coefficients, we try to find a solution of the form

$$f(n) = k \cdot 2^n.$$

Thus

$$k \cdot 2^n - 4k \cdot 2^{n-1} - 5k \cdot 2^{n-2} = 2^n$$
$$k - 4k/2 - 5k/4 = 1$$
$$k = -4/9.$$

Thus, by Theorem 6.10, we conclude that our general solution is

$$f(n) = c_1 5^n + c_2(-1)^n - (4/9)2^n. \quad \blacksquare$$

Although other functions do occasionally arise on the right-hand side of recurrence equations, exponential functions, constant functions, and polynomials are very common. We will not discuss nonhomogeneous equations for other types of functions here. Another difficulty arises when the right-hand side of a nonhomogeneous, second-order, linear recurrence equation is of the form c^n, where c is a root of the characteristic equation. It can be shown that if c is a nonrepeated root of the characteristic equation, $f(n) = knc^n$ is a solution. See the Exercises.

Higher-Order Equations

The methods we have developed here can be easily extended to equations of degree 3 and higher. For instance, suppose we have a recurrence equation of the form

$$f(n) + af(n-1) + bf(n-2) + cf(n-3) = 0.$$

This is a homogeneous, linear recurrence equation with constant coefficients of order 3. If we assume that it has a solution of the form $f(n) = t^n$, we will find that it has a characteristic equation

$$t^3 + at^2 + bt + c = 0.$$

If this has three distinct real roots, t_1, t_2, and t_3, we can conclude that the equation has the solution

$$f(n) = c_1 t_1{}^n + c_2 t_2{}^n + c_3 t_3{}^n.$$

Similarly if two roots are equal and one distinct, the solution is

$$f(n) = c_1 t_1{}^n + c_2 t_2{}^n + c_3 n t_2{}^n.$$

If all three roots are equal, the solution is

$$f(n) = c_1 t_1{}^n + c_2 n t_1{}^n + c_3 n^2 t_1{}^n.$$

With a cubic equation, at least one of the roots must be real, since a cubic curve must always cross the x-axis at least once. The others must be either both real or complex conjugates. If they are complex conjugates, the solution is of the form

$$f(n) = c_1 t_1{}^n + r^n(k_1 \cos n\theta + k_2 \sin n\theta)$$

for values of r, θ, k_1, and k_2. Similar results hold for equations of fourth and higher orders.

TERMINOLOGY

method of characteristic roots   multiple root of a characteristic equation

characteristic equation   method of undetermined coefficients

characteristic root

SUMMARY

1. Second-order, linear recurrence equations can be classified as homogeneous versus nonhomogeneous and as having constant or non-constant coefficients.

2. The homogeneous equations with constant coefficients can be solved by the method of characteristic roots. If the equation is written

$$f(n) + af(n-1) + bf(n-2) = 0,$$

the characteristic equation is

$$t^2 + at + b = 0$$

and the characteristic roots are the roots of this equation.

3. If the roots t_1 and t_2 of the characteristic equation are

   a) *real and unequal,* the homogeneous equation has the solution

   $$f(n) = c_1 t_1{}^n + c_2 t_2{}^n;$$

   b) *real and equal,* the homogeneous equation has the solution

   $$f(n) = c_1 t^n + c_2 n t^n;$$

   c) *complex with real initial conditions,* the homogeneous equation has the solution

   $$f(n) = r^n(k_1 \cos n\theta + k_2 \sin n\theta).$$

4. The nonhomogeneous equation

   $$f(n) + af(n-1) + bf(n-2) = h(n)$$

   can be solved by the method of undetermined coefficients if h is a constant, a polynomial, or an exponential function whose base is not a root of the characteristic equation. This means assuming a solution of the same form as $h(n)$ but with unknown coefficients, substituting this solution into the equation, and finding the coefficients. We have not addressed the situation in which h is of other forms.

5. Linear recurrence equations with constant coefficients of order higher than 2 can be solved by generalizations of these methods.

EXERCISES 6.4

For each of the following recurrence equations, find the characteristic equation and use it to find the general solution of the given equation. No complex numbers should appear in your answers. In each case, check the correctness of your answer by substitution.

1. $f(n) = 3f(n-1) - 2f(n-2)$

2. $f(n) - f(n-1) - 2f(n-2) = 0$

3. $f(n+1) = 6f(n) - f(n-1)$

4. $f(n+1) - 7f(n-1) + 6f(n) = 0$

5. $2f(n) = 6f(n-2) - f(n-1)$

6. $f(n) + 2f(n-1) + f(n-2) = 0$

7. $f(n+1) = f(n) - f(n-1)$

8. $2f(n) - 2f(n-1) + 3f(n-2) = 0$

Each of the following refers to one of the preceding exercises. Find the solution with the given initial condition.

9. Exercise 1, $f(0) = 1, f(1) = 1$

10. Exercise 2, $f(0) = 1, f(1) = 0$

11. Exercise 3, $f(0) = 0, f(1) = 1$

12. Exercise 4, $f(0) = 1, f(1) = 2$

13. Exercise 5, $f(0) = 0, f(1) = 1$

14. Exercise 6, $f(0) = 0, f(1) = 1$

15. Exercise 7, $f(0) = 0, f(1) = 1$

Find a particular solution for each of the following by the method of undetermined coefficients.

16. $f(n) - f(n-1) - 2f(n-2) = 1$

17. $f(n) = 6f(n-2) - f(n-1) + 2$

18. $f(n) - f(n-1) - 2f(n-2) = n$

19. $f(n) = 6f(n-2) - f(n-1) + 2n$

20. $f(n) - f(n-1) - 2f(n-2) = 3^n$

21. $f(n) = 6f(n-2) - f(n-1) + 3^n$

22. $f(n) - f(n-1) - 2f(n-2) = 2n + 1$

23. $f(n) = 6f(n-2) - f(n-1) + n^2 + 1$

24. $f(n) + 2f(n-1) + f(n-2) = 2$

25. $f(n) + 2f(n-1) + f(n-2) = n + 3$

26. $f(n) + 2f(n-1) + f(n-2) = n^2$

27. $f(n) + 2f(n-1) + f(n-2) = 2^n$

Each of the following refers to Exercises 16–27. In each case, find the *general* solution and use it to solve for $f(n)$ when $f(0) = 0$ and $f(1) = 1$.

28. Exercise 16

29. Exercise 17

30. Exercise 18

31. Exercise 19

32. Exercise 20   **33.** Exercise 21

34. Exercise 22   **35.** Exercise 23

36. Exercise 24   **37.** Exercise 25

38. Exercise 26   **39.** Exercise 27

40. Find the general solution for the deer population example (Example 8) of Section 6.1.

41. Prove Theorem 6.9.

42. Prove that if a recurrence equation of the form

$$f(n) + af(n-1) + bf(n-2) = c^n$$

has a characteristic equation for which c is a nonrepeated root, then $f(n) = knc^n$ is a solution of the equation for some k.

43. Prove that if a recurrence equation of the form

$$f(n) + af(n-1) + bf(n-2) = c^n, \quad a \neq 0, c \neq 0,$$

has a characteristic equation for which c is a repeated root with multiplicity two, then $f(n) = kn^2c^n$ is a solution of the equation for some k.

6.5
RECURSIVE ALGORITHMS

In the previous four sections, we examined recurrence equations—equations in which a function $f(n)$ is defined in terms of $f(n-1)$ and possibly $f(n-2)$ and $f(n-3)$ as well. Not only functions but also algorithms may be defined recursively. Algorithms and functions are similar; both have sets of acceptable inputs, outputs corresponding to those inputs, and a rule or set of rules for transforming the inputs into the outputs. Thus what applies to functions will often apply to algorithms as well. A **recursive algorithm** is similar to a recurrence equation. Suppose we have a set of n data items. A recursive algorithm typically states what the algorithm is supposed to do if n equals 1. It also specifies what the algorithm does if $n > 1$ in terms of two things—what it does to sets of fewer than n items and what additional activity is needed to handle the nth item. Thus, if $n > 1$, it is like an equation of the form $f(n) = f(n-1) + 2$. That is, $f(n)$ is specified in terms of $f(n-1)$ and an action (add 2) to handle n. For instance, here is a typical example. It computes the function $f(n) = a^n$ using the recursive definition of a^n presented in Section 5.1. In its standard form, the programming language Pascal does not have an exponential function. Thus the algorithm could be used to compute exponentials in Pascal. Unlike the other algorithms in this section (which are written in pseudo-code), this is actual

Pascal code. The algorithm assumes that $n \geq 1$; if $n < 1$, the algorithm will not return the correct value as its output.

EXAMPLE 1

ALGORITHM 16   Computing Exponentials

Function exp(a,n : integer): integer;

{Computes the value of the exponential function directly from its definition. n is an integer such that $n \geq 1$, a is any real number.}

begin

1.   if $n \leq 1$

2.   then exp := a

3.   else exp := $a * \exp(a,n-1)$

end; ∎

Note that the algorithm computes a^n (denoted $\exp(a,n)$) by specifying that $\exp(a,1) = a$ and $\exp(a,n) = a * \exp(a,n-1)$. This is the recurrence equation, $f(1) = a$, $f(n) = a \cdot f(n-1)$, $n > 1$, expressed in algorithmic language.

Tracing Recursive Algorithms

Recall that in mathematical induction we *first* verify what needs to be checked with a small number such as $n = 1$. Then we assume that the statement is true for n and prove that it must be true for $n + 1$. Tracing a recursive algorithm is very similar to mathematical induction. We start with a small number such as 1. Before going to the induction step, however, we usually try a few small numbers until we understand how the algorithm works. That is, we try the algorithm with 1, then 2, then 3, and perhaps a few more. Then move to the induction step. Here is a trace of Algorithm 16 when $n = 1$, 2, 3, and an arbitrary value.

Initial value	Step	Action
$n = 1$	1	$1 = 1$: test is true; advance to step 2.
	2	Set exp $= a$ and then halt.
$n = 2$	1	$2 \neq 1$: test is false; advance to step 3.
	3	Set exp $= a * \exp(a,1)$. We already know that $\exp(a,1) = a$ from the $n = 1$ case. Thus exp $= a * a = a^2$.

$n = 3$	1	$3 \neq 1$: test is false; advance to step 3.
	3	Set exp $= a * \exp(a,1)$. We already know that $\exp(a,2) = a^2$ from the $n = 2$ case. Thus exp $= a^2 * a = a^3$.
$n > 1$ arbitrary	1	$n \neq 1$: test is false; advance to step 3.
	3	Set exp $= a * \exp(a, n - 1)$. In this case, we assume the induction hypothesis, that $\exp(a, n - 1) = a^{n-1}$. Thus $\exp(a,n) = a * \exp(a, n - 1) = a * a^{n-1} = a^n$.

The key to this approach is that when we get to $n = 2$, we have already computed the result of the recursive call when $n = 1$; with $n = 3$, we have already computed the result of the call when $n = 2$, etc.

The area where students usually go astray in dealing with recursive algorithms is this: Suppose we try to trace the preceding algorithm by *first* letting n be an arbitrary natural number greater than 1. The test in line 1 fails, so we advance to line 3. This requires us to go back to the beginning and trace the algorithm again, with $n - 1$ replacing n. The test in line 1 fails if n was originally greater than 2. Thus, in step 3, we must go back to the beginning, with $n - 2$ replacing n. This process must continue until the algorithm is called with 1 replacing n. This approach quickly becomes very confusing, and students who try it are very likely to say that they can't understand recursion. This is why it is better to start with $n = 1$ than with an arbitrary n.

At this point, someone usually says, "But a *computer* starts with an arbitrary value of n. I want to understand how a computer carries out a recursive algorithm." This is a reasonable question, and it is worth pursuing *once* here. In carrying out a recursive algorithm, a computer uses a special structure called a **stack,** which is used to keep track of each of the unfinished problems that accumulate as it works its way down toward $n = 1$. Think of it as a stack of pancakes—the last one on the plate is the first one taken off. Let's trace this algorithm, showing intuitively what happens on the stack. We will ignore the technical details of how a computer actually places the current problem on the stack. To use a concrete example, let $a = 5$ and $n = 3$.

Initial value	Step	Action	Stack
$n = 3$	1	Test fails; advance to step 3.	empty
	3	Put the problem on the stack and call the procedure again with $n = 2$.	exp $= 5 * \exp(5,2)$

$n = 2$	1	Test fails; advance to step 3.	
	3	Put the problem on the stack and call the procedure again with $n = 1$.	$\exp = 5 * \exp(5,1)$ $\exp = 5 * \exp(5,2)$
$n = 1$	1	Test succeeds; advance to step 2.	
	2	Set $\exp = 5$; after this END is encountered The unfinished problem, $\exp = 5 * \exp(5,1)$, is removed from the top of the stack.	$\exp = 5 * \exp(5,2)$
		$\exp = 5 * \exp(5,1) = 5 * 5 = 25$, since $\exp(5,1)$ was found earlier; this completes step 3, so END is reached. Remove the next unfinished problem from the stack.	
		$\exp = 5 * \exp(5,2) = 5 * 25 = 125$, END is reached.	empty
		The stack is now empty, so the computation is finished.	

Notice what happens here. A list of problems is stacked up, each with a value of n that is one smaller than the previous one. This continues until $n = 1$. Then problems in the list are solved, starting from $n = 1$, going to $n = 2$, etc. That is, the computer must first work down from n to 1 and then reverse itself, working up from 1 back to n. A human being, when tracing a recursive algorithm, can shorten the process by skipping the part that uses the stack and starting at $n = 1$, using induction to verify the pattern after the first few have been computed.

EXAMPLE 2 Trace the following recursive algorithm and verify that it finds the largest element of an array.

ALGORITHM 17   **Finding the Largest Element in an Array**

Procedure Find_largest(n,A; var largest);

{A is an array of integer or real numbers, n is the length of the array, largest is an output variable that gets the value of the largest element of the array.}

begin

1.   if $n = 1$

2.   then largest := $A[1]$

else begin

3.   find_largest($n - 1$, A,largest);

4.   if $A[n] >$ largest

5.   then largest := $A[n]$

end

end;

To illustrate the action of this algorithm, we will trace it with the array $A = [-1, 5, 2, 3, 6, 8, 10]$. Using the process described previously, we will start with $n = 1$ and work our way up.

Initial value	Step	Action
$n = 1$	1	Test succeeds; advance to step 2.
	2	Largest := -1; the algorithm then halts.
$n = 2$	1	Test fails; advance to step 3.
	3	Calls *Find_largest* with $n = 1$; this was done previously so largest becomes -1.
	4	$A[2]$ is compared to largest; since $5 > -1$, advance to step 5.
	5	Largest becomes 5.
$n = 3$	1	Test fails; advance to step 3.
	3	Calls *Find_largest* with $n = 2$; largest becomes 5.
	4	$A[3]$ compared to largest; $2 < 5$, so skip step 5 and halt with largest equal to 5.
$n = 7$	{we will skip to this case to show the last step}	
	1	Test fails; advance to step 3.
	3	Calls *Find_largest* with $n = 6$; we assume that it works up to 6, and thus largest becomes 8.

4          $A[7]$ compared to largest; $10 > 8$, so
           advance to step 5.

5   Largest becomes 10 and halt.

We now generalize to a verification that the algorithm works correctly. We proceed by induction. Let

$$A = [a_1, a_2, \ldots, a_n],$$

where $n \geq 1$. Our precondition is that $[a_1, \cdots, a_n]$ is an array of n integers. Our postcondition is that at the end of the algorithm, the variable *largest* equals the largest of $a_1, \ldots, a_n$.

First, let $n = 1$ and begin to trace the algorithm. The test in step 1 succeeds, so *largest* becomes a_1. The algorithm then halts, and it indeed gets the correct value for *largest* in this case. Thus we have completed the basis step for our induction.

Now suppose $n > 1$ and assume that *Find_largest(n − 1, A, largest)* sets the variable *largest* equal to some a_i such that $largest \geq a_i$ for all i, $1 \leq i \leq n − 1$. Now begin the trace. In line 1, the test fails and we advance to line 3. In line 3, by the induction hypothesis, we can assume that *largest* becomes a value such that $largest \geq a_i$ for all i, $1 \leq i \leq n − 1$. In line 4, a_n is compared to *largest*. If the test fails, then $largest \geq a_i$ for all i, $1 \leq i \leq n$. If it succeeds, *largest* takes on the value of a_n, and thus the new value of *largest* is greater than or equal to a_i for all i, $1 \leq i \leq n$. Thus we have verified that, in either case, the value of *largest* becomes the value of the greatest element in the array. ∎

In practice, the processes of writing and verifying an algorithm are not entirely separable; that is, we are continually attempting to verify the parts of an algorithm we have written and are modifying it accordingly as we write.

For the next example, we will need the following algorithm. It is a slight variation of Algorithm 4 of Section 4.3.

ALGORITHM 18   Move the Largest Item in an Array to the End

Procedure Bubble_down(n; var A);

{n is the length of the array; A is an array of integers or reals; i is an integer; temp is of the same data type as A.}

begin

    for $i := 1$ to $n − 1$ do

      if $A[i] > A[i + 1]$ then begin

        temp $:= A[i]$

        $A[i] := A[i + 1]$;

$$A[i + 1] := \text{temp}$$
$$\text{end}$$
$$\text{end};$$

EXAMPLE 3   The following algorithm is a recursive version of *Bubble_sort* that we saw in Chapter 4. It uses the procedure *Bubble_down*.

ALGORITHM 19   **Sort an Array from Smallest to Largest**

Procedure Bubble_sort(n; var A);

{A is the array to be sorted, n is its length.}

begin

1.   if $n > 1$

then begin

2.   bubble_down(n,A);

3.   bubble_sort ($n - 1,A$)

end;

end;

Verify that it sorts an array of size n from smallest to largest.

SOLUTION   Our precondition is that A is an array of length n. Our post-condition is that A is sorted from smallest to largest. As usual, we proceed by induction. First, let $n = 1$. Then the test in line 1 fails and the algorithm halts. That is, it does nothing, which is exactly what an algorithm that sorts an array of length 1 should do. Rather than going directly to the induction step, let's continue to trace for $n = 2$ and $n = 3$.

Initial value	Line	Action
$n = 2$	1	Test fails; advance to step 2.
	2	The larger of the two elements is placed in $A[2]$.
	3	*Bubble_sort* is called with $n = 1$; this does nothing, as we saw previously; so the algorithm halts. $A[1]$ is smaller and $A[2]$ larger; hence the array is sorted.
$n = 3$	1	Test fails; advance to step 2.
	2	The largest of the three elements is moved to $A[3]$.

3   *Bubble_sort* is called with $n = 2$; as we saw, it sorts from smallest to largest. The algorithm then halts and the array is sorted.

Now let n be an arbitrary integer greater than 1 and assume that after completion of the algorithm for a data set of size $n - 1$, $A[1], \ldots, A[n - 1]$ are sorted in increasing order.

Arbitrary n 1   Test fails; advance to step 2.

2   The largest element in the array is moved to $A[n]$.

3   *Bubble_sort* is called to sort the first $n - 1$ elements of the array. By the induction hypothesis, we can assume that this results in $A[1]$ to $A[n - 1]$ being sorted in increasing order. Since $A[n]$ is greater than all of these values, the algorithm halts, with the array sorted in the proper order. ∎

EXAMPLE 4   Recursive algorithms provide a natural way to compute a table of values for a recurrence equation. For instance, consider the equation $f(n) = (n - 1) + f(n - 1)$, $f(2) = 1$, which we saw at the beginning of Section 6.1. The following algorithm can be used to compute a list of values of f.

ALGORITHM 20   Computing a Table of Values for the Recurrence Equation $f(n) = (n - 1) + f(n - 1)$, $f(1) = 0$

Function $f(n)$;

{n is an integer greater than or equal to 1; f is also an integer.}

begin

  if $n \leq 1$

  then $f := 0$

  else $f := (n - 1) + f(n - 1)$

end;

We will leave the tracing of this algorithm for the Exercises. ∎

EXAMPLE 5   The following algorithm is a recursive version of Algorithm 11 of Section 4.4, which listed all the subsets of $\{1, \cdots, n\}$. Trace and verify this algorithm.

ALGORITHM 21   List All Subsets of $\{1, \cdots, n\}$

Procedure Subsets(n; var number_of_subsets, list_of_subsets);

begin

1. initialize list_of_subsets to the empty list;
2. if $n = 0$

   then begin
3.   add $\varnothing$ to list_of_subsets;
4.   number_of_subsets := 1

   end

   else begin
5.   Subsets($n - 1$, number_of_subsets,list_of_subsets);
6.   for $i := 1$ to number_of_subsets do begin
7.   let $S_i = i$th subset in list_of_subsets;
8.   add $\{n\} \cup S_i$ to the end of list_of_subsets

      end;
9.   number_of_subsets := number_of_subsets $*$ 2

   end;

end;

SOLUTION   As usual, we proceed by induction. The precondition is that n is an integer greater than or equal to zero. The postcondition is that *number_of_subsets* $= 2^n$ and *list_of_subsets* contains all of the subsets of $\{1, \ldots, n\}$. First, let $n = 0$. The condition in line 2 is true, so lines 3 and 4 are executed, while lines 5 through 8 are not. Thus the empty set is added to the list of subsets and *number_of_subsets* is set equal to 1. The algorithm then halts. Thus the *list_of_subsets* and the *number_of_subsets* are correct for a set with zero elements.

At this point, we could go directly to the induction step. However, we will trace the algorithm for $n = 1$ and $n = 2$ in order to illustrate more clearly how it works. First, let $n = 1$. The condition in line 2 is false, so we move to line 5. The procedure, *Subsets,* is called with $n = 0$, and this line is completed with *number_of_subsets* equal to 1 and *list_of_subsets* containing only the empty set. Line 6 initiates a loop that will be executed once. In line 7, S_1 becomes $\varnothing$ and in line 8, $\{1\}$ is added to *list_of_subsets*. In line 9, *number_of_subsets* becomes 2. This completes the algorithm. Its output is correct, as there are two subsets of $\{1\}$, namely, $\varnothing$ and $\{1\}$.

Now let $n = 2$. The condition in line 2 is false, so we advance to line 5. In line 5 the procedure *Subsets* is called with $n = 1$. Hence, at the end of this line, *number_of_subsets* is 2 and *list_of_subsets* is $\{\varnothing, \{1\}\}$. In line 6, a loop is begun that will be executed twice. Thus, at line 7, S_1 becomes $\varnothing$ and at line 8, {2} is added to the end of *list_of_subsets*. At the next execution of line 7, S_2 becomes {1} and at line 8, {1,2} is added to the list. At line 9, *number_of_subsets* becomes 4. Again, this result is correct, since there are four subsets of {1,2}, namely, $\varnothing$, {1}, {2}, and {1,2}.

Now let n be an arbitrary integer greater than 1 and assume that the algorithm correctly lists all of the subsets of $\{1, \ldots, n-1\}$ and sets *number_of_subsets* to 2^{n-1}. In line 2, the condition is false, so we advance to line 5. This line, by the induction hypothesis, places all of the subsets of $\{1, \cdots, n-1\}$ in *list_of_subsets* and sets *number_of_subsets* to 2^{n-1}. In lines 6 through 8, a loop is executed 2^{n-1} times. At each pass, one subset of $\{1, \ldots, n-1\}$ is retrieved, {n} is included in it, and the resulting new subset is added to the end of the list. We have the following situation:

List of Subsets Before the Loop	New Subsets Added to the List
$\varnothing$	{n}
{1}	{1,n}
{2}	{2,n}
{1,2}	{1,2,n}
{3}	{3,n}
{1,3}	{1,3,n}
.	.
.	.
.	.
$\{1,2,3, \ldots, n-1\}$	$\{1,2,3, \ldots, n-1,n\}$

After completion of the loop, *number_of_subsets* becomes 2^n, which verifies that part of the postcondition. To verify that the new *list_of_subsets* is correct, we can argue as follows: First, a set of n elements has 2^n distinct subsets. The new list has exactly 2^n entries, since the list before the loop had 2^{n-1} and one additional subset has been added for each of the original 2^{n-1}. Also, each of the 2^n sets is distinct, since the original 2^{n-1} are distinct. That is, the new sets are all distinct from the old ones, since the new ones include n. The new ones are distinct from each other, since each is the result of including n with a list of distinct sets. Thus the new list has exactly 2^n distinct subsets of $\{1, \ldots, n\}$ which is all that exist. Thus the algorithm has correctly listed the subsets of $\{1, \ldots, n\}$. ∎

Complexity of Recursive Algorithms

Analyzing the time complexity of recursive algorithms is very similar to analyzing the time complexity of nonrecursive algorithms except for one factor. If $f(n)$ is the number of computations required for a data set of size n, counting the number of computations will normally lead us to a recurrence equation. This equation can often be solved by the methods presented in Sections 6.1 through 6.4. Thus, for computer scientists, analysis of recursive algorithms is one of the most important applications of recurrence equations. For instance, consider Example 6.

EXAMPLE 6 Determine the complexity of Algorithm 16.

SOLUTION   The algorithm is as follows:

ALGORITHM 16   Computing Exponentials
Function exp(a,n : integer): integer;

{Computes the value of the exponential function directly from its definition. n is an integer such that $n \geq 1$, a is any real number.}

begin

1.   if $n \leq 1$
2.   then exp := a
3.   else exp := $a * $exp($a,n - 1$)
     end;

To analyze the complexity of this algorithm, we count each comparison, each assignment, and each arithmetic operation as one computation. We also let $f(n)$ denote the total number of computations needed to find a^n.

	$n = 1$		$n > 1$
Step	**Number of Computations**	**Step**	**Number of Computations**
1	1	1	1
2	1	2	0
3	0	3	$2 + f(n - 1)$

Thus we get

$$f(1) = 2, f(n) = 3 + f(n - 1),$$

which has the solution

$$f(n) = 3n - 1.$$

Hence this algorithm has a complexity of $O(n)$. ∎

EXAMPLE 7   Compute the complexity of Algorithm 17—*Find_largest.*

SOLUTION   The body of the algorithm is

```
        begin
1.      if n = 1
2.          then largest := A[1]
            else begin
3.              find_largest(n − 1,A,largest);
4.              if A[n] > largest
5.                  then largest := A[n]
            end
        end;
```

We get

	n = 1		n > 1
Step	**Number of Computations**	**Step**	**Number of Computations**
1	1	1	1
2	1	2	0
3	0	3	$f(n-1)$
4	0	4	1
5	0	5	$x(n) \quad 0 \le x(n) \le 1$

Thus $f(1) = 2$ and $f(n) = 2 + x(n) + f(n-1)$. In the worst case, the test in line 4 succeeds every time, so $x(n) = 1$ for all n. Thus the equation becomes

$$f(1) = 2, f(n) = 3 + f(n-1).$$

Thus $f(n) = 3n - 1$. In the best case, the test fails every time, and thus $x(n) = 0$ for all n. In this case, the equation is

$$f(1) = 2, f(n) = 2 + f(n-1),$$

and thus $f(n) = 2n$. The average case is between the two, and hence, in every case, the complexity of this algorithm is in $O(n)$.   ∎

EXAMPLE 8   Compute the complexity of Algorithm 21—listing subsets.

SOLUTION   The body of the algorithm is

> begin
> **1.**   initialize list_of_subsets to the empty list;
> **2.**   if $n = 0$
>   then begin
> **3.**   add $\emptyset$ to list_of_subsets;
> **4.**   number_of_subsets := 1
>   end
>   else begin
> **5.**   Subsets($n - 1$, number_of_subsets,list_of_subsets);
> **6.**   for $i := 1$ to number_of_subsets do begin
> **7.**   let $S_i = i$th subset in list_of_subsets;
> **8.**   add $\{n\} \cup S_i$ to the end of list_of_subsets
>   end;
> **9.**   number_of_subsets := number_of_subsets $* 2$
>   end;
>   end;

Suppose at first that each set operation can be counted as one computation. We get

	$n = 0$		$n > 1$
Step	**Number of Computations**	**Step**	**Number of Computations**
1	1	1	1
2	1	2	1
3	1	3	0
4	1	4	0
5	0	5	$f(n - 1)$
6	0	6	$2^{n-1} + 1$
7	0	7	2^{n-1}
8	0	8	$2 \cdot 2^{n-1}$
9	0	9	2

Thus

$$f(0) = 4, f(n) = 5 + 4 \cdot 2^{n-1} + f(n-1)$$
$$f(0) = 4, f(n) = 5 + 2^{n+1} + f(n-1).$$

By Theorem 6.6, the equation

$$f(n+1) = Af(n) + B(n), f(0) = c$$

has the unique solution

$$f(n) = cA^n + \sum_{i=0}^{n-1} (B(i)A^{n-i-1}).$$

Rewriting the equation to put it in this form, we get,

$$f(0) = 4, f(n+1) = 5 + 2^{n+2} + f(n).$$

Thus,

$$f(n) = 4 \cdot 1^n + \sum_{i=0}^{n-1} (5 + 2^{i+2})1^{n-i-1} = 4 + 5n + 4\sum_{i=0}^{n-1} 2^i$$
$$= 4 + 5n + 4(2^n - 1) = 4 \cdot 2^n + 5n$$
$$f(n) \in O(2^n).$$

If, in fact, each set operation requires n steps rather than one, the number of times that some of the steps (such as step 8) will be executed will be multiplied by a factor $O(n)$. Since those steps are executed $O(2^n)$ times, the resulting complexity will be $O(n \cdot 2^n)$. ■

EXAMPLE 9 Compute the complexity of the following algorithm for finding the nth Fibonacci number:

ALGORITHM 22 Finding the nth Fibonacci Number
   Function Fib(n);
   {n is an integer ≥ 1; Fib is the nth Fibonacci number.}
   begin
1.   if $n = 1$
2.   then Fib := 0
3.   else if $n = 2$
4.   then Fib := 1
5.   else Fib := Fib($n-1$) + Fib($n-2$)
   end;

SOLUTION   We first tabulate the number of steps required:

$n = 1$		$n = 2$		$n > 2$	
Step	Number of Computations	Step	Number of Computations	Step	Number of Computations
1	1	1	1	1	1
2	1	2	0	2	0
3	0	3	1	3	1
4	0	4	1	4	0
5	0	5	0	5	$2 + f(n-1) + f(n-2)$
Total:	2		3		$4 + f(n-1) + f(n-2)$

Thus we have the recurrence equation

$$f(n) = 4 + f(n-1) + f(n-2), n > 2,$$
$$f(1) = 2,$$
$$f(2) = 3,$$

which is a second-order, linear, nonhomogeneous recurrence equation. By the method of undetermined coefficients and Example 2 of Section 6.4, we can conclude that

$$f(n) = c_1 \left(\frac{1 + \sqrt{5}}{2} \right)^n + c_2 \left(\frac{1 - \sqrt{5}}{2} \right)^n - 4.$$

Note that

$$\left| \frac{(1 - \sqrt{5})}{2} \right| < 1,$$

and hence the second term of $f(n)$ is in $O(1)$, as is the term -4. Thus

$$f(n) \in O\left(\left(\frac{1 + \sqrt{5}}{2} \right)^n \right),$$

which is of exponential time. ∎

TERMINOLOGY

recursive algorithm

stack

SUMMARY

1. In recursive algorithms, the action of the algorithm on a data set of size n is often specified in two ways. First, what happens when $n = 1$ or some other starting value is specified. Second, what happens for arbitrary n is specified in terms of what happens to the nth item and a call to the algorithm itself to handle the first $n - 1$ items.

2. Recursive algorithms are best traced by first assuming that $n = 1$, and then letting $n = 2$, $n = 3$, etc., until the pattern is clear. Verification of the algorithm then proceeds by mathematical induction.

3. Analysis of the complexity of recursive algorithms typically involves the solution of recurrence equations.

EXERCISES 6.5

Trace each of the specified algorithms with the given set of data. Start with the smallest possible value of n and work up to the largest value.

1. Algorithm 17, $A = [2, 3, -7, 14, 5]$
2. Algorithm 17, $A = [17, 3, 29, 11, 18, 21, 13]$
3. Algorithm 20, $n = 4$
4. Algorithm 20, $n = 6$
5. Algorithm 21, $n = 3$
6. Algorithm 22, $n = 7$

Trace each of the specified algorithms. Start with the largest value of n and show the contents of the stack at each recursive call to the algorithm.

7. Algorithm 17, $A = [3, -1, 6]$
8. Algorithm 17, $A = [14, 2, -12, 6]$
9. Algorithm 19, $A = [3, -1, 6]$
10. Algorithm 19, $A = [14, 2, -12, 6]$
11. Algorithm 20, $n = 4$
12. Algorithm 21, $n = 3$

ALGORITHM 23   Counting Down
    Procedure Countdown(n);
    {n is any positive integer.}
    begin
        if $n \neq 0$

```
      then begin
          write(n);
          Countdown(n − 1)
      end
  end;
```

13. Trace Algorithm 23 with $n = 3$.

14. Trace Algorithm 23 with $n = 5$.

ALGORITHM 24   Counting Up
```
  Procedure Countup(n);
  {n is any positive integer.}
  if n ≠ 0
    then begin
        Countup (n − 1);
        write(n)
    end
  end;
```

---

15. Trace Algorithm 24 with $n = 3$.

16. Trace Algorithm 24 with $n = 5$.

ALGORITHM 25   Computing *n*!
```
  Function fact(n);
  {n is any nonnegative integer.}
  begin
  if n = 0
    then fact := 1
    else fact := n * fact(n − 1)
  end;
```

---

17. Trace Algorithm 25 with $n = 3$.

18. Trace Algorithm 25 with $n = 5$.

Verify that:

19. Algorithm 23 lists the integers from n down to 1.

20. Algorithm 24 lists the integers from 1 up to n.

21. Algorithm 25 computes $n!$.

Compute the complexity of the following algorithms.

22. Algorithm 19 (First, compute the complexity of Algorithm 18.)

23. Algorithm 20   **24.** Algorithm 23

25. Algorithm 24   **26.** Algorithm 25

Modify the following algorithms in the way indicated and verify your answer.

27. Algorithm 17 so that it finds the smallest element in an array.

28. Algorithm 18 so that it moves the smallest element to the bottom.

29. Algorithm 19 so that it sorts from largest to smallest instead of from smallest to largest.

Write recursive algorithms to compute $f(n)$ for the following recurrence equations.

30. $f(n) = (n-1)f(n-1); f(0) = 1$

31. $f(n) = (n-1) + f(n-1); f(2) = 1$

32. $f(n) = f(n-1) + f(n-2); f(0) = 0, f(1) = 1$

33. $f(n) = 2f(n-1) + 2; f(0) = 1$

34. Write a nonrecursive algorithm to compute the nth Fibonacci number that has a time complexity of $O(n)$ and verify that its complexity is $O(n)$.

6.6
MORE RECURSIVE ALGORITHMS

In the last section, we considered recursion as it is applied to algorithms. We looked at several recursive algorithms and analyzed their complexity. In this section, we will focus on two particularly important problems—searching and sorting—and examine a recursive algorithm for each. Both of these algorithms are very efficient and are therefore of great importance to computer scientists.

Binary Search

We have already examined binary search in Section 4.5. **Binary search** is a divide and conquer algorithm. It starts with a sorted array, examines the middle to see which half of the array contains the item being sought, and searches only the appropriate half array. Thus it follows a pattern typical of recursive algorithms: A problem is broken down into smaller problems that are of the same form as the original problem. Before writing the algorithm, we must clarify the problem it is supposed to solve. In this case, we assume that we have an array A of length n that is sorted in increasing order, although duplicate entries are allowed. We let X denote the item we are searching for. Imagine the array written as follows:

$$[A[1], A[2], \ldots, A[n-1], A[n]].$$

We will use two variables, *lower* and *upper*, which will be subscripts for the array. *Lower* will initially be 1 and *upper* will initially be n. We will find a subscript midway between *lower* and *upper* and change either *lower* or *upper*, depending on how X compares to the middle entry. The key to designing the algorithm is to note that *lower* and *upper* keep moving closer to each other whenever X is between them. Once they meet, we either have found X or have found where it would go if it were in the array. Thus the algorithm will include a conditional in which the condition *lower* < *upper* is checked. Once that condition is violated, the algorithm will terminate with *lower* equal to *upper*. The output of the algorithm is the final common value of *lower* and *upper*. Here is the algorithm.

ALGORITHM 26   Searching a Sorted Array*

Procedure Binary_search(X,A;var lower, upper);

{X is the item being searched for; A is the array; lower and upper are the lowest and highest subscripts of A. Thus the procedure that calls Binary_search must initialize lower to 1 and upper to n.}

begin

1.   if upper $\neq$ lower then begin

2.   mid := (lower + upper) div 2;

3.   if $X \leq A[\text{mid}]$

4.   then upper := mid

5.   else lower := mid + 1;

6.   binary_search(X, A, lower, upper)

      end

end;

---

* Several of the ideas used in this algorithm and its discussion were suggested in a talk by David Gries at the 1986 meeting of the Association of Computing Machinery's Special Interest Group on Computer Science Education (ACM/SIGCSE).

The algorithm has the following properties, which we will verify. We let N denote that common value of *lower* and *upper* that *binary_search* returns.

P1: If $X \leq A[i]$ for all i, $N = 1$.

P2: If $X > A[i]$ for all i, $N = n$.

P3: If $A[1] < X \leq A[n]$, it returns a value N such that $A[N - 1] < X \leq A[N]$.

P1 says that if X is less than or equal to all entries in the array, N is set to 1. P2 says that if X is greater than all entries in the array, N is set to the length of the array. P3 says, in essence, that if X is not in the array but is between $A[1]$ and $A[n]$, the algorithm returns the location where X would be inserted if it were placed in the array. It also says that if there are multiple copies of X in the array, the algorithm finds the location of the "leftmost" copy, i.e., the one with the smallest subscript.

Before attempting to verify the correctness of the algorithm, we must trace it for a few examples. As usual with recursive algorithms, we start with $n = 1$, then move to $n = 2$, etc.

EXAMPLE 1 Trace *binary_search* for an array of length 1.

SOLUTION   Let $A = [a_0]$ for some a_0. Then *binary_search* is called with *lower = upper =* 1. Thus, when the algorithm is executed, the test in line 1 fails and the algorithm halts with $N = 1$. Note that the answer is the same whether $X = a_0$ or not. In terms of the three properties, P3 does not apply. If $X < a_0$, P1 is satisfied; if $X \geq a_0$, P2 is satisfied.   ∎

EXAMPLE 2 Trace *binary_search* for the array $A = [3,6]$ with

a) $X = 6$

b) $X = 2$.

SOLUTION

a) In this case, *binary_search* is called with *lower =* 1 and *upper =* 2. Thus the test in line 1 succeeds; mid becomes 1 in line 2, and in line 3 the question "Is $6 \leq 3$?" is asked. The answer is no, so *binary_search* is called with *lower = upper =* 2. This halts immediately and $N = 2$, which is the location of X.

b) In line 1, the question becomes "Is $3 \geq 2$?" The answer is yes, so the algorithm is called again, with *lower = upper =* 1; hence, as we saw previously, it halts with $N = 1$. This is not the location of X, since 2 is not in the array. But this is consistent with P1, since 2 is less than both elements of the array.   ∎

EXAMPLE 3 Trace *binary_search* with $X = 8$ and $X = 11$ for the array $A = [3, 4, 7, 9, 11]$.

SOLUTION   Here is a line-by-line trace with $X = 8$. When the algorithm is initially called, *lower* = 1 and *upper* = 5.

Step	Action
1	$1 \neq 5$, so advance to line 2.
2	mid := 3.
3	$A[3] = 7$. The test, $8 \leq 7$, fails, so advance to line 5.
5	lower := 4.
6	Call *binary_search* with *lower* = 4, *upper* = 5. If we were to trace this, we would follow the same pattern as in the second part of the last example; i.e., the algorithm will halt with $N = 4$.

Note again that 8 is not in the array. The location 4 is the place where it would be inserted if it were placed in the array.

Next, we trace the algorithm with $X = 11$. Again, *lower* = 1 and *upper* = 5 initially.

Step	Action
1	$1 \neq 5$, so advance to line 2.
2	mid := 3.
3	$A[3] = 7$. The test, $11 \leq 7$, fails. Advance to line 5.
5	lower := 4.
6	Call *binary_search* with *lower* = 4, *upper* = 5. This is the same as the first part of example 2. Thus the algorithm halts with $N = 5$. ∎

Verification of *binary_search*

Like any algorithm, *binary_search* must be rigorously verified. Thus let

$$A = [A[1], A[2], \ldots, A[n]], n \geq 1$$

be an array of integers. Assume that $A[i] \leq A[i + 1]$, $1 \leq i \leq n - 1$, and let X be an integer. This assumption is our precondition. The proof proceeds by the second principle of mathematical induction—the rule of inference:

$$\frac{\begin{array}{l} p(1) \\ \forall n \in \mathbf{N} \ (p(1), p(2), \ldots, p(n)) \Rightarrow p(n + 1) \end{array}}{\forall n \in \mathbf{N} \ p(n).}$$

We must show that at the completion of the algorithm, properties P1 through P3 are satisfied, our postcondition. First, let $n = 1$. Then, when the algorithm is called, *lower* = *upper* = 1. Hence the test at the beginning of the loop fails and the algorithm halts with $N = 1$. If $X < A[1]$, P1 is satisfied. If $X \geq A[1]$, P2 is satisfied. P3 does not apply, since when $n = 1$, there is no "otherwise."

Now let $n = 2$. Since P3 was not tested when $n = 1$, we must also consider this case before the induction step. This time $A = [A[1], A[2]]$; *lower* initially equals 1 and *upper* initially equals 2. Thus, when the algorithm is carried out, the test in line 1 succeeds and in line 2, mid becomes 1. Thus, in line 3, $A[1]$ is compared to X. If $X \leq A[1]$, *binary_search* is called with *lower* = *upper* = 1. We already know that this halts with $N = 1$, and thus condition P1 is satisfied. If $X > A[1]$, *binary_search* is called with *lower* = *upper* = 2, and this halts with $N = 2$. Thus, if $X > A[2]$, $N = 2$ and P2 is satisfied. If $X \leq A[2]$, we have $A[1] < X \leq A[2]$ and condition P3 is satisfied. Note that if $A[1] = A[2] = X$, N becomes 1. That is, the leftmost of the two duplicates of X is selected.

Now let $n > 2$. The induction hypothesis is that for any array of length less than n, *binary_search* will halt with N satisfying conditions P1–P3. *Lower* is initially 1 and *upper* is initially n. The test in line 1 fails and mid := $(n + 1)$div 2, which is greater than 1 and less than n. The former is obvious and the latter can be easily shown:

$$n > 1 \Rightarrow 2n > n + 1 \Rightarrow n > (n + 1)/2 \geq (n + 1) \text{ div } 2.$$

Thus, if $A[\text{mid}] \geq X$, *binary_search* is called with *lower* = 1 and *upper* = mid. By the induction hypothesis, if $X \leq A[1]$, N will become 1. By assumption, $X > A[\text{mid}]$ is impossible, so P2 does not apply. Again by induction, if $A[1] < X \leq A[\text{mid}]$, N will be such that $A[N - 1] < X \leq A[N]$. Thus, if $A[\text{mid}] \geq X$, the algorithm is verified. Suppose instead that $A[\text{mid}] < X$. Then *binary_search* is called with *lower* = mid + 1 and *upper* = n. If $X \leq A[\text{mid} + 1]$, $N = \text{mid} + 1$. This satisfies P2, since $A[\text{mid}] < X \leq A[\text{mid} + 1]$. If $X > A[n]$, $N = n$, which satisfies P3. Lastly, if $A[\text{mid} + 1] < X \leq A[n]$, by the induction hypothesis, P2 will be satisfied. Thus *binary_search* is verified.

Complexity of Binary Search

If we let $f(n)$ denote the number of steps required to search a sorted array of length n, we get the following number of computations for binary search, where each comparison, assignment, or arithmetic operation is considered a computation. Note that if $n = 1$, $f(1) = 1$. Assuming $n > 1$,

Step	Number of Computations
1	1
2	3
3	1
4–6	at most, $1 + f(n/2)$

Thus $f(n) \leq 6 + f(n/2)$. If we solve the equation $f(n) = 6 + f(n/2)$, $f(1) = 1$, we get a value for $f(n)$ that is the upper bound for the actual value. Suppose for the moment that n is of the form $n = 2^k$ for $k \geq 1$ an integer. Denoting $f(2^k)$ by $g(k)$, we get $g(k) = 6 + g(k - 1)$ and $g(0) = 1$. This is a recurrence equation, and its solution is $g(k) = 6k + 1$. But $k = \log_2 n$. Thus $f(n) = 6 \log_2 n + 1$ and $f(n) \in O(\log_2 n)$. If n is not of the form 2^k, there is a smallest integer k such that $n < 2^k$. Since f is an increasing function of n,

$$f(n) \leq f(2^k) = 6k + 1 = 6\lceil \log_2 n \rceil + 1.$$

Thus $f \in O(\log_2 n)$ in this case as well.

Quicksort

There are many algorithms available for sorting arrays. We examined *bubble_sort* in Chapter 4 and a recursive version of it in the previous section. As we saw, *bubble_sort* has a complexity of $O(n^2)$. There are a number of sorting algorithms that have an average behavior of $O(n \log n)$. Experimental testing has shown one of these, **Quicksort,** to be consistently one of the fastest. It is a recursive algorithm and is an excellent example of an elegant algorithm. It is based on a very simple idea. One element in the array, usually the first, is selected. The algorithm then decides where it should go in the final sorted array. This is done by comparing all elements to it, shifting all elements smaller than it to the front of the array, and shifting all elements larger than it to the rear. Lastly, the element is shifted to the location where the front and the rear meet. Once this has been done, the array has been divided into two subarrays, both of which are shorter than the original array. The first subarray contains only elements that are smaller than the selected element; the second contains only larger elements. Quicksort is then called (recursively) to sort each of these subarrays. Thus Quicksort is also an example of a divide and conquer algorithm.

ALGORITHM 27   Sorting an Array

Procedure Quicksort(var A, m, n);

{A is the array to be sorted; when called initially, $m = 1$ and n is the length of A. On subsequent calls, m and n are subscripts such that the subarray $A[m], \ldots, A[n]$ is to be sorted. An extra element, $A[n + 1]$, greater than all elements of the array, is assumed to have been appended to it.}

begin

1.   if $n > m$
      then begin

2. $k := A[m]$;   {k denotes the value of the first element of the array}

3. $i := m$;   {i will count up from the left}

4. $j := n + 1$;   {j will count down from the right}
          Repeat

5.   Repeat $i := i + 1$ until $A[i] > k$;   {find an $A[i]$ that must be moved, if any}

6.   Repeat $j := j - 1$ until $A[j] \leq k$;   {find an $A[j]$ that must be moved, if any}

7.   if $i < j$ then switch $A[i]$ and $A[j]$;

8.   until $i \geq j$;

9.   switch $A[m]$ and $A[j]$;

10.   Quicksort(A, m, $j - 1$);

11.   Quicksort(A, $j + 1$, n)

      end

end;

EXAMPLE 4   Trace Quicksort for the one-element array $A = [3]$.

SOLUTION   Quicksort is called with both $m = 1$ and $n = 1$. Thus the test in line 1 fails and the algorithm halts immediately. Since a one-element array is already sorted, this is exactly what we want it to do. ∎

EXAMPLE 5   Trace Quicksort for the array $A = [7, 4]$.

SOLUTION   Initially $m = 1$ and $n = 2$. Assume that the extra element added is $A[3] = 100$.

Step	Action
1	Test succeeds; proceed to step 2.
2–4	$k := 7$; $i := 1$; $j := 3$.
5	$i := 3$ since $A[2] \leq 7$ and $A[3] > 7$.
6	$j := 2$ since $A[2] < 7$.
7	Test fails, so continue to step 8.
8	Test succeeds, so continue to step 9.
9	$A[1]$ and $A[2]$ are switched. $A[1] := 4$ and $A[2] := 7$. Thus $A = [4, 7]$.
10	Quicksort is called with $m = 1$ and $n = 1$. This simply halts.
11	Quicksort is called with $m = 3$ and $n = 2$. This also simply halts. ∎

EXAMPLE 6 Trace Quicksort with $A = [4, 1, 5, 3, 7]$.

SOLUTION Quicksort is called with $m = 1$ and $n = 5$ initially. Assume also that $A[6] - 100$ is appended to the array. The question of why the extra element needs to be appended will be left for the Exercises.

Step	Action
1	$5 > 1$; advance to step 2.
2–4	$k := 4$; $i := 1$; $j := 6$.
5	$i := 3$ since $A[3]$ is the first element from the left to exceed the value of k, namely, 4.
6	$j := 4$ since $A[4]$ is the first element from the right whose value is less than k.
7	$3 < 4$; thus $A[3]$ and $A[4]$ are switched. A becomes $[4, 1, 3, 5, 7]$.
8	$3 < 4$, so this test fails. Return to step 5.
5	$i := 4$ since now $A[4] = 5$, which is greater than k.
6	$j := 3$ since $A[3] = 3 < k$.
7	$4 > 3$, so this test fails. Do not switch $A[3]$ and $A[4]$.
8	$3 \leq 4$; the test succeeds, so proceed to step 9.
9	$A[1]$ and $A[3]$ are switched; A becomes $[3, 1, 4, 5, 7]$.
10	Quicksort is called with $m = 1$ and $n = 2$. Thus the subarray $[3, 1]$ is sorted as we saw in Example 2, yielding $[1, 3]$. Hence A becomes $[1, 3, 4, 5, 7]$.
11	Quicksort is called with $m = 4$, $n = 5$. Thus the subarray $[5, 7]$ is sorted, yielding $[5, 7]$. Thus A becomes $[1, 3, 4, 5, 7]$. ∎

Verification of Quicksort

Let A be an array of integers,

$$A = [A[1], A[2], \ldots, A[n]], \text{ where } n \geq 1.$$

We must show that at the completion of the algorithm, $A[1] \leq A[2] \leq \cdots \leq A[n]$. Once again, the proof will proceed by the second principle of mathematical induction. Our induction hypothesis will be that for any subarray of length 1 up to and including length $n - 1$,

$$A[m] \leq A[m + 1] \leq \cdots \leq A[m + n - 2]$$

at the completion of the algorithm.

First, let $n = 1$. Then $A = [A[1]]$. Thus Quicksort is called with $m = n = 1$, the test in line 1 fails, and the algorithm halts immediately. Thus the condition is satisfied.

Now assume that $n > 1$. The algorithm is called with $m = 1$ and $n > 1$; we also assume that

$$A[n + 1] = M > A[i], \text{ for all } i, 1 \leq i \leq n.$$

The test in line 1 succeeds, so k becomes $A[1]$, i becomes 1, and j becomes $n + 1$. In line 5, i is incremented until $A[i]$ exceeds k. Thus, at the completion of line 5, $A[i] > A[1]$ and $2 \leq i \leq n + 1$. Line 6 is similar; at its completion, $A[j] \leq A[1]$ and $1 \leq j \leq n$. Thus, at line 7, $A[i] > A[j]$.

If, at line 7, $j \leq i$, then j has two properties: (1) all array elements up to the jth are less than or equal to k (since all elements up to the ith are less than k) and (2) all elements after the jth exceed k. No switching is done in this case and the repeat loop halts. Thus, at line 9, $A[1]$ is switched to the jth position and hence $A[i] \leq A[j]$, $i < j$ by property (1). Also, $A[i] > A[j]$, $i > j$ by property (2). By the induction hypothesis, lines 10 and 11 correctly sort the two subarrays and the algorithm is verified.

If $j > i$ at line 7, $A[j]$ and $A[i]$ are switched. Thus, after the switch, all entries from the second up to and including the ith are less than or equal to $A[1]$ and all entries from the jth to the $n + 1$th are greater than $A[1]$. This process continues until $j \leq i$, at which point the argument of the previous paragraph applies. Thus, in either case, the algorithm is verified.

Complexity of Quicksort

The proof that the average case behavior of Quicksort is $O(n \log n)$ is beyond the scope of this text. The worst case is $O(n^2)$; the proof of this is left for the Exercises. In spite of this relatively slow worst case behavior, though, Quicksort is still faster than most other algorithms in experimental tests. Thus it is a good example of the limitations of worst case analysis, i.e., the worst case behavior of Quicksort is worse than that of some other algorithms, such as Mergesort. In spite of this, in experimental tests it consistently does better. Thus it is not sufficient to consider only the worst case behavior of an algorithm when comparing it to another algorithm.

TERMINOLOGY

binary search
Quicksort

SUMMARY

1. Binary search can be written recursively. The version of binary search considered here returns a value N with the following properties:

P1: If $X \leq A[i]$ for all i, $N = 1$.
P2: If $X > A[i]$ for all i, $N = n$.
P3: If $A[1] < X \leq A[n]$, it returns a value N such that
$A[N - 1] < X \leq A[N]$.

These properties imply that when there are duplicate copies of X in the array, the algorithm finds the location of the leftmost copy, i.e., the one with the smallest subscript.

2. Binary search has a complexity of $O(\log_2 n)$; as such, it is very efficient.

3. Quicksort is a recursive algorithm for sorting arrays; it has an average case complexity of $O(n \log n)$. While its worst case performance is $O(n^2)$, it generally performs as well as or better than other algorithms in trials on randomly selected arrays.

EXERCISES 6.6

Trace binary search for the following sets of data.

1. $A = [2]$, $X = 10$
2. $A = [2, 5]$, $X = 10$
3. $A = [2, 5]$, $X = 4$
4. $A = [2, 5]$, $X = 1$
5. $A = [2, 2, 4]$, $X = 2$
6. $A = [2, 2, 4]$, $X = 3$
7. $A = [-1, -1, 3, 5, 5, 5]$, $X = -1$
8. $A = [-2, 3, 11, 14, 17, 23, 27, 31]$, $X = 3$

Algorithm 28 is an alternative form of *Binary_search* that gives the rightmost element if there are duplicates in an array and that satisfies the following properties:

P1$'$: $X < A[1] \Rightarrow N = 1$
P2$'$: $X \geq A[n] \Rightarrow N = n$
P3$'$: otherwise $A[N] \leq X < A[N + 1]$

Trace Algorithm 28 for the following data.

9. $A = [3]$, $X = 3$
10. $A = [3]$, $X = 2$
11. $A = [3, 6]$, $X = 6$
12. $A = [3, 6]$, $X = 2$

ALGORITHM 28   Searching an Array

Procedure Binary_search(X,A; var lower, upper)

{X is the item being searched for; A is the array; lower and upper are the lowest and highest subscripts of A; thus, when initially called, lower $= 1$ and upper $= n$.}

begin

1.   if upper $\neq$ lower then begin
2.   mid := (lower + upper + 1) div 2;
3.   if $X \geq A[\text{mid}]$
4.   then lower := mid
5.   else upper := mid $-$ 1;
6.   binary_search(X, A, lower, upper)
             end
       end;

---

13. $A = [3, 6], X = 4$
14. $A = [3, 4, 7, 9, 11], X = 8$
15. $A = [3, 4, 7, 9, 11], X = 11$
16. $A = [3, 4, 7, 9, 11], X = 10$
17. $A = [2, 7, 7, 9, 13], X = 7$
18. $A = [-1, -1, 3, 5, 5, 5], X = 5$
19. $A = [-1, -1, 3, 5, 5, 5], X = -1$
20. $A = [-1, -1, 3, 5, 5, 5], X = 2$
21. Verify that Algorithm 28 satisfies properties P1'–P3'.
22. Compute the complexity of Algorithm 28. Which is more efficient, Algorithm 26 or Algorithm 28?

In Exercises 23–26, a possible modification of Algorithm 26 is presented. Each of these modifications will cause the algorithm to fail. Show that the modified algorithm fails for the given data set in the way stated.

23. Change line 2 to "mid := (lower + upper + 1) div 2." Let $A = [3, 6]$ and $X = 3$. The algorithm goes into an infinite loop.
24. Change line 3 to "If $X < A[\text{mid}]$." Let $A = [3, 5, 8]$ and $X = 5$. The algorithm sets $N = 3$ instead of $N = 2$.
25. Change line 4 to "then upper := mid -1." Let $A = [3, 6]$ and $X = 3$. The algorithm tries to search from a lower value of 1 to an upper value of 0.
26. Change line 5 to "then lower := mid." Let $A = [3, 5, 8]$ and $X = 6$. The algorithm goes into an infinite loop.

Trace Quicksort for the following arrays.

27. $A = [5]$ 28. $A = [13, 6]$ 29. $A = [5, 3, 1]$ 30. $A = [8, 3, 13, 5, 1]$

Quicksort can be modified to sort from largest to smallest by (1) adding an element $A[n + 1]$ that is less than $A[i]$ for all i between 1 and n and (2) modifying lines 5 and 6 of the algorithm so that they read:

> 5. Repeat $i := i + 1$ until $A[i] < k$
>
> 6. Repeat $j := j - 1$ until $A[j] \geq k$

Trace this version of Quicksort for the following arrays.

31. $A = [3]$ **32.** $A = [7, 4]$ **33.** $A = [13, 6]$ **34.** $A = [3, 1, 5]$

35. $A = [6, 4, 2]$ **36.** $A = [8, 3, 13, 5, 1]$

37. $A = [4, 1, 5, 3, 7]$ **38.** $A = [15, 13, 4, 3, 5, -1]$

39. Verify that the revised version of Quicksort does indeed sort from largest to smallest.

40. Suppose that the extra element $A[n + 1]$ is not added. Construct an example to show why this can cause the algorithm to fail.

41. Suppose line 6 of Quicksort is modified to read "Repeat $j := j + 1$ until $A[j] < k$." Construct an example to show why this can cause the algorithm to fail.

42. Suppose $A = [A[1], \ldots, A[n]]$ is already sorted from smallest to largest. Suppose also that switching two elements of the array requires three computations. Show that the complexity, $f(n)$, of Quicksort is in $O(n^2)$.

REVIEW EXERCISES—CHAPTER 6

1. Write a recurrence equation and an initial condition to model the following situation. Suppose the time required for an algorithm to evaluate n items is the time required to sort the first n plus the time to handle the nth. The time required for the nth item is proportional to $\log n$.

2. Write a difference equation for the functions whose values are given in the following table. Also, rewrite the difference equation as a recurrence equation.

n	$f(n)$
1	2
2	5
3	10
4	17

3. Modify Example 8 of Section 6.1 to reflect the assumption that deer live for 3 years rather than 2 years.

4. Modify your answers to Exercise 3 to reflect the assumption that 3-year-old females give birth to only one fawn. [*Hint:* Write an equation for z_n rather than f_n.]

5. Suppose that the amount of revenue a business produces after n periods of operation is directly proportional to the value of the business at the beginning of that period. Suppose also that all revenues are invested in the business to make it grow as rapidly as possible. Write an equation for u_{n+1}, the value of the business after $n + 1$ periods.

6. Modify your answer to Exercise 5 to allow for the withdrawal of a certain fraction k of the revenues earned in each period for the stockholders' profit.

7. A family is saving money for a child's college tuition. An amount, R, is set aside each year at a fixed rate, r, of interest. Write a recurrence equation for the amount t_n that will be saved after n years.

8. Suppose that after 10 years the family begins to withdraw an amount, T, from the investment each year. Modify your answer to Exercise 7 to reflect this change.

In Exercises 9 and 10, you are given recurrence equation and a possible solution. In each case, test whether the given function is actually a solution.

9. $f(n) = 3 f(n - 1); f(n) = 3^{n-1}$ 10. $f(n) = 2 f(n - 1) + 2; f(n) = 2^n + 2$

Classify each of the following equations as linear or nonlinear. For the linear equations, identify the order, state whether the coefficients are constant, and state whether the equation is homogeneous.

11. $f(n + 2) - 2 f(n - 1) - 4 f(n - 2) = 6$ 12. $f(n) = n! f(n - 1)$

13. $f(n + 3) = 3 f(n + 2) - f(n)$ 14. $f(n) = 2 + 1/(f(n - 1) + f(n - 2))$

Apply Theorem 6.1 to decide whether each of the following linear recurrence equations has a unique solution.

15. $f(n + 1) = 2 + f(n); f(0) = 0, f(1) = 1$

16. $f(n + 3) - 2 f(n - 2) + f(n) = 1; f(0) = f(1) = 1$

17. Show that the principle of superposition does not apply if the equation is non-linear by considering the equation $f(n) = [f(n - 1)]^2$ and its solutions, $f_1(n) = 1$ and $f_0(n) = 0$.

18. Generalize Theorem 6.3 by showing that if g is a solution of the nonhomogeneous equation and f is a solution of its reduced equation, $cf + g$ is a solution for any constant c.

Solve each of the following by the iteration method, and in each case verify your solution.

19. $f(n) = (n - 1) f(n - 1); f(1) = 1$ 20. $f(n) = n + f(n - 1); f(1) = 1$

21. $f(n) + f(n - 1) = 1; f(0) = 1$

Solve the following recurrence equations by the method of substitution.

22. $f(n) = 2 + f(n - 1); f(0) = 0$ 23. $f(n) = 3 + (1/2) f(n - 1); f(0) = 1$

24. Solve the recurrence equation for compound interest (Example 4, Section 6.1).

25. Solve the recurrence equation for simple interest (Example 4, Section 6.1).

Solve the recurrence equations developed in each of the following exercises from Section 6.1.

26. Exercise 19   27. Exercise 23

Using Theorem 6.6 of Section 6.3, solve the following and verify that your solution is correct.

28. Exercise 13, Section 6.1   **29.** Exercise 14, Section 6.1

30. Exercise 20, Section 6.1

For each of the following recurrence equations, find the characteristic equation and use it to find the general solution of the given equation. No complex numbers should appear in your answers. In each case, check the correctness of your answer by substitution. Then find the solution with the specified initial condition.

31. $f(n) - (1/2)f(n-1) - 5f(n-2) = 0, f(0) = 1, f(1) = -1$

32. $f(n) = f(n-1) - (1/4)f(n-2), f(0) = 1, f(1) = 0$

33. $f(n) + 2f(n-1) + 2f(n-2) = 0, f(0) = f(1) = 1$

Find a particular solution for each of the following by the method of undetermined coefficients. Then find the general solution and use it to solve the equation when $f(0) = 0$ and $f(1) = 1$.

34. $f(n) = f(n-1) - f(n-2) + 3$ **35.** $f(n) = f(n-1) - f(n-2) + 2n - 5$

36. $f(n) = f(n-1) - f(n-2) + n^2 + 2n + 1$

37. $f(n) = f(n-1) - f(n-2) + 3^n$

38. In general, the method of undetermined coefficients fails for the equation $f(n) + af(n-1) + bf(n-2) = k^n$ when k is a root of the characteristic equation. Show that the method fails for $f(n) - 3f(n-1) + 2f(n-2) = 2^n$. Find a particular solution for this equation.

Trace each of the specified algorithms with the given set of data. Start with the smallest possible value of n and work up to the largest value. Repeat the trace, starting with the largest value of n and showing the contents of the stack at each recursive call.

39. Algorithm 22, $n = 2$ **40.** Algorithm 22, $n = 4$

ALGORITHM 29   **Drawing a Spiral**

     Procedure Spiro(size, angle, n);

     {This procedure could be used in computer graphics. Size is measured in "pixels"; i.e., dots on a screen. Angle is measured in degrees, and n is a positive integer. A dot starts at the middle of a screen; Forward (size) moves it ahead size pixels and illuminates each pixel it encounters while moving; Left(angle) rotates the direction the dot will move angle degrees counterclockwise.}

     begin

       if $n \neq 0$

       then begin

          forward(size);

          left(angle);

$$\text{Spiro(size} + 10, \text{angle}, n - 1)$$
$$\text{end}$$
$$\text{end;}$$

41. Trace the output of Spiro when size $= 10$, angle $= 10$, $n = 1$.

42. Trace the output of Spiro when size $= 10$, angle $= 10$, $n = 2$.

43. Trace the output of Spiro when size $= 10$, angle $= 20$, $n = 3$.

44. Trace the output of Spiro when size $= 10$, angle $= 30$, $n = 6$.

45. Verify that Algorithm 29 does form a counterclockwise spiral.

46. Compute the complexity of Algorithm 29.

47. Modify Algorithm 29 so that it forms a clockwise rather than a counterclockwise spiral.

Trace binary search for the following sets of data.

48. $A = [-1, -1, 3, 5, 5, 5]$, $X = 5$

49. $A = [-2, 3, 11, 14, 17, 23, 27, 31]$, $X = 9$

Trace Algorithms 26 and 28 for the following data.

50. $A = [1, 2, 7, 12, 13, 14]$, $X = 8$ **51.** $A = [1, 1, 4, 9, 9, 9]$, $X = 9$

52. $A = [1, 1, 4, 9, 9, 9]$, $X = 1$

Trace Quicksort for the following arrays.

53. $A = [5]$ **54.** $A = [13, 6]$

55. $A = [5, 3, 1]$ **56.** $A = [8, 3, 13, 5, 1]$

57. Verify that Quicksort still sorts as desired even if line 5 is modified to read "Repeat $i := i + 1$ until $A[i] \geq k$."

58. Construct an array to illustrate that if Quicksort is modified as suggested in Exercise 57, then the resulting algorithm may be less efficient than the original because it carries out unnecessary switches.

REFERENCES

Aho, Alfred, John Hopcroft and Jeffrey Ullman, *Data Structures and Algorithms.* Reading, Mass.: Addison-Wesley, 1983.

Goldberg, Samuel, *Introduction to Difference Equations.* New York: Wiley, 1958.

Grimaldi, Ralph, *Discrete and Combinatorial Mathematics.* Reading, Mass.: Addison-Wesley, 1985.

Liu, C. L., *Elements of Discrete Mathematics.* New York: McGraw-Hill, 1985.

Stanat, Donald, and David McAlister, *Discrete Mathematics in Computer Science.* Englewood Cliffs, N.J.: Prentice Hall, 1977.

Wilf, Herbert, *Algorithms and Complexity.* Englewood Cliffs, N.J.: Prentice-Hall, 1986.

Combinatorics and Discrete Probability

7

This chapter deals primarily with counting. That sounds simple enough, since counting is something we have all done since childhood. But counting is not always easy, particularly when the set of things we need to count is large. For instance, at the beginning of Chapter 6, we examined the following problem:

Suppose n *diplomats are at a party, and during the festivities every diplomat shakes hands with every other diplomat exactly once. How many handshakes occur?*

One way to solve this problem would be to get a guest list for the party and attempt to enumerate every handshake that occurs. But this is not a very good strategy; it is strongly error prone, and since the solution depends on the particular guest list, it is not general. We would like to have a means of solving problems like this one without listing every possibility, i.e., to count without enumerating. Such means fall within the realm of **combinatorics,** which can be described as the study of methods used to count the number of ways in which an action or sequence of actions can be performed. Most of the actions we will consider involve selecting items of a specified type from a specified finite set.

Combinatorics has important applications in the analysis of algorithms, coding theory, probability, statistics, quality control, genetics, medicine, agriculture, quantum mechanics, communications, and many other disciplines as well. In Sections 7.1 through 7.4, we will examine the fundamental counting principles and a number of applications. Sections 7.5 through 7.7 will focus on one of the most important applications: discrete probability models.

7.1

THE ADDITION AND MULTIPLICATION PRINCIPLES

The Addition Principle

The **addition principle** is the most basic and simplest of the counting principles. It is formally stated as follows:

Addition principle   Let $X = S_1 \cup S_2 \cup \cdots \cup S_n$, where $S_1, S_2, \ldots, S_n$ are disjoint subsets of some finite universal set. Then $|X| = |S_1| + |S_2| + \cdots + |S_n|$ (see Figure 7.1).

EXAMPLE 1   In how many ways can a heart or a diamond be drawn from an ordinary deck of cards?

SOLUTION   Let X denote the set of cards that are hearts or diamonds, let S_1 denote the set of hearts, and let S_2 denote the set of diamonds. Then S_1

Figure 7.1 X is the disjoint union of S_1, S_2, S_3, S_4, S_5, and S_6.

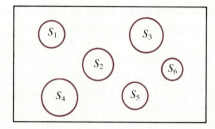

and S_2 are disjoint, since no card is both a heart and a diamond. Since there are 13 hearts and 13 diamonds, $|X| = 13 + 13 = 26$. ∎

EXAMPLE 2 How many ways are there to roll a number between 6 and 8 inclusive with two six-sided dice, one of which is red and the other green?

SOLUTION If we write "12," for instance, we mean that the green die was 1 and the red one was 2. There are five ways to roll a sum of 6 (15, 24, 33, 42, 51), six ways to roll a sum of 7 (16, 25, 34, 43, 52, 61), and five ways to roll a sum of 8 (26, 35, 44, 53, 62). Note that each of these lists forms a disjoint subset of the set of all possible outcomes of rolling the two dice. Thus there are $5 + 6 + 5 = 16$ different ways to roll a number between 6 and 8. ∎

If the dice were not distinguishable, the answer would be different. There would be only three ways to roll a 6—15, 24, 33—since 15 and 51 and 24 and 42 are not distinguishable. Similarly, there are three ways to roll a sum of 7 and three ways to roll a sum of 8. Thus there are $3 + 3 + 3 = 9$ ways to roll a sum between 6 and 8 in this case. ∎

Note that in both of these problems the word *Or* is either used or implied. The card had to be a heart or a diamond, and the sum had to be a 6 or a 7 or an 8. Whenever we have a *single* action to perform, but that action must satisfy one condition or another *when the conditions are mutually exclusive,* we normally use the addition principle. Situations that involve processes consisting of *successive* actions are more likely to involve the multiplication principle.

The Multiplication Principle

Suppose an action consists of a sequence of steps (e.g., rolling a die, then another die, then a third die). We say that the steps are **independent** if the number of ways each step can be done does not depend on the number of ways any of the other steps can be done. (A more precise definition of independence in terms of probability is given in Section 7.7.)

Multiplication principle   Suppose that an action consists of k independent steps. Suppose also that the first step can be done in n_1 ways, the second in n_2 ways, etc. Then the whole action can be done in $n_1 \times n_2 \times \cdots \times n_k$ ways.

If $k = 2$, $n_1 = 3$, and $n_2 = 2$, this can be described by the following decision tree (Figure 7.2). (We first saw decision trees in Section 3.3 in connection with truth tables.)

Figure 7.2   Tree diagram for a two-step decision.

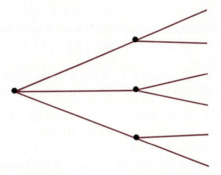

EXAMPLE 3   Suppose that three distinguishable dice are rolled. How many possible outcomes are there?

SOLUTION   Since the dice are distinguishable, we can label them *first, second,* and *third* and can treat the roll as an action with three successive steps. Thus we can look at this experiment as consisting of three successive steps, each with six possible outcomes. The number of possibilities is $6 \times 6 \times 6 = 216$. ∎

Note that if the dice in Example 3 are not distinguishable, the answer would be different. For instance, it would be impossible to distinguish an outcome of 112 from an outcome of 121.

EXAMPLE 4   How many license plates are possible if a license number consists of three letters followed by three digits?

SOLUTION   Let n denote the number of possible license plates. Then

$$n = 26 \times 26 \times 26 \times 10 \times 10 \times 10 = 17{,}576{,}000.$$

Note that if a state has 20 million residents and one car per resident, this scheme would not provide enough license numbers for the state. ∎

EXAMPLE 5   How many seven-digit phone numbers are possible, keeping in mind that neither the first nor the second digit can be a 0 or a 1?

SOLUTION   The number of possibilities is $8 \times 8 \times 10 \times 10 \times 10 \times 10 \times 10 = 64 \times 10^5 = 6.4$ million. This, then, is the maximum number of distinct phone numbers possible in one area code. It explains why there is more than one area code for New York City. ∎

EXAMPLE 6   We can use the multiplication principle to explain why there are $(n - 1)!$ distinct routes for the n-city traveling salesman problem. The salesman chooses any of the $n - 1$ cities to visit first, then any of the $n - 2$ that remain to visit next, then chooses from $n - 3$, etc., until there is only one choice. Hence the total number of possibilities is $(n - 1)(n - 2)(n - 3) \cdots (2)(1) = (n - 1)!$. Note that when counting in this way, the *order* in which the cities are visited is important, i.e., a route such as ABCDEFGA is distinct from AGFEDCBA. Treating routes traversed in opposite orders as the same would reduce the number of possibilities to $(n - 1)!/2$. ∎

At this point, the following theorem is fairly obvious. But it is very useful.

THEOREM 7.1   Suppose r successive decisions are made and there are n choices for each decision. Then there are n^r possible choices for the sequence of decisions.

PROOF   By the multiplication principle, the number of possibilities is $n \times n \times n \times \cdots \times n = n^r$. ☐

EXAMPLE 7   Suppose a number is r digits in length. How many such numbers are possible in base 2? In base 3?

SOLUTION   In base 2, $n = 2$, and thus there are 2^r such integers. In base 3, $n = 3$. Hence there are 3^r such integers. ∎

Situations Involving Both the Multiplication and Addition Principles

Some situations require the application of both principles. For instance, consider Example 8.

EXAMPLE 8   An algorithm begins with an IF ... THEN ... ELSE conditional. Selection of the ELSE option causes the algorithm to halt immediately. Selection of the THEN option causes the algorithm to execute a sequence of two independent conditionals, both of which could result in either the THEN or the ELSE option being chosen. How many sets of test data are needed to test this algorithm?

SOLUTION   The algorithm has the structure of Figure 7.3. We cannot use the multiplication principle directly, since the steps are not indepen-

Figure 7.3   Tree diagram representation of an algorithm.

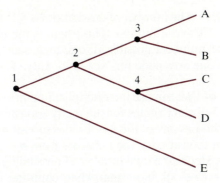

dent, i.e., the number of possibilities after step 1 depends on whether the first condition is true or false. However, if we assume that the first conditional was satisfied, the other two conditionals are independent. Thus there are four possibilities if condition 1 is true by the multiplication principle. There is only one possibility if condition 1 is false. By the addition principle, there are five possible routes through this algorithm, and hence five sets of test data are needed. ∎

The previous example is typical of many counting situations. Actions involve sequences of steps, only some of which are independent. The number of possibilities for parts of the tree diagram consisting of an independent sequence of steps can be computed by using the multiplication principle. Afterward, the number of possibilities for the separate branches of the tree diagram must be added.

EXAMPLE 9   How many license plates are possible that consist of three letters followed by three letters or three digits?

SOLUTION   We have already found that there are 17,576,000 possibilities if three letters are followed by three digits. If three letters are followed by three letters, there are $26^6 = 308,915,776$ possibilities. By the addition principle, the total is 326,491,776 different possible license plates. ∎

EXAMPLE 10   Consider a traveling salesman problem with five cities, A, B, C, D, and E, and with base city A. Suppose also that the salesman cannot visit E until after he has visited B or C. How many possible routes are there?

SOLUTION   If B is visited first, the salesman has three choices for his next visit, two for the next, and one for the last. By the multiplication principle, this is $3! = 6$. Similarly, if C is visited first, there are six choices. If D is visited first and then B, there are two choices, and if D is visited first, followed by C, there are two choices. Thus, by the addition principle, there are 16 possible routes.

Another solution method is this: By the multiplication principle, there are 4! = 24 different routes if we disregard the restriction that B or C must precede E. We can then subtract from 24 the number of routes that violate the restriction. There are 3! = 6 routes that start with E and are thus illegal and two more routes (DEBC and DECB) that start with D and are illegal. Hence again there are 16 routes. ∎

The Inclusion-Exclusion Principle

The addition principle, which we looked at earlier, states that if X is the disjoint union of a collection of sets, S_1, S_2, ... , S_n, then $|X| = |S_1| + |S_2| + \cdots + |S_n|$. However, we must often find the number of elements of a set X that is the union of a collection of sets S_1, S_2, ..., S_n that are *not* disjoint. The inclusion-exclusion principle tells us how to do this in terms of the number of members of the individual sets S_1, S_2, ..., S_n. We will study the principle with $n = 2$ and $n = 3$ in this section and save the general case for Section 7.4.

THEOREM 7.2 (inclusion-exclusion principle for two sets) Let A and B be subsets of $\cup$. Then

$$|A \cup B| = |A| + |B| - |A \cap B|.$$

Figure 7.4 Venn diagram for $A \cup B$.

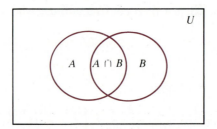

We can intuitively justify this theorem by examining Figure 7.4. When we add together the number of elements in A and the number of elements in B, the elements in $A \cap B$ are counted twice. Thus, to find $|A \cup B|$, we must add $|A|$ to $|B|$ and subtract $|A \cap B|$. The proof is as follows:

PROOF

$$A = (A \setminus B) \cup (A \cap B)$$
$$B = (B \setminus A) \cup (A \cap B)$$

Since $(A \setminus B)$, $(B \setminus A)$, and $(A \cap B)$ are all disjoint and their union is

$A \cup B$, we have, by the addition principle,

$$|A| = |A \setminus B| + |A \cap B|$$
$$|B| = |B \setminus A| + |A \cap B|$$
$$|A \cup B| = |A \setminus B| + |B \setminus A| + |A \cap B|.$$

Thus

$$|A \cup B| = |A| - |A \cap B| + |B| - |A \cap B| + |A \cap B|$$
$$|A \cup B| = |A| + |B| - |A \cap B|. \quad \square$$

EXAMPLE 11   Find the number of cards in an ordinary deck that are spades or aces.

SOLUTION   Let S denote the set of cards that are spades and let A denote the set of cards that are aces. Then $|S| = 13$, $|A| = 4$, and $|A \cap S| = 1$ (namely, the ace of spades). Thus

$$|A \cup S| = |S| + |A| - |A \cap S| = 13 + 4 - 1 = 16. \quad \blacksquare$$

EXAMPLE 12   Of a group of 62 programmers, 35 are familiar with manufacturer A's computers and 41 are familiar with manufacturer B's. If 16 are familiar with neither, how many are familiar with both?

SOLUTION   Let P denote the set of all programmers, let A denote those familiar with type A computers, and let B denote those familiar with type B computers. Then $A \cap B$ consists of those who are familiar with both (see Figure 7.5). Thus

$$|A \cup B| = 62 - 16 = 46$$
$$|A| = 35$$
$$|B| = 41$$
$$|A \cup B| = |A| + |B| - |A \cap B|$$
$$46 = 35 + 41 - |A \cap B|$$
$$|A \cap B| = 30. \quad \blacksquare$$

Figure 7.5   Venn diagram for Example 12.

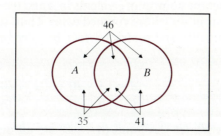

THEOREM 7.3 (inclusion-exclusion principle for three sets; see Figure 7.6) Let A, B, and C be subsets of U. Then

$$|A \cup B \cup C| = |A| + |B| + |C| - |A \cap B|$$
$$- |A \cap C| - |B \cap C| + |A \cap B \cap C|.$$

Figure 7.6 $A \cup B \cup C$ as a disjoint union of subsets.

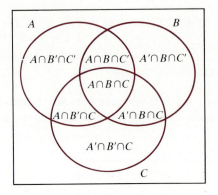

PROOF This proof is a generalization of Theorem 7.2; the details are left for the Review Exercises. Note that the sum $|A| + |B| + |C|$ counts $|A' \cap B \cap C|$, $|A \cap B' \cap C|$, and $|A \cap B \cap C'|$ twice and $|A \cap B \cap C|$ three times. □

EXAMPLE 13 How many numbers are there between 1 and 1000 inclusive that are neither perfect squares, perfect cubes, nor perfect fourth powers?

SOLUTION Let the universal set U be the integers between 1 and 1000. A denotes the set of perfect squares in U, B the set of perfect cubes, and C the set of perfect fourth powers. First, we find the number of integers that are perfect squares ($|A|$), cubes ($|B|$), or fourth powers ($|C|$). Then

$$|A| = 31, \text{ since } 31^2 = 961 \text{ but } 32^2 = 1024.$$
$$|B| = 10, \text{ since } 10^3 = 1000.$$
$$|C| = 5, \text{ since } 5^4 = 625 \text{ but } 6^4 = 1296.$$

We must also find $|A \cap B|$, $|A \cap C|$, $|B \cap C|$, and $|A \cap B \cap C|$. $A \cap B$ consists of integers that are both squares and cubes, i.e., are of the form n^6 for some n. Since $3^6 = 729$ and $4^6 = 4096$, $|A \cap B| = 3$. Since every fourth power is also a square, $A \cap C = C$. $B \cap C$ consists of integers that are both cubes and fourth powers, i.e., are of the form n^{12}. Since every twelfth power is also a square, $B \cap C = A \cap B \cap C$. Since $2^{12} = 4096$,

$|B \cap C| = |A \cap B \cap C| = 1.$ Hence

$$|A \cap B| = 3$$
$$|A \cap C| = 5$$
$$|B \cap C| = |A \cap B \cap C| = 1$$

and thus

$$|A \cup B \cup C| = |A| + |B| + |C| - |A \cap B|$$
$$- |A \cap C| - |B \cap C| + |A \cap B \cap C|$$
$$= 31 + 10 + 5 - 3 - 5 - 1 + 1$$
$$= 38.$$

Hence there are 962 integers between 1 and 1000 that are not perfect squares, cubes, or fourth powers (see Figure 7.7). ∎

Figure 7.7 Numbers of squares, cubes, and fourth powers between 1 and 1000.

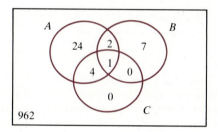

TERMINOLOGY

combinatorics   multiplication principle
addition principle   inclusion-exclusion principle
independent

SUMMARY

1. *Addition principle* Let $X = S_1 \cup S_2 \cup \cdots \cup S_n$ where $S_1, S_2, \ldots, S_n$ are disjoint subsets of some universal set. Then $|X| = |S_1| + |S_2| + \ldots + |S_n|$.

2. *Multiplication principle* Suppose that an action consists of k independent steps. Suppose that the first step can be done in n_1 ways, the second in n_2 ways, etc. Then the whole action can be done in $n_1 \times n_2 \times \cdots \times n_k$ ways.

3. If there are n choices for each of r successive decisions, there are n^r possible choices for the sequence of decisions.

4. Let A and B be subsets of U. Then $|A \cup B| = |A| + |B| - |A \cap B|$.

5. Let A, B, and C be subsets of U. Then
$$|A \cup B \cup C| = |A| + |B| + |C| - |A \cap B|$$
$$- |A \cap C| - |B \cap C| + |A \cap B \cap C|.$$

EXERCISES 7.1

Find the number of ways to do each of the following.

1. Roll two distinguishable dice and get a sum of 7 or 11.

2. Roll two indistinguishable dice ad get a sum of 7 or 11.

3. Roll two distinguishable dice and get doubles.

4. Roll two indistinguishable dice and get doubles.

5. Select a face card or an ace from a deck of 52 playing cards.

6. Select a card that is a 5 or less, assuming an ace is treated as a 1.

A die in the shape of a tetrahedron has four sides numbered 1 through 4. How many ways are there to roll two such dice and get

7. a sum of 3 if the dice are distinguishable? Of 4? Of 5 or less?

8. a sum of 3 if the dice are indistinguishable? Of 4? Of 5 or less?

How many different possibilities are there in each of the following situations?

9. Rolling four distinguishable six-sided dice.

10. Rolling three distinguishable four-sided dice.

11. A license plate with no more than seven characters. The first cannot be blank, but any of the rest can be either a letter or a blank.

12. Another possible license plate pattern with up to six characters, each of which may be a blank, a letter, a digit, or one of the four special characters—!, *, #, |. The only restriction is that the license plate cannot consist of all blanks.

13. A combination lock that consists of three successive turns of the knob, halting on any of 40 different numbers.

14. The same as Exercise 13, assuming that each number in the combination is different.

15. A true-false test consisting of 12 questions.

16. In early versions of the computer language BASIC, identifiers consisted of a single letter followed by an optional digit followed by an optional dollar sign. How many such identifiers are possible?

17. Later versions of BASIC allowed more freedom in specifying identifiers. The second character is still optional, but it can be a letter or a digit. Similarly, the third character can be a dollar sign, a pound sign (#), an exclamation point, or a percentage sign, or it can be omitted. Now how many identifiers are possible?

18. A traveling salesman is visiting cities B, C, D, and E, starting from A. Each city is joined by two roads, not just one. A route consists of a sequence of roads that visit each city once and return to A. How many routes are possible?

19. This is the same as Exercise 18, except that there are three roads joining A and B.

Answer each of the following questions.

20. A combination lock for a bicycle consists of four digits, each of which is between 1 and 6. You have forgotten the combination and decide to try all of the combinations in order. If it takes an average of 2 seconds to test a combination, what is the maximum time needed to open the lock?

21. An automobile manufacturer advertises that no two cars of that make are identical. Sixteen different colors are possible, as well as three engine sizes, four body types, and a choice between cloth and plastic seats. In addition, the following options may be included: radio, tape deck, air conditioning, tinted glass, rear window defogger, automatic transmission, and cruise control. The manufacturer sells about 100,000 cars per year. Is the claim reasonable?

22. Construct a tree diagram to illustrate the solution to Example 10.

23. An algorithm consists of a series of five conditionals. If any one of them is false, the algorithm terminates. If one is true, the algorithm proceeds to the next conditional. Illustrate this situation with a tree diagram and count the number of paths through the algorithm.

Answer the following questions, using the inclusion-exclusion principles.

24. Guinea pigs are either long- or short-haired and golden in color or not. Of 492 guinea pigs, 401 are short-haired and 42 are golden; 10 are both long-haired and golden. How many are neither long-haired nor golden?

25. A biologist studying the effect of pollution in the Great Lakes nets 242 salmon. It is found that 147 have excessive chemical pollutants in their tissues. Of these, 97 have an excess of polychlorinated biphenyls and 106 have an excess of other pollutants. How many fish have an excess of both? How many have an excess of neither?

26. How many digits are there between 1 and 600 inclusive that are divisible by either 3 or 5?

27. How many digits are there between 1 and 1000 that are not divisible by either 2 or 3?

28. A year is a leap year if it is divisible by 4 and not divisible by 100 except when it is divisible by 400. How many leap years are there between year 1 and year 2000 inclusive?

29. Show that the inclusion-exclusion principles both reduce to the addition principle when the sets involved are disjoint.

7.2
PERMUTATIONS

How many different ways are there to order a deck of cards, for a group of people to order themselves in a waiting line, or for a collection of cities to be listed in the order in which they are to be visited? These are all situations that involve successive independent steps, and thus all involve applications of the multiplication principle. But they have an added feature: Each step is a selection of one item from a specified set such that once an item has been selected, it is not replaced. That is, once an item has been selected to be first, it is no longer available for any subsequent position. A second item is then selected, and so on, until all items have been selected. Such situations occur so often that they have been extensively studied. They are modeled mathematically by the concept of the permutation, which we define formally as follows:

Definition   A **permutation** is a one-to-one function from a finite set A onto itself.

Less formally, a permutation can be defined as an arrangement of the items of A. To understand the connection between the arrangements and the formal definition, suppose $|A| = n$. We can think of the elements of A as occupying each of n slots or positions, numbered from 1 to n. An arrangement is a function, then, because no element can occupy two positions. It is one-to-one, since no two elements can occupy the same position, and it is onto since each position has to be filled. Going back to our original problems, any order of the cards in a deck is a permutation, since it is an arrangement of the cards.

Counting the number of permutations of a set of n items is a common task.

THEOREM 7.4   If $|A| = n$, there are $n!$ permutations of the elements of A.

PROOF   Imagine the elements of A arranged in the order $\{a_1, a_2, \ldots, a_n\}$. There are n positions that a_1 could be mapped to, $n - 1$ positions that a_2 could be mapped to (since one position was used by a_1), $n - 2$ possible positions for a_3, etc. Finally, there is one possible position for a_n. Using the multiplication principle, there are thus $n \cdot (n - 1) \cdot (n - 2) \cdots (1) = n!$ permutations.   ▢

One way to visualize this situation is to list the n positions that each item could be placed in horizontally and then over each position list the number of possibilities that could be placed in it.

$$\frac{n}{1\text{st}} \quad \frac{n-1}{2\text{nd}} \quad \frac{n-2}{3\text{rd}} \quad \frac{\cdot}{\cdot} \quad \frac{\cdot}{\cdot} \quad \frac{\cdot}{\cdot} \quad \frac{1}{n\text{th}}$$
$$\text{Position}$$

Thus, applying the multiplication principle, we can again see that the total number of possibilities is $n!$.

EXAMPLE 1   In scrambled word games, one is asked to find the unscrambled word that could be constructed from a collection of letters. For instance, RECOV, when unscrambled, is COVER. How many arrangements of the letters of RECOV are possible?

SOLUTION   This is a classic example of a permutation problem; each letter must be used and put in one of the five places. Hence the number of permutations is 5! = 120. ∎

EXAMPLE 2   How many arrangements of LINEAR are possible?

SOLUTION   The number of permutations is 6! = 720. Comparing this answer to the previous one, it is easy to see why six-letter words are so much harder to unscramble than five-letter words. ∎

EXAMPLE 3   We have said from the beginning that in the traveling salesman problem with n cities, there are $(n - 1)!$ distinct routes starting at a given city and returning to the same city. If we number the cities from 1 through n and start and end at city 1, any list of the cities in the order visited contains a permutation of the set $\{2, 3, \ldots, n\}$. Thus permutations give us an easy way to see why there are $(n - 1)!$ routes. ∎

EXAMPLE 4   How many different ways are there to arrange a 52-card deck?

SOLUTION   There are 52! ways. This is approximately 8.07×10^{67}. ∎

EXAMPLE 5   Seven administrators are to be seated in a single row on a platform at a commencement ceremony. In how many different ways can their seats be assigned?

SOLUTION   Again, the answer is 7!, which equals 5040. ∎

Permutations and Recursion

The permutations of $A = \{1, \ldots, n\}$ can be listed recursively. Suppose $n = 1$. Then $\{1\}$ has only one permutation, namely, itself. Now suppose $n > 1$ and consider $A_1 = \{1, \ldots, n - 1\}$. Suppose $i_1, i_2, \ldots, i_{n-1}$ is a permutation of A_1. Then n permutations of A can be constructed in a manner similar to that of Example 3 of Section 6.1—by inserting n in each possible location in $i_1, i_2, \ldots, i_{n-1}$:

$$n, i_1, i_2, \ldots, i_{n-1}$$
$$i_1, n, i_2, \ldots, i_{n-1}$$
$$\cdot$$
$$\cdot$$
$$\cdot$$
$$i_1, i_2, \ldots, i_{n-1}, n.$$

This can be done for each of the permutations of A_1, yielding the permutations of A. The proof that this process yields all of the permutations of A is left for the Exercises. For instance, the only permutation of $\{1\}$ is 1. Thus the permutations of $\{1, 2\}$ are

$$12 \qquad 21.$$

Similarly, when inserting 3 in each possible location in the list of permutations of $\{1, 2\}$, the permutations of $\{1, 2, 3\}$ are

$$312 \quad 132 \quad 123 \quad 321 \quad 231 \quad 213.$$

This process also gives us a recurrence equation that can be solved for $f(n)$, the number of permutations of $\{1, \ldots, n\}$. From the previous discussion, we have

$$f(1) = 1$$
$$f(n) = n \cdot f(n-1).$$

Solving this recurrence equation with the given initial condition gives $f(n) = n!$, the same formula we derived using the multiplication principle (see Example 6, Section 6.3).

r-Permutations

In one typical application of permutations, we have an n-element set and we select all n, one after another. But there are a number of situations in which we do not want to select all n, but rather some number of items, r, that is less than n. This situation is modeled by the concept of the r-permutation.

Definition   Suppose $|A| = n$ and $1 \leq r \leq n$. An **r-permutation** of A is a permutation of any r-element subset of A. An r-permutation is also called a **permutation of n things taken r at a time**. The number of r-permutations of a set of size n is denoted **$P(n,r)$**.

EXAMPLE 6   Suppose $A = \{1, 2, 3, 4\}$. Find all r-permutations of A for $r = 1, 2,$ and 3.

SOLUTION   We can list the r-permutations as follows:

  1–permutations: 1, 2, 3, 4

  2–permutations: 12, 13, 14, 21, 23, 24, 31, 32, 34, 41, 42, 43

  3–permutations: 123, 124, 132, 134, 142, 143, 213, 214, 231, 234,
           241, 243, 312, 314, 321, 324, 341, 342,
           412, 413, 421, 423, 431, 432. ∎

The following theorem is used to count r-permutations.

THEOREM 7.5   Let $|A| = n$ and suppose $1 \le r \le n$. Then the number of r-permutations of A is

$$P(n,r) = \frac{n!}{(n-r)!}.$$

PROOF   By the multiplication rule, there are $n \cdot (n-1) \cdot \cdots \cdot (n-r+1)$ possibilities. This can be written

$$\frac{n(n-1)\cdots(n-r+1)(n-r)\cdots(1)}{(n-r)(n-r-1)\cdots(1)} = \frac{n!}{(n-r)!}. \quad \square$$

This can also be visualized by a list of positions with the number of possibilities for each position written above it:

$$\frac{n}{\text{1st}} \quad \frac{n-1}{\text{2nd}} \quad \frac{n-2}{\text{3rd}} \quad \frac{\cdot}{\cdot} \quad \frac{\cdot}{\cdot} \quad \frac{\cdot}{\cdot} \quad \frac{(n-r+1)}{r\text{th}}.$$

$$\text{Position}$$

Definition   In order to be consistent with the formula in Theorem 7.5, we adopt the convention that $P(n,0) = 1$ for every $n \ge 0$. We can think of $P(n,0)$ as counting 0-permutations—one-to-one, onto functions from the empty set to itself. But there is only one such function—the empty function.

EXAMPLE 7   Find the number of 1-, 2-, and 3-permutations of $\{1, 2, 3, 4\}$.

SOLUTION

$$P(4,1) = \frac{4!}{3!} = \frac{4 \cdot 3!}{3!} = 4$$

$$P(4,2) = \frac{4!}{2!} = \frac{4 \cdot 3 \cdot 2!}{2!} = 12$$

$$P(4,3) = \frac{4!}{1!} = \frac{24}{1} = 24.$$

Note that these numbers are consistent with the lists given in Example 6. The values of $P(n,r)$ verify the correctness of the preceding lists; for instance, consider the list of 2-permutations. We know that $P(4,2) = 12$. Since our list has 12 2-permutations and they are all distinct, we have a complete list. ∎

EXAMPLE 8   Consider a traveling salesman with a base city and seven additional cities to visit. He decides to visit four cities one day and three the next. How many different routes are possible for the first day? How many are possible for the second day without regard to which cities were visited the first day?

SOLUTION   On the first day, the salesman can follow $P(7,4) = 7!/3!$ $= 7 \cdot 6 \cdot 5 \cdot 4 = 840$ different itineraries. On the second day, he can follow $P(7,3) = 7!/4! = 210$ itineraries.

Note that the product of these two numbers, 176,400, is *not* the number of routes he could follow to visit all seven cities, namely, $7! = 5040$. To see why, consider a specific route that the salesman might follow the first day. After visiting four cities, there are three left, and hence only $3! = 6$ possible itineraries from that point. But in the preceding computation, we looked at the number of possible routes for the second day *without regard for what had happened the first day.* Hence the number of possibilities is substantially increased.   ∎

EXAMPLE 9   A school board consists of nine members. If there are four officers—president, vice-president, secretary, and treasurer—how many different slates of officers are possible?

SOLUTION   The school board is a nine-member set. Four members are to be selected in a certain order. Thus there are $P(9,4) = 9!/5! = 3024$ slates of officers.   ∎

EXAMPLE 10   How many different three-letter words can be made from the English alphabet if we do not insist that the words have vowels or make sense, but do insist that all letters be distinct?

SOLUTION   There are $P(26,3) = 26!/23! = 26 \cdot 25 \cdot 24 = 15,600$.   ∎

Circular Arrangements

One important variation on the notion of counting permutations or *r*-permutations is finding the number of arrangements when the members of a set are arranged in a circle. Typically in this case, the actual positions do not matter but only the relative positions, i.e., which items are next to each other. For instance, consider Example 11.

EXAMPLE 11   Twenty members of the diplomatic corps are invited to a state dinner and are to be seated around a circular table. How many different arrangements are possible?

SOLUTION   What is important in this situation is who sits next to whom. The first diplomat selected can be placed anywhere. Starting from that diplomat and moving around the table clockwise or counterclockwise, there are $P(19,19) = 19! \cong 1.22 \times 10^{17}$ possibilities.

An alternative way to solve this problem is first to compute the number of arrangements as if the exact position mattered. This would yield 20! different arrangements. Then consider one of these arrangements. It could be rotated through 20 different positions without changing the relative locations of anyone. Thus every arrangement is equivalent to 20 others. Hence there are 20!/20 arrangements that are not equivalent, and this is 19!, as before. ■

EXAMPLE 12   Suppose the 20 individuals in the previous example consist of 10 diplomats and their spouses. How many arrangements are possible that alternate men and women?

SOLUTION   Any individual can be placed anywhere at the start. The 10 individuals of the opposite sex are placed in the next 10 alternating seats. Thus there are 10! possibilities for them. The remaining nine individuals are placed in the remaining seats, and there are 9! possibilities. By the multiplication principle, there are $10! \cdot 9!$ possibilities overall, which is approximately 1.32×10^{12}. ■

Permutations Involving Duplicates

When we write $A = \{a_1, a_2, \ldots a_n\}$, we assume that all of the a_i values are distinct. Thus, when we speak about the permutations of an n-element set, we are assuming that the n elements are all different. However, there are a number of situations in which we must count the number of arrangements that are possible for a collection of n objects that are not all distinct. For the moment, we will consider only permutations, not r-permutations. In this case, we must modify some of our earlier procedures. For instance, consider Example 13.

EXAMPLE 13   Find the number of distinct four-letter words that can be formed from the letters SEED.

SOLUTION   If the E's were different letters, we know that there would be $4! = 24$ possibilities. If we list all 24 of them, distinguishing the two E's by making one lowercase, we get

SEeD	SEDe	SeED	SeDE	SDEe	SDeE
ESDe	ESeD	EeSD	EeDS	EDSe	EDeS
DESe	DEeS	DeES	DeSE	DSEe	DSeE
eESD	eEDS	eDSE	eDES	eSED	eSDE

If we now replace the lowercase e with an uppercase E, SEeD and SeED, for instance, would both become SEED. Similarly, SEDe and SeDE would become SEDE. That is, every entry in the table has one other entry equivalent to it if the E's are taken as identical. Thus there are 12 distinct possible words that can be formed from SEED. ∎

THEOREM 7.6 Suppose that among n objects there are k distinct types of objects. Suppose also that there are n_1 objects of type 1, n_2 of type 2, up to n_k of type k, and $n_1 + n_2 + \cdots n_k = n$. Then the number of distinct permutations of the n objects is

$$\frac{n!}{n_1!\, n_2! \cdots n_k!}.$$

PROOF There are $n!$ permutations of the n objects if all are distinguished. Consider any one specific permutation. Since there are $n_1!$ permutations of the type 1 objects, there are $n_1!$ permutations of the n objects that have the objects not of type 1 all in the same place (see Figure 7.8).

Figure 7.8 There are $3! = 6$ arrangements of the x_i values that leave the rest of the permutation the same.

$a\; b\; c\; x_1\; d\; e\; x_2\; x_3\; f\; g$

Thus there are only $\dfrac{n!}{n_1!}$ permutations if the type 1 objects are not distinguished. Proceeding by mathematical induction, we can assume that there are

$$\frac{n!}{n_1!\, n_2! \cdots n_{k-1}!}$$

different permutations if objects of types 1 through $k-1$ are not distinguished. By the same argument used with the type 1 objects, we can conclude that there are

$$\frac{n!}{n_1!\, n_2! \cdots n_k!}$$

different permutations if the objects of type k are not distinguished either. □

Definition The expression

$$\frac{n!}{n_1!\, n_2! \cdots n_k!}$$

is often denoted $P(n; n_1, n_2, \ldots, n_k)$.

EXAMPLE 14   How many distinct six-letter words can be formed from the word LETTER?

SOLUTION   There are six letters: one L, two E's, two T's, and one R. Thus there are

$$P(6;1,2,2,1) = \frac{6!}{1!2!2!1!} = \frac{720}{4} = 180$$

possible words. ∎

EXAMPLE 15   How many distinct arrangements are there of the letters in MISSISSIPPI?

SOLUTION   There are 11 letters: one M, four I's, four S's, and two P's. Thus there are

$$P(11;1,4,4,2) = \frac{11!}{1!4!4!2!}$$

possibilities, totaling 34,650. ∎

EXAMPLE 16   How many different arrangements are there of a deck of cards if the suits are distinguished and the denominations of the cards within the suits are not?

SOLUTION   The number of arrangements is $P(52;13, 13, 13, 13) =$

$$\frac{52!}{13!13!13!13!}$$

which is approximately 5.36×10^{28}. ∎

EXAMPLE 17   A literature student has five novels by Dickens, two by Melville, two by Tolstoy, and four by Dostoevsky. In how many ways can these novels be arranged on a shelf if the student doesn't distinguish different books by the same author?

SOLUTION   There are 13 novels altogether. Thus there are $P(13;5,2,2,4) =$

$$\frac{13!}{5!2!2!4!}$$

possibilities, totaling 540,540. ∎

r-Permutations of Sets with Duplicate Elements

What about *r*-permutations of sets that involve duplicates? Unfortunately, this is quite a difficult problem. For instance, consider a fairly easy example: How many 2-permutations are there of the letters XXYZ? If the letters are all distinct, the answer is simple—$P(4,2) = 12$. But there are two X's, so some of the 12 permutations will be identical. Thus the result will be less than 12. One way to find out how much less than 12 this result will be is to list the 2-permutations of XXYZ. Doing so, we get XX, XY, XZ, YZ, YX, ZX, ZY; that is, there are seven 2-permutations of XXYZ. Although this method works in this case, listing as a means of counting is not a good technique; it is too easy to omit items and is unworkable for large numbers. Listing can be improved on for this problem, but the method is still cumbersome. It is as follows: Since the X is duplicated, we ask how many 2-element subcollections of XXYZ there are that contain two X's, that contain one X, and that contain no X's. These are XX, XY, XZ, and YZ, respectively. We now count the number of permutations of each of these subcollections and then use the addition principle to find our answer. In this case, the answer is $1 + 2 + 2 + 2 = 7$. This method can be generalized to larger problems, although, for r-permutations, listing the r-element subsets of the original set may prove difficult. Once they are listed, though, the methods presented here for calculating the number of permutations of the r elements with possible duplicates can be applied. The methods presented in the next section will help us to find the subsets of size r.

Generating Permutations

At this point, it should be clear that the number of permutations of even a relatively small set is quite large. Also, increasing the size of a set causes the number of its permutations to grow very rapidly. However, there are occasional situations in which we may want to list all of the permutations of a set, may want a computer to sort through them to determine which satisfy some given condition, or may simply want to generate *one* permutation of a set. The recursive process we examined earlier can be used to generate all of the permutations of $\{1, \ldots, n\}$, although it is very inefficient. The algorithm considered in this section can be used to list all of the permutations of an array $A = [A[1], A[2], \ldots, A[n]]$, but it can also be used to accept one permutation as input and generate another as output. We will examine the idea behind it by first looking at the following list of all permutations of $\{1, 2, 3, 4\}$.

1234	1423	2314	3124	3412	4213
1243	1432	2341	3142	3421	4231
1324	2134	2413	3214	4123	4312
1342	2143	2431	3241	4132	4321

If this table is read vertically, starting at the left, the permutations appear in increasing numerical order. More generally, this order is called the **lexicographic order**. For numbers, lexicographic order is numerically increasing order; for strings, it is alphabetic order. The following algorithm takes one permutation as an input and generates the next one in lexicographic order as its output.

This algorithm works as follows: It starts with a given permutation and then selects two elements of the permutation to switch. After these elements are selected, the remaining elements are rearranged. Specifically, the first of the two elements is selected by starting at the right side of a permutation and finding the first $A[k]$ such that $A[k] < A[k+1]$. This one is selected, since all elements from $A[k+1]$ to $A[n]$ are in decreasing order from left to right, and switching any of them would not give a permutation that *follows* the current one in lexicographic order. Then start again at the right and find the first element $A[m]$ such that $A[m] > A[k]$. This element is the smallest number to the right of $A[k]$ that can be switched with $A[k]$, yielding a permutation following the current one. Switch $A[m]$ and $A[k]$. The entries to the right of the new $A[k]$ will now be in decreasing order, since the new $A[m]$ is less then the old one but is greater than $A[m+1]$ to $A[n]$. Thus the entries from $A[k+1]$ to $A[n]$ are reversed in order and the algorithm is done. The algorithm is formally stated as Algorithm 30.

ALGORITHM 30   Generating the Next Permutation in Lexicographic Order

Procedure Get_next_perm(n; var A);

{A is an array of length n. Its data can be of any type that are ordered lexicographically. If $n = 1$ or A is the last permutation, the algorithm will fail. Thus these conditions need to be checked for in the calling program.}

begin

1. $k := n - 1$;

2. while $A[k] > A[k+1]$ do   {find first $A[k]$ such
      $k := k - 1$;   that $A[k] < A[k+1]$}

3. $m := n$;

4. while $A[m] < A[k]$ do   {find first $A[m]$
      $m := m - 1$;   greater than $A[k]$}

5. switch $A[m]$ and $A[k]$;

6. reverse the order of $A[k+1]$ to $A[n]$

end;

EXAMPLE 18   Trace *get_next_perm* for $n = 4$ and $A = [3, 1, 4, 2]$.

SOLUTION

Step	Action
1	$k := 3$.
2	$k := 2$, since 1 is the first entry not in increasing order from the right.
3	$m := 4$.
4	$A[4]$ (which equals 2) exceeds $A[2]$ (which equals 1), so m is unchanged.
5	Switch $A[4]$ and $A[2]$, giving $A = [3, 2, 4, 1]$.
6	Reverse the order of $A[3]$ to $A[4]$, giving $A = [3, 2, 1, 4]$. ∎

EXAMPLE 19 Trace *get_next_perm* for $n = 4$ and $A = [3, 4, 2, 1]$.

SOLUTION

Step	Action
1	$k := 3$.
2	$k := 1$, since 3 is the first entry not in increasing order from the right.
3	$m := 4$.
4	$A[4] = 1$, $A[3] = 2$, both less than $A[1]$. $A[2] = 4 > A[1]$, so $m := 2$.
5	Switch $A[1]$ and $A[2]$, giving $A = [4, 3, 2, 1]$.
6	Reverse the order of $A[2]$ to $A[4]$, giving $A = [4, 1, 2, 3]$. ∎

EXAMPLE 20 Trace *get_next_perm* with $n = 7$, $A = [7, 3, 6, 5, 4, 2, 1]$.

SOLUTION

Step	Action
1	$k := 6$.
2	$k := 2$, since 3 (i.e., $A[2]$) is the first entry less than the one to its right.
3	$m := 7$.
4	$m := 5$, since 4 (i.e., $A[5]$) is the first entry greater than $A[2]$.
5	A becomes $[7, 4, 6, 5, 3, 2, 1]$.
6	A becomes $[7, 4, 1, 2, 3, 5, 6]$. ∎

Get_next_perm has a worst case complexity of $O(n)$. However, when it is used in a loop to generate all permutations of A, the resulting loop will have a complexity between $n!$ and $n \cdot n!$. Thus, for any practical problem, such a loop cannot be used unless n is 13 or 14 at most.

TERMINOLOGY

permutation	$P(n,r)$
r-permutation	$P(n; n_1, n_2, \ldots, n_k)$
permutation of n things taken r at a time	lexicographic order

SUMMARY

1. If $|A| = n$, there are $n!$ permutations of the elements of A.

2. Let $|A| = n$ and suppose $0 \le r \le n$. Then the number of r-permutations of A is

$$P(n,r) = \frac{n!}{(n-r)!}.$$

3. Suppose that among n objects there are k distinct types of objects. Suppose also that there are n_1 objects of type 1, n_2 of type 2, up to n_k of type k, and $n_1 + n_2 + \cdots n_k = n$. Then the number of distinct permutations of the n objects is

$$\frac{n!}{n_1! n_2! \cdots n_k!}.$$

4. Algorithm 30 can be used to generate the permutations of $\{1, \cdots, n\}$ in lexicographic order.

EXERCISES 7.2

Evaluate the following:

1. $P(10,10)$ 2. $P(10,4)$
3. $P(10,0)$ 4. $P(8,1)$
5. $P(10;3,3,4)$ 6. $P(9;2,4,4,1)$

How many different ways are there to do the following?

7. Scramble the letters of SCRAMBLE.

8. Arrange the letters of COURSE.

9. Arrange 20 different books on a shelf.

10. Seat 12 trustees in a row on a platform at a commencement ceremony.

Solve the following problems.

11. Using recursion, list all of the permutations of $\{a, b, c\}$.

12. Using recursion, list all of the permutations of $\{1, 2, 3, 4\}$.

13. In Exercise 9, suppose that two of the books are by the same author. How many arrangements are there that place the two together?

14. Nine horses are in a race. How many possible finishes are there, not counting ties?

15. Nineteen students write computer programs and submit them to a printer queue at about the same time. The queue handles them one at a time. How many different arrangements of the queue are there?

16. A baseball team has 20 players but only 9 can start a game. How many starting lineups are possible when a lineup is a list of nine players arranged in order by position?

17. A total of 110 students are competing for first, second, and third prize in an art contest. How many different outcomes are possible?

18. How many seven-digit phone numbers are there in which all of the digits are different? How many are there in which digits may be used more than once?

19. Users of a computer's operating system are often placed in a circular queue, a list through which the operating system continuously cycles, serving each person in succession. How many different such queues can be formed of 18 users if relative position in the queue matters but absolute position does not?

20. In Exercise 19, how many queues can be formed in which Marie and Frank are placed next to each other? How many queues can be formed in which not only Marie and Frank but Hank and Judy as well are placed next to each other?

21. How many distinct three-letter sequences can be formed from the letters of SCRAMBLE?

22. Each of three countries sends six diplomats to a conference. They are seated around a circular table. How many seating arrangements are possible if the individuals are designated by their country rather than by their individual names?

23. In Exercise 22, how many arrangements are possible if they are seated in the order country A, country B, country C, country A, country B, etc.?

24. In Exercise 22, how many arrangements are possible in which all of the diplomats from each country sit next to each other?

How many arrangements are there of the letters of the following words?

25. DISCRETE

26. MATHEMATICS

27. ELIGIBLE

28. ARRANGEMENT

29. In how many ways can four copies of *War and Peace,* three copies of *Anna Karenina,* three copies of *The Brothers Karamazov,* and two copies of *Notes from the Underground* be arranged on a shelf?

30. In Exercise 29, another copy of *Anna Karenina* has been donated to the library. By what factor does this increase the number of possible arrangements?

31. A bridge hand consists of five spades, four hearts, two diamonds, and two clubs. In how many ways can the cards be arranged?

32. A different bridge hand consists of four hearts, three spades, three diamonds, and three clubs. Now how many arrangements are possible?

33. A closet has eight light bulbs, four of which work and four of which do not. How many different orders of selection are there for the bulbs if all eight are removed from the closet and tried one at a time?

Trace Algorithm 30 with each of the following permutations as input.

34. 1234

35. 1243

36. 1432

37. 41235

38. 43215

39. 243156

40. Find a permutation that will cause the algorithm to execute in a minimum number of steps and determine how many steps it will take. Let $n = 4$.

41. Show that the recursive method for listing permutations of $A = \{1, \ldots, n\}$ does not list any permutation twice.

42. Show that the recursive method for listing permutations of $A = \{1, \ldots, n\}$ lists all $(n!)$ permutations of A.

7.3

COMBINATIONS

Once again, let's look at this problem:

Suppose n *diplomats are at a party, and during the festivities, every diplomat shakes hands with every other diplomat exactly once. How many handshakes occur?*

This is a counting problem, and we solved it one way in the previous chapter. It is similar to the permutation problems in Section 7.2 in that it involves a set of size n (the diplomats) and the selection of some number (in this case, two) of elements from that set. But there is a significant difference: If two diplomats are labeled A and B, the handshake AB is the same as the handshake BA. With permutations, AB and BA are different. Thus, in this section, we will focus on situations that involve counting selections of r elements from a set of size n, but in which the order of selection does *not* matter. Such a selection is called a **combination.** This concept is more formally defined as follows.

Definition   Let S be a set with n elements. Any subset of S is called a **combination.** A subset with r elements is called an **r-combination** or a **combination of n things taken r at a time.** The number of r-combinations of S is denoted $C(n,r)$ or $\binom{n}{r}$.

$C(n,r)$ is also called the **binomial coefficient** for reasons that will become clear in Section 7.4.

Note that a combination is simply a subset; the order in which its elements appear does not matter. A permutation, on the other hand, is like an array; it is a list in which the order of appearance of its elements is of critical importance. Thus whether order matters is the principal factor that distinguishes combinations from permutations. Whether order matters is a key factor in much of discrete mathematics. For instance, in subsequent chapters we will see that order is what distinguishes digraphs (Chapters 8 and 10) and graphs (Chapters 9 and 10).

One of the most commonly asked counting questions is, "Given a set S, how many subsets of S are there that have r elements?" This question arises in many different forms in mathematics, computer science, statistics, biology, sociology, and many other fields. We will answer it in the following two theorems and then apply the answer to a number of examples.

THEOREM 7.7 $C(n,r) = \dfrac{n!}{r!(n-r)!}$.

PROOF We know from the previous section that there are

$$P(n,r) = \frac{n!}{(n-r)!}$$

r-permutations of the elements of S. If we list all of the r-permutations of S, we could organize them into collections consisting of those that have the same elements. (For instance, $s_1 s_2 \cdots s_r$ and $s_2 s_1 \cdots s_r$ would be in the same collection.) Consider one such permutation—$s_1 s_2 \cdots s_r$. There are $r!$ permutations of the elements of this r-permutation. Each such collection has $r!$ elements. The number of r-element subsets of S is the same as the number of these collections. Thus there are $P(n,r)/r!$ subsets. That is, the number of subsets is

$$\frac{n!}{r!(n-r)!} \quad \square$$

An alternative and perhaps easier proof of this theorem is as follows: Think of a list of n letters, each of which is either an I (for *include*) or an E (for *exclude*). Suppose there are exactly r I's. A permutation of these n letters is a sequence such as IIEIEEEI $\cdots$ Each such permutation corresponds to a subset of size r; that is, it corresponds to going through the set $S = \{a_1, \ldots, a_n\}$ and for the ith element, either including it if the ith letter is an I or excluding it if the ith letter is an E. Since there are $n - r$ E's, by the previous section there are

$$\frac{n!}{r!(n-r)!}$$

lists of n elements where r are I's and $n - r$ are E's. But each such list corresponds to a subset of size r.

EXAMPLE 1 List all of the 2-permutations and 2-combinations of the set {1, 2, 3, 4}.

SOLUTION We know that there are $P(4,2) = 4!/2! = 12$ 2-permutations and $C(4,2) = 4!/2!2! = 6$ 2-combinations of the set. They are as follows:

2-permutations		2-combinations
12	21	{1, 2}
13	31	{1, 3}
14	41	{1, 4}
23	32	{2, 3}
24	42	{2, 4}
34	43	{3, 4}

Note that the 2-permutations are organized into pairs made up of the same elements. Each such pair corresponds to a 2-combination. This illustrates why the number of r-combinations is $P(n,r)/r!$. ∎

EXAMPLE 2 Suppose S has 10 elements. Determine how many subsets it has of size r where $r = 0, 1, 2, \cdots, 10$.

SOLUTION The number of subsets is given by $C(10,r)$ for each r. Thus

$$C(10,0) = \frac{10!}{0!10!} = \frac{10!}{10!} = 1$$

$$C(10,1) = \frac{10!}{1!9!} = \frac{10!}{9!} = 10$$

$$C(10,2) = \frac{10!}{2!8!} = \frac{10 \times 9}{2} = 45$$

$$C(10,3) = \frac{10!}{3!7!} = \frac{10 \times 9 \times 8}{3 \times 2 \times 1} = 120$$

$$C(10,4) = \frac{10!}{4!6!} = \frac{10 \times 9 \times 8 \times 7}{4 \times 3 \times 2 \times 1} = 210$$

$$C(10,5) = \frac{10!}{5!5!} = \frac{10 \times 9 \times 8 \times 7 \times 6}{5 \times 4 \times 3 \times 2 \times 1} = 252$$

$$C(10,6) = \frac{10!}{6!4!} = \frac{10 \times 9 \times 8 \times 7}{4 \times 3 \times 2 \times 1} = 210$$

$$C(10,7) = \frac{10!}{7!3!} = \frac{10 \times 9 \times 8}{6} = 120$$

$$C(10,8) = \frac{10!}{8!2!} = \frac{10 \times 9}{2} = 45$$

$$C(10,9) = \frac{10!}{9!1!} = \frac{10!}{9!} = 10$$

$$C(10,10) = \frac{10!}{10!0!} = \frac{10!}{10!} = 1$$

Note that the sum of these 11 numbers is 1024, which is 2^{10}. This observation is generalized in Theorem 7.8 ∎

THEOREM 7.8 $\sum_{r=0}^{n} C(n,r) = 2^n$.

PROOF $C(n,0)$ is the number of zero-element subsets of a set S with n elements, $C(n,1)$ is the number of one-element subsets, etc. Thus

$$\sum_{r=0}^{n} C(n,r)$$

is the total number of subsets of S, namely, 2^n. ☐

EXAMPLE 3 Solve the handshake problem using combinations.

SOLUTION A handshake is a subset of size two. Thus there are $C(n,2) = n(n-1)/2$ possible handshakes. ∎

EXAMPLE 4 A poker hand consists of five cards selected from a standard deck. A bridge hand consists of 13 cards. How many distinct poker hands are there? How many distinct bridge hands are there?

SOLUTION There are $C(52,5)$ poker hands and $C(52,13)$ bridge hands.

$$C(52,5) = \frac{52!}{5!47!} = \frac{52 \times 51 \times 50 \times 49 \times 48}{5 \times 4 \times 3 \times 2 \times 1} = 2{,}598{,}960$$

$$C(52,13) = \frac{52!}{13!39!} = \frac{52 \times 51 \times \cdots \times 40}{13 \times \cdots 2 \times 1} \cong 6.35 \times 10^{11}. ∎$$

EXAMPLE 5 The study of n-person games is largely a study of the different coalitions that can be formed, how they are made, and how they are broken. Consider seven people playing the game of Risk. How many different coalitions of four could be formed?

SOLUTION A coalition is simply a subset. Thus $C(7,4) = 35$ different coalitions are possible. ∎

EXAMPLE 6   One variation of poker is the game of seven-card stud. In this game, an individual is dealt two cards face down, four cards face up, and one more card face down. A hand consists of these seven cards, and the best hand wins at the end. However, if betting occurs after each card that is dealt face up and after the last card, how many different ways are there to be dealt a seven-card stud hand?

SOLUTION   For the first two cards, the order in which they are dealt does not matter. Thus there are $C(52,2)$ different pairs. For the next four cards, the order does matter, so there are $P(50,4)$ possibilities. At this point 46 cards are left in the deck, so there are 46 possibilities for that card. Hence, by the multiplication principle, there are

$$C(52,2) \times P(50,4) \times 46 = 1326 \times 5,527,200 \times 46 \cong 3.37 \times 10^{11}$$

possible seven-card stud hands.   ∎

Just as with permutations, combinations can be both listed and counted recursively. We will discuss this aspect of combinations in the next section.

Selection with Repetition Allowed

Suppose four couples go out to dinner together, and the menu has three soups that can be ordered as an appetizer—French onion, cream of broccoli, and clam chowder. How many different soup orders are possible?

This situation is typical of certain types of counting problems. There is a set, S, of n elements (in this case, the three kinds of soup) and r selections have to be made from S (the eight people, each selecting one kind of soup). But it differs from the combination questions we dealt with earlier in significant ways. First, r can be greater than n, which is impossible if we are selecting an r-element subset of a set of size n. Second, elements of S can be selected more than once, whereas with subsets, each element must be distinct. But it is not a permutation problem either, since the order of selection does not matter. What constitutes an order is the number of bowls of French onion, cream of broccoli, and clam chowder needed, not who chose which one.

As with many problems, the solution to this one depends on finding the right way to look at it. Once it is seen in the right light, the solution is relatively easy. Until it is seen this way, it can be very confusing. One simple way to visualize it is to imagine the waiter's order pad and suppose that the soup section is divided into three areas separated by vertical bars as follows:

$$\underline{\hspace{2cm}} | \underline{\hspace{2cm}} | \underline{\hspace{2cm}}.$$

Each time a customer orders one of the three soups, the waiter marks an X in the appropriate area on the pad. Thus, if the first person orders French onion soup, the pad will look like this:

$$\text{\_\_\_\_X} \,|\, \text{\_\_\_\_\_} \,|\, \text{\_\_\_\_\_}.$$

If this is followed by an order for cream of broccoli soup, the mark on the pad will become

$$\text{\_\_\_\_X} \,|\, \text{\_\_\_\_X} \,|\, \text{\_\_\_\_\_}.$$

A complete order for this table (assuming everyone orders soup) will consist of eight X's and two vertical bars arranged in some order. Some sample orders are

XXXX | XXX | X (four onion, three broccoli, one chowder).

| XXXX | XXXX (no onion, four broccoli, four chowder).

XXXXXXXX | | (everyone ordered onion).

Looked at this way, we now have the means to solve our problem. We have 10 "slots" in which to place either an X or a vertical bar. We have exactly eight X's and two bars. Hence the number of distinct soup orders is precisely the number of ways that exist to select the two slots in which to place the two bars (which is, of course, equal to the number of ways to select the eight slots in which to place the X's). Thus the number of distinct soup orders is the number of ways to arrange eight X's and two | 's, namely, $C(10,2)$, which equals 45.

This can be generalized as follows:

THEOREM 7.9 Suppose a set S has n elements, $\{s_1, s_2, \ldots, s_n\}$, and r elements are to be selected from S where each element can be selected an arbitrary number of times but the order of selection does not matter. Then $C(n + r - 1, r)$ different selections can be made.

PROOF The problem can be modeled as follows. Suppose a string of r X's is written

XXXXXX....XXXXX.

$n - 1$ vertical bars are placed among the X's, separating the X's into n subcollections. Then the leftmost subcollection can be looked upon as the s_1's selected, the second subcollection from the left as the s_2's, etc. Thus the problem is equivalent to the problem of how many ways exist to select places for $n - 1$ bars among $r + n - 1$ possible positions. Hence there are $C(n + r - 1, n - 1)$ possibilities. But this equals $C(n + r - 1, r)$. □

EXAMPLE 7 Suppose there are four desserts for the eight people to choose from. How many different dessert orders are possible?

SOLUTION In problems of this type, we typically let n denote the number of elements in the set and let r denote the number of selections to be made. Thus $n = 4$, $r = 8$. Hence there are $C(11,8) = 165$ different possibilities. ∎

There is another important alternative way to look at Example 7. Suppose we let x_1 denote the number of customers ordering dessert 1, x_2 denote the number ordering dessert 2, x_3 denote the number ordering dessert 3, and x_4 denote the number ordering dessert 4. Then

$$x_1 + x_2 + x_3 + x_4 = 8.$$

The question of how many different dessert orders there are is really the same as the question "How many different solutions are there of this equation in which each of the x_i's is a nonnegative integer?" The answer, as we have already shown, is $C(4 + 8 - 1,8) = C(11,8)$.

The result is generalized in the following corollary to Theorem 7.9.

COROLLARY 7.10 The equation

$$i_1 + i_2 + \cdots + i_n = r, \qquad n \geq 1,$$

where r and all i_j are nonnegative integers, has $C(n + r - 1,r)$ distinct solutions.

PROOF This proof is left for the Exercises. □

EXAMPLE 8 How many solutions are there to the equation

$$x_1 + x_2 + x_3 + x_4 + x_5 = 12,$$

in which all of the x_i's are nonnegative integers?

SOLUTION In this case, $n = 5$ and $r = 12$. Thus there are $C(16,12) = 1820$ solutions. The equation would model the preceding restaurant problem if there were five choices of soups and 12 customers. ∎

EXAMPLE 9 How many different outcomes are possible when flipping six indistinguishable coins?

SOLUTION Let six X's denote the six coins. Imagine a vertical bar with the X's on its left denoting heads and the X's on its right denoting tails. Thus $n = 2$, since there are the two choices (heads and tails) and $r = 6$ for the six flips. Thus there are $C(6 + 2 - 1,6) = C(7,6) = 7$ possibilities. Alternatively, we could write the equation

$$H + T = 6$$

and seek nonnegative integer solutions of this equation. This would yield the same answer. ∎

The previous example may seem to be the hard way to do an easy problem; the outcomes could be listed easily, namely, zero heads, one head, ... , six heads. But the principle can be applied even when listing is difficult, as in the next example.

EXAMPLE 10 How many different outcomes are possible when rolling six indistinguishable dice?

SOLUTION   Again, we let six X's denote the six dice. We need five bars to separate the six possible outcomes. An X to the left of all the bars denotes an outcome of 1; an X between the first and second bars on the left denotes a 2, etc. Thus $n = 6$ and $r = 6$. Thus there are $C(11,6) = 462$ possible outcomes. Letting i_1 denote the number of 1's, i_2 denote the number of 2's, etc., the corresponding equation is

$$i_1 + \cdots + i_6 = 6. \quad \blacksquare$$

EXAMPLE 11 How many different ways are there to place 15 indistinguishable balls in five boxes?

SOLUTION   Each ball has to "choose" which box it will occupy. Thus we have 15 X's and four bars. Hence $n = 5$ and $r = 15$. There are $C(19,15) = 3876$ different ways. In this case, the corresponding equation is

$$i_1 + \cdots + i_5 = 15. \quad \blacksquare$$

EXAMPLE 12 How many different ways are there to place 15 indistinguishable balls in five boxes if each box must have at least 2 balls?

SOLUTION   Since 10 balls must be distributed to the five boxes, 2 to a box, we need only consider the 5 remaining balls. These can be distributed in $C(5 + 5 - 1,5) = C(9,5) = 126$ ways. The corresponding equation could be written as follows:

$$n_1 + \cdots + n_5 = 5.$$

Note that this could also have been solved using the equation in Example 11. That is, we can let $i_1 = n_1 + 2$, $i_2 = n_2 + 2$, ... where n_j is the excess over two in each cell. The equation of Example 11 then becomes

$$n_1 + 2 + n_2 + 2 + n_3 + 2 + n_4 + 2 + n_5 + 2 = 15$$

or

$$n_1 + n_2 + n_3 + n_4 + n_5 = 5,$$

which is the same equation we derived at the beginning of this exercise. $\quad \blacksquare$

Generating Combinations

In the last section, we looked at an algorithm that generates the next permutation of a set of n elements, where *next* refers to lexicographic order. In this section, we will examine an algorithm that lists the next r-combination of the set $\{1, 2, \ldots, n\}$. Since an r-combination is merely a subset, we will assume that all elements of each subset are listed in increasing numerical order from left to right. It is this ordering that is the basis of the algorithm. To illustrate how it works, suppose $n = 5$ and we are looking for the next 3-combination. Following is a list of all 10 3-combinations of $\{1, 2, 3, 4, 5\}$ in increasing order:

$$123 \quad 124 \quad 125 \quad 134 \quad 135 \quad 145 \quad 234 \quad 235 \quad 245 \quad 345.$$

Note that if the last element of some particular r-combination is not 5, we can get the next combination by adding 1 to it. If it is 5, we go back to the second element and, if possible, add 1 to it. Then the third element has to become one more than the second. For instance, consider 125. We cannot add 1 to 5, so we add to the 2, giving 3. We then make the third entry 4, i.e., one more than the second. If it is impossible to add 1 to it, go back to the first element and add 1 to it. Then the second element becomes one more than the first and the third, one more than the second. This process is generalized in the following algorithm.

ALGORITHM 31   Generating the Next r-Combination of $\{1, \cdots, n\}$

Procedure Get_next_combination(n,r; var X);

{n is any integer greater than or equal to 1; $1 \leq r \leq n$; X is an array of length r. Before execution of the algorithm X is any r-combination; after completion it is the next r-combination. If X is the last r-combination, the algorithm fails. Thus this condition needs to be checked for in the calling procedure.}

begin

1. $j := 0$;
2. while $(j < r)$ and $(x_{r-j} = n - j)$ do
3. $\quad j := j + 1$;   {find an x_i that is not at its maximal value}
4. $x_{r-j} := x_{r-j} + 1$;   {increment that x_i}
5. for $i := r - j + 1$ to r do
6. $\quad x_i := x_{i-1} + 1$ {each subsequent x_i should be one more than the previous one}

end;

EXAMPLE 15 Trace the previous algorithm for $n = 6$ and each of the following 4-combinations:

a) 1235

b) 1236

c) 2356

SOLUTION

a) Let $X = [1, 2, 3, 5]$. We have $n = 6$ and $r = 4$.

Step	Action
1	$j := 0$.
2	$0 < 4$ but $x_4 \neq 6$, so loop test fails. Advance to step 4.
4	$x_4 := 6$. Thus X becomes $[1, 2, 3, 6]$.
5	$r - j + 1 = 5$ while $r = 4$. Thus the body of the for loop is skipped and the algorithm halts.

b) Let $X = [1, 2, 3, 6]$.

Step	Action
1	$j := 0$.
2	$0 < 4$ and $x_4 = 6$, so advance to step 3.
3	$j := 1$.
2	$1 < 4$ but $x_3 = 3 \neq 5$, so skip to step 4.
4	$x_3 := 4$.
5	$r - j + 1 = 4$, so execute the body of the loop once.
6	$x_4 := 5$. The algorithm halts with $X = [1, 2, 4, 5]$.

c) Let $X = [2, 3, 5, 6]$.

Step	Action
1	$j := 0$.
2	$0 < 4$ and $x_4 = 6$, so advance to step 3.
3	$j := 1$.
2	$1 < 4$ and $x_3 = 5$, so again advance to step 3.
3	$j := 2$.
2	$2 < 4$ but $x_2 \neq 4$, so skip to step 4.
4	$x_2 := 4$.
5	$r - j + 1 = 3$, so start the loop with $i = 3$.

6        $x_3 := x_2 + 1 = 5.$

5   i is incremented to 4.

6        $x_4 := x_3 + 1 = 6.$ The algorithm halts with $X = [2, 4, 5, 6]$.

The maximum number of steps this algorithm will take can be easily shown to be in $O(r)$ (see the Exercises). But if we were to use it to generate all r-combinations of a set of size n, it would have to be called $C(n,r)$ times. Depending on the relative values of n and r, this can be quite inefficient. ∎

TERMINOLOGY

combination
r-combination
combination of n things taken r at a time
$C(n,r)$

$\binom{n}{r}$
binomial coefficient

SUMMARY

1. Suppose $|S| = n$. Then S has

$$C(n,r) = \frac{n!}{r!(n-r)!}$$

subsets of S of size r.

2. $\displaystyle\sum_{r=0}^{n} C(n-r) = 2^n$

3. One of the most common counting problems is that of counting the number of ways of selecting r elements from a set of n distinct elements. These problems can be classified according to whether the order of selection matters and whether elements of the set may be selected more than once. The number of ways are tabulated in the following table, where "repeated selection" refers to being allowed to select an element of the set more than once.

	Repeated Selection Not Allowed	Repeated Selection Allowed
Order matters	$P(n,r)$	n^r
Order doesn't matter	$C(n,r)$	$C(n+r-1,r)$

If the underlying set itself has duplicates, this table does not apply.

4. Algorithm 31 can be used to generate r-combinations of the set $\{1, 2, \ldots, n\}$.

EXERCISES 7.3

Let $S = \{1, 2, 3, 4, 5\}$. List all of the r-combinations and r-permutations of S for the following values of r.

1. $r = 2$

2. $r = 1$

3. $r = 3$

4. $r = 4$

Let $S = \{1, \ldots, n\}$. For the specified n and for each r, $0 \leq r \leq n$, find the number of subsets of S of size r and verify that the total number of the subsets adds to 2^n.

5. $n = 4$

6. $n = 5$

7. $n = 7$

8. $n = 9$

Solve the following problems.

9. Given a school board of 12 members, how many distinct 3-person executive committees can be selected from the board?

10. If nine people are playing the game of Risk, how many five-person coalitions can be formed? How many can be formed that include both Paul and Judy?

11. A gin rummy hand consists of 10 cards. How many different hands are possible?

12. How many distinct gin rummy hands exist that include exactly one ace?

13. How many distinct gin rummy hands exist that include one or more aces?

14. Four students are selected at random from a class of 25. How many different selections are possible?

15. There are m distinct balls in a jar, and k are to be selected. How many different selections are possible?

16. A nine-member school board elects its own president. There are three nominees. If each member of the board votes for one nominee, a tally of the votes is a count of the number of votes each nominee received. (For instance, 9-0-0 is one possible tally; 0-9-0 is another.) How many different tallies are possible?

17. Repeat Exercise 16 with four nominees.

18. In Exercise 16, in how many of the tallies does one nominee receive a majority of the votes? (Hint: consider separately the cases in which one nominee gets five, six, seven, eight, or nine votes.)

19. In Exercise 17, in how many of the tallies does one candidate receive a majority of the votes?

20. There are 12 entrees on a menu. How many different orders are possible from a party of six people?

21. There are six beverages on a menu. How many different beverage orders are possible from a party of five people?

22. How many different ways are there to put 12 indistinguishable balls into five boxes?

23. How many distinct ways are there to put 20 balls in 12 boxes if no box is allowed to remain empty?

24. Prove Corollary 7.10.

How many solutions are there to the following equations, in which each variable is a nonnegative integer?

25. $x_1 + x_2 + x_3 = 11$ **26.** $w + x + y + z = 16$

27. $a + b + c + d + e = 25$ **28.** $x_1 + x_2 + x_3 + x_4 = 18$

Trace Algorithm 31 for the specified values of n and r and for each of the following input combinations.

29. $n = 6, r = 4, 1245$ **30.** $n = 6, r = 4, 1346$

31. $n = 6, r = 4, 1256$ **32.** $n = 5, r = 3, 123$

33. $n = 5, r = 3, 245$ **34.** $n = 5, r = 3, 345$

35. With $n = 6$ and $r = 4$, find an input combination that results in Algorithm 30 taking the minimum number of steps to execute. Find this number and explain why it is minimal.

36. With $n = 6$ and $r = 4$, find an input combination that results in Algorithm 30 taking the maximum number of steps to execute. Find this number and explain why it is maximal.

7.4

APPLICATIONS AND PROPERTIES OF THE BINOMIAL COEFFICIENT

Since the binomial coefficient, $C(n,r)$, tells us the number of r-element subsets of an n-element set, S, and since sets are present in all of mathematics, it should not be surprising that the binomial coefficient occurs in a variety of situations. Thus facility in working with the $C(n,r)$ notation is very helpful. We will begin this section by studying a number of identities involving $C(n,r)$ and $P(n,r)$; mastery of the proofs of these identities can help build that facility. We will then discuss three important applications of the binomial coefficient: the binomial theorem, the multinomial theorem, and the generalized inclusion-exclusion principle.

Combinatorial Identities

The binomial coefficient, $C(n,r)$, has both a combinatorial interpretation, as the number of r-element subsets of an n-element set, and an algebraic interpretation, as $\dfrac{n!}{(r!(n-r)!)}$. Both are important and both need to be understood if we are to work with $C(n,r)$ easily. The following identities summarize some of the more commonly used properties of $C(n,r)$. They will be interpreted both algebraically and in terms of combinations whenever possible.

THEOREM 7.11 Suppose n and r are integers, $n \geq 0$, and $0 \leq r \leq n$. Then the following are true.

a) $C(n,0) = C(n,n) = 1$

b) $C(n,1) = C(n,n-1) = n$

c) $C(n,r) = C(n,n-r)$

d) $C(n,r) = C(n-1,r) + C(n-1,r-1)$, $1 \leq r \leq n-1$

e) $rC(n,r) = nC(n-1,r-1)$, $n \geq 1$, $r \geq 1$

f) $P(n,0) = 1$

g) $P(n,n) = n!$

h) $P(n,1) = n$

i) $P(n,r) = nP(n-1,r-1)$, $n \geq 1$, $r \geq 1$

PROOF

a) $C(n,0) = \dfrac{n!}{0!n!} = 1$

$C(n,n) = \dfrac{n!}{n!0!} = 1$

This says that a set, S, of size n has only one zero-element subset and only one n-element subset.

b) This is left for the Exercises.

c) This says that the number of ways of selecting r elements from n elements is the same as the number of ways of designating $n - r$ elements not to select. Algebraic verification of this identity is left for the Review Exercises.

d) Designate one of the n elements by x. There are $C(n-1,r-1)$ ways to select sets of size r that include x, and there are $C(n-1,r)$ ways to select r elements that do not include x. Thus

$$C(n,r) = C(n-1,r-1) + C(n-1,r).$$

Algebraically,

$$C(n-1,r-1) + C(n-1,r)$$

$$= \frac{(n-1)!}{r!(n-1-r)!} + \frac{(n-1)!}{(r-1)!(n-r)!}$$

$$= \frac{(n-1)!}{(r-1)!(n-r-1)!} \left(\frac{1}{r} + \frac{1}{n-r} \right)$$

$$= \frac{(n-1)!}{(r-1)!(n-r-1)!} \frac{n}{r(n-r)}$$

$$= \frac{(n)!}{(r)!(n-r)!}$$

$$= C(n,r).$$

We will comment more on this identity later. Note that together with parts (a) and (b), the equation in part (d) forms a recursive definition of $C(n,r)$. The preceding combinatorial analysis can be modified to generate r-combinations recursively, as well as to count them (see the Exercises).

e) This is left for the Exercises.

f) $P(n,0) = n!/(n-0)! = n!/n! = 1$. This is interpreted to mean that there is only one way to select no elements from an n-element set.

g) This is left for the Review Exercises.

h) $P(n,1) = n!/(n-1)! = n$. This says that there are n ways to select one element from an n-element set.

i) $P(n,r) = \dfrac{n!}{(n-r)!} = \dfrac{n(n-1)!}{((n-1)-(r-1))!} = nP(n-1,r-1).$ ☐

Pascal's Triangle

One way to display the binomial coefficients is called **Pascal's triangle,** after the French philosopher and mathematician Blaise Pascal (1623–1662), for whom the programming language Pascal is named (see Figure 7.9). Note that this triangle illustrates several of the relationships proven in Theorem 7.10 if we interpret each of its entries as being $C(n,r)$, where n is the row number counting down from the top and r is the entry in the row starting from the left. Both n and r start from 0. Thus the fact that the outermost entries are all 1 is consistent with Theorem 7.10(a): $C(n,0) = C(n,n) = 1$. The diagonal just inside the outer ones on each side

Figure 7.9 Pascal's triangle.

```
              1
            1   1
          1   2   1
        1   3   3   1
      1   4   6   4   1
    1   5  10  10   5   1
  1   6  15  20  15   6   1
```

has the entries 1, 2, 3, 4, ... This is consistent with Theorem 7.10(b): that $C(n,1) = C(n,n-1) = n$. Each entry (other than the outside ones) also has the property of being the sum of the two immediately above it. This is consistent with Theorem 7.10(e): that $C(n,r) = C(n-1,r-1) + C(n-1,r)$. These relationships can also be combined into a recursive algorithm for computing $C(n,r)$ (see the Exercises).

The Binomial Theorem

Expressions like $(x + y)^2$ or $(a + b)^4$ arise fairly often in algebra. $x + y$ and $a + b$ are called **binomials.** It is often useful to be able to express a binomial raised to the nth power as a **polynomial** involving two variables. For instance,

$$(x + y)^0 = 1$$
$$(x + y)^1 = 1 \cdot x + 1 \cdot y$$
$$(x + y)^2 = 1 \cdot x^2 + 2 \cdot xy + 1 \cdot y^2$$
$$(x + y)^3 = 1 \cdot x^3 + 3 \cdot x^2 y + 3 \cdot xy^2 + 1 \cdot y^3$$
$$(x + y)^4 = 1 \cdot x^4 + 4 \cdot x^3 y + 6 \cdot x^2 y^2 + 4 \cdot xy^3 + y^4.$$

Note that the coefficients of the polynomials on the right are precisely the numbers appearing in the first five rows of Pascal's triangle. Thus, if we use the symbol a_{nr} to denote the coefficient of $x^{n-r}y^r$ in the polynomial expansion of $(x + y)^n$, the preceding expansion shows that $a_{nr} = C(n,r)$ for any r as long as $n \le 4$. The assertion that $a_{nr} = C(n,r)$ for any n and any r is the **binomial theorem,** which we will now prove. Note that this explains why $C(n,r)$ is called the *binomial coefficient.*

THEOREM 7.12 (binomial theorem)   Let n be a nonnegative integer. Then for all x and y,

$$(x + y)^n = C(n,0)x^n + C(n,1)x^{n-1}y^1 + \cdots + C(n,n-1)x^1 y^{n-1} + C(n,n)y^n.$$

PROOF   We will prove this theorem two ways, first by a combinatorial argument and then by an algebraic argument.

Consider first $(x + y)^2$. This can be written $(x + y) \cdot (x + y)$, which equals $x^2 + xy + yx + y^2$, which in turn equals $x^2 + 2xy + y^2$. The xy term has a coefficient of 2, since it can arise in two ways—x from the first $(x + y)$ term, y from the second, or y from the first, x from the second. x^2, by contrast, has a coefficient of 1 because it can arise in only one way—x from the first and x from the second. Similarly, $(x + y)^n$ can be written $(x + y) \cdot (x + y) \cdot \cdots \cdot (x + y)$. The term $x^{n-r}y^r$ will arise as the product of $n - r$ x's and r y's. Its coefficient will be the number of ways of selecting the $n - r$ elements of this product from which the x will come and hence the r elements from which the y will come. Thus the coefficient must be $C(n,r)$.

Algebraically, the theorem can be proven by mathematical induction. First, let $n = 1$. Then $(x + y)^1 = 1 \cdot x + 1 \cdot y$ and thus the coefficients are $C(1,0)$ and $C(1,1)$. Now assume that the theorem holds for $n - 1$. That is, assume that

$$(x + y)^{n-1} = C(n - 1,0)x^{n-1} + C(n - 1,1)x^{n-2}y + \cdots$$
$$+ C(n - 1,n - 2)xy^{n-2} + C(n - 1,n - 1)y^{n-1}.$$

Now multiply both sides of this equation by $(x + y)$. We get

$$(x + y)^n = C(n - 1,0)x^n + C(n - 1,1)x^{n-1}y + \cdots$$
$$+ C(n - 1,n - 2)x^2y^{n-2} + C(n - 1,n - 1)xy^{n-1}$$
$$+ C(n - 1,0)x^{n-1}y + C(n - 1,1)x^{n-2}y^2 + \cdots$$
$$+ C(n - 1,n - 2)x^1y^{n-1} + C(n - 1,n - 1)y^n$$
$$= C(n - 1,0)x^n + [C(n - 1,1) + C(n - 1,0)]x^{n-1}y + \cdots$$
$$+ [C(n - 1,n - 2) + C(n - 1,n - 1)]xy^{n-1} + C(n - 1,n - 1)y^n$$
$$= C(n,0)x^n + C(n,1)x^{n-1}y + \cdots + C(n,n - 1)x^1y^{n-1} + C(n,n)y^n.$$

by Theorem 7.10(e). □

EXAMPLE 1   Use the binomial theorem to express the following as polynomials:

a) $(x + y)^4$

b) $(x + 1)^5$

c) $(2x - 1)^4$

SOLUTION

a) $(x + y)^4 = C(4,0)x^4y^0 + C(4,1)x^3y^1 + C(4,2)x^2y^2 + C(4,3)x^1y^3$
$$+ C(4,4)x^0y^4$$
$$= x^4 + 4x^3y + 6x^2y^2 + 4xy^3 + y^4.$$

b) $(x + 1)^5 = C(5,0)x^51^0 + C(5,1)x^41^1 + C(5,2)x^31^2 + C(5,3)x^21^3$
$$+ C(5,4)x^11^4 + C(5,5)x^01^5$$
$$= x^5 + 5x^4 \cdot 1 + 10x^31^2 + 10x^21^3 + 5x \cdot 1^4 + 1^5$$
$$= x^5 + 5x^4 + 10x^3 + 10x^2 + 5x + 1.$$

c) $(2x - 1)^4 = C(4,0)(2x)^4(-1)^0 + C(4,1)(2x)^3(-1)^1 + C(4,2)(2x)^2(-1)^2$
$$+ C(4,3)(2x)^1(-1)^3 + C(4,4)(2x)^0(-1)^4$$
$$= 16x^4 - 32x^3 + 24x^2 - 8x + 1. \quad ■$$

The Multinomial Theorem

The **multinomial theorem** is a generalization of the binomial theorem. Instead of expressions of the form $(x + y)^n$, it deals with expressions of the form $(x_1 + x_2 + \cdots + x_k)^n$. A polynomial expansion of this latter expression consists of a sum of terms of the form

$$x_1^{n_1} \cdot x_2^{n_2} \ldots x_k^{n_k},$$

where $n_1 + n_2 + \cdots + n_k = n$ and each $n_i \geq 0$. Each of these terms has a coefficient $a_{n_1, n_2, \ldots, n_k}$. What the multinomial theorem gives us is a formula for each of these coefficients.

THEOREM 7.13 (multinomial theorem) Let n be any integer greater than 0. Then for all $x_1, x_2, \ldots, x_k, k \geq 1$,

$$(x_1 + x_2 + \cdots + x_k)^n = \Sigma \frac{n!}{n_1! n_2! \cdots n_k!} x_1^{n_1} x_2^{n_2} \ldots x_k^{n_k}$$

where the sum is over all nonnegative integer solutions of the equation $n_1 + n_2 + \cdots + n_k = n$. In addition, the number of terms in this sum is $C(k + n - 1, n)$.

PROOF   The pattern of this proof is very similar to the combinatorial proof of the binomial theorem. We imagine $(x_1 + x_2 + \cdots + x_k)^n$ written as

$$(x_1 + x_2 + \cdots + x_k)(x_1 + x_2 + \cdots + x_k) \cdots (x_1 + x_2 + \cdots + x_k).$$

If this product were expanded term by term, each term would result from a selection of one of the x_i's from the first sum, one of the x_i's (possibly the same, possibly different) from the second sum, etc. Hence it will consist of a sum of products of the form $x_1^{n_1} x_2^{n_2} \ldots x_k^{n_k}$, where $n_1 + n_2 + \cdots + n_k = n$. The number of these terms that can be combined by virtue of having the same exponents for every x_i will be the resulting coefficient of $x_1^{n_1} x_2^{n_2} \ldots x_k^{n_k}$ in the final polynomial expression. That is, to get a term of the form $x_1^{n_1} x_2^{n_2} \ldots x_k^{n_k}$, we need n_1 x_1's, n_2 x_2's, etc. These terms are a permutation of the list $x_1 \ldots x_1 x_2 \ldots x_2 \ldots x_k \ldots x_k$, where each x_i is repeated n_i times. There are $P(n; n_1, n_2, \ldots, n_k)$ such permutations, and thus this number is the coefficient of $x_1^{n_1} x_2^{n_2} \ldots x_k^{n_k}$ in the statement of the theorem. As for the number of terms in the sum, we can apply Corollary 7.10, where k replaces n and n replaces r.   □

EXAMPLE 2   Use the multinomial theorem to compute the following:

a) $(x + y + z)^2$

b) $(x - y - 1)^3$

SOLUTION

a) The solution consists of six terms of the form $x^{n_1} y^{n_2} z^{n_3}$, where $n_1 + n_2 + n_3 = 2$.

$$(x + y + z)^2 = \frac{2!}{2!0!0!} x^2 y^0 z^0 + \frac{2!}{1!1!0!} x^1 y^1 z^0 + \frac{2!}{1!1!0!} x^1 y^0 z^1$$

$$+ \frac{2!}{1!1!0!} x^0 y^1 z^1 + \frac{2!}{0!2!0!} x^0 y^2 z^0 + \frac{2!}{0!0!2!} x^0 y^0 z^2$$

$$= x^2 y^0 z^0 + 2x^1 y^1 z^0 + 2x^1 y^0 z^1 + 2x^0 y^1 z^1 + x^0 y^2 z^0$$
$$+ x^0 y^0 z^2$$
$$= x^2 + 2xy + 2xz + 2yz + y^2 + z^2.$$

b) This solution consists of 10 terms of the form $x^{n_1} (-y)^{n_2} 1^{n_3}$, where $n_1 + n_2 + n_3 = 3$.

$$(x - y - 1)^3 = \frac{3!}{3!0!0!}x^3(-y)^0 1^0 + \frac{3!}{2!1!0!}x^2(-y)^1 1^0$$

$$+ \frac{3!}{2!0!1!}x^2(-y)^0 1^1 + \frac{3!}{1!2!0!}x^1(-y)^2 1^0$$

$$+ \frac{3!}{1!1!1!}x^1(-y)^1 1^1 + \frac{3!}{1!0!2!}x^1(-y)^0 1^2$$

$$+ \frac{3!}{0!3!0!}x^0(-y)^3 1^0 + \frac{3!}{0!2!1!}x^0(-y)^2 1^1$$

$$+ \frac{3!}{0!1!2!}x^0(-y)^1 1^2 + \frac{3!}{0!0!3!}x^0(-y)^0 1^3$$

$$= x^3 - 3x^2 y + 3x^2 + 3xy^2 - 6xy + 3x - y^3 + 3y^2$$
$$- 3y + 1. \quad \blacksquare$$

The Generalized Inclusion-Exclusion Principle

In Section 7.1, we proved the **inclusion-exclusion principle** for two and three sets. It can be generalized to n sets, but doing so requires the binomial coefficient. Hence it is another application of this particular counting tool.

THEOREM 7.14 (the generalized inclusion-exclusion principle)   Let A_1, $\ldots$, A_n be subsets of some universal set **U**. Then

$$|A_1 \cup A_2 \cup \cdots \cup A_n| = \sum_{i=1}^{n} |A_i| - \sum_{i,j} |A_i \cap A_j| + \sum_{i,j,k} |A_i \cap A_j \cap A_k|$$

$$+ \cdots + (-1)^{n-1} |A_1 \cap A_2 \cap \cdots \cap A_n|,$$

where the sum over i,j extends over all 2-combinations of $\{1, 2, \ldots, n\}$, the sum over i,j,k extends over all 3-combinations of $\{1, 2, \ldots, n\}$, etc. $\square$

The proof is by induction on n. We have already proven the basis step by showing that the theorem is true for $n = 2$. The details of the induction step are omitted, as they are quite complex and not very instructive.

TERMINOLOGY

Pascal's triangle   polynomial

binomial   multinomial theorem

binomial theorem   inclusion-exclusion principle

SUMMARY

1. Suppose n and r are integers, $n \geq 0$, and $0 \leq r \leq n$. Then the following are true.

   a) $C(n,0) = C(n,n) = 1$

   b) $C(n,1) = C(n,n-1) = n$

   c) $C(n,r) = C(n,n-r)$

   d) $C(n,r) = C(n-1,r) + C(n-1,r-1)$, $1 \leq r \leq n-1$

   e) $rC(n,r) = nC(n-1,r-1)$, $n \geq 1$, $r \geq 1$

   f) $P(n,0) = 1$

   g) $P(n,n) = n!$

   h) $P(n,1) = n$

   i) $P(n,r) = nP(n-1,r-1)$, $n \geq 1$, $r \geq 1$

2. (Binomial theorem)   Let n be a nonnegative integer. Then for all x and y,

$$(x + y)^n = C(n,0)x^n + C(n,1)x^{n-1}y^1 + \cdots + C(n,n-1)x^1y^{n-1} + C(n,n)y^n.$$

3. (Multinomial theorem)   Let n be any integer greater than 0. Then for all x_1, $x_2, \ldots, x_k$, $k \geq 1$,

$$(x_1 + x_2 + \cdots + x_k)^n = \sum \frac{n!}{n_1!n_2! \ldots n_k!} x_1^{n_1} x_2^{n_2} \ldots x_k^{n_k}$$

4. (The generalized inclusion-exclusion principle)   Let $A_1, \ldots, A_n$ be subsets of some universal set U. Then

$$|A_1 \cup A_2 \cup \cdots \cup A_n| = \sum_{i=1}^{n} |A_i| - \sum_{i,j} |A_i \cap A_j| + \sum_{i,j,k} |A_i \cap A_j \cap A_k|$$
$$+ \cdots + (-1)^{n-1}|A_1 \cap A_2 \cap \cdots \cap A_n|.$$

EXERCISES 7.4

Expand the following binomials using the binomial theorem.

1. $(1 + x)^4$

2. $(1 - x)^3$

3. $(2 + x)^6$

4. $(x - 1/2)^4$

5. $(x - y)^3$

6. $(x - y)^4$

Expand the following multinomials using the multinomial theorem.

7. $(x + y + z)^3$ **8.** $(x - y + 1)^2$

9. $(2x - y + 1)^3$ **10.** $(x - 1 + 1/x)^3$

Prove the following, using an algebraic argument, and give a combinatorial interpretation of each.

11. Theorem 7.11(b)   **12.** Theorem 7.11(e)

13. $C(n,2) = \sum_{i=1}^{n-1} i$ **14.** $C(n, n-2) = C(n,2)$

15. If n is odd, $C(n,0) + C(n,1) + \cdots + C(n,(n-1)/2) = 2^{n-1}$.

16. $C(n,i) = C(n, i-1) \cdot \dfrac{(n - i + 1)}{i}$

17. $P(n,r) = r!\, C(n,r)$

18. $P(n, r-1) = \dfrac{P(n,r)}{(n - r + 1)}$

19. Give a combinatorial proof that $\sum_{i=k}^{n} C(i,k) = C(n + 1, k + 1)$.

20. Show that $\sum_{k=0}^{n} C(2n + 1, k) = 2^{2n}$. (Hint: use Theorem 7.8.)

21. Show by both algebraic and combinatorial arguments that $C(2n,2) = 2C(n,2) + n^2$.

22. Show by both algebraic and combinatorial arguments that $C(3n,3) = 3C(n,3) + 6nC(n,2) + n^3$.

23. Write out the next three lines of Pascal's triangle.

24. Write a recursive algorithm for computing $C(n,r)$ using Pascal's identity.

25. Show how the combinatorial argument of Theorem 7.11(d) can be used to generate r-combinations recursively. Use your process to generate all of the 2-combinations of $\{1, 2, 3, 4\}$.

26. Show that the generalized inclusion-exclusion principle reduces to the principle proven in Section 7.1 when $n = 2$.

27. Show that the generalized inclusion-exclusion principle reduces to the principle proven in Section 7.1 when $n = 3$.

7.5

PROBABILITY MODELS

One of the most important applications of combinatorics is in discrete probability. In this section, we will focus on fundamental concepts of probability and, in the next two, see how combinatorics can be applied to them. Probability theory is one of the most widely applied branches of mathematics. It is used in weather forecasting, economics and finance, game theory, statistics, analysis of algorithms, genetics, sociology, physics, psychology, quality control, operations research, chemistry, and many other disciplines.

Introduction to Probability

Probability theory is the study of mathematical models of situations that involve uncertainty. For instance, consider the following five situations:

Flipping a coin

Spinning a roulette wheel

Predicting tomorrow's weather

Observing the number of user interactions with a computer's operating system in some time interval

Throwing darts at a dart board

These situations have in common the fact that the outcome is uncertain. Thus, the basic concepts of probability should abstract common characteristics from these situations (and others similar to them) and symbolize them in some cogent fashion. These concepts, once formulated, can be used to prove theorems that, in turn, will allow us to say a great deal about these and other similar situations. Thus, as a starting point, we must identify the common characteristics that these situations share.

The most obvious common characteristic is that, while we may not know what is going to happen in each situation, we can at least identify the possibilities. Thus, when flipping a coin, we assume that the possibilities are head or tail. Note that in saying this, though, we are making the assumption that no other possibilities are allowed. Once when I was at a high school wrestling match, the referee flipped a coin to decide which wrestler would start on top, and it landed on the mat balanced on its edge. Generally, though, in flipping coins we ignore this possibility because it is so unlikely. Thus, in describing any situation that involves uncertainty, the critical first step is to decide which of the possible outcomes are to be considered and which are to be ignored. With the roulette wheel, we would list the 37 numbers, 0, . . . , 36 and, possibly, 00, which some roulette wheels have and others do not. We would normally ignore the possibility of the ball becoming stuck or flying out of the wheel. With the weather, we could make a simple list ({sunny, partly cloudy, cloudy, precipitation}) or we could refine it and replace precipitation by snow, rain, sleet, and hail. We could also add other unusual prospects such as tornadoes, depending on the season of the year, the location, and the purpose of the study. With the computer interactions, our possibilities would be the integers starting from zero. The dart board could be looked at two ways—as the infinite set of all points within a circle or as the score earned when the dart hits the board. Of course, the possibility of missing the board completely could also be included.

The second characteristic is that each of these possible outcomes has a certain probability or likelihood associated with it. In flipping a coin, we typically assume that heads and tails are equally likely, i.e., that each will come up half of the time. In spinning a roulettte wheel that does not have a 00, we assume that each of the numbers will occur 1/37th of the time. With weather, a reporter says something like "There is a 90 percent chance of rain tomorrow." By the probability or likelihood or chance of something, we normally mean a number between 0 and 1 inclusive. It it is 1 (as in "There is a 100 percent chance of rain tomorrow"), that tells us an event is certain. It it is 0, the event is impossible. Otherwise, the closer to 0 it is, the less likely the possibility, and the closer to 1 it is, the more likely the possibility.

The Probability Model

The preceding observations are summarized in the following definitions.

Definition An **experiment** is any situation in which it is conceivable that more than one outcome may occur. A **sample space** is the set of possible outcomes we have chosen to consider in some experiment. An **event** is any subset of the sample space. An event that consists of a single outcome is called a **simple event**. An event that consists of more than one single outcome is called a **compound event.**

EXAMPLE 1 In flipping a coin, getting heads is an event; in fact, it is a simple event. In spinning the roulette wheel, spinning a 0 is an event; spinning any number between 1 and 9 and spinning an odd number are also events. Spinning a 0 is a simple event, whereas spinning a number between 1 and 9 and spinning an odd number are compound events. With weather, the statement "It will not be sunny tomorrow" describes an event that can consist of various possibilities, depending on which ones were listed in setting up the sample space. ∎

Definition Let S be a sample space, $E \subset S$. Recall that $P(S)$ denotes the power set of S, the set of all subsets of S. Also, recall that $[0, 1]$ denotes $\{x \in \mathbf{R} \mid 0 \le x \le 1\}$. Let $p : P(S) \to [0, 1]$. Then p is a **probability measure*** if

1. $p(\varnothing) = 0$
2. $p(S) = 1$
3. $0 \le p(E) \le 1$
4. $p(E_1 \cup E_2) = p(E_1) + p(E_2)$ if E_1 and E_2 are disjoint subsets of S.

---

* If S is infinite, it is possible that some subsets of S cannot be assigned a probability measure. Defining the class of such sets is beyond our scope here; however, this will not be a problem since our focus will be on finite sets.

EXAMPLE 2   Show how flipping a coin can be modeled using the concepts of sample space, events, and a probability measure.

SOLUTION   Let $S = \{\text{heads, tails}\}$. Then $P(S) = \{\varnothing, \{\text{heads}\}, \{\text{tails}\}, S\}$. If we let $p(\{\text{heads}\}) = 1/2$, $p(\{\text{tails}\}) = 1/2$, $p(\varnothing) = 0$, and $p(S) = 1$, we have a function with domain $P(S)$ that satisfies each of the four properties and hence is a probability measure. Note that if we let $p(\{\text{heads}\}) = 1/3$ and $p(\{\text{tails}\}) = 2/3$, p still satisfies the axioms. These probabilities could occur if the coin were weighted in some fashion, what is commonly called a *biased* coin. Thus the axioms are general enough to model both fair and biased situations. ∎

EXAMPLE 3   Show that the roulette wheel can be modeled using the preceding concepts.

SOLUTION   We will use the roulette wheel without a 00. Thus the sample space is $S = \{0, \cdots, 36\}$. If $E \subset S$, let $p(E) = \dfrac{|E|}{37}$. Thus the probability of each simple event is 1/37 (which is consistent with our intuitive notion of how a fair roulette wheel should behave). Clearly, $p(\varnothing) = 0$, $p(S) = 1$, and $0 \le p(E) \le 1$ for any E. Property iv follows from the addition principle of Section 7.1; that is, if E_1 and E_2 are disjoint,

$$|E_1| + |E_2| = |E_1 \cup E_2|.$$

Dividing this equation by 37, we get $p(E_1) + p(E_2) = p(E_1 \cup E_2)$.

An unfair wheel could also be modeled. For instance, if 0 were twice as likely to come up as any other number, we could set $p(\{0\}) = \dfrac{2}{38}$ and $p(\{x\}) = \dfrac{1}{38}$ for any x other than 0. For any event, E, we would then set $p(E)$ equal to the sum of the probabilities of its elements. ∎

The unbiased roulette wheel can be generalized to the following theorem, which is the basis of most of what is called *finite probability*.

THEOREM 7.15   Let S be a sample space with $|S| = n > 0$. Let $E \subset S$, and let $p(E) = \dfrac{|E|}{|S|}$. Then p is a probability measure.

PROOF   We must check that conditions i–iv are satisfied.

i) $p(\varnothing) = \dfrac{|\varnothing|}{|S|} = 0.$

ii) $p(S) = \dfrac{|S|}{|S|} = 1.$

iii) If $E \subset S$, then $0 \le |E| \le |S|$. So, dividing by $|S|$, $0 \le p(E) \le 1$.

iv) If $E_1 \cap E_2 = \emptyset$, then $|E_1| + |E_2| = |E_1 \cup E_2|$. Dividing this equation by $|S|$, we get $p(E_1) + p(E_2) = p(E_1 \cup E_2)$. □

The function p of Theorem 7.15 is called the **equiprobable measure.** It is the basis for most of our intuitive notions about probability. If we are flipping a coin, since there are two possibilities, we expect each to come up half of the time. That is, we expect the probability of each to be one-half. If that doesn't happen, we believe that the coin is unfair. Similarly, a roulette wheel is expected to have each outcome come up 1/37th of the time. *In everyday experience, we expect probabilities to be the same as proportions.* That is, we normally expect the probability of an event occurring in an unbiased situation to be the same as the relative size (or proportion) of that event in the sample space. The notion of probability as proportion will be the basis of most of the examples we examine in the next section.

Another way to interpret the previous theorem is to say that it tells us that our model of uncertainty—the concepts of sample space, event, and probability measure—is adequate to conceptualize any situation with a finite set of possible outcomes in which each one is equally likely to occur. But our model goes beyond that situation. It is appropriate for situations in which the simple events are not equally likely, such as the unfair coin, the unbalanced roulette wheel, or predicting tomorrow's weather. It can also model situations in which the sample space is not finite, as seen in the next two examples.

EXAMPLE 4   Show that recording the number of user interactions with a computer's operating system within a given interval of time can be described by a probability model.

SOLUTION   We model this situation by letting S be the set $\{0, 1, 2, \ldots\}$, although we recognize that if x is in S and x is large, $p(x)$ will either be 0 or very close to 0. We cannot use a finite subset of S, $\{0, 1, \ldots, N\}$, for the model, since we normally cannot identify a particular N such that no more than N interactions can occur in the given interval. As for the probability measure, suppose g is any function on S such that $g_n \geq 0$ for all $n \in S$ and

$$\sum_{n=0}^{\infty} g_n = 1.$$

Then if $E \subset S$, $p(E)$ is defined as the sum of the g_n's for each x_n in E. In an application of the model, we could compute the g_n's by recording the number of user interactions with a particular computer within several time intervals of the given length. g_0 would be the proportion of intervals in which there were no interactions, g_1 the proportion of intervals in which there was one interaction, etc. However, we don't need the actual g_n values to show how this situation can be modeled.

To show that the preceding definition for p fits our probability model, we must check that each of the four properties of a probability measure is satisfied:

i) $p(\varnothing) = 0$ since there are no x_n's in $\varnothing$.

ii) $p(S) = 1$ since we have defined the sum in such a way that all of the g_n's add up to 1.

iii) This follows from the fact that each $g_n \geq 0$ and the sum of all of the g_n's is 1.

iv) This follows from the fact that $p(E_1 \cup E_2)$ is the sum of the g_i for each x_i in $E_1 \cup E_2$. If E_1 and E_2 are disjoint, this sum can be split into two sums, one consisting only of elements of E_1 and one consisting only of elements of E_2. Thus $p(E_1 \cup E_2) = p(E_1) + p(E_2)$ if E_1 and E_2 are disjoint subsets of S. ∎

EXAMPLE 5 Show that the dart board example (Figure 7.10) can be described by a probability model.

SOLUTION   One way to look at this situation is to list the sample space by the scores that could be earned: $\{0, 25, 50, 75, 100\}$. Based on a player's record, the proportion of times an individual has scored 100, 75, etc. could be assigned as the probability of each outcome. This proportion is called the **experimental probability** of each outcome. The proof that experimental probabilities fit the probability model will be left for the Exercises.

Figure 7.10   A dart board.

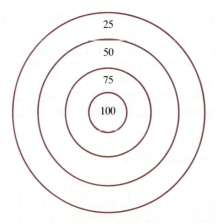

Another way to look at this situation, however, is to assume that the player is sure to hit the board but that every point on the board is equally likely to be hit. This leads us to the paradoxical situation of having infinitely many points, each of which is equally likely; hence each point has a probability of 0. Yet, when the dart is thrown, it has to hit *some* point. This paradox can be resolved by looking at regions on the board rather than at individual points. That is, the probability of scoring 25 is the area of the region labeled 25 divided by the area of the whole board. A similar pattern follows for the other regions. Again, all four properties are easily satisfied. The probability of $\varnothing$ is 0 since its area is zero; the probability of S is 1. Any region will have an area between 0 and the area of S, and hence will have a probability between 0 and 1. If E_1 and E_2 are disjoint subsets of S, the area of $E_1 \cup E_2$ will be the area of E_1 plus the area of E_2. Property iv follows from this if we divide by the area of S. ∎

We will not discuss probability for infinite sample spaces further. In the dart board example, finding the areas of more than the most elementary regions can involve advanced calculus, which is outside the domain of discrete mathematics. Also, defining exactly what we mean by area for complicated regions and proving even apparently obvious properties such as the additivity of the areas of disjoint regions involves mathematics that are considerably beyond the scope of this book. In the computer user example, the sample space was the natural numbers and thus does fall within the domain of discrete mathematics. However, it involves infinite series, and we have dealt only with finite series in this book.

Further Properties of Probability Measures

At this point, we have examined the basic concept of a probability model, with several examples, and have examined the equiprobable measure. We will now prove three further useful properties of probability measures. These will serve as the bases for computing probabilities in Section 7.6 and will complete our examination of the elementary properties of probability measures. In the preceding examples, we verified that certain functions were probability measures. Now we will assume that p is a probability measure and derive some properties from that assumption.

THEOREM 7.16   Let S be a sample space, E be an event, and p be any probability measure. Then

$$p(E) + p(E') = 1.$$

PROOF   Since $E' = \{x \mid x \text{ is not in } E\}$, we know that $E \cup E' = S$ and $E \cap E' = \varnothing$. Thus,

$$p(E) + p(E') = p(E \cup E') = p(S) = 1. \quad \square$$

EXAMPLE 6   On the roulette wheel that does not have a 00, find the probability of not spinning a 0, both for a fair wheel and a wheel biased to make 0 twice as likely as the other numbers.

SOLUTION   Let E be the event, spinning a 0. Then E' is the event of not spinning a 0. Thus, if the wheel is fair,

$$p(E') = 1 - p(E) = 1 - \frac{1}{37} = \frac{36}{37}.$$

If the wheel is biased in the way described here.

$$p(E') = 1 - \frac{2}{38} = \frac{36}{38}. \quad \blacksquare$$

THEOREM 7.17   Let S be a sample space, E_1 and E_2 events, not necessarily disjoint, and p a probability measure. Then

$$p(E_1 \cup E_2) = p(E_1) + p(E_2) - p(E_1 \cap E_2).$$

PROOF   From basic properties of sets, we see that

$$E_1 \cup E_2 = (E_1 \setminus E_2) \cup (E_2 \setminus E_1) \cup (E_1 \cap E_2),$$
$$E_1 = (E_1 \setminus E_2) \cup (E_1 \cap E_2),$$
$$E_2 = (E_2 \setminus E_1) \cup (E_1 \cap E_2).$$

In each of these three equations, the sets combined via the union operation are disjoint. Thus

$$p(E_1) = p(E_1 \setminus E_2) + p(E_1 \cap E_2),$$
$$p(E_2) = p(E_2 \setminus E_1) + p(E_1 \cap E_2),$$
$$p(E_1 \cup E_2) = p(E_1 \setminus E_2) + p(E_2 \setminus E_1) + p(E_1 \cap E_2).$$

Combining these, we get

$$p(E_1 \cup E_2) = p(E_1) + p(E_2) - p(E_1 \cap E_2). \quad \square$$

Theorem 7.17 can be illustrated by the familiar Venn diagram of Figure 7.11. We think of the entire rectangle as having area 1 and each of E_1, E_2, $E_1 \cup E_2$, and $E_1 \cap E_2$ having an area equal to its probability.

Figure 7.11   Venn diagram representing Theorem 7.17.

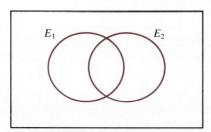

EXAMPLE 7   Assume that a deck of cards is well shuffled so that each of the 52 cards is equally likely to be selected. Find the probability of selecting a spade or an ace when one card is picked at random.

SOLUTION   Let E_1 be the event of picking a spade and E_2 the event of picking an ace. Since each card is equally likely to be chosen, we use the equiprobable measure. Thus

$$p(E_1) = \frac{13}{52} = \frac{1}{4} \quad \text{and} \quad p(E_2) = \frac{4}{52} = \frac{1}{13}.$$

Also, $p(E_1 \cap E_2) = \frac{1}{52}$ since there is only one ace of spades. Thus

$$p(E_1 \cup E_2) = p(E_1) + p(E_2) - p(E_1 \cap E_2) = \frac{13}{52} + \frac{4}{52} - \frac{1}{52} = \frac{16}{52} = \frac{4}{13}. \quad \blacksquare$$

THEOREM 7.18   Let S be a sample space and $A_1, \ldots, A_n$ be events. Then

$$p(A_1 \cup A_2 \cup \cdots \cup A_n) = \sum_{i=1}^{n} p(A_i) - \sum_{i,j} p(A_i \cap A_j) + \sum_{i,j,k} p(A_i \cap A_j \cap A_k)$$
$$+ \cdots + (-1)^{n-1} p(A_1 \cap A_2 \cap \cdots \cap A_n)$$

where the sum over i,j extends over all 2-combinations of $\{1, 2, \ldots, n\}$, the sum over i,j,k extends over all 3-combinations of $\{1, 2, \ldots, n\}$, etc.

PROOF   If S is finite and p is the equiprobable measure, this follows directly from the generalized inclusion-exclusion principle. Otherwise, it can be proven by induction. We will not discuss the details here. Note, however, that we have already verified the basis step in Theorem 7.17.   □

TERMINOLOGY

experiment	compound event
sample space	probability measure
event	equiprobable measure
simple event	experimental probability

SUMMARY

1. Situations involving uncertainty can be modeled by the concepts of sample space (the possible outcomes), event (subsets of the sample space), and probability measure (a function whose domain is the power set of the sample space).

2. Given a sample space S and events E_1 and E_2, the properties of a probability measure are as follows:

   i) $p(\emptyset) = 0$.

   ii) $p(S) = 1$.

   iii) $0 \le p(E) \le 1$.

   iv) $p(E_1 \cup E_2) = p(E_1) + p(E_2)$ if E_1 and E_2 are disjoint subsets of S.

3. If S is finite and all simple events of S have equal probability, p is called the *equiprobable measure*. Many common probability questions fit this pattern.

4. Let S be a sample space. Three basic properties of a probability measure, p, are as follows:

   i) $p(E) + p(E') = 1$ for any event E.

   ii) $p(E_1 \cup E_2) = p(E_1) + p(E_2) - p(E_1 \cap E_2)$ for any events E_1 and E_2.

   iii) $p(A_1 \cup A_2 \cup \cdots \cup A_n) = \sum_{i=1}^{n} p(A_i) - \sum_{i,j} p(A_i \cap A_j) + \sum_{i,j,k} p(A_i \cap A_j \cap A_k) + \cdots + (-1)^{n-1} p(A_1 \cap A_2 \cap \cdots \cap A_n)$

   where the sum over i,j extends over all 2-combinations of $\{1, 2, \ldots, n\}$, the sum over i,j,k extends over all 3-combinations of $\{1, 2, \ldots, n\}$, etc.

EXERCISES 7.5

In each of the following, identify an appropriate sample space.

1. Spinning a roulette wheel with 00
2. Selecting one card from an ordinary deck
3. Rolling one die
4. Flipping 50 coins and recording the heads
5. Feeding rats various diets for 3 months
6. Observing the number of trials required for a rat to learn to run a T maze
7. The number of phone calls arriving at a switchboard in one minute
8. Surveying people regarding their political affiliation
9. Asking people how many cups of coffee per day they drink
10. Measuring the length and weight of salmon caught in Lake Michigan

Construct a probability model for each of the following experiments. In each case, show that the probabilities you define form a probability measure.

11. Spinning a roulette wheel with 00
12. Selecting one marble from a box containing five red, four blue, and three yellow marbles
13. Flipping a coin until it comes up heads (Hint: $1/2 + 1/4 + 1/8 + \cdots = 1$)

14. Selecting one student at random from a class of 25

15. Selecting two light bulbs at random for testing from a batch of 100, of which 10 are defective.

16. Playing a game in which a spinner can point to any of 12 different areas, each of which has a vertex angle of $30°$; four are colored white, two green, two blue, one orange, one brown, one yellow, and one red

17. Rolling a weighted die

18. Flipping a coin weighted so that heads comes up 60% of the time

Use the properties of probability measures to solve each of the following problems. Indicate which property you are using.

19. If the probability of rain today is 0.4, find the probability that it won't rain.

20. If the probability that ten or more users interact with the operating system in one minute is 0.75, what is the probability that fewer than ten interact?

21. Three distinguishable dice are rolled. For each die, the probability that it is odd is $\frac{1}{2}$. Also, suppose that for any event A involving only one die and any event B involving only one other die, $p(A \cap B) = p(A) \cdot p(B)$. Find the probability that at least one die is odd.

22. Three questions are asked in a survey: Did you vote in the last election? Do you live within the city limits? Are you female? Nineteen percent of those surveyed responded no to all three questions. Forty-three percent of the whole sample voted, 23% were females who voted, and 14% were city residents who voted. Thirty-four percent lived in the city, of whom exactly 50% were female. In the whole sample, 52% were female. What is the probability that an individual selected at random answered yes to all three questions?

Let S be a sample space and p be a probability measure. Prove the following.

23. If $X \subset S$ and $p(x_i) > 0$ for all i, then $p(X) = 1$ if and only if $X = S$.

24. If $A \subset B$, then $p(A) \leq p(B)$.

25. $p(A_1 \cup A_2 \cup \cdots \cup A_n) \leq \sum_{i=1}^{n} p(A_i)$.

26. If the A_i are disjoint, then $p(A_1 \cup A_2 \cup \cdots \cup A_n) = \sum_{i=1}^{n} p(A_i)$.

27. Prove Theorem 7.17 in the case where $n = 3$.

28. Derive Theorem 7.17 from Theorem 7.2 in the case where p is the equiprobable measure.

29. Derive Theorem 7.18 from Theorem 7.14 in the case where p is the equiprobable measure.

30. Suppose $S = \{1, \ldots, n\}$ is a sample space. Suppose also an experiment is performed N times with i occurring as outcome f_i times. Define $p(A)$ to be $\Sigma f_i/N$ for any event A where the sum is carried out over each i in A. $p(A)$ is the experimental probability of A. Show that p is a probability measure.

7.6

COMPUTING DISCRETE PROBABILITIES

In the last section, we discussed the equiprobable measure, part of a probability model in which the sample space is finite and each simple event is equally likely to occur. We defined the probability of an event, E, to be $\frac{|E|}{|S|}$ and denoted it as $p(E)$. This is equivalent to saying that the probability of E is the same as the proportion E is of the entire sample space, S. In this section, we will calculate the probabilities of a number of events, E, by means of the equiprobable measure. Keep in mind that our answers will be good estimates of the proportion of times an arbitrary selection from the given sample space will be in E. Such estimates are the basis of many practical applications of probability theory.

Examples Using Enumeration

EXAMPLE 1 Calculate the probability that when a die is rolled, the number shown is

a) a 4

b) even

c) greater than 4 or less than 3.

SOLUTION $S = \{1, 2, 3, 4, 5, 6\}$; thus $|S| = 6$.

a) $E = \{4\}$; hence $|E| = 1$ and $p(E) = 1/6$.

b) $E = \{2, 4, 6\}$; hence $|E| = 3$ and $p(E) = 1/2$.

c) $E = \{1, 2, 5, 6\}$; thus $p(E) = 4/6 = 2/3$. ∎

Note that in each case the number of elements in E was found by listing them and counting. The same was true for S. Part(c) is an application of the addition principle, since $E = E_1 \cup E_2$, where $E_1 = \{x \in S \mid x > 4\}$ and $E_2 = \{x \in S \mid x < 3\}$.

Perhaps the single most common source of errors in finite probability problems is ambiguity regarding the identity of the members of the sample space. This information must be made explicit. Often this can be done by listing the sample space as in Example 1 or Example 2.

EXAMPLE 2 Suppose two distinguishable dice are rolled. Find the probability of rolling

a) a sum of 7

b) a sum of 7 or 11.

SOLUTION　The sample space is listed in Figure 7.12.

Figure 7.12　Sample space for the roll of two dice.

$$
\begin{array}{cccccc}
11 & 12 & 13 & 14 & 15 & 16 \\
21 & 22 & 23 & 24 & 25 & 26 \\
31 & 32 & 33 & 34 & 35 & 36 \\
41 & 42 & 43 & 44 & 45 & 46 \\
51 & 52 & 53 & 54 & 55 & 56 \\
61 & 62 & 63 & 64 & 65 & 66
\end{array}
$$

Let E_7 denote the event of rolling a sum of 7 and let E_{11} denote the event of rolling a sum of 11.

a) $p(E_7) = \dfrac{6}{36} = \dfrac{1}{6}$, since $E_7 = \{61, 52, 43, 34, 25, 16\}$.

b) $E_{11} = \{56, 65\}$. Thus $p(\text{rolling a sum of 7 or 11}) = p(E_7 \cup E_{11}) = \dfrac{8}{36} = \dfrac{2}{9}$. ∎

If the two dice are not distinguishable, we can write the sample space as

$$
S' = \{11, 12, 13, 14, 15, 16, 22, 23, 24, 25, 26, \\
33, 34, 35, 36, 44, 45, 46, 55, 56, 66\}.
$$

Note that $|S'| = 21$, not 36. The difference between S and S' occurs because an outcome such as 12, for instance, is not distinguished from 21. One common error in dealing with finite probabilities is this: The sample space for rolling two dice is written as S'. To find the probability of rolling a sum of 7, the number of outcomes in S' that sum to 7 are counted, namely, three, and it is concluded that the probability of rolling a 7 is $3/21 = 1/7$ when actually it is $1/6$. The reason this is an error is that the simple events of S' are not equally likely. For instance, to get an outcome of 11, both dice must come up 1, but to get an outcome of 12, either can be the 1 and the other the 2. Thus 12 is twice as likely to occur as 11. So S' does not satisfy the assumption of equally likely simple events, and thus the probability of an event cannot be calculated by finding $|E| / |S|$.

Examples Using the Multiplication Principle

When sample spaces grow large, it is no longer possible to list all of their elements, as we did in Examples 1 and 2. However, with care, we can use the principles of combinatorics to count both $|S|$ and $|E|$. The following examples illustrate some situations in which this can be done using only the multiplication principle.

EXAMPLE 3 A license plate consists of three letters followed by three digits. What is the probability that a particular plate starts with the letters VT and ends with either a 1 or a 7?

SOLUTION There are 26 possibilities for the third letter, 10 for each of the next two digits, and 2 for the last digit. Hence there are $26 \times 10 \times 10 \times 2$ plates that satisfy the condition. Overall there are $26 \times 26 \times 26 \times 10 \times 10 \times 10$ plates. Hence the probability that a particular plate chosen at random satisfies the condition is

$$\frac{26 \times 10 \times 10 \times 2}{26 \times 26 \times 26 \times 10 \times 10 \times 10} = \frac{2}{26 \times 26 \times 10} = 0.000296. \quad \blacksquare$$

EXAMPLE 4 Find the probability that a coin flipped 10 times comes up heads every time.

SOLUTION If S is the set of all possible outcomes, $|S| = 2^{10} = 1024$. To satisfy this condition, each flip has only one possibility. Thus $|E| = 1$ and hence

$$p(E) = \frac{1}{1024}. \quad \blacksquare$$

Examples Using Permutations

EXAMPLE 5 Seven administrators will be seated on a platform side by side during a commencement ceremony. Two are having a feud. If seats are assigned randomly, what is the probability that they will be seated next to each other?

SOLUTION There are 7! possible seating arrangements. If we treat the two feuding administrators as one unit (which we can do if we consider only the case in which they sit side by side), there are 6! different arrangements. However, we have not yet distinguished the cases in which administrator A is on the left of administrator B from the cases in which A is on the right. When we make this distinction, there are $2! \cdot 6!$ arrangements with A and B together. Hence the probability of their being together is

$$\frac{2! \cdot 6!}{7!} = \frac{2 \cdot 6!}{7!} = \frac{2}{7}. \quad \blacksquare$$

EXAMPLE 6 Ten diplomats and their spouses are invited to a state dinner. If they are seated randomly around a circular table, what is the probability that the men and women are seated alternately?

SOLUTION In Examples 11 and 12 of Section 7.2, S denotes the set of all possible arrangements and E denotes the set of arrangements in which men and women alternate. We found that $|S| = 19!$ and $|E| = 10!9!$. Thus

$$p(E) = \frac{10!9!}{19!} = \frac{1}{92378} \cong 1.08 \cdot 10^{-5}. \quad \blacksquare$$

EXAMPLE 7 Suppose the letters of the word MISSISSIPPI are rearranged randomly. What is the probability that all four S's are together?

SOLUTION Let $W =$ the set of all possible arrangements. From Example 15 of Section 7.2, we know that

$$|W| = \frac{11!}{1!4!4!2!} = 34,650.$$

If all of the S's are together, we can treat them as one letter and count the arrangements of an eight-letter word with four I's, one M, two P's, and one S. Call the set of these arrangements E. Then

$$|E| = \frac{8!}{4!1!2!1!} = 840,$$

and thus

$$p(E) = \frac{840}{34,160} = 0.024. \quad \blacksquare$$

EXAMPLE 8 A literature student has five novels by Dickens, two by Melville, two by Tolstoy, and four by Dostoevsky. If these books are placed randomly on a shelf, what is the probability that each author's books are together?

SOLUTION From Example 17 of Section 7.2, there are $13!/5!2!2!4!$ = 540,540 distinct possible arrangements on the shelf. If each author's books are to be together, we can treat each such collection of books as one item and ask "How many ways are there to arrange the four groups?" The answer is $P(4,4) = 4!$. Thus the required probability is

$$\frac{4!}{540,540} \cong 0.0000444. \quad \blacksquare$$

Arguments like the preceding one are often used as the basis for a form of circumstantial evidence. That is, in a random arrangement, the probability of having each author's books all together is very low. Thus, if we find them all together, we conclude that it is very unlikely that they were placed randomly. In other words, it seems far more likely that they were deliberately placed that way. Such arguments have been used in court cases against defendants who claim that a given collection of ties to a crime is coincidental.

Examples Involving Combinations

EXAMPLE 9 In five-card draw poker, the possible winning hands are ranked in order from most valuable to least valuable. In order, the hands are

1. royal flush (ace, king, queen, jack, 10, all in the same suit)
2. straight flush (five cards in sequence order, all in the same suit; the highest is not an ace)
3. four of a kind (four cards of the same denomination; the fifth can be anything)
4. full house (three of one denomination; two of another)
5. flush (all of the same suit)
6. straight (all five in sequential order, of any suit)
7. three of a kind
8. two pairs
9. one pair
10. a "bust" hand, i.e, none of the preceding.

Find the probability of being dealt each possible hand.

SOLUTION In Example 4 of Section 7.3, we found that there are $C(52,5) = 2,598,960$ possible hands. This number will serve as $|S|$.

There are only four ways to get a royal flush. Thus p(royal flush) $= 4/2,598,960$, so a royal flush really is close to a "one in a million" chance.

For four of a kind, there are $C(13,1)$ ways to pick the denomination, $C(4,4)$ ways to pick the four cards of that denomination, and $C(48,1)$ ways to pick the other card. Thus

$$p(\text{four of a kind}) = C(13,1) \cdot C(4,4) \cdot C(48,1)$$

$$= \frac{13 \cdot 1 \cdot 48}{2,598,960} \cong 0.00024.$$

For a flush, any 5 cards can be selected out of 13. Thus there are $C(13,5)$ distinct flushes possible in each suit and four suits. Hence

$$p(\text{flush}) = \frac{4 \cdot C(13,5)}{C(52,5)} \cong 0.00198.$$

Note that this value includes both the royal flushes and the straight flushes. To find the probability of flushes that are not straight or royal flushes, they would have to be subtracted.

For one pair, any denomination can be selected. The other three cards must all be different from the pair and from each other. Thus there are 13 ways to select the denomination to be used as the pair and $C(12,3)$ ways to select the other denominations. We use combinations here, since it does not matter which one is selected to be the first other card, the second other card, etc. Once the four denominations have been selected, we have $C(4,2)$ ways to select the pair and $4 \times 4 \times 4$ ways to select the other three cards. Putting these conditions together, we have

$$p(\text{one pair}) = \frac{13 \cdot C(12,3) \cdot C(4,2) \cdot 64}{2,598,960} \cong 0.423.$$

Computation of the remaining probabilities is left for the Exercises. Since these probabilities are so small, it is not hard to understand why people who play poker often use wild cards. ∎

EXAMPLE 10   Four couples go to dinner together. The menu has three soups that can be ordered as an appetizer—French onion, cream of broccoli, and clam chowder. What is the probability that all eight people, choosing randomly, will select the same soup?

SOLUTION   From Section 7.3, we know that there are $C(8 + 3 - 1,8) = C(10,8) = 45$ possible distinct orders. Three of these orders will have all soups alike. Hence the probability is $3/45 = 1/15$. ∎

In the previous example, if all eight people did order the same soup, we might be tempted to infer that the decision was really not random; perhaps they were all following the lead of the person who ordered first. With a probability of $1/15$ of such an occurrence happening by chance, this inference is plausible but shaky. The question "How small must a probability be before we can infer that the event is probably not based on a random process?" is at the heart of inferential statistics. We won't pursue it further here, but there are many excellent texts on the subject.

SUMMARY

1. For finite sample spaces in which each simple event is equally likely to occur, we can define the probability of an event, E, to be $|E| / |S|$.

2. For small sample spaces, explicit listing of the sample space is often very helpful in computing probabilities. For larger sample spaces, careful application of counting principles is often needed. This can involve the use of the addition principle, the multiplication principle, permutations, and/or combinations.

EXERCISES 7.6

A roulette wheel without a 00 is spun. Find the probability that the outcome is

1. between 1 and 9.   2. an odd number.

3. 0 or even.   4. odd and less than 24.

5. not between 18 and 36 inclusive.   6. evenly divisible by 4.

Two distinguishable four-sided dice with sides 1, 2, 3, and 4 are rolled. Find the probability of

7. a sum of 7.   **8.** doubles.

9. not rolling a sum between 3 and 6.   **10.** a sum that is odd.

11. both dice being odd.   **12.** one die being odd and one even.

13. Marbles numbered 1, 2, 3, 4, and 5 are placed in a box. Three are selected at random. What is the probability that the three are in numerical order?

14. See Exercise 13. What is the probability that the marbles are in numerical order and the first marble is a 1?

Eight coins are flipped in succession. What is the probability that they are

15. all alike?   **16.** alternating heads and tails?

17. starting and ending with heads?

18. alternating two heads and two tails?

A traveling salesman is visiting cities B, C, D, and E and returning to A. Suppose he selects a route at random.

19. A critical piece of information he needs at C is in city B, although he doesn't know it. What is the probability that he will visit B before C?

20. Suppose he needs information from both B and C before visiting D. What is the probability that he will visit them before visiting D?

Suppose nine horses are racing in this year's Kentucky Derby.

21. You have predicted the first-, second-, and third-place finishers. What is the probability that you would be correct if all outcomes were equally likely?

22. You have predicted the first three finishers without specifying which is first, second, or third. What is the probability that you are correct?

23. Twelve jobs are submitted to a computer's printer queue. Two are long and the others are short. What is the probability that the two long ones are at the head of the queue?

24. See Exercise 23. What is the probability that the two long jobs are next to each other in the queue?

A byte consists of 8 bits. What is the probability that a randomly selected byte will have the following characteristics?

25. The number of 1's and 0's is the same.

26. There are five 1's and three 0's.

27. There are six 1's and two 0's.

28. There are more 1's than 0's.

29. There are exactly four 0's together.

30. There are four or more 0's together.

A box contains 25 marbles, of which 10 are blue and the rest yellow. Find the probabilities of each of the following events.

31. A marble selected at random is blue.

32. Two marbles selected at random are both blue.

33. Of five marbles selected at random, three are blue.

34. Of five marbles selected at random, none are blue.

Three couples go to a restaurant together. The dessert menu has seven selections. Find the probabilities that if each of the six people choose independently,

35. all choose the cherry cheesecake.

36. all choose alike.

37. three choose the cherry cheesecake and three the key lime pie.

38. all choose something different.

39. three choose one item and three choose another.

40. four choose the key lime pie and the others choose something different.

Compute the probabilities of being dealt each of the following poker hands.

41. A straight flush   **42.** A full house

43. A straight (excluding straight and royal flushes)   **44.** Three of a kind

45. Two pairs   **46.** A bust hand

7.7

CONDITIONAL PROBABILITY, INDEPENDENCE, AND EXPECTATION

In this section, we will continue to address the question raised in Section 7.6—how to compute discrete probabilities. We will also discuss an important application—computation of the average behavior in a situation in which the outcome is uncertain.

Conditional Probability

Suppose a fair coin is flipped and comes up heads four times in a row. What is the probability that the fifth flip also comes up heads? The concept of conditional probability enables us to address such questions.

Definition Suppose B is an event such that $p(B) \neq 0$. We define the **conditional probability of A given B,** denoted $p(A \mid B)$, by

$$p(A \mid B) = \frac{p(A \cap B)}{p(B)}.$$

Note that if we are using the equiprobable measure, this simplifies to (Figure 7.13)

$$p(A \mid B) = \frac{|A \cap B| \, / \, |S|}{|B| \, / \, |S|} = \frac{|A \cap B|}{|B|}.$$

Figure 7.13 $p(A \mid B)$ is the measure of $A \cap B$ divided by the measure of B.

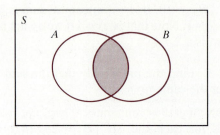

The intuitive idea behind this definition is as follows: Suppose we want to find the probability that an event, A, occurs when using the equiprobable measure. If we have no additional information, we must find the size of the sample space, S, the size of A, and divide. But suppose we have more information, namely, that another event, B, has definitely occurred. Then we no longer need to count all elements in S, only those in B. Also, we do not need to count every element in A, only those in $A \cap B$. Hence, with this additional information, we can find the probability of A by computing $\frac{|A \cap B|}{|B|}$. This is the conditional probability of A given B.

EXAMPLE 1 Suppose a fair coin comes up heads four times in a row. What is the probability that it will come up heads on the fifth flip as well?

SOLUTION Let A be the event "five successive flips are heads" and let B be the event "four successive flips are heads." We are looking for $p(A \mid B)$, the probability of five successive heads given that four successive heads have occurred:

$$
\begin{aligned}
p(A \mid B) &= \frac{p(A \cap B)}{p(B)} \\
&= \frac{p(\text{five successive heads})}{p(\text{four successive heads})}, \qquad \text{since } A \cap B = A \\
&= \frac{(\frac{1}{2})^5}{(\frac{1}{2})^4} \\
&= \frac{1}{2}.
\end{aligned}
$$

Note that this is exactly the same as the probability of one flip coming up heads. No matter how many heads have come up in a row, the probability of flipping another head remains the same. ∎

EXAMPLE 2 Find the probability that the sum of two dice is a 7, given that one of the dice came up a 2.

SOLUTION Let B be the event that one die is a 2. Then

$$B = \{12, 32, 42, 52, 62, 21, 23, 24, 25, 26\},$$

and hence $|B| = 10$. If $A = \{$outcomes $|$ sum is 7$\}$, $|A \cap B| = 2$. Hence $p(A \mid B) = \frac{2}{10} = \frac{1}{5}$. Note that $p(A) = \frac{1}{6}$ if we do not have the information that one of them is a 2. ∎

EXAMPLE 3   Find the probability that the sum of two dice is a 5, given that neither die came up a 1.

SOLUTION   Let $B = \{$outcomes $|$ 1 was not rolled$\}$. Then there are five possibilities for each die, so $|B| = 25$. If $A = \{$outcomes $|$ sum is 5$\}$, $A \cap B = \{23, 32\}$ and hence $|A \cap B| = 2$. Thus

$$p(A \mid B) = \frac{2}{25} = 0.08.$$

Note that $p(A) = \frac{4}{36} = \frac{1}{9} = 0.111 \ldots .$ As in the previous example, $p(A) \neq p(A \mid B)$. ∎

EXAMPLE 4   Suppose two dice are rolled. Find the probability that the first die is a 6, given that the second one is a 6.

SOLUTION   Let A be the event that the first die is a 6 and let B be the event that the second die is a 6. Then $|B| = 6$ and $|A \cap B| = 1$. Hence $p(A \mid B) = \frac{1}{6}$. Note that this time $p(A) = \frac{1}{6}$ and thus $p(A) = p(A \mid B)$. ∎

Independence

In Example 4, $p(A) = p(A \mid B)$, while in Examples 2 and 3, $p(A) \neq p(A \mid B)$. That is, in Example 4, the fact that B occurs has no influence on the probability of A, while in Examples 2 and 3, the fact that B occurs does influence the probability of A. It is this absence of influence of one event on the probability of another that we want to capture in the notion of independence. Thus we could give a formal definition of independence of events by saying that two events, A and B, are independent if and only if $p(A) = p(A \mid B)$. However, this would mean that independence would not be defined if $p(B) = 0$, since $p(A \mid B)$ is defined only when $p(B) \neq 0$. Instead, we use a different definition and then show in Theorem 7.19 that if $p(B) \neq 0$, the formal definition is equivalent to the notion that independence means that $p(A) = p(A \mid B)$. Our formal definition will also prove helpful because it provides an easy way to compute the probability of both A and B occurring if the individual probabilities of A and B are known and if it is known that A and B are independent.

Definition   Two events, A and B, are said to be **independent** if $p(A \cap B) = p(A) \cdot p(B)$.

Figure 7.14 is a Venn diagram representing two events, A and B. If A and B are independent, our definition says that the measure of $A \cap B$ is the product of the measures of A and B. Theorem 7.19 also tells us that if A and B are independent, the proportion that A is of S is the same as the proportion $A \cap B$ is of B;

$$\frac{p(A)}{p(S)} = \frac{p(A \cap B)}{p(B)}.$$

Figure 7.14 A and B are independent if $p(A \cap B)$ $= p(A) \cdot p(B)$.

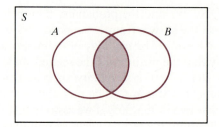

THEOREM 7.19 Suppose S is a sample space and A and B are events such that $p(B) \neq 0$. Then A and B are independent if and only if $p(A) = p(A \mid B)$.

PROOF First, suppose that $p(A) = p(A \mid B)$. Then

$$p(A) = \frac{p(A \cap B)}{p(B)}.$$

Hence

$$p(A \cap B) = p(A) \cdot p(B).$$

Conversely, suppose that $p(A \cap B) = p(A) \cdot p(B)$. Then

$$p(A \mid B) = \frac{p(A \cap B)}{p(B) = p(A)}. \quad \square$$

EXAMPLE 5 The probability that an electronic switching circuit fails during testing is 0.01. If we regard the test of each circuit as independent, what is the probability that two successive circuits will fail? What is the probability that two out of three will fail?

SOLUTION The probability of two circuits failing in a row is $0.01 \times 0.01 = 0.0001$. Since there are $C(3,2)$ ways to pick the two out of three that fail, the probability of two out of three failing is $C(3,2) \times 0.0001 = 0.0003$. ∎

Computations like those done in Example 5 are basic to quality control. Suppose that the assumption of independence leads us to conclude that the probability of two failures in a row is extremely small. When two failures in a row do occur, we strongly suspect that the tests are not independent. That is, we suspect the possibility of something being systematically wrong with the production process, and we investigate and, if necessary, correct the situation.

EXAMPLE 6 A class consists of 23 students. What is the probability that two have a birthday on the same date?

SOLUTION We find the probability that their birthdays are all different and subtract the answer from 1. Suppose the students are arranged in some order from 1 to 23. The birthday of the first student can be any date. The probability that the birthday of the second student is different from that of the first is 364/365. (For simplicity, suppose that all February 29 birthdays are celebrated on March 1.) The probability that the third birthday differs from the first two is 363/365. If we assume that there are no twins, triplets, quadruplets, quintuplets, sextuplets, or septuplets in the class, each of these probabilities is independent. Thus, if we let p denote the probability that all of the birthdays differ,

$$p = \frac{365}{365} \cdot \frac{364}{365} \cdot \frac{363}{365} \cdots \frac{343}{365} \cong 0.4927.$$

Thus the probability that two students do have the same birthday is $1 - p \cong 0.5063$, which is greater than one half.

An alternative way to compute this answer is to say that $|S| = 365^{23}$, where S is the set of all possible birthdays for the 23 students. $|E| = P(365,23)$, since E is the set of distinct ordered selections of 23 birthdays out of 365 days. This gives the same value for p as before. ∎

The previous result is usually very surprising. At first, it seems that far more than 23 students would be required to have a 50% chance of two having the same birthday. This conclusion is probably due to the fact that we are really thinking about a different question: In a class of n students, what is the probability of someone having the same birthday as me? To find this answer, we again assume independence, compute the probability of all n being different from me—$(364/365)^n$—and subtract that number from 1. To find out how large n must be to have this probability equal 0.5, we set

$$\left(\frac{364}{365}\right)^n = 0.5$$

$$n \log_{10}\left(\frac{364}{365}\right) = \log_{10} 0.5$$

$$n = 252.65.$$

Thus n must be 253 or greater for there to be a chance of 50% or more of having someone else in the class with the same birthday as myself.

The previous example shows that in dealing with probability, the results will often seem inconsistent with our intuitive beliefs. Most people, when asked what either of the probabilities computed in the previous example are, will answer somewhat as follows: "There are 365 days in the year. We want a probability of 0.5. So we would need about 182 people." This is not close to the correct answer to either question, indicating why careful analysis is critical in solving probability problems. Our intuition improves with experience, but as soon as we encounter a new and unfamiliar situation, it is likely to let us down.

EXAMPLE 7   A computer has 40 incoming lines but only 20 ports to accommodate them. If the probability that any given line is in use at a given time is 1/3, and whether a given line is in use is independent of whether others are in use, what is the probability that exactly 20 lines are in use at a specified time?

SOLUTION   There are $C(40,20)$ ways to select the 20 busy lines. Since usage of lines is independent, we can say that for each such selection, the probability that any particular selection of lines is in use is $(\frac{1}{3})^{20} (\frac{2}{3})^{20}$. By property 4 of probability measures, this number can be added to itself $C(40,20)$ times to get the probability that exactly 20 lines are in use. Thus the required probability is

$$C(40,20) \cdot \left(\frac{1}{3}\right)^{20} \cdot \left(\frac{2}{3}\right)^{20} \cong 0.0119. \quad \blacksquare$$

Expectation

Suppose the members of a sample space $\{s_1, \ldots, s_n\}$ are real numbers, such as the sample space for rolling a single die—$\{1, \ldots, 6\}$. If we also have a probability measure on S, we can define the average or expected value of the s_i's as follows:

Definition   Let $S = \{s_1, \ldots, s_n\}$ be a sample space with probability measure p such that each $s_i \in \mathbf{R}$. The **expectation** or **expected value** is

$$E = \sum_{i=1}^{n} s_i \cdot p(s_i).$$

This definition can be generalized to sample spaces whose events are not in **R** but have a real value associated with them. Such associations are called **random variables.** We will not pursue the study of random variables further.

EXAMPLE 8   Suppose a fair coin is flipped. What is the expected value of the number of heads?

SOLUTION $S = \{1,0\}$, the number of heads in each possible outcome. $p(1) = \frac{1}{2}$ and $p(0) = \frac{1}{2}$. Thus

$$E = 1 \cdot \left(\frac{1}{2}\right) + 0 \cdot \left(\frac{1}{2}\right) = \frac{1}{2}. \quad \blacksquare$$

EXAMPLE 9   Suppose two fair dice are rolled and their result is summed. Find the expected value of their sum.

SOLUTION   The sample space consists of the 36 pairs $\{11, \ldots, 66\}$. Thus the possible sums are 2, 3, $\ldots$, 12. If we let p_x denote the probability of each possible sum, we have

$$p_x(2) = \frac{1}{36} \qquad p_x(8) = \frac{5}{36}$$

$$p_x(3) = \frac{2}{36} \qquad p_x(9) = \frac{4}{36}$$

$$p_x(4) = \frac{3}{36} \qquad p_x(10) = \frac{3}{36}$$

$$p_x(5) = \frac{4}{36} \qquad p_x(11) = \frac{2}{36}$$

$$p_x(6) = \frac{5}{36} \qquad p_x(12) = \frac{1}{36}$$

$$p_x(7) = \frac{6}{36}$$

(See Figure 7.15.)

Figure 7.15 Probabilities of each possible sum when rolling two dice.

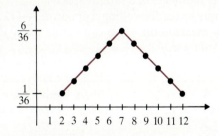

Thus we have

$$E = 2 \cdot \frac{1}{36} + 3 \cdot \frac{2}{36} + 4 \cdot \frac{3}{36} + 5 \cdot \frac{4}{36} + 6 \cdot \frac{5}{36} + 7 \cdot \frac{6}{36}$$

$$+ 8 \cdot \frac{5}{36} + 9 \cdot \frac{4}{36} + 10 \cdot \frac{3}{36} + 11 \cdot \frac{2}{36} + 12 \cdot \frac{1}{36} = \frac{252}{36} = 7. \quad \blacksquare$$

One of the most important computer science applications of expectation is in computing the average-case behavior of algorithms. Our next example illustrates one algorithm for which the average-case behavior can be found fairly simply.

EXAMPLE 10   Consider Algorithm 12, linear search, which was discussed in Section 4.5. In that algorithm, an array, A, of length n was searched for an element, K. The location where it was found was returned or a 0 was returned if it was not found. Compute the average number of items in A that must be examined by linear search. Assume that all elements of A are distinct.

SOLUTION   We will do this twice, once assuming that K is definitely in the array A and once assuming that there is a probability, p_0, that K will not be found in A. We will also assume that if K is in A, each location where it could be found is equally likely.

If K is in A, we can consider the sample space $\{1, \ldots, n\}$ of positions where K could be found. Thus

$$E = 1 \cdot \frac{1}{n} + 2 \cdot \frac{1}{n} + \cdots + i \cdot \frac{1}{n} + \cdots + n \cdot \frac{1}{n}$$

$$= \frac{1}{n}(1 + 2 + \cdots + n)$$

$$= \frac{1}{n} \cdot \frac{n(n+1)}{2}$$

$$= \frac{(n+1)}{2}.$$

If K is not in A, there is a probability of $1 - p_0$ that all n items will be examined without K being found and a probability of p_0/n that K will be found in any specified location in the array. Thus

$$E = 1 \cdot \frac{p_0}{n} + 2 \cdot \frac{p_0}{n} + \cdots + i \cdot \frac{p_0}{n} + \cdots + n \cdot \frac{p_0}{n} + n \cdot (1 - p_0)$$

$$= \frac{p_0(n+1)}{2} + (1 - p_0)n$$

$$= \frac{n \cdot p_0}{2} + \frac{p_0}{2} + n - np_0$$

$$= n\left(\frac{1 - p_0}{2}\right) + \frac{p_0}{2}.$$

Thus, in either case, the average-case behavior of linear search is in $O(n)$. ■

TERMINOLOGY

conditional probability of A given B expectation

$p(A \mid B)$ expected value

independent events   random variable

SUMMARY

1. Suppose A and B are events in the same sample space and $p(B) \neq 0$. Then the conditional probability of A given B is the probability that A happens if it is known that B happens. It is given by the formula $p(A \mid B) = \dfrac{p(A \cap B)}{p(B)}$.

2. Two events, A and B, are independent if $p(A \cap B) = p(A) \cdot p(B)$. If A and B are independent, the probability that one occurs does not affect the probability of the other occurring.

3. The average value of the elements of a sample space, S, with a probability measure, p, can be computed by means of the expected value. This value can also be used to compute the average-case behavior of algorithms.

EXERCISES 7.7

Suppose a deck of cards is shuffled and two cards are selected at random. Find the probability requested and determine whether the two events are independent.

1. Find the probability that the second card is an ace, given that the first is an ace.

2. Repeat Exercise 1, assuming that the first ace was reinserted and the deck was reshuffled before the second card was drawn.

3. Find the probability that the second card is black, given that the first is an ace.

4. Repeat Exercise 3, assuming that the first ace was reinserted and the deck was reshuffled before the second card was drawn.

In Exercises 5–8, find the required probability.

5. Seven administrators are to be seated in a line on a platform at a commencement ceremony. Administrators A and B are having a feud. Given that A must sit at one end of the platform and the rest are placed randomly, what is the probability that A and B will be seated next to each other?

6. This is the same as Exercise 5, except that either A or B must sit at one end of the platform. What is the probability that the two will be seated together?

7. A class consists of 15 students. What is the probability that two of them have a birthday on the same date?

8. This is the same as Exercise 7, but with a class of 30 students.

The switchboard of a small company has five outside lines. There are 40 phones on the employees' desks, each of which is capable of making an outside call. If at any instant the likelihood of an outside call being made from one of the phones is 0.05, find the probability that at any given moment

9. no one is making an outside call.

10. exactly five people are making outside calls.

11. five or fewer people are making outside calls.

12. more than five people are attempting to make outside calls.

Two dice are rolled. Determine if the following events are independent.

13. The first die is a 1; the second die is a 1.

14. The first die is a 1; the sum of the dice is 7.

15. The first die is a 1; the sum of the dice is 8.

16. The sum of the dice is 8; both dice are odd.

In a batch of computer chips, 10% are defective. What is the probability that

17. four selected at random are all defective?

18. four selected at random have no defects?

19. one of four selected at random is defective?

20. one or more of four selected at random is defective?

Prove the following.

21. If $p(A \mid B) = p(A)$, then $p(B \mid A) = p(B)$.

22. $p(A)p(B \mid A) = p(B)p(A \mid B)$.

23. If A_1, A_2, and A_3 are any events, then $p(A_1 \cap A_2 \cap A_3) = p(A_1)p(A_2 \mid A_1)p(A_3 \mid A_1 \cap A_2)$.

24. If A_1, A_2, and A_3 are independent events, then $p(A_1 \cap A_2 \cap A_3) = p(A_1)p(A_2)p(A_3)$.

Compute the expectation of each of the following.

25. The outcome of rolling a single four-sided die.

26. Flipping three coins and recording the number of heads.

27. The matching pennies game. The rules are that both players hide a penny. Both expose their pennies simultaneously. If both are heads or both are tails, player A takes both pennies. If they differ, player B takes both pennies.

28. Being dealt one card in the card game Twenty-One. The rules are that face cards have a value of 10 points and all other cards have their face value. For this question, consider the value of aces as 1 point.

29. Repeat Exercise 28, counting aces as having 11 points.

30. Linear search is applied to an array of length n where there is a probability of 0.5 that the item being searched for is in the array.

The average-case behavior of binary search can be approximated as follows: Suppose that the element being searched for is X and suppose that the array being searched has exactly 2^k distinct elements, each equally likely to be X. Assume that X is in the array. When the search begins, the middle of the array is examined. If the item in the middle is X, the search halts in one step. Otherwise the array is divided in half and either of the two halves is searched. Thus there are two ways to find X on the second step, and the probability of each of those being X is approximately 2^{-k}. Continue in this fashion for k steps.

31. Identify the sample space, S, for this situation.

32. Show that the expectation, E, can be written

$$E = \frac{1}{2^{k+1}} \sum_{i=1}^{k} i 2^i.$$

33. Show that $2^{k+1}E$ satisfies the recurrence equation $S_{k+1} - S_k = (k + 1)2^{k+1}$, and hence show that $E = (k - 1)$ by substituting in the equation. (Note that the average number of steps is only one less than that of the worst case.)

REVIEW EXERCISES—CHAPTER 7

1. Prove Theorem 7.3.

2. In FORTRAN, an identifier consists of *up to* six characters. The first must be a letter; others may be a letter or a digit. How many distinct identifiers are there in FORTRAN?

3. How many seven-digit numbers exist that do not include the sequence 00 anywhere? (Note that a seven-digit number cannot start with 0.)

4. A set X has n elements. How many functions are there from X to itself?

5. A set X has n elements. How many one-to-one functions are there from X to itself?

6. How many digits are there between 1 and 1000 that are not divisible by 3, 5, or 7?

7. Of a sample of restaurant customers who ate dessert, 93 tried the blintzes, 37 the cheesecake, and 25 the chocolate tort. Thirteen tried both the blintzes and the cheesecake and seven tried the blintzes and the chocolate tort. Nobody tried both the cheesecake and the chocolate tort. How many customers were surveyed?

8. Sixteen people are waiting to enter a theater. How many ways are there to place them in line?

9. A basketball team has 11 players and only 5 can start. How many starting line-ups are there?

10. A library restricts book borrowing to a maximum of five books at a time. There are 16 books you would like to read. In how many ways can five be selected and placed in order for reading this semester?

11. See Exercise 10. In how many ways could five or fewer books be selected and placed in order?

How many arrangements are there of the letters of the following words?

12. BASEBALL   13. PRACTICAL

Trace Algorithm 30 with each of the following permutations as input.

14. 125364   15. 654321

16. Replace step 5 of Algorithm 30—switch $A[m]$ and $A[k]$—by pseudo-code that will carry out the switch using only assignment statements.

17. Replace step 6 of Algorithm 30—reverse the order of $A[k + 1]$ to $A[n]$—by pseudo-code that will carry out the reversal using only assignment statements.

18. A basketball team consists of 11 players. How many different starting teams of five are possible? (Note that this is not the same as starting lineups.)

19. How many different outcomes are possible for five indistinguishable dice where an outcome is a list of the number of 1's, 2's, etc.?

20. Repeat Exercise 19, assuming that the dice are dodecahedrons (i.e., 20-sided).

How many solutions are there to the following equations, in which each variable is a nonnegative integer?

21. $x_1 + x_2 + x_3 = 13$ 22. $u + v + w + x + y + z = 16$

23. $a + b + c + d + e = 10$ 24. $x_1 + x_2 + x_3 = 20$

Trace Algorithm 31 for the specified values of n and r and for each of the following input combinations.

25. $n = 6, r = 4,$ 3456   26. $n = 5, r = 3,$ 125

27. Give an algebraic proof, using induction, that $\sum_{r=0}^{n} C(n,r) = 2^n$.

28. Verify Theorem 7.11(c) algebraically.

29. Verify Theorem 7.11(g) algebraically.

30. Show by a combinatorial argument that

$$C(n + m,k) = C(n,0)C(m,k) + C(n,1)C(m,k - 1) + \cdots + C(n,k)C(m,0).$$

(*Hint:* the left-hand side counts the number of k-element subsets of an $n + m$-element set. If we split the $n + m$ elements into two subsets, one of size n and one of size m, the right-hand side does the same.)

31. Use the result of Exercise 30 to prove that

$$C(n,0)^2 + C(n,1)^2 + \cdots + C(n,n)^2 = C(2n,n).$$

32. Two individuals are throwing horseshoes. The probability that each throws a ringer is $\frac{1}{10}$ on any given toss, and the probability that both throw a ringer on the same toss is $\frac{1}{100}$. What is the probability that one or the other throws a ringer on any given toss?

33. On a roulette wheel, numbers are coded as red and black. Suppose $p(\text{red}) = \frac{18}{37}$,

$p(\text{odd}) = \frac{18}{37}$, and the probability that a spin is both red and odd is $\frac{5}{37}$. Find the probability that a spin is red or odd.

34. Let S be any finite set and let f be any function on S with the property that $f(x) > 0$ for all x in S. Let $R = \Sigma f(x)$ where the sum is taken over all x in S. We define $p(x) = f(x)/R$. Show that $p(x)$ is a probability measure.

Construct a probability model for each of the following experiments. In each case, show that the probabilities you define form a probability measure.

35. Rolling a die until a 6 comes up (*Hint:* $1 + \frac{5}{6} + (\frac{5}{6})^2 + \cdots = 6$)

36. Selecting a student at random from a student body that is 52% female and 48% male

Suppose the letters in MISSISSIPPI are arranged randomly. What is the probability that

37. the P's are together?

38. the P's are together and the I's are together?

39. the letters are in alphabetical order?

40. all duplicate letters are together?

Two four-sided dice are rolled. Find the probability of the events specified and decide in each case whether the two events are independent.

41. The sum is 5, given that one die is a 1.

42. The sum is 5, given that the first die is a 1.

43. The sum is 6, given that neither die is an odd number.

44. The second die is 4, given that the first die is odd.

45. The first die is even, given that the second die is also even.

46. The sum is 6, given that the first die is a 1.

47. If $A_1, A_2, \ldots, A_n$ are events in some sample space, S, then

$$p(A_1 \cap A_2 \cap \cdots \cap A_n) = p(A_1)p(A_2 | A_1) \cdots p(A_n | A_1 \cap A_2 \cap \cdots \cap A_{n-1}).$$

48. If $A_1, A_2, \ldots, A_n$ are events in some sample space, S,

$$p(A_1 \cap A_2 \cap \cdots \cap A_n) = p(A_1)p(A_2) \cdots p(A_n).$$

49. Compute the expectation for one spin of a roulette wheel with a 00 if the value of 00 is 0.

50. A computer user's request to have a job printed is placed in a printer queue. If there are four positions in the queue and the probability that the user is in position i is $\frac{i}{10}$, find her expected position.

51. An organism dies with probability p and splits in half, yielding two identical organisms with probability $1 - p$. What is the probability that one such organism produces an eternal colony? (From Donald J. Newman, A Problem Seminar, p. 3, © 1982. Reprinted by permission of Springer-Verlag, New York.)

REFERENCES

Bauer, Heinz, *Probability Theory and Elements of Measure Theory.* New York: Holt, Rinehart, & Winston, 1972.

Feller, William, *An Introduction to Probability Theory and Its Applications.* 3rd ed. New York: Wiley, 1968.

Grimaldi, Ralph P., *Discrete and Combinatorial Mathematics.* Reading, Mass.: Addison-Wesley, 1985.

Knuth, Donald, *The Art of Computer Programming,* Vol. 1., *Fundamental Algorithms.* Reading, Mass.: Addison-Wesley, 1973.

Larson, Harold J., *Introduction to Probability Theory and Statistical Inference.* New York: Wiley, 1969.

Liu, Chung L., *Elements of Discrete Mathematics.* New York: McGraw-Hill, 1977.

Mott, Joe L., Abraham Kandel, and Theodore P. Baker. *Discrete Mathematics for Computer Scientists.* Reston, Va.: Reston, 1983.

Newman, Donald J., *A Problem Seminar.* New York: Springer-Verlag, 1982.

Spiegel, Murray R., *Probability and Statistics.* New York: McGraw-Hill, 1975.

Stanat, Donald, and David McAllister, *Discrete Mathematics in Computer Science.* Englewood Cliffs, N.J.: Prentice-Hall, 1977.

RELATIONS

Suppose you were given a set of names—{Boaz, Obed, Jesse, David, Eliab, Abinadab, Absolom, Amnon, Tamar, Solomon}. The set has 10 members and 2^{10} subsets, and subsets of various sizes can be listed and counted. Beyond this, there is not much we can say about it. But if we also know that the names are part of the family tree of the biblical king David, and that Boaz was Obed's father, Obed was Jesse's father, Jesse was David's, Eliab's, and Abinadab's father, and David was Absolom's, Amnon's, Tamar's, and Solomon's father, we would know far more about the set. Instead of being merely a collection, these names are now related to each other. This gives the set a structure and that structure gives the set a great deal more meaning than it had without the structure. See Figure 8.1.

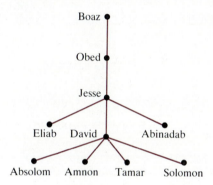

Figure 8.1   Part of a family tree.

The remaining chapters of this book deal with some of the possible structures that sets can have; we will learn more specifically what the word *structure* means as we go along. This chapter is about **relations,** which is the mathematical term for relationships. Relations have a wide variety of applications. In computer science, they are important in the study of databases, state diagrams, and automata. In mathematics, they are widely used in number theory, topology, set theory, and as an object of study in their own right. They also have applications in a number of other disciplines as well. For instance, in sociology, they are very important in the study of kinship relations. In fact, any kind of comparison between pairs of items establishes a relationship between them that can be modeled by the concept of the relation. We will define the word *relation* in a precise way in Section 8.1 and examine some of the properties that relations can have in Sections 8.2 through 8.4. Our focus will be primarily on finite sets, although most of the ideas we will examine can be generalized to infinite sets as well.

8.1
BASIC CONCEPTS

The concept of the function was introduced in Chapter 2. Functions can be considered as special cases of relations; alternatively, relations can be considered as a generalization of the concept of the function. So, before defining relations, we must briefly review the concept of the function.

Given two sets, A and B, we defined the **cross-product,** $A \times B$, of the sets as

$$A \times B = \{(x,y) \mid x \in A, \ y \in B\}.$$

Thus, for instance, if $A = \{1, 2, 3\}$ and $B = \{4, 5, 6\}$,

$$A \times B = \{(1,4), (1,5), (1,6), (2,4), (2,5), (2,6), (3,4), (3,5), (3,6)\}.$$

We then said that a **function,** f, with **domain** A and **codomain** B is a subset of $A \times B$ with the property that every x in A has precisely one y in B associated with it. Thus $\{(1,4), (2,5), (3,6)\}$ is a function but $\{(1,4), (1,5), (1,6)\}$ and $\{(2,5), (3,6)\}$ are not functions.

For relations, the restriction that every x in A must have precisely one y in B associated with it is removed. The number of y's associated with a particular x can be anything, even zero.

Definition Let A and B be any two sets. Then a **binary relation** from A to B is any subset, R, of $A \times B$. We continue to call A and B the *domain* and the *codomain,* respectively, of the relation. If (x,y) is a member of the relation, we say that $\boldsymbol{x}$ **is related to** $\boldsymbol{y}$ and write $\boldsymbol{xRy}$. If $A = B$, we call R a **relation on** $\boldsymbol{A.}$ The ordered pair $(\boldsymbol{A,R})$ denotes the relation R on the set A.

Examples of Binary Relations

EXAMPLE 1 Let $A = \{1, 2, 3\}$ and $B = \{4, 5, 6\}$. Then the following are all binary relations with domain A and codomain B.

a) $\{(1,4), (2,5), (3,6)\}$

b) $\{(1,4), (1,5), (1,6)\}$

c) $\{(2,5), (3,6)\}$

d) $\varnothing$

e) $\{(1,4), (1,5), (1,6), (2,4), (2,5), (2,6), (3,4), (3,5), (3,6)\}$

f) $\{(1,6), (2,6), (3,6)\}$

Note that (a) and (f) are functions, but the rest are not. ∎

EXAMPLE 2   Consider the equation $x = y^2$, mapping **R** into **R** with x in the domain and y in the codomain (see Figure 8.2). This is not a function, since for all $x > 0$ there are two y's corresponding to each x. But it is a binary relation. Relations provide us with the mathematical tool we need to deal with equations like this that are not functions. ■

Figure 8.2   Graph of $x = y^2$.

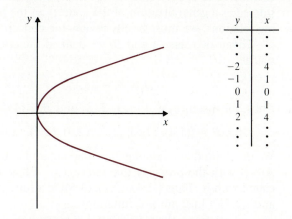

y	x
⋮	⋮
-2	4
-1	1
0	0
1	1
2	4
⋮	⋮

EXAMPLE 3   Show that the family tree presented at the beginning of this chapter represents a relation.

SOLUTION   Let $S = \{$Boaz, Obed, Jesse, David, Eliab, Abinadab, Absolom, Amnon, Tamar, Solomon$\}$. Let R denote the phrase "is the parent of." We can write

> Boaz R Obed
>
> Obed R Jesse
>
> Jesse R David
>
> Jesse R Eliab
>
> Jesse R Abinadab
>
> David R Absolom
>
> David R Amnon
>
> David R Tamar
>
> David R Solomon

Since this collection of ordered pairs is a subset of $S \times S$, it is a relation. ■

EXAMPLE 4   There are a number of important and interesting binary relations on the real numbers, **R**. For instance, let $R = \{(x,y) \mid x \in \mathbf{R},\ y \in \mathbf{R},\ x < y\}$. Then $R \subset \mathbf{R} \times \mathbf{R}$ and thus *less than* is a relation. *Equals, less than or equal to, greater than,* and *greater than or equal to* are all relations on **R**. ■

Relations are an underlying concept that tie together many of the concepts we have examined earlier in this text. The next two examples and several of the exercises provide a demonstration.

EXAMPLE 5 Let n be any positive integer. In Section 5.5 we defined $x \equiv (y \bmod n)$ to mean that $x - y$ is divisible by n. If $x \equiv (y \bmod n)$, we say that x is congruent to y mod n. Show that congruent mod n defines a relation on **Z**.

SOLUTION If x and y are integers, then xRy if and only if $x - y = kn$ for some integer k. The set of pairs that satisfies this condition is a subset of $\mathbf{Z} \times \mathbf{Z}$ and is thus a relation. Note that the elements of the set $\{\ldots, -2n, -n, 0, n, 2n, \ldots\}$ are all related. The elements of $\{\ldots -2n + 1, -n + 1, 1, n + 1, 2n + 1, \ldots\}$ are also related. These sets are denoted $\bar{0}, \bar{1}$, etc. (See Section 5.5 for more details.) ∎

EXAMPLE 6 Suppose P denotes the set of all propositions. Then logical equivalence of propositions defines a relation on P. That is, if p and q are propositions, they are logically equivalent if they have the same truth table. The set of all pairs of propositions (p,q) in which p is logically equivalent to q is a subset of $P \times P$ and hence is a relation. ∎

EXAMPLE 7 Let S be any set. Then we can say that two subsets, A and B, of S are related if $A \subset B$. This defines a binary relation on $P(S)$, the power set of S. Note that the elements of $p(S)$ are themselves sets, so the ordered pairs here are pairs of sets. ∎

EXAMPLE 8 Suppose a computer program is organized into a collection of procedures, as in Figure 8.3. We can say that procedure A is related to procedure B if A calls B. Then *calls* is a binary relation on the collection of procedures that make up the program. ∎

EXAMPLE 9 Consider a set, V, of points and a set, E, of directed arcs joining these points, as in Figure 8.4. Such a pair, (V,E), of sets is called a **digraph.** The points are called **vertices** (plural of **vertex**) or **nodes.** The elements of E are called **edges.** If each pair of vertices, (v_i,v_j), has at most one edge associated with it, a digraph is a binary relation on V. That is, if $V = \{v_1, v_2, \ldots, v_n\}$, we can label the edges by their starting and ending vertices—for instance, (v_1, v_2) in Figure 8.4(a). In drawing digraphs, the convention is that if (v_1, v_2) is in R, then the tail of the arrow is at v_1 and the head of the arrow is at v_2. Digraphs are an extremely useful mathematical tool, having applications in biology, engineering, sociology, computer science, and many other disciplines. We will see more of digraphs in this chapter and in Chapter 10. ∎

Figure 8.3 Structure
chart for a computer
program.

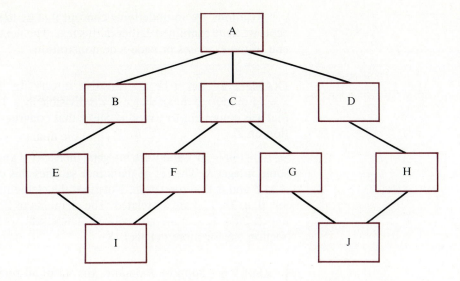

Figure 8.4 Digraphs.

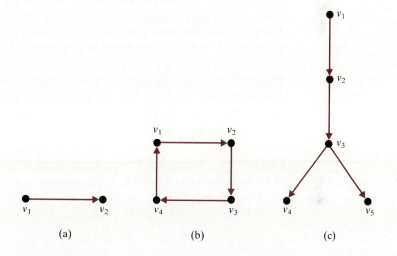

(a)   (b)   (c)

n-ary Relations

Binary relations are subsets of $A \times B$ for two sets, A and B. But we can
just as well consider subsets of the cross-product of n sets, written
$A_1 \times A_2 \times \cdots \times A_n$. Such subsets are called **_n_-ary relations.** Although not
as widely used as binary relations, n-ary relations have come to be of critical
importance in the study of databases. If $A_1, A_2, \ldots, A_n$ are all the same set,
A, we speak of an **_n_-ary relation on _A_.**

EXAMPLE 10   Consider the so-called unit cube—$\{(x,y,z) \mid 0 \leq x \leq 1, 0 \leq
y \leq 1, 0 \leq z \leq 1\}$. Then the statement "$(x,y,z)$ is a point in the unit cube"

defines a ternary relation on **R,** since the set of points that satisfy the statement is a subset of $\mathbf{R} \times \mathbf{R} \times \mathbf{R}$. The same approach could be used to define ternary relations based on any subset of $\mathbf{R} \times \mathbf{R} \times \mathbf{R}$. ∎

EXAMPLE 11   A teacher's grade book for a course consists of the student's name, the grades on three tests, and the grade on the final exam. Show that this collection of data is a 5-ary relation.

SOLUTION   Suppose each test and the final are graded on a scale from 0 to 100. Let N denote the set of names of students in the class and let U denote the integers from 0 to 100. Then the set of entries in the grade book is a subset of $N \times U \times U \times U \times U$. ∎

EXAMPLE 12   The parts department of an automobile dealership classifies parts according to part name, part number, cost, and quantity on hand. Show that this classification can be modeled by a 4-ary relation.

SOLUTION   Let P be the set of possible part names and suppose that part numbers are all natural numbers. Then the current inventory is a subset of $P \times \mathbf{N} \times \mathbf{R} \times \mathbf{Z}$. Note that the part number, cost, and quantity on hand are not a *function* of the part name, since it is possible to have several different carburetors, for instance. However, if part numbers are well chosen, the name, cost, and quantity on hand will be a function of the part number, i.e., every part number has a unique name, cost, and quantity. In database theory, the part number would be called a **key.** Note also that every time a new part or an additional inventory of existing parts arrives, the relation will change, i.e., it will become a new subset of $P \times \mathbf{N} \times \mathbf{R} \times \mathbf{Z}$. ∎

Representing Relations

Besides presenting the concept of relation, a second objective of this section is to present two methods that are commonly used to represent relations, i.e., to present them in a way that makes them easier to visualize. These are digraphs and matrices. Both become unwieldy when the size of A or B is more than a small finite number. In spite of this limitation, both are very helpful in visualizing a number of useful binary relations. In the next sections, they will also help to explain several important properties of relations.

EXAMPLE 13   The family tree, Figure 8.1, and the hierarchy chart for the computer program, Figure 8.3, are digraphs representing relations. ∎

EXAMPLE 14   Represent the relation *greater than* on the set $\{1, 2, 3, 4, 5\}$ by a digraph.

SOLUTION   Consider 2 and 1. Since $2 > 1$, we can write $2R1$; this relationship is represented by an arrow with the tail at 2 and the head at 1. The complete digraph is in Figure 8.5.

Figure 8.5   Digraph representing *greater than* on $\{1, 2, 3, 4, 5\}$.

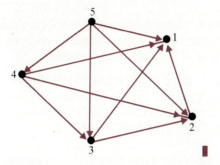

EXAMPLE 15   Represent the relation *proper subset* by a digraph for the subsets of $\{1, 2, 3\}$.

SOLUTION   See Figure 8.6.

Figure 8.6   Digraph representing the subset relation on $\{1, 2, 3\}$.

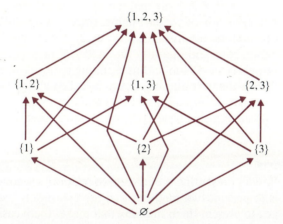

The vertices are the subsets. A directed edge from A_1 to A_2 implies that $A_1 \subset A_2$ and $A_2 \not\subset A_1$. ∎

A binary relation on any set, A, gives rise to a digraph. The vertices are the elements of A. The directed edges are those pairs (a,b) such that a is related to b.

An alternative way to represent relations is with the use of Boolean matrices—matrices in which all of the entries are either 0 or 1. The rows correspond to the elements of A and the columns to the elements of B. An entry is 1 if (a,b) is in the relation; otherwise it is 0. Such a matrix is called the **adjacency matrix** of the relation. (See Appendix B for a more extensive discussion of matrices.)

EXAMPLE 16   Represent the relations of Examples 1(a), (b), and (d) by Boolean matrices.

SOLUTION   Let the rows correspond to 1, 2, and 3, respectively, and the columns to 4, 5, and 6, respectively. Example 1(a) consisted of the pairs, {(1,4), (2,5), (3,6)}. Its matrix representation is

$$\begin{bmatrix} 1 & 0 & 0 \\ 0 & 1 & 0 \\ 0 & 0 & 1 \end{bmatrix}.$$

Example 1(b) consisted of the pairs, {(1,4), (1,5), (1,6)}. Its matrix representation is

$$\begin{bmatrix} 1 & 1 & 1 \\ 0 & 0 & 0 \\ 0 & 0 & 0 \end{bmatrix}.$$

Example 1(d) was the empty set, and hence its matrix representation is

$$\begin{bmatrix} 0 & 0 & 0 \\ 0 & 0 & 0 \\ 0 & 0 & 0 \end{bmatrix}. \quad \blacksquare$$

EXAMPLE 17   Represent the relation *less than* on {1, 2, 3, 4, 5} by a matrix.

SOLUTION   Let row 1 correspond to 1 in the set, column 1 to 1, etc. Then an entry in the matrix will be 1 if its row number is less than its column number and will be 0 otherwise.

$$\begin{bmatrix} 0 & 1 & 1 & 1 & 1 \\ 0 & 0 & 1 & 1 & 1 \\ 0 & 0 & 0 & 1 & 1 \\ 0 & 0 & 0 & 0 & 1 \\ 0 & 0 & 0 & 0 & 0 \end{bmatrix} \quad \blacksquare$$

TERMINOLOGY

relation	x is related to y	node
cross-product	xRy	edge
function	relation on A	n-ary relation

domain	(A, R)	key
codomain	digraph	adjacency matrix
binary relation	vertex	

SUMMARY

1. A binary relation is any subset of $A \times B$ for sets A and B. Binary relations are a generalization of the concept of function and can be used to describe almost any comparison or association.

2. n-ary relations are subsets of $A_1 \times A_2 \times \cdots \times A_n$. They are especially important in the study of databases.

3. Binary relations on finite sets can be represented either by digraphs or by Boolean matrices.

EXERCISES 8.1

Let $A = \{1, 2, 3\}$, $B = \{a, b, c, d\}$, and $C = \{n, m\}$. List all elements in each of the following cross-products.

1. $A \times B$

2. $A \times C$

3. $B \times C$

4. $B \times A$

5. $C \times A$

6. $C \times B$

Decide whether each of the following equations is a function, with x as the independent variable and domain $\mathbf{R}$. Show that each is a relation.

7. $x = y^3$

8. $x = |y|$

9. $x = 1 - y^2$

10. $x = \begin{cases} y^2 & y \geq 0 \\ -y^2 & y < 0 \end{cases}$

Each of the following situations can be modeled using relations. In each case, indicate an appropriate domain and codomain and informally describe the set of ordered pairs.

11. Given any set, S, subsets are compared to see if one is a superset of another.

12. English words are compared to see which is first in alphabetical order.

13. The question is asked about a collection of 10 small cities in a state, "Which ones have direct airline connections between them?"

14. The students at a college are listed, along with the courses each is taking in the current semester.

15. A class list has each student's grade next to his or her name.

16. A rumor is spreading through the student body.

Draw digraphs to represent each of the following relations.

17. {(1,2), (2,3), (3,4), (4,5)}   **18.** {(1,1), (2,2), (3,3)}

19. {(1,2), (2,1), (2,3), (3,2), (3,1)}   **20.** {(a,b), (b,c), (b,d), (c,e), (d,e)}

21. Let $S = \{1, 2, 3, 4, 6, 8\}$. Then iRj if and only if i divides j.

22. This is the same as Exercise 21, but with $S = \{3, 5, 7, 11, 13\}$.

Represent each of the following digraphs by a Boolean matrix.

23.

24.

25.

26.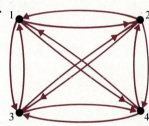

27.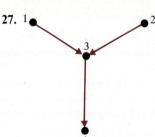

28.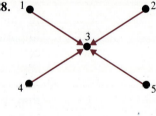

Represent each of the following Boolean matrices by a digraph.

29. $\begin{bmatrix} 1 & 1 \\ 1 & 1 \end{bmatrix}$ **30.** $\begin{bmatrix} 1 & 0 \\ 0 & 1 \end{bmatrix}$

31. $\begin{bmatrix} 1 \end{bmatrix}$ **32.** $\begin{bmatrix} 0 \end{bmatrix}$

33. $\begin{bmatrix} 0 & 1 & 0 \\ 0 & 0 & 1 \\ 1 & 0 & 0 \end{bmatrix}$ **34.** $\begin{bmatrix} 0 & 0 & 1 \\ 1 & 0 & 0 \\ 0 & 1 & 0 \end{bmatrix}$

35. $\begin{bmatrix} 1 & 0 & 1 & 0 \\ 0 & 1 & 0 & 1 \\ 1 & 0 & 1 & 0 \\ 0 & 1 & 0 & 1 \end{bmatrix}$ **36.** $\begin{bmatrix} 1 & 1 & 0 & 0 & 0 \\ 0 & 1 & 1 & 0 & 0 \\ 0 & 0 & 1 & 1 & 0 \\ 0 & 0 & 0 & 1 & 1 \\ 1 & 0 & 0 & 0 & 1 \end{bmatrix}$

Show that each of the following situations can be represented by an *n*-ary relation.

37. Three integers, *x, y,* and *z,* are related if $x^2 + y^2 + z^2 \le 1$.

38. A compiler for a computer language collects all of the symbols the program uses into a symbol table. This table includes the symbol itself, its type (integer, real, boolean, etc.), and its location in memory.

39. A class list contains a student ID number, the student's name, and a single letter code containing special information such as whether the student is auditing or repeating the course.

40. A list of library books contains the book's classification number, title, author, publisher, and date of publication.

8.2

PROPERTIES OF RELATIONS

In the previous section, we examined the concept of relation and two important ways to represent relations—by digraphs and by boolean matrices. Two types of relations arise especially often in mathematics and its applications—relations that express the notion that items are equivalent to each other and relations that express the idea that things are arranged in some order. We will consider equivalence in this section and order in Section 8.3. In this and subsequent sections, we will assume familiarity with matrix notation and the properties of matrices as presented in Appendix B. We start by defining four fundamental properties that relations may or may not possess.

Definition   Let *A* be any set and let *R* be a binary relation on *A*.

If for all *a* in *A, aRa,* then *R* is **reflexive.**

If *aRb* implies *bRa,* then *R* is **symmetric.**

If *aRb* and *bRa* implies *a = b,* then *R* is **antisymmetric.**

If *aRb* and *bRc* implies *aRc,* then *R* is **transitive.**

Recall that *R* is a subset of $A \times A$. Thus the previous properties can be expressed in terms of ordered pairs of elements of *R*. That is, *reflexive* means that for all $a \in A$, $(a,a) \in R$. *Symmetric* means that $(a,b) \in R$ implies that $(b,a) \in R$. *Antisymmetric* means that $(a,b) \in R$ and $(b,a) \in R$ implies that $(a,a) \in R$. *Transitive* means that $(a,b) \in R$ and $(b,c) \in R$ implies that $(a,c) \in R$.

Examples

EXAMPLE 1 Let S be any set and suppose that for any x and y in S, xRy if and only if $x = y$. Which of the previously discussed properties hold for R?

SOLUTION In any set, every element is equal to itself. Thus, for all a in S, aRa, and hence R is reflexive. Suppose aRb; that is, $a = b$. Then $b = a$ also and hence aRb implies bRa. Thus R is symmetric. Suppose aRb and bRa. Then $a = b$ and hence R is antisymmetric. Lastly, if $a = b$ and $b = c$, we can conclude that $a = c$. Thus R is also transitive. ∎

Note that symmetric and antisymmetric are not opposites. While this relation is both symmetric and antisymmetric, there are other relations that are neither. We will see one such relation in Example 6.

EXAMPLE 2 Which of the preceding properties hold for $(R, \leq)$?

SOLUTION For any real number a, $a \leq a$. Thus $\leq$ is reflexive. $a \leq b$ does not imply that $b \leq a$, however, and thus $\leq$ is not symmetric. It is antisymmetric, though, since $a \leq b$ and $b \leq a$ imply that $b = a$. It is also transitive, since $a \leq b$ and $b \leq c$ imply that $a \leq c$. ∎

EXAMPLE 3 Which of the properties hold for $(\mathbf{R}, <)$?

SOLUTION $<$ is not reflexive, since for any a, $a < a$ is false. $<$ is not symmetric; however, it is transitive. Verification of these last two statements is left for the Exercises. That $<$ is antisymmetric is vacuously true, since it is impossible to have both $a < b$ and $b < a$ be true simultaneously. ∎

EXAMPLE 4 A political scientist, William Riker, has given an interesting example of a nontransitive relation. Suppose an issue is being debated by a three-member committee and there are three proposals, A, B, and C, to be voted on. A > B will mean that a majority of the committee prefers proposal A to proposal B. If one committee member prefers, say, A to B, we write A ≥ B. Suppose that for one committee member, A ≥ B ≥ C. Suppose that for a second member, B ≥ C ≥ A and for a third, C ≥ A ≥ B. If all three proposals are presented at once, the committee will be divided, each proposal getting one vote. So it is decided to vote on just two of them. The committee will then choose between the winner of that vote and the one proposal that is left. Thus, for instance, if A and B are voted on first, A will win. In the runoff between A and C, C will win. In fact, voting on two proposals at a time, A will be preferred to B, B to C, and C to A. That is, A > B > C > A. This violates transitivity, since A > B and B > C would imply that A > C if the relation were transitive. Political scientists have noted that skillful politicians have occasionally observed this phenomenon and have used it to bring about the victory of their own proposals. ∎

EXAMPLE 5   Let S be a set of people and let R_1 be the relation "a is the parent of b." Then this relation is not reflexive, since no one can be his or her own parent. It is also not symmetric, although it is antisymmetric, since aRb and bRa is impossible. It is also not transitive, since a being the parent of b and b being the parent of c implies that a is the *grandparent,* not the parent, of c. If R_2 is the relation "a is the sibling of b," where sibling means both having the same parents, then R_2 is reflexive, symmetric, and transitive, although it is not antisymmetric. Verification of these properties of R_2 will be left for the Exercises. ∎

EXAMPLE 6   What properties must the matrix of a relation have for the relation it represents to be reflexive, symmetric, antisymmetric, or transitive?

SOLUTION   To be reflexive, a relation must satisfy aRa for every a. Thus the diagonal entries of the matrix must be all nonzero. For instance,

$$\begin{bmatrix} 1 & 0 & 0 \\ 0 & 1 & 1 \\ 0 & 0 & 1 \end{bmatrix}$$

represents a reflexive relation.

To be symmetric, aRb must imply that bRa. Thus the matrix must be symmetric. For instance,

$$\begin{bmatrix} 1 & 0 & 1 & 0 \\ 0 & 0 & 0 & 1 \\ 1 & 0 & 0 & 1 \\ 0 & 1 & 1 & 1 \end{bmatrix}$$

represents a symmetric relation.

In order for a relation to be antisymmetric, aRb and bRa must imply that $a = b$. aRb means that the entry in the ath row, bth column is a 1. For entries on the diagonal of the matrix, $a = b$. Thus 1's on the diagonal satisfy the condition for antisymmetry. If there is a 1 off the diagonal, however, in, say the abth position, the corresponding entry in the bath position must be a 0, since otherwise the condition for antisymmetry would be violated. For instance, the relation represented by the following matrix is *not* antisymmetric, since both M_{14} and M_{41} are nonzero:

$$\begin{bmatrix} 1 & 0 & 0 & 1 \\ 0 & 1 & 0 & 0 \\ 0 & 0 & 0 & 1 \\ 1 & 0 & 0 & 0 \end{bmatrix}$$

Note that the relation represented by this matrix is not symmetric either.

A relation is transitive if aRb and bRc imply that aRc. In terms of matrices, this says that if the abth entry and the bcth entry are 1, then the acth entry must be 1. For instance,

$$\begin{bmatrix} 0 & 1 & 1 & 0 \\ 0 & 0 & 1 & 0 \\ 0 & 0 & 0 & 0 \\ 0 & 0 & 0 & 0 \end{bmatrix}$$

represents a transitive relation. Since M_{12} and M_{23} are both 1, for the relation to be transitive, the lines across the first row and down the third column must meet in a 1. Note that this does *not* say that for any pair of 1's in the matrix, a rectangle formed using them as corners must have all of its corners with a value of 1. It applies only to the case where the abth and bcth entries are 1's, and then it requires that only the acth entry be a 1. ∎

EXAMPLE 7   Which of the four properties do the digraphs of Figure 8.7 possess?

SOLUTION

a) To be reflexive, every node of a digraph must possess a **loop,** i.e., an edge that both begins and ends at that node. Thus Figure 8.7(a) is reflexive. Symmetry (aRb implies bRa) requires that every edge be *bidirectional*; i.e., if there is an edge from a to b, there must also be one from b to a. Thus Figure 8.7(a) is not symmetric. Antisymmetry (aRb and bRa imply that $a = b$) means that there are no bidirectional edges, although loops are allowed. Thus Figure 8.7(a) is antisymmetric. A digraph is transitive if, whenever there is an edge joining a and b and an edge joining b and c, there is an edge joining a and c. Since there are only three edges, it can be easily checked that Figure 8.7(a) is transitive.

b) This one is not reflexive, since two nodes do not have loops. It is not symmetric, since none of the edges are bidirectional, although it is antisymmetric, since there are no bidirectional edges. It is not transitive, since there is an edge from a to b and one from b to c, but no edge from a to c.

c) This one is not reflexive but it is symmetric, since each edge is bidirectional. It is neither antisymmetric nor transitive.

d) This one is neither reflexive nor symmetric. But it is antisymmetric and transitive. ▮

Figure 8.7 Four digraphs having various properties.

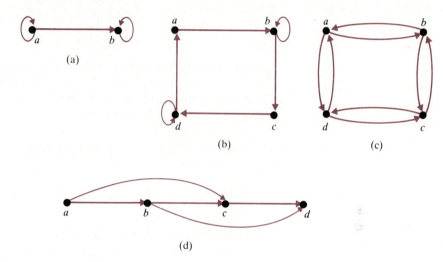

(a)

(b)

(c)

(d)

Equivalence Relations

Equivalence is a concept we constantly encounter. If a store does not have the particular item we want, we may be asked, "Would you accept an equivalent substitute?" In Chapter 3, we asked whether two Boolean expressions were logically equivalent. Logical circuits and computer programs are considered equivalent if the same inputs always give the same outputs. In Chapter 4, we studied asymptotic equivalence of functions. Note that *equivalent* is not the same as *equal*. The similarities and differences between them will become clearer in the following discussion.

Definition Let A be any set. By an **equivalence relation** on A, we mean a relation that is reflexive, symmetric, and transitive. If two elements, a and b, of A are related by an equivalence relation, we say that a and b are **equivalent** and we write $a \equiv b$.

This statement formalizes three properties we intuitively associate with equivalence: (1) every object is equivalent to itself; (2) if a is equivalent to b, then b is equivalent to a; and (3) if a is equivalent to b and b is equivalent to c, then a is equivalent to c. As we saw in Example 1, equality in any set is reflexive, symmetric, and transitive. Thus *equals* is always an equivalence relation. But there are many relations other than *equals* that are equivalence relations. Thus the notion of an equivalence relation is a generalization of the concept of equality that preserves all of the properties we intuitively associate with equivalence.

EXAMPLE 8 Two Boolean expressions are said to be logically equivalent if they have the same truth table. Show that this notion of equivalence is an equivalence relation.

SOLUTION To give a specific example, $p \rightarrow q$ is logically equivalent to $\neg p \vee q$, since they have the same truth table. To show in general that logical equivalence is an equivalence relation, let a, b, and c denote boolean expressions and let R denote the notion of logical equivalence. Then aRa is true, since any logical expression has the same truth table as itself. Also aRb implies bRa, since if a has the same truth table as b, then b has the same truth table as a. Lastly, logical equivalence is transitive, since if a has the same truth table as b and b has the same truth table as c, then a has the same truth table as c. This is a good illustration of the difference between equality and equivalence. Certainly, both expressions would have to use the same variables in order to have the same truth table. But $a = b$ can also mean that the expressions are identical, symbol by symbol, while $a \equiv b$ can mean that they have the same truth table even though the connectives used and the order in which the symbols appear may differ. ∎

EXAMPLE 9 Show that congruence mod 5 is an equivalence relation on **Z.**

SOLUTION Two integers, a and b, are congruent mod 5 if 5 divides $a - b$. This relation is reflexive, since for any integer a, $a - a = 0$ and 5 divides 0. It is symmetric, since if 5 divides $a - b$, it also divides $b - a$. Lastly, it is transitive, since if 5 divides $a - b$ and $b - c$, then 5 divides $(a - b) + (b - c) = a - c$. ∎

EXAMPLE 10 Which of the following matrices represent equivalence relations?

a) $\begin{bmatrix} 1 & 1 & 0 & 0 & 0 \\ 1 & 1 & 0 & 0 & 0 \\ 0 & 0 & 1 & 0 & 0 \\ 0 & 0 & 0 & 1 & 1 \\ 0 & 0 & 0 & 1 & 1 \end{bmatrix}$
b) $\begin{bmatrix} 1 & 1 & 0 & 0 \\ 0 & 1 & 0 & 1 \\ 0 & 0 & 0 & 0 \\ 0 & 1 & 0 & 1 \end{bmatrix}$
c) $\begin{bmatrix} 1 & 1 & 0 & 0 \\ 1 & 1 & 1 & 0 \\ 0 & 1 & 1 & 0 \\ 0 & 0 & 0 & 1 \end{bmatrix}$

SOLUTION

a) This one does represent an equivalence relation. It is reflexive, since all diagonal entries are 1, and it is obviously symmetric. It is transitive, since each of the four nonzero entries off the main diagonal satisfies the condition that $a_{ij} = 1$ and $a_{jk} = 1$ implies that $a_{ik} = 1$.

b) This is neither symmetric nor reflexive, so it does not represent an equivalence relation.

c) This is reflexive and symmetric, but not transitive. For instance, a_{12} and a_{23} are both 1, but $a_{13} = 0$. ∎

Equivalence Classes and Partitions

The next two concepts are quite important in our study of equivalence.

Definition   Let R be an equivalence relation on a set A. Let a be an element of A. By the **equivalence class of a,** denoted **[a],** we mean that $\{x \in A \mid x \equiv a\}$. Let B be any set. By a **partition** of B, we mean a collection, Γ, of nonempty subsets, $\{B_i\}$, such that $U\,B_i = B$ and $B_i \cap B_j = \varnothing$.

Equivalence classes allow us to break sets down into subsets, all of whose elements are equivalent to each other. A partition divides a set into a collection of disjoint subsets whose union is the entire set. See Figure 8.8.

Figure 8.8   A partition of a set.

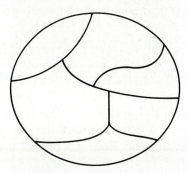

EXAMPLE 11   Find the equivalence classes for the following equivalence relations:

a) equality on any set A.

b) the set $\{a, b, c, d, e\}$ where the relation is given by the matrix of Example 10(a).

c) the integers with the relation of congruence mod 5.

SOLUTION

a) Under this relation, two elements are equivalent if and only if they are equal to each other. Thus, for each element a, $[a] = \{a\}$. That is, each element is its own equivalence class.

b) Under this relation, $a \equiv b$ and $d \equiv e$. No other elements ar equivalent to any other elements. Thus the equivalence classes are $\{a, b\}$, $\{c\}$, and $\{d, e\}$. Note that these form a partition of the set $\{a, b, c, d, e\}$.

c) The equivalence classes are

$$\overline{0} = \{\ldots -10, -5, 0, 5, 10, \ldots\}$$
$$\overline{1} = \{\ldots -9, -4, 1, 6, 11, \ldots\}$$
$$\overline{2} = \{\ldots -8, -3, 2, 7, 12, \ldots\}$$
$$\overline{3} = \{\ldots -7, -2, 3, 8, 13, \ldots\}$$
$$\overline{4} = \{\ldots -6, -1, 4, 9, 14, \ldots\}$$

That is, each class is a set of the form $\{i + 5n \mid n \in \mathbf{Z}\}$ where $i = 0, 1, 2, 3,$ or 4. These sets form a partition of the set $\mathbf{Z}$. ∎

The following theorem summarizes the fundamental properties of equivalence classes.

THEOREM 8.1 Let (A,R) be an equivalence relation. Then for all a, b, and c in A,

a) $a \in [a]$.

b) $b,c \in [a] \Rightarrow b \equiv c$.

c) $[a] = [b] \Leftrightarrow a \equiv b$.

d) either $[a] = [b]$ or $[a] \cap [b] = \varnothing$.

PROOF

a) Since R is an equivalence relation, R is reflexive. Thus $a \equiv a$ and hence $a \in [a]$.

b) If b and c are in $[a]$, then $b \equiv a$ and $c \equiv a$. Since R is transitive and symmetric, $b \equiv c$.

c) First, suppose that $[a] = [b]$. By part (a), $a \in [a]$ and $b \in [b]$. Hence a and b are both elements of $[a]$, and thus, by part (b), $a \equiv b$. Now suppose $a \equiv b$. If x is any element of $[a]$, then $x \equiv a$. Hence $x \equiv b$ by transitivity and x is an element of $[b]$. Thus $[a] \subset [b]$. Similarly, $[b] \subset [a]$ and thus $[a] = [b]$.

d) Suppose there exists $x \in [a] \cap [b]$. Then $x \equiv a$ and $x \equiv b$ by parts (a) and (b) of this theorem. Thus $a \equiv b$ and, by part (c), $[a] = [b]$. ☐

The next theorem presents two of the most commonly used properties of equivalence relations.

THEOREM 8.2

a) Let (A,R) be an equivalence relation. Then the set of equivalence classes of R forms a partition of A.

b) Let A be any set and let $\Gamma = \{A_i\}$ be a partition of A. Define a relation, R, on A by the rule that aRb if and only if a and b are both elements of the same A_i. Then R is an equivalence relation.

PROOF

a) For each element, *a,* of *A,* there exists an equivalence class [*a*]. Thus the union of all of the equivalence classes will be *A.* By Theorem 8.1(d), two equivalence classes, [*a*] and [*b*], are either equal or disjoint. Thus the equivalence classes form a partition of *A.*

b) Since Γ is a partition, $U\,A_i = A$. Hence, each *a* in *A* will be in some A_i. Thus *aRa* is true for all *a*. If *a* is in the same A_i as *b,* then it is also true that *b* is in the same A_i as *a.* Hence *R* is symmetric. Lastly, suppose that *a* and *b* are related, i.e., both are in A_i for some *i.* Also, suppose that *b* and *c* are related, i.e., both are in A_j for some *j.* Thus, since *b* is a member of both A_i and A_j, $A_i = A_j$ and thus *a* is related to *c.* Thus *R* is transitive as well. ☐

Theorem 8.2 is important for two reasons. (1) An equivalence relation breaks down a set into subsets consisting of equivalent elements. Thus, if we want an equivalent alternative to an element, any element in its equivalence class will do. (2) The equivalence classes provide a complete listing of all of the possible nonequivalent values the relation could take on in the original set. The set of those values is often quite important. It is formally defined as follows.

Definition   If (A,R) is an equivalence relation, the set of equivalence classes of *R* is called *A* **modulo *R*** or the **quotient set** and is denoted by *A/R*.

EXAMPLE 12   If *R* is the relation *congruence mod 5,* then

$$\frac{\mathbf{Z}}{R} = \{\overline{0},\ \overline{1},\ \overline{2},\ \overline{3},\ \overline{4}\}. \quad \blacksquare$$

EXAMPLE 13   If *S* is any set and *R* is the relation *equals,* then

$$S/R = S. \quad \blacksquare$$

EXAMPLE 14   A concept often used in physics is that of the vector. Equivalence classes provide a nice way to define this concept. By a **directed line segment,** we mean (1) the set of points that lie on the straight line segment joining two specific points and (2) one of the two points is designated as the head of the segment and the other as the tail (see Figure 8.9). Two directed line segments are considered equivalent if they have the same length, if they are parallel, and if line segments joining the heads together and the tails together are also parallel. Then this notion of equivalence can be shown to be an equivalence relation. If we let *N* denote the set of all line segments in a plane and *R* the equivalence relation, by a **vector** we mean an element of *N/R*. That is, all directed line segments that are parallel have the same length, and "point in the same direction" are regarded as equivalent. A vector is this class of equivalent line segments. One common way to look at vectors that is easy to visualize is to consider a vector as one representative element of this equivalence class, typically the directed line segment that has its tail at the origin. ∎

Figure 8.9 A collection of directed line segments.

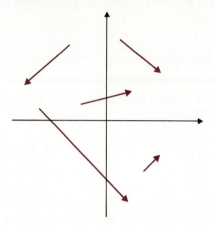

One additional question we might ask at this point is, "Given a set, *A*, with $|A| = n$, how many equivalence relations are there on *A*?" This question is addressed in Appendix F.

TERMINOLOGY

reflexive	equivalence class
symmetric	[*a*]
antisymmetric	partition
transitive	*A* modulo *R*
loop in a digraph	quotient set
equivalence relation	*A/R*
equivalent	directed line segment
$a \equiv b$	vector

SUMMARY

1. Four fundamental properties that relations may have are reflexivity, symmetry, antisymmetry, and transitivity.

2. Relations that are reflexive, symmetric, and transitive are called *equivalence relations*. Equivalence is a generalization of the notion of equality.

3. Equivalence relations partition sets into equivalence classes—disjoint subsets whose union is the entire original set. Similarly, any partition gives rise to an equivalence relation.

4. The set of equivalence classes of a set is called the *quotient set*.

EXERCISES 8.2

For each of the relations represented by the following matrices, decide if it is reflexive, symmetric, antisymmetric, or transitive.

1.
$$\begin{bmatrix} 1 & 0 & 0 \\ 0 & 1 & 0 \\ 0 & 0 & 1 \end{bmatrix}$$

2.
$$\begin{bmatrix} 1 & 1 & 1 \\ 0 & 1 & 1 \\ 0 & 0 & 1 \end{bmatrix}$$

3.
$$\begin{bmatrix} 1 & 1 & 0 & 0 \\ 1 & 1 & 0 & 0 \\ 0 & 0 & 1 & 1 \\ 0 & 0 & 1 & 1 \end{bmatrix}$$

4.
$$\begin{bmatrix} 0 & 1 & 1 & 1 \\ 1 & 0 & 1 & 1 \\ 1 & 1 & 0 & 1 \\ 1 & 1 & 1 & 0 \end{bmatrix}$$

For each of the relations represented by the following digraphs, decide if it is reflexive, symmetric, antisymmetric, or transitive.

5.

6.

7.

8.

For each of the following relations, decide if it is reflexive, symmetric, antisymmetric, or transitive.

9. $(\mathbf{Z}, \leq)$

10. $(\mathbf{R}, <)$

11. Two sets, A and B, are related if both are subsets of the same set, **U**, and $|A| = |B|$.

12. If f is a function from $\mathbf{R} \rightarrow \mathbf{R}$, two real numbers, x and y are related if $f(x) = f(y)$.

13. Two people are related if they are siblings, where *sibling* is defined to mean both having the same parents.

14. Two real numbers, x and y, are related if $x = y^2$.

15. $\{(1,2), (2,3), (3,2), (2,1), (1,3), (3,1), (4,4)\}$

16. $\{(1,1), (2,2), (3,4), (4,3)\}$

Show that the following relations are equivalence relations.

17. 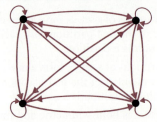   18.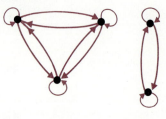

(Note that this diagram is all one digraph.)

19. Two triangles are related if they are congruent.

20. Two integers are related if they are congruent mod 7.

21. Two sets are related if there is a one-to-one correspondence between them.

22. Two functions from **Z** to **R** are related if they are asymptotically equivalent.

23. Two events in a sample space are equivalent if they are equally likely.

24. Two combinational circuits are equivalent if they have the same truth tables.

25. Two bytes are equivalent if they have the same number of 1's.

26. Two fractions, a/b and c/d, are equivalent if $ad = bc$.

Give examples of relations that satisfy the following properties.

27. Both symmetric and antisymmetric.

28. Neither reflexive, symmetric, nor transitive.

29. An equivalence relation that is antisymmetric.

30. Reflexive and transitive but not symmetric.

Identify the equivalence classes for the equivalence relations of the following exercises.

31. Exercise 17   32. Exercise 18

33. Exercise 20   34. Exercise 21

35. Exercise 22   36. Exercise 26

Test if the following are partitions of the given set, S.

37. $\{\{1,2\}, \{3,4,5\}, \{6\}, \{7,8\}\}; S = \{1, \ldots, 8\}$.

38. $\{\{1, 2, 3\}, \{3, 4, 5\}, \{6, 7, 8\}\}$; $S = \{1, \ldots, 8\}$.

39. $A = \{$Subsets of $\{1, \ldots, n\}$ with less than $n/2$ elements$\}$ and $B = \{$Subsets of $\{1, \ldots, n\}$ with $n/2$ or more elements$\}$; $S = $ power set of $\{1, \ldots, n\}$.

40. $\{S_L\}$, where L is a line in the Cartesian plane containing the origin and S_L denotes the set of all lines in the plane parallel to L; $S = \{$all lines in the plane$\}$.

Describe the quotient set of the given set for the equivalence relation of the specified exercise.

41. {all people}; Exercise 13

42. **Z**; Exercise 20

43. {all sets}; Exercise 21

44. {all fractions}; Exercise 26

Prove or disprove each of the following assertions.

45. Given two partitions of a set, S, form a new collection of subsets by intersecting each member of the first partition with each member of the second partition. Such a collection is a partition.

46. Given two partitions of a set, S, form a new collection of subsets by forming the union of each member of the first partition with each member of the second partition. Such a collection is a partition.

47. Any relation that is both symmetric and transitive is reflexive.

48. If R is an equivalence relation, R is not antisymmetric.

Solve the following problems.

49. If $|A| = n$, how many relations are there on A that are reflexive?

50. If $|A| = n$, how many relations are there on A that are symmetric?

51. If $|A| = n$, how many relations are there on A that are antisymmetric?

52. Given the partition $\{\{1,2\}, \{3\}, \{4,5\}\}$, represent the equivalence relation corresponding to it with both a matrix and a digraph.

8.3

ORDER RELATIONS

At the beginning of Section 8.2, we discussed the two main questions underlying our study of the properties of relations, namely, "What does it mean for things to be equivalent?" and "What does it mean for things to be arranged in order?" Having discussed equivalence, we are now ready to consider order, a concept we see in several contexts in discrete mathematics. This section is of particular importance in computer science, since sequential algorithms execute instructions in a specified order. Thus the process of ordering items and the kinds of order relationships that exist between items take on special importance in that discipline.

Partial Orders

Some sets are like the real numbers in that every element can be compared to every other element and all of the elements are arranged in a certain order. Some sets are not like the real numbers; they contain mixtures of elements, some of which can be compared and some of which cannot. We will examine several examples of such sets. In this section, we will discuss two main concepts that will clarify this distinction—partial order and total order. We begin with the notion of partial order.

Definition Let R be a relation on set A. We say that R is a **partial order** if R is reflexive, transitive, and antisymmetric. If R is a partial order on A, we say that (A,R) is a **partially ordered set** or a **poset.** We will denote a relation that is a partial order by $\lesssim$. Thus we will use the notation $(A,\lesssim)$ to denote a poset where A is the set and $\lesssim$ is a partial order on A. If A is a poset and for two elements, a and b of A, either $a \lesssim b$ or $b \lesssim a$, then a and b are **comparable.**

The notion of partial order is a generalization of the relation $\leq$ on **R** and **Z**. It differs from an equivalence relation in that a partial order is antisymmetric, while an equivalence relation is symmetric. Thus, with an equivalence relation, $R,$ it is possible to have distinct elements, a and $b,$ such that aRb and bRa. With a partial order, $\lesssim$, this is impossible; if a and b are distinct, comparable elements, then $a \lesssim b$ or $b \lesssim a,$ but not both. It is this one-directional property of the relation that makes it an "ordering."

EXAMPLE 1 Which of the digraphs in Figure 8.10 represent partial orders?

Figure 8.10 Of the various digraphs (a) through (d), only (d) is a poset.

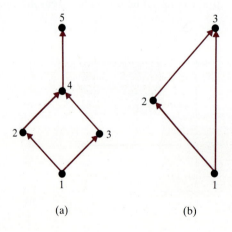

(a)　　　　　　　　(b)

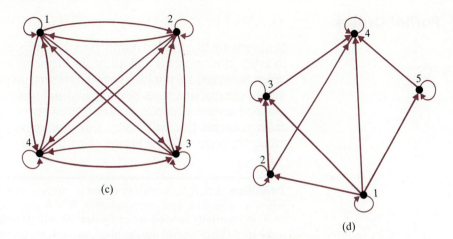

(c)

Figure 8.10   *(continued)*

(d)

SOLUTION   Figures 8.10(a), 8.10(b), and 8.10(c) are not partial orders. Figure 8.10(a) is neither transitive nor reflexive. Figure 8.10(b) is not reflexive. Figure 8.10(c) is not antisymmetric. Figure 8.10(d), however, is a partial order, as it is reflexive, transitive, and antisymmetric. Note that in Figure 8.10(d), there are pairs of elements that are not comparable, for instance, 2 and 5.   ■

EXAMPLE 2 $(\mathbf{R}, \leq)$ is a partial order, since $\leq$ is transitive, reflexive, and antisymmetric. However, $(\mathbf{R}, <)$ is not a partial order, since it is not reflexive. If we let $\mathbf{R}^2$ represent the Cartesian plane and say that for two points, (x_1, y_1) and (x_2, y_2), $(x_1, y_1) \leq (x_2, y_2)$ if and only if $x_1 \leq x_2$ and $y_1 \leq y_2$, then $\leq$ is a partial order. Note that many pairs of points such as (1,2) and (2,1), for instance, are not comparable (see Figure 8.11). The shaded area in the upper right part of this figure represents all points greater than or equal to (a, b), and the shaded area in the lower left part represents all points less than or equal to (a, b). The unshaded areas are those points not comparable to (a, b).

Figure 8.11   Shaded points are comparable to (a, b); unshaded points are not.

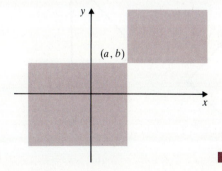

EXAMPLE 3   Let A be any community of people and let S be the "ancestor" relation, i.e., for individuals a and b, aSb if and only if a and b denote the same person or b is a direct descendant of a. S is reflexive, transitive, and antisymmetric, and thus is a partial order. Cousins are people who share a common grandparent. The relation "cousin" is not a partial order, since it is not transitive and not antisymmetric. ∎

EXAMPLE 4   A carpentry project typically requires many steps that must be executed sequentially. For instance, boards, nails, and glue have to be purchased. Boards have to be measured and then cut. Pieces have to be assembled. These steps can be arranged in a form called a **PERT (Program Evaluation and Review Technique) chart** as shown in Figure 8.12.

Figure 8.12   PERT chart for a carpentry project.

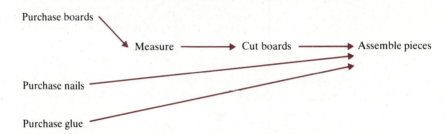

Let $\precsim$ denote the relation where $a \precsim b$ means that either $a = b$ or a must be done before b. Then $\precsim$ is a partial order. PERT charts are a scheduling tool used in many different types of projects. ∎

EXAMPLE 5   Let S be any set. Then $\subset$ is partial order on $P(S)$. It is reflexive since, for any set A, $A \subset A$. It is antisymmetric, since $A \subset B$ and $B \subset A$ implies that $A = B$. And it is transitive, since $A \subset B$ and $B \subset C$ imply that $A \subset C$. ∎

EXAMPLE 6   A relation, $\precsim$, on **N** is defined as follows. We say that $n \precsim m$ if $n \mid m$. Then $\precsim$ is a partial order. First, $\precsim$ is reflexive, since $n \mid n$ for all n. Also, if $n \mid m$ and $m \mid n$, $m = \pm n$. But since the domain of $\precsim$ is restricted to the positive integers, $m = n$. Thus $\precsim$ is antisymmetric. Lastly, if $m \mid n$ and $n \mid p$, $m \mid p$. Hence $\precsim$ is transitive. ∎

Hasse Diagrams

Since every relation can be represented by a digraph, so can every partial order. But digraphs of partial orders are often unnecessarily complicated. For instance, consider the digraph of Figure 8.10(d). Knowing that this represents a poset, there are two unnecessary sets of edges. (1) Since a partial order must be reflexive, the loops are not needed. (2) Since a partial order must be transitive, edges such as 13 and 14 are not needed either. Further-more, since partial orders are antisymmetric, all connections between nodes must be one-directional. If we adopt the convention that the arrows will always point up, we can also drop the arrowheads. Such a reduced dia-gram is called a **Hasse diagram.** It is not a digraph but is a simplification of one in the case where the relation is a partial order. The Hasse diagram of the relation represented in Figure 8.10(d) is presented in Figure 8.13.

Figure 8.13   Hasse dia-gram for the digraph of Figure 8.10(d).

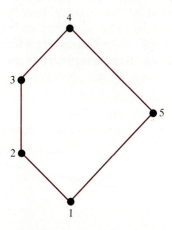

EXAMPLE 7   Represent the "divides" relation on the set {1, 2, 3, 4, 5, 6} by a digraph and by a Hasse diagram.

SOLUTION   See Figure 8.14.   ∎

Figure 8.14   Digraph and Hasse diagram for the divides relation.

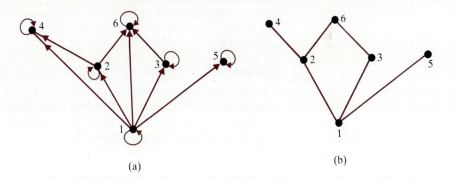

(a)

(b)

EXAMPLE 8   Represent the subset relation on the subsets of {1, 2, 3} by a Hasse diagram.

SOLUTION   Recall that we have already represented this relation by a digraph in Figure 8.6. The Hasse diagram is simpler (Figure 8.15).   ∎

Figure 8.15 Hasse diagram for the subset relation.

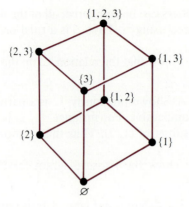

Total Orders

Definition   Let A be any set and let R be a relation on A. Then R is a **total order** if R is a partial order and all elements of A are comparable.

EXAMPLE 9   Show that the relation $\le$ is a total order on $\mathbf{R}$ and on $\mathbf{N}$.

SOLUTION   We have already shown that $\le$ is a partial order on $\mathbf{R}$; also, it is a property of $\mathbf{R}$ that for any two numbers, a and b, either $a \le b$ or $b \le a$. Thus $\le$ is a total order on $\mathbf{R}$; the same arguments apply on $\mathbf{N}$. ∎

EXAMPLE 10   Let A be the set of all possible words in the English language and let $\lesssim$ denote lexicographic (i.e., dictionary) order. Show that $\lesssim$ is a total order.

SOLUTION   First, we must clarify more precisely what we mean by lexicographic order. Words in English are sequences of one to n characters where n is any positive integer. The allowable characters are the 26 characters of the English alphabet arranged in their normal order, a to z. Suppose $a_1 a_2 \ldots a_m$ and $b_1 b_2 \ldots b_n$ are two words. We define $\lesssim$ inductively as follows. If a_1 precedes b_1 in the alphabet, $a_1 a_2 \ldots a_m \lesssim b_1 b_2 \ldots b_n$. If $a_1 a_2 \ldots a_k = b_1 b_2 \ldots b_k$ and either $m = k$ or a_{k+1} precedes b_{k+1}, then $a_1 a_2 \ldots a_m \lesssim b_1 b_2 \ldots b_n$.

Then $\lesssim$ is reflexive, since $a_1 a_2 \ldots a_m \lesssim a_1 a_2 \ldots a_m$. For any distinct letters, a_i and b_i, we cannot have both a_i preceding b_i in the alphabet and b_i preceding a_i in the alphabet. Thus $a_1 a_2 \ldots a_m \lesssim b_1 b_2 \ldots b_n$ and $b_1 b_2 \ldots b_n \lesssim a_1 a_2 \ldots a_m$ implies that for each i less than or equal to the smaller of m and n, $a_i = b_i$. Similarly, $m = n$, since otherwise the shorter word would follow the longer one in lexicographic order. Hence lexicographic order is antisymmetric. Verification that $\lesssim$ is transitive will be left for the Exercises.

Since all letters can be compared, all of the words made from them can also be compared using $\lesssim$. Thus $\lesssim$ is a total order. ∎

EXAMPLE 11   Show that the relation $\subset$ on the subsets of $\{1, 2, 3\}$ is not a total order.

SOLUTION   As shown previously, $\subset$ is a partial order. However, not all elements are comparable. For instance, $\{1, 2\}$ is not a subset of $\{1, 3\}$ and $\{1, 3\}$ is not a subset of $\{1, 2\}$. Thus this relation is not a total order. ∎

EXAMPLE 12   Draw the Hasse diagram for the $\leq$ relation on the set $\{1, 2, 3, 4, 5, 6\}$.

SOLUTION   The diagram in Figure 8.16 is typical of the Hasse diagrams for total orders and illustrates why total orders are sometimes called **chains.** Note that all posets contain chains.

Figure 8.16   Hasse diagram for $\leq$ on $\{1, 2, 3, 4, 5, 6\}$.

6

5

4

3

2

1 ∎

Maximal and Minimal Elements

In a totally ordered set like $\{1, 2, 3, 4, 5, 6\}$, it is obvious what the largest and smallest elements are. But in a partially ordered set such as the one in the Hasse diagram of Figure 8.14, it is not so obvious. The concepts of maximal and minimal element, which we are about to define, enable us to clarify these notions. Since many mathematical models are created to allow us to optimize some factor, these concepts arise frequently in applications.

Definition   Let $(A, \lesssim)$ be a poset. By a **maximal element** of A, we mean an element, a, of A such that $a \lesssim b$ does not hold for any $b \in A$ distinct from a. Similarly, a **minimal element** of A is an element, a, of A such that $b \lesssim a$ does not hold for any $b \in A$ distinct from a.

EXAMPLE 13   Find the maximal and minimal elements for the posets represented by the Hasse diagrams of Figure 8.17.

Figure 8.17   Posets having varying numbers of maximal and minimal elements.

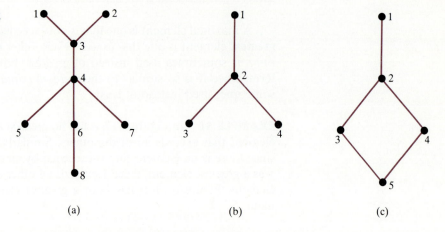

(a)   (b)   (c)

SOLUTION   Recall that a Hasse diagram represents a poset with the arrows pointing up; i.e., if $a \lesssim b$, a will be lower on the diagram than b. Thus every element with no other element above it satisfies the requirements for a maximal element and every element with no other element below it satisfies the requirement for a minimal element. Hence

a) Both 1 and 2 are maximal elements. 5, 7, and 8 are minimal.

b) Only 1 is maximal. 3 and 4 are both minimal.

c) Only 1 is maximal. Only 5 is minimal.   ∎

EXAMPLE 14   Find the maximal and minimal elements for the $\subset$ relation on the power set of any set S.

SOLUTION S itself contains all of its subsets and hence is the only maximal element. $\varnothing$ is a subset of every other subset of S. Hence $\varnothing$ is the only minimal element. Note that if the question were reworded to ask for the minimal element of $(P(S) \setminus \{\varnothing\}, \subset)$, every singleton, $\{s\}$, for each s in S would be minimal.   ∎

EXAMPLE 15   What are the maximal and minimal elements for $\lesssim$ on **Z** and on **N**?

SOLUTION   Since $\mathbf{N} = \{1, 2, 3, \ldots\}$, the minimal element is 1. There is no maximal element in **N**. **Z** has neither maximal nor minimal elements.   ∎

Definition The **greatest** element of a poset, A, is an element a such that $x \lesssim a$ for all x in A. Similarly the **least** element of a poset is an element, b, such that $b \lesssim x$ for all x in A.

A maximal element is one that is not exceeded by any other element. A **greatest** element is one that exceeds every other element. The term *maximum* is sometimes used instead of *greatest*; however, we will avoid this term, since it is so similar to the word *maximal*. Similarly, *minimum* is sometimes used instead of **least.**

EXAMPLE 16 In Figure 8.17, *a* has no greatest element, since there is no element that exceeds all of the others. Similarly, it has no least element, since there is no element that is exceeded by all others. Figure 8.17(b) has 1 as a greatest element, since 1 exceeds all other elements but has no least element. Figure 8.17(c) has both a greatest element, 1, and a least element, 5. ∎

EXAMPLE 17 In Example 14, the subset example, S is a greatest element, since it contains all of the other subsets of S. $\varnothing$ is a least element, since it is contained in every subset. ∎

The following theorems clarify the basic properties of greatest and least elements, on the one hand, and maximal and minimal elements on the other.

THEOREM 8.3 In any poset, if greatest or least elements exist, they are unique.

PROOF Let $(A, \lesssim)$ be a poset and assume that a and b are greatest elements. Then $x \lesssim a$ for all x in A and $x \lesssim b$ for all x in A. Thus $b \lesssim a$ and $a \lesssim b$. Since $\lesssim$ is antisymmetric, $a = b$ and thus there is, at most, one greatest element. The argument for least elements is similar. ☐

THEOREM 8.4 Every finite poset has at least one maximal and at least one minimal element.

PROOF Let $(A, \lesssim)$ be a finite poset and let $|A| = n$. Let a_1 be any element of A. If there are no elements, a, of A distinct from a_1 such that $a_1 \lesssim a$, then a_1 is maximal. If there are elements of A other than a_1 that exceed a_1, let a_2 be one of them. If no element of A distinct from a_2 exceeds a_2, then a_2 is maximal. Continuing this process yields a sequence of elements $a_1 \lesssim a_2 \lesssim \ldots \lesssim a_n$. Note that such a sequence cannot contain any repeated elements. If a_1, for instance, appeared again subsequent to a_2, by transitivity we would have $a_2 \lesssim a_1$. Since $a_1 \lesssim a_2$, by antisymmetry we would have $a_1 = a_2$.

But a_1 and a_2 are distinct. Hence our sequence must terminate in, at most, n steps, since $|A| = n$. The last element in this list has the property that no element of A exceeds it and hence is maximal. The argument for minimal elements is similar but is left for the Exercises. ⬜

THEOREM 8.5   If a finite poset, $(A, \precsim)$, has a unique maximal (minimal) element, that element is the greatest (least) element.

PROOF   What follows is a sketch of the proof. The details are left for the Exercises. Let a_0 denote the unique maximal element. We claim that any other element, $a \in A$, is comparable to a_0. Since a cannot exceed a_0, it follows from this claim that a_0 exceeds a and thus a_0 is maximal. Thus we suppose that there exists an $a \in A$ that is not comparable to a_0. By an argument similar to that used in Theorem 8.4, we can conclude that there is an element, $b \in A$, that is not comparable to a_0 and is maximal. Our theorem follows from this contradiction. ⬜

The previous theorem provides an easy way to find the greatest or least element in a poset or to show that a poset does not have such elements.

EXAMPLE 18   We can see immediately that Figure 8.17(a) has neither a greatest nor a least element, since maximal and minimal elements are not unique. Figure 8.17(b) has a greatest element, 1, since 1 is a unique maximal element, but no least element. Figure 8.17(c) has a greatest element, 1, and a least element, 5. ∎

Note also that in totally ordered sets such as $(\mathbf{R}, \le)$, there is no difference between maximal and greatest. That is, if some element, a_0, is maximal, no other element exceeds it. Thus all other elements are exceeded by a_0, since all elements are comparable. In posets, however, $\neg(a \precsim b)$ does not imply that $(b \precsim a)$. It is this latter property that particularly distinguishes partially ordered and totally ordered sets.

Topological Sorting

Partially ordered sets arise quite often. However, in computing, data must be handled sequentially, one item at a time. That is, the sequence in which the data are processed imposes a total order on them. Thus it is not uncommon to have to impose a total order on a poset that is consistent with its partial order in the sense that if a precedes b in the partial order, a also precedes b in the total order. The process of selecting such an order is called **topological sorting.** Given a finite poset, there is an algorithm for topologically sorting it.

ALGORITHM 32   Topological Sort

Procedure Top_sort(P; var T);

{P is any finite poset with a partial order denoted $\precsim$. T is a set containing the same elements as P; the sequence of its elements form a total order consistent with $\precsim$.}

begin

1.   while P is not empty do begin

2.   select a in P such that a is minimal with respect to $\precsim$;

3.   make a the next element of T;

4.   delete a from P

        end

end;

That is, we simply keep selecting minimal elements of P until the set is empty. Since we know that every finite poset has at least one minimal element, we know that the algorithm always works.

EXAMPLE 19   Consider the poset represented by the Hasse diagram of Figure 8.18. Use Algorithm 32 to sort this poset topologically.

Figure 8.18   Poset to be topologically sorted.

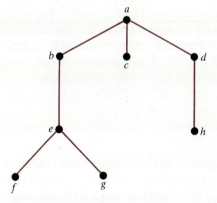

SOLUTION   The set has four minimal elements—f, g, c, and h. Thus one possible sequence is {h, g, f, e, d, c, b, a}. But there are many others, for instance, {f, g, e, b, c, h, d, a}.   ■

EXAMPLE 20 Topologically sort the poset of Example 4 in this section.

SOLUTION Purchasing boards, purchasing nails, and purchasing glue are all minimal elements of this set. Thus any one can be done first. Note that the nails and glue don't have to be purchased until just before the pieces are assembled, however. Thus two possible topological sorts of this poset are {purchase boards, measure boards, cut boards, buy nails, buy glue, assemble pieces} and {buy glue, buy nails, buy boards, measure boards, cut boards, assemble pieces}. On large projects, a careful analysis of a PERT chart can lead to substantial financial savings. Some items can be purchased together to take advantage of quantity discounts, and other purchases can be put off until just before the items are needed so that money that could be used elsewhere is not tied up in inventory. ∎

TERMINOLOGY

partial order	PERT chart	minimal element
partially ordered set	Hasse diagram	greatest element
poset	total order	least element
$(A, \preceq)$	chain	topological sorting
comparable	maximal element	

SUMMARY

1. A partial order on a set A is a relation that is reflexive, antisymmetric, and transitive.

2. Hasse diagrams are simplifications of the digraphs for partially ordered sets (posets); loops and paths implied by the transitivity of the relation are omitted. The direction of the relation is always up.

3. A totally ordered set is a poset in which all elements are comparable.

4. Maximal elements of a poset are those that are not exceeded by any other element. A greatest element is one that exceeds every other element. If a greatest element exists, it is unique. In finite posets, maximal elements always exist; they may not be unique, however. Similar properties hold for minimal and least elements.

5. Topological sorting is the process of placing a total order on a poset that is consistent with its partial order. For finite posets, this can be done using Algorithm 32.

EXERCISES 8.3

Which of the following digraphs represent partial orders and why?

1.

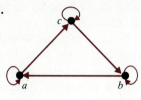

2.

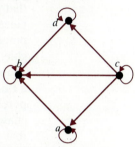

3.

4.

5.

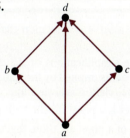

6.

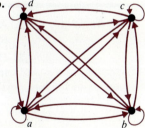

Which of the following relations are partial orders and why?

7. Two English words are related if the length of one is less than or equal to the length of the other.

8. Two bytes are related if the number of 1's in the first does not exceed the number of 1's in the second.

9. Two points in the planes, (x_1, y_1) and (x_2, y_2), are related if $x_1 \geq x_2$ and $y_1 \geq y_2$.

10. Two individuals are related if one is the cousin of the other. *Cousin* means that they have a common grandparent but not a common parent.

11. Two sets, A and B, are related if $A \supset B$.

12. A calendar of events is set up. Events A and B are related iff B does not precede A in the calendar.

13. Two individuals are related if they are siblings, where *sibling* means that they both have the same parents.

Find Hasse diagrams to represent the following posets.

14. The subsets of $\{1, 2\}$ ordered by $\subset$.

15. The subsets of $\{1, 2, 3, 4\}$ ordered by $\subset$.

16. The set $\{1, 2, 3, 4, 5, 6, 7, 8\}$. m and n are related if $m \mid n$.

17. The set $\{1, 2, 3, 4, 5, 6, 7, 8, 9, 10, 11, 12\}$. m and n are related if $m \mid n$.

Which of the following are total orders?

18. The subset relation on a set, S, where $|S| > 1$.

19. The relation $\leq$ on **N**.

20. The relation of Exercise 7.

21. The relation of Exercise 9.

22. The relation $<$ on **R**.

Find all maximal and minimal elements for the following posets, if possible.

23. The relation of Exercise 14.

24. The relation of Exercise 15.

25. The relation of Exercise 16.

26. The relation of Exercise 17.

27. $\{x \subset \mathbf{R} \mid 0 < x < 1\}$.

Using Algorithm 32, topologically sort the posets represented by the following Hasse diagrams.

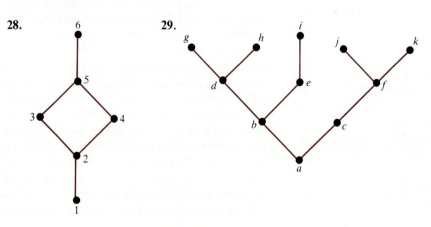

28.

29.

30. The relation of Exercise 14.

31. The relation of Exercise 15.

32. The relation of Exercise 16.

33. The relation of Exercise 17.

34. Show that lexicographic order is transitive.

35. Show that any finite poset has a minimal element.

36. Find a set with three maximal but no minimal elements.

37. Find a set with a least element but no maximal elements.

38. Find an example of a set in which the divides relation is a total order.

39. List three distinct partial orders possible on the set {1, 2, 3}. (*Hint:* consider the possible matrix representations of relations on {1, 2, 3}.)

40. What is the result of topologically sorting a totally ordered set?

41. Fill in the details in the proof of Theorem 8.6.

42. Modify Algorithm 32 so that it sorts a set from maximal to minimal elements.

8.4
OPERATIONS ON RELATIONS

Consider the relations represented by the two digraphs shown in Figure 8.19. They have a great deal in common; both are relations on the same set, $A = \{1, 2, 3, 4\}$. Both consist of two ordered pairs; Figure 8.19(a) contains (1, 2) and (2, 4) and Figure 8.19(b) contains (1, 2) and (3, 4). Neither is symmetric or reflexive. Both are antisymmetric. And yet there are obvious and significant differences between them. First, Figure 8.19(b) is transitive, whereas Figure 8.19(a) is not. Second, if we imagine the arrows as representing routes that can be followed, in Figure 8.19(a) it is possible to go from 1 to 2 and then from 2 to 4. In Figure 8.19(b), one can go from 1 to 2 or from 3 to 4, but no further steps are possible. That is, in Figure 8.19(a), 1 and 4 are not directly related, but there is an indirect connection between them that does not exist in Figure 8.19(b). Yet nothing that we have done so far with relations provides us with a way to make this distinction; up to this point, elements are either related or not. We haven't provided adequate mathematical tools to model relationships that are removed by a "generation," such as Figure 8.19(a) or that of a child to its grandparent. The purpose of this section is to develop such tools. Our results will parallel some similar results about graphs that we will see in the next chapter. Before we do so, we must apply a few basic concepts from set theory to relations. The principal concept we will use is the union of relations; however, the others are useful at times and are included for completeness.

Figure 8.19 Two different relations on {1, 2, 3, 4}.

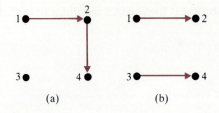

(a)   (b)

Set Operations on Relations

Definition   Let R_1 and R_2 be relations on $A \times B$. Then we define the following:

$$\mathbf{R_1} \cup \mathbf{R_2} = \{(x,y) \mid (x,y) \in R_1 \text{ or } (x,y) \in R_2\}$$
$$\mathbf{R_1} \cap \mathbf{R_2} = \{(x,y) \mid (x,y) \in R_1 \text{ and } (x,y) \in R_2\}$$
$$\mathbf{R_1}^c = \{(x,y) \mid (x,y) \text{ is not a member of } R_1\}$$
$$\mathbf{R_1}^{-1} = \{(x,y) \mid (x,y) \in R_1\}$$

R_1^c is read as R_1 *complement* and R_1^{-1} as R_1 *inverse*.

Note that neither union nor intersection is meaningful unless both R_1 and R_2 are subsets of the same universal set $A \times B$.

EXAMPLE 1   Consider the relations on $A \times A$ given by the digraphs shown in Figure 8.20. Find their union, intersection, complement, and inverse.

Figure 8.20   Two digraphs on {a, b, c}.

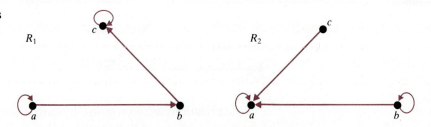

SOLUTION   The union consists of all edges present in either digraph. See Figure 8.21.

Figure 8.21   The union of two relations.

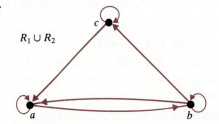

The intersection is those edges common to both digraphs (see Figure 8.22).

Figure 8.22 The intersection of two relations.

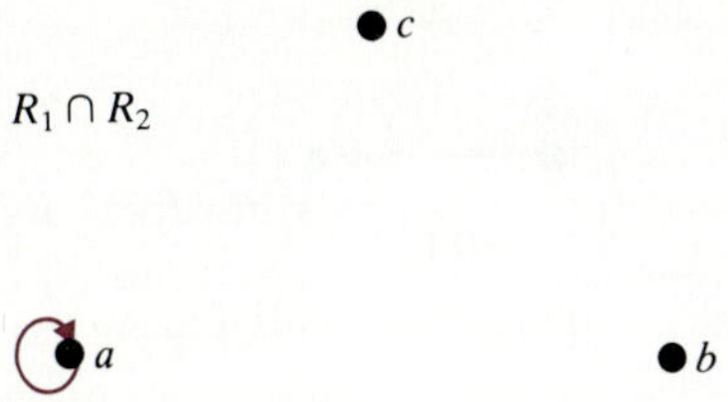

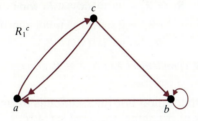

$R_1 \cap R_2$

The complement is all edges not present in a particular digraph (see Figure 8.23).

Figure 8.23 Complement of a relation.

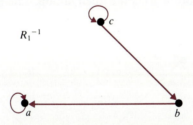

$R_1{}^c$

The inverse is the digraph with the arrows reversed (see Figure 8.24).

Figure 8.24 Inverse of a relation.

$R_1{}^{-1}$

When relations are represented by matrices, the four set operations can be described in terms of those matrices. Let $m(R)$ stand for the matrix representation of R.

THEOREM 8.6 Matrix representation of the operations of union, intersection, complement, and inverse on relations has the following properties:

M1: $m(R_1 \cup R_2) = m(R_1) + m(R_2)$, where addition is a logical OR: $1 + 1 = 1$, $1 + 0 = 1$, $0 + 1 = 1$, $0 + 0 = 0$.

M2: $m(R_1 \cap R_2) = m(R_1) * m(R_2)$, where $*$ denotes an element-by-element logical AND, —i.e., if $W = S * T$, then $w_{ij} = s_{ij} * t_{ij}$ where $1 * 1 = 1$, $1 * 0 = 0 * 1 = 0 * 0 = 0$.

M3: $m(R^c)$ is the matrix that has 1's where $m(R)$ had 0's and 0's where

$m(R)$ had 1's.

M4: $m(R^{-1})$ is the transpose of $m(R)$.

PROOF

M1: If $m(R_1 \cup R_2)_{ij} = 1$, then the ith element of A is related to the jth element of A in the relation $R_1 \cup R_2$. Thus the ith element must be related to the jth element in R_1 or R_2. Hence $m(R_1)_{ij} = 1$ or $m(R_2)_{ij} = 1$. Hence $(m(R_1) + m(R_2))_{ij} = 1$. If $m(R_1 \cup R_2)_{ij} = 0$, then the ith element of A is not related to the jth element of A in the relation $R_1 \cup R_2$. Thus the ith element is not related to the jth element in either R_1 or R_2. Hence $m(R_1)_{ij} = 0$ and $m(R_2)_{ij} = 0$. Hence $(m(R_1) + m(R_2))_{ij} = 0$.

We will leave justification of M2–M4 for the Exercises. □

EXAMPLE 2 Represent the relations given in Figure 8.20 by matrices and find the matrices of their union, intersection, complements, and inverses.

SOLUTION The matrices of the original relations are

$$m(R_1) = \begin{bmatrix} 1 & 1 & 0 \\ 0 & 0 & 1 \\ 0 & 0 & 1 \end{bmatrix} \qquad m(R_2) = \begin{bmatrix} 1 & 0 & 0 \\ 1 & 1 & 0 \\ 1 & 0 & 0 \end{bmatrix}$$

Then

$$m(R_1 \cup R_2) = \begin{bmatrix} 1 & 1 & 0 \\ 1 & 1 & 1 \\ 1 & 0 & 1 \end{bmatrix} \qquad m(R_1 \cap R_2) = \begin{bmatrix} 1 & 0 & 0 \\ 0 & 0 & 0 \\ 0 & 0 & 0 \end{bmatrix}$$

$$m(R_1{}^c) = \begin{bmatrix} 0 & 0 & 1 \\ 1 & 1 & 0 \\ 1 & 1 & 0 \end{bmatrix} \qquad m(R_1{}^{-1}) = \begin{bmatrix} 1 & 0 & 0 \\ 1 & 0 & 0 \\ 0 & 1 & 1 \end{bmatrix}$$

$R_2{}^c$ and $R_2{}^{-1}$ are left for the Exercises. Note that these matrices are consistent with Figures 8.21–8.24 ∎

EXAMPLE 3 Let $A = \mathbf{R}$, let R_1 be the "less than" relation, and let R_2 be the "equality" relation. Find $R_1 \cup R_2$, $R_1 \cap R_2$, $R_1{}^c$, and $R_1{}^{-1}$.

SOLUTION $R_1 \cup R_2$ consists of all pairs of real numbers that satisfy either relation. Thus it is the "less than or equal to" relation. $R_1 \cap R_2$ is the empty relation. $R_1{}^c$ is the "greater than or equal to" relation, and $R_1{}^{-1}$ is the "greater than" relation. ∎

EXAMPLE 4　Let S be a finite set and let R and S be relations on A.

a) If R and S are reflexive, is $R \cap S$ reflexive?

b) If R is symmetric, is R^{-1}?

c) If R is transitive, is R^c?

SOLUTION

a) The matrices for R and S will have 1's in each location on the main diagonal. Thus the matrix of $R \cap S$ will also have 1's in each location on the main diagonal, and thus $R \cap S$ will also be reflexive.

b) The matrix for R^{-1} will simply have 1's where the matrix for R had 0's and vice versa. Thus, since the matrix for R is symmetric, the matrix for R^{-1} will also be symmetric. Hence R^{-1} is a symmetric relation.

c) R^c is not necessarily transitive. See Figure 8.25 for an example. Note that in R^c, b relates to a and a relates to c but b does not relate to c. ■

Figure 8.25　R is transitive, but R^c is not.

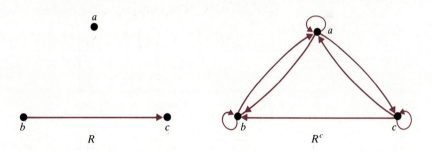

Composition of Relations

Returning now to our original question of developing tools to deal with indirect relations, we will define the concept of composition, our principal tool.

Definition　Let R_1 be a relation on $A \times B$ and R_2 a relation on $B \times C$. Then we define a relation $\boldsymbol{R_2 \circ R_1}$ on $A \times C$, called the **composite of R_2 and R_1**, as follows: $(x,z) \in R_2 \circ R_1$ if and only if there is a $y \in B$ such that $(x,y) \in R_1$ and $(y,z) \in R_2$.

Note that this definition is consistent with the notation for the composition of two functions; when we write $(f \circ g)(x)$, we mean $f(g(x))$. That is, g acts on x first, and then f acts on $g(x)$. With $R_2 \circ R_1$, R_1 acts first, followed by R_2.

EXAMPLE 5　Suppose that both R_1 and R_2 are the parent–child relation of Section 8.1. Find $R_2 \circ R_1$.

SOLUTION   The relation will be the grandparent–grandchild relation. Thus, since Boaz is the parent of Obed and Obed is the parent of Jesse, the relation will contain the pair

(Boaz, Jesse).

Similarly, since Obed is the parent of Jesse and Jesse is the parent of Eliab, it will contain

(Obed, Eliab).

Besides the two pairs already listed, the relation will include Obed and each of his grandchildren and Jesse and each of his grandchildren.   ∎

EXAMPLE 6   Suppose a database consists of two files—R_1, listing each student's name with his or her identification number, and R_2, listing each student's identification number along with each course for which the student is registered in the current semester. Each of these files is a set of ordered pairs and hence a relation. Then $R_2 \circ R_1$ is a relation consisting of all registrations for the current semester, i.e., all ordered pairs of the form (student name, course) in which that student is taking that particular course.   ∎

EXAMPLE 7   Find a digraph representing $R_1 \circ R_1$ where R_1 is given by the digraph of Figure 8.20.

SOLUTION   The digraphs will consist of edges representing all possible two-step connections between elements of A. See Figure 8.26. Note that the matrix of $R_1 \circ R_1$ is

$$\begin{bmatrix} 1 & 1 & 1 \\ 0 & 0 & 1 \\ 0 & 0 & 1 \end{bmatrix}$$

In property M2 of Theorem 8.6, we saw that the matrix of the relation $R_1 \cap R_2$ was $m(R_1) * m(R_2)$, where multiplication of matrices was an element-by-element logical AND. If we use standard matrix multiplication (see Appendix B) instead of element-by-element logical AND and take

Figure 8.26   Composition of relations.

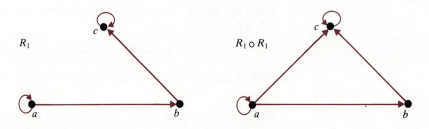

addition to be a logical OR, we get the following:

$$m(R_1) * m(R_1) = \begin{bmatrix} 1 & 1 & 0 \\ 0 & 0 & 1 \\ 0 & 0 & 1 \end{bmatrix} * \begin{bmatrix} 1 & 1 & 0 \\ 0 & 0 & 1 \\ 0 & 0 & 1 \end{bmatrix}$$

$$= \begin{bmatrix} 1+0+0 & 1+0+0 & 0+1+0 \\ 0+0+0 & 0+0+0 & 0+0+1 \\ 0+0+0 & 0+0+0 & 0+0+1 \end{bmatrix}$$

$$= \begin{bmatrix} 1 & 1 & 1 \\ 0 & 0 & 1 \\ 0 & 0 & 1 \end{bmatrix}$$

$$= m(R_1 \circ R_1). \quad \blacksquare$$

Note that the addition in this case is the same as ordinary addition, since no term had more than one 1. This is not always the case. For instance, if one of the terms had been $0 + 1 + 1$, the sum would still be 1.

Properties of Composition

There are two important properties of composition of relations that we will use.

THEOREM 8.7 Composition of relations is associative. If $R_1 \subset A \times B$, $R_2 \subset B \times C$, and $R_3 \subset C \times D$, then $R_3 \circ (R_2 \circ R_1) = (R_3 \circ R_2) \circ R_1$.

PROOF If both the left- and righthand sides of the equation are empty relations, they are equal. If not, suppose $(a,d) \in (R_3 \circ R_2) \circ R_1$. Then there exists b such that $(a,b) \in R_1$ and $(b,d) \in R_3 \circ R_2$. Hence there exists c such that $(b,c) \in R_2$ and $(c,d) \in R_3$. Thus $(a,c) \in R_2 \circ R_1$, and hence $(a,d) \in R_3 \circ (R_2 \circ R_1)$. Thus $(R_3 \circ R_2) \circ R_1 \subset R_3 \circ (R_2 \circ R_1)$. Proof of the opposite inclusion follows the same pattern, and hence the left- and right-hand sides are equal. □

THEOREM 8.8 Suppose A, B, and C are all finite sets. Let $R_1 \subset A \times B$ and $R_2 \subset B \times C$. Then

$$m(R_2 \circ R_1) = m(R_1) * m(R_2),$$

where $*$ denotes standard matrix multiplication and addition is the logical OR.

PROOF Since both $m(R_2 \circ R_1)$ and $m(R_1) * m(R_2)$ are boolean matrices, we only need to show that they have 1's in the same places.

Thus suppose first that $m(R_2 \circ R_1)$ has a 1 in the ijth position. For simplicity, suppose that the elements of A are denoted $\{1, \ldots, m\}$, those of B are denoted $\{1, \ldots, n\}$ and those of C are denoted $\{1, \ldots, p\}$. Then $(i,j) \in R_2 \circ R_1$ implies that there is at least one k in B such that $(i,k) \in R_1$ and $(k,j) \in R_2$ (see Figure 8.27a). Thus $m(R_1)$ has a 1 in the ikth position and $m(R_2)$ has a 1 in the kjth position. Hence $m(R_1) * m(R_2)$ has a 1 in the ijth position, since the dot product of the ith row of $m(R_1)$ and the jth column of $m(R_2)$ will be 1.

Now suppose that $m(R_1) * m(R_2)$ has a 1 in the ijth position. Then the dot product of the ith row of R_1 and the jth column of R_2 is 1, and hence there is at least one k between 1 and n such that the kth entry of that row and the kth entry of that column are both 1 (see Figure 8.27b). Thus iR_1k and kR_2j and thus $i(R_2 \circ R_1)j$. Hence $m(R_2 \circ R_1)$ also has a 1 in the ijth position. ☐

Figure 8.27 Part of the proof that $m(R_2 \circ R_1) = m(R_1)m(R_2)$.

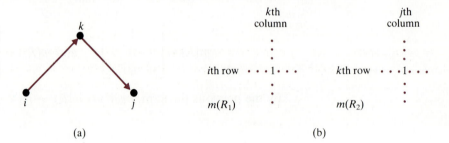

(a)　　　　　　　　　　　　　　　　(b)

See Examples 7 and 8 for applications of Theorem 8.8.

Composition of a Relation with Itself

When we have a relation, R, on a set A, we denote $R \circ R$ by R^2. R^2 represents pairs of vertices connected by paths of length 2. Similarly R^n represents pairs of vertices connected by paths of length n. Thus R^n will be a relation in which aR^nb means that there is an indirect connection of length n between a and b. This notion can be defined more formally for R^n as follows:

Definition　Let $n \geq 1$. Then we define

$$R^n = \begin{cases} R & \text{if } n = 1 \\ R \circ R^{n-1} & \text{if } n > 1 \end{cases}$$

Note that this is a recursive definition. We also define $\boldsymbol{R^\infty}$ to be $\displaystyle\bigcup_{n=1}^{\infty} R^n$.

R^∞ represents all pairs of vertices that can be connected by at least one path of some finite length.

We can find the matrix of R^n from the original matrix of R by matrix multiplication as follows.

THEOREM 8.9   Let R be a relation on a finite set A and let $n \geq 1$. Then

$$m(R^n) = m(R)^n.$$

PROOF   This follows directly from Theorem 8.8 by induction. The details are left for the Exercises.   ☐

EXAMPLE 8   Find R^n for all $n \geq 1$ and R^∞ for the relation given by the digraph in Figure 8.20(a).

SOLUTION   We already know that

$$m(R) = \begin{bmatrix} 1 & 1 & 0 \\ 0 & 0 & 1 \\ 0 & 0 & 1 \end{bmatrix} \quad \text{and } m(R^2) = \begin{bmatrix} 1 & 1 & 1 \\ 0 & 0 & 1 \\ 0 & 0 & 1 \end{bmatrix}.$$

By the previous theorem, $m(R^3) = m(R)^3 = m(R)m(R)^2$. Thus

$$m(R^3) = \begin{bmatrix} 1 & 1 & 1 \\ 0 & 0 & 1 \\ 0 & 0 & 1 \end{bmatrix}.$$

That is, $m(R)^3 = m(R)^2$ and hence $m(R)^4 = m(R)m(R)^3 = m(R)m(R)^2 = m(R)^3 = m(R)^2$. Thus $m(R^n) = m(R)^n = m(R)^2$ for all $n \geq 2$. Thus the digraph of R^n is also the same as Figure 8.26 for all $n \geq 2$. Thus $m(R^\infty) = m(R) + m(R^2) = m(R^2)$.   ■

EXAMPLE 9   Find R^n for all $n \geq 1$ and R^∞ for the relation represented by the digraph shown in Figure 8.28.

Figure 8.28 Digraph for Example 9.

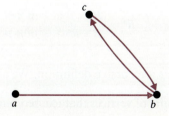

Figure 8.29 Composition of the digraph in Figure 8.28 with itself.

SOLUTION   First, we find $m(R)$, which is

$$m(R) = \begin{bmatrix} 0 & 1 & 0 \\ 0 & 0 & 1 \\ 0 & 1 & 0 \end{bmatrix}.$$

Then it follows that

$$m(R^2) = m(R)^2 = \begin{bmatrix} 0 & 0 & 1 \\ 0 & 1 & 0 \\ 0 & 0 & 1 \end{bmatrix},$$

which is represented by the digraph shown in Figure 8.29. Further matrix multiplication will show that $m(R) = m(R^3) = \ldots = m(R^{2n+1})$ for all n and $m(R^2) = m(R^4) = \ldots = m(R^{2n})$ for all n. Thus Figure 8.28 represents R^n whenever n is odd and Figure 8.29 represents R^n whenever n is even. Note that Figure 8.29 represents all connections between nodes of the original digraph that can be made in an even number of steps and Figure 8.28 represents all connections that can be made in an odd number of steps. Then

$$m(R^\infty) = \begin{bmatrix} 0 & 1 & 1 \\ 0 & 1 & 1 \\ 0 & 1 & 1 \end{bmatrix},$$

and this represents all connections that can be made in any number of steps, even or odd. The column of zeroes tells us that nothing is related to a, no matter how many steps are taken.   ∎

The next theorem presents some additional properties of R^n. Part (a) will be used subsequently; part (b) is included for completeness.

THEOREM 8.10   Let R be a relation on a set A, $m \geq 1$, $n \geq 1$. Then

a) $R^n \circ R^m = R^{n+m}$

b) $(R^n)^m = R^{nm}$

PROOF

a) Let m be any fixed positive integer. The proof is by induction on n. Let $n = 1$. Then, by the definition of R^n, $R \circ R^m = R^{m+1}$. Now assume that $n > 1$ and that $R^{n-1} \circ R^m = R^{n+m-1}$. By definition, $R^n = R \circ R^{n-1}$. Using Theorem 8.7, $R^n \circ R^m = (R \circ R^{n-1}) \circ R^m = R \circ (R^{n-1} \circ R^m) = R \circ R^{n+m-1} = R^{n+m}$. ▯

b) (Left for the Exercises.)

Transitive Closure

Given a relation like the one in Figure 8.19(a), which is not transitive, it is often useful to be able to add whatever relationships are needed to make it transitive. That is, for every indirect connection of any length, we will change the relation by adding a direct connection (of length 1). The result will be a transitive relation in which every indirect relation is supplemented by a direct one. This idea is formalized as follows.

Definition   Let S denote the set of all relations with domain A and codomain B for sets A and B. Let R be a relation in S. The **transitive closure** of R is the smallest transitive relation that includes R as a subset. ("Smallest" is taken relative to the poset $(S, \subset)$.)

THEOREM 8.11   For any relation R, R^∞ is its transitive closure.

PROOF   First, we show that R^∞ is transitive. Let (x,y) and $(y,z) \in R^\infty$. Then $(x,y) \in R^n$ for some n and $(y,z) \in R^m$ for some m. Hence $(x,z) \in R^n \circ R^m = R^{n+m}$ and thus $(x,z) \in R^\infty$. Thus R^∞ is transitive.

We will now show that R^∞ is the transitive closure of R. Let $(x,y) \in R^\infty$. Then $(x,y) \in R^m$ for some m; since R^m consists of all of the m-step connections between elements of A, there exists a sequence $x_1, x_2, \ldots, x_{m-1}$ such that $(x,x_1) \in R$, $(x_1,x_2) \in R$, $\ldots$, $(x_{m-1}, y) \in R$. Since the transitive closure must itself be transitive, (x,x_2) is contained in it. Thus (x,x_3) is also in it. By induction, it follows that (x,y) is in the transitive closure of R as well. Thus R^∞ is a subset of the transitive closure of R. The transitive closure, $T(R)$, is the smallest transitive relation containing R. Since $T(R) \supset R^\infty \supset R$ and we have shown R^∞ to be transitive, it follows that $R^\infty = T(R)$. ▯

COROLLARY 8.12   Let R be a relation on a set A. Then R is transitive if and only if $R = R^\infty$.

PROOF   Suppose R is transitive. Then it is itself the smallest transitive set containing R and hence is its own transitive closure. Thus $R = R^{\infty}$.

Suppose $R = R^{\infty}$. Since R^{∞} is transitive, so is R.   □

EXAMPLE 10   Find the transitive closure of the relation given by the digraph of Figure 8.20(a).

SOLUTION R_1^{∞} will be the transitive closure. From Example 8, its matrix is

$$\begin{bmatrix} 1 & 1 & 1 \\ 0 & 0 & 1 \\ 0 & 0 & 1 \end{bmatrix}$$

See Figure 8.26 for its digraph.   ∎

Further properties of R^n and R^{∞} are left for the Exercises.

TERMINOLOGY

$R_1 \cup R_2$	composite of R_2 and R_1
$R_1 \cap R_2$	R^n
R_1^c	R^{∞}
R_1^{-1}	transitive closure
$R_2 \circ R_1$	

SUMMARY

1. Relations are sets and, as such, the normal set operations—union, intersection, and complement—can be applied to them. Another operation, the inverse, is also useful.

2. Using boolean addition, the set operations on relations can be described in terms of the matrices of the relations involved, namely,

   M1: $m(R_1 \cup R_2) = m(R_1) + m(R_2)$, where $+$ denotes the logical OR.

   M2: $m(R_1 \cap R_2) = m(R_1) * m(R_2)$, where $*$ is an element-by-element logical AND.

   M3: $m(R^c)$ is the matrix that has 1's where R had 0's and 0's where R had 1's.

   M4: $m(R^{-1})$ is the transpose of $m(R)$.

3. The concept of the composite of relations enables us to handle indirect relations—situations where a relates to b and b relates to c but a does not relate
directly to c.

4. Some properties of composition of relations are as follows:

   Composition is associative.

   $m(R_2 \circ R_1) = m(R_1) * m(R_2)$, where the matrix multiplication here is standard matrix multiplication, but using the logical OR for addition.

   $m(R^n) = m(R)^n$

   $R^n \circ R^m = R^{n+m}$

   $(R^n)^m = R^{nm}$

5. R^∞ is the transitive closure of R.

EXERCISES 8.4

Questions 1 through 8 refer to the relations given by the following digraphs. In each
case, find the digraph of the relations specified.

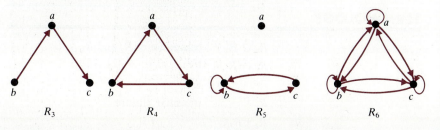

R_3 R_4 R_5 R_6

1. $R_3 \cup R_4,\ R_3 \cap R_4$

2. $R_3 \cup R_5,\ R_3 \cap R_5$

3. $R_6 \cup R_4,\ R_6 \cap R_4$

4. $R_4 \cup R_5,\ R_4 \cap R_5$

5. $R_3{}^c,\ R_3{}^{-1}$

6. $R_4{}^c,\ R_4{}^{-1}$

7. $R_5{}^c,\ R_5{}^{-1}$

8. $R_6{}^c,\ R_6{}^{-1}$

Prove the following properties from Theorem 8.7.

9. M2

10. M3

11. M4

In Exercises 12 through 19, assume that R and S are relations on a set A. In each
exercise, a property and a set are specified. Assume that R and S have the specified
property. Prove or disprove that the specified set also has that property.

12. reflexive; $R \cup S$

13. transitive; $R \cup S$

14. symmetric; $R \cap S$

15. antisymmetric; $R \cap S$

16. antisymmetric; R^c

17. reflexive; R^c

18. transitive; R^{-1}

19. symmetric; R^{-1}

Describe the elements of each of the following composite relations.

20. A database consists of two relations. R contains part names and their corresponding part numbers and P contains part numbers and their quantities on hand. Describe $P \circ R$.

21. R is the parent–child relation. Describe R^3.

22. D is the relation aDb iff $a \mid b$. Describe D^2.

23. L is the relation xLy iff $x \le y - 1$. Describe L^n.

The next six questions relate to R_1 and R_2 of Example 1 in the text.

24. Find the digraph for $R_2 \circ R_2$ and represent it by a matrix.

25. Using your answer to Exercise 24, verify that $m(R_2 \circ R_2) = m(R_2)m(R_2)$.

26. Find $R_1 \circ R_2$ by computing its matrix and then constructing the digraph.

27. Find $R_2 \circ R_1$ by computing its matrix and then constructing the digraph.

28. Using R_3 from the beginning of this exercise set, verify that the composition of R_1, R_2, and R_3 is associative.

29. Verify that $m(R_1)^3 = m(R_1^3)$.

For each of the following matrices, find R^n for all n and R^∞. Also draw the digraph and find the transitive closure directly from the digraph. Verify that the matrix of the transitive closure you have drawn equals R^∞.

30. $\begin{bmatrix} 1 & 0 & 1 \\ 0 & 0 & 0 \\ 0 & 0 & 1 \end{bmatrix}$

31. $\begin{bmatrix} 1 & 0 & 1 \\ 1 & 1 & 0 \\ 0 & 0 & 1 \end{bmatrix}$

32. $\begin{bmatrix} 1 & 1 & 0 & 0 \\ 0 & 0 & 1 & 0 \\ 0 & 0 & 0 & 1 \\ 0 & 1 & 0 & 0 \end{bmatrix}$

33. $\begin{bmatrix} 0 & 1 & 0 & 0 \\ 0 & 0 & 1 & 0 \\ 0 & 0 & 0 & 1 \\ 1 & 0 & 0 & 0 \end{bmatrix}$

Let R, S, and T be relations on a set A. Prove each of the following.

34. $(S \cup T) \circ R = (S \circ R) \cup (T \circ R)$

35. $(S \cap T) \circ R = (S \circ R) \cap (T \circ R)$

36. $(R \cup S)^c = R^c \cap S^c$

37. $(R \cap S)^c = R^c \cup S^c$

38. $(R \cup S)^{-1} = R^{-1} \cup S^{-1}$

39. $(R \cap S)^{-1} = R^{-1} \cap S^{-1}$

Answer the following.

40. Fill in the details of the proof of Theorem 8.9.

41. Prove Theorem 8.10(b).

42. Suppose that R_1 and R_2 are relations on A and $R_1 \supset R_2$. Show that $R_1^\infty \supset R_2^\infty$.

43. In Exercise 42, is it possible for R_1 to properly contain R_2, but $R_1^\infty = R_2^\infty$? Why?

REVIEW EXERCISES—CHAPTER 8

Model the following situations using relations.

1. Employees of a company are listed with their titles.
2. A new strain of influenza is spreading through a community.

Represent the following relations by both a digraph and a boolean matrix.

3. The $<$ relation on $\{(1,2), (2,1), (3,3), (0,0), (2,3), (0,2)\}$, which is defined by the rule $(a,b) < (c,d)$ if and only if both $a < c$ and $b < d$.
4. The same as Exercise 3, but using $\leq$ instead of $<$ throughout.
5. The subset relation on $\{1, 2, 3\}$.

Decide if the following relations are symmetric, reflexive, transitive, or antisymmetric.

6. Two positive integers are related if they are relatively prime.
7. Two triangles, A and B, are related if one angle of A equals one angle of B.

Show that each of the following are equivalence relations.

8. Two individuals are related if they were born in the same year.
9. Two integers are related if they are congruent mod 7.
10. Two $n \times n$ matrices are equivalent if there exists an invertible matrix, X, such that $A = BX$.
11. Two lines in the plane are equivalent if they are parallel.
12. Two triangles are equivalent if they are similar.

Identify the equivalence classes for the equivalence relations of the following exercises.

13. Exercise 8
14. Exercise 12

Modify each of the following to make them (a) reflexive, (b) symmetric, (c) antisymmetric, (d) transitive.

15. $\begin{bmatrix} 1 & 0 & 1 \\ 0 & 1 & 0 \\ 0 & 0 & 0 \end{bmatrix}$

16. $\begin{bmatrix} 1 & 1 & 0 & 0 \\ 1 & 0 & 0 & 0 \\ 0 & 0 & 1 & 1 \\ 0 & 0 & 0 & 1 \end{bmatrix}$

17. $\begin{bmatrix} 1 & 1 & 1 \\ 1 & 1 & 1 \\ 1 & 1 & 1 \end{bmatrix}$

18. $\begin{bmatrix} 0 & 0 & 0 \\ 0 & 0 & 0 \\ 0 & 0 & 0 \end{bmatrix}$

19.   **20.**

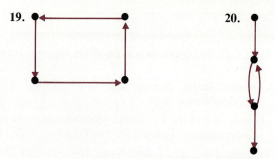

For each of the following, indicate whether the given relation is a partial order and/or a total order and why.

21. Two points in the plane (x_1, y_1) and (x_2, y_2) are related if $x_1 \geq x_2$ and $y_1 \leq y_2$.

22. Two individuals are related if they are siblings, where sibling is defined to mean that both have the same parents.

23. Consider the relation $\Rightarrow$ on predicates $p(x)$ and $q(x)$. Consider as equal two predicates that are logically equivalent.

24. The integers appearing in Pascal's triangle (Section 7.4) satisfy the relation $C(n,r) = C(n-1, r-1) + C(n-1, r)$. Two positive integers, a and b, are related if a appears on the left of such an expression and b is one of the integers on the right.

Find all maximal and minimal elements for the following poset, if possible.

25. $\{(x,y) \mid x \text{ and } y \text{ are both in } \mathbf{N}\}.$ $(x_1, y_1) \leq (x_2, y_2)$ if $x_1 \leq x_2$ and $y_1 \leq y_2$.

Answer the following questions.

26. Under what conditions is the subset relation a total order?

27. How would one recognize from its matrix representation that a relation is a total order?

Let R_1 and R_2 be relations on $\mathbf{Z}$ given by xR_1y iff x is congruent to y mod 3 and xR_2y iff x is congruent to y mod 4. Find the following.

28. $R_1 \cup R_2$ **29.** $R_1 \cap R_2$

30. R_1^c **31.** R_1^{-1}

We define the reflexive closure, $r(R)$, of a relation to be the smallest reflexive relation containing R and the symmetric closure, $s(R)$, to be the smallest symmetric relation containing R. If R is a relation on a set A with $|A| = n$, show that

32. $m(r(R)) = m(R) + 1_n$, where 1_n denotes the $n \times n$ identity matrix and $+$ denotes logical OR.

33. $s(R) = m(R) + m(R)^t$, where $+$ denotes logical OR.

34. $s(r(R))^\infty$ is an equivalence relation.

35. Explain why $s(r(R))^\infty$ can be regarded as the smallest equivalence relation containing R.

REFERENCES

Date, C.J., *An Introduction to Database Systems,* 3rd ed. Reading, Mass.: Addison-Wesley, 1982.

Grimaldi, Ralph, *Discrete and Combinatorial Mathematics.* Reading, Mass.: Addison-Wesley, 1985.

Hillman, Abraham P., Gerald Alexanderson, and Richard Grassl, *Discrete and Combinatorial Mathematics.* San Francisco, Calif.: Dellen Publishing Co., 1987.

Kemeny, John, and Laurie Snell, *Mathematical Models in the Social Sciences.* Cambridge, Mass.: MIT Press, 1972.

Knuth, Donald, *The Art of Computer Programming,* Vol. 1., *Fundamental Algorithms.* Reading, Mass.: Addison-Wesley, 1973.

Lewis, H.R., and C.H. Papadimitriou, *Elements of the Theory of Computation.* Englewood Cliffs, N.J.:Prentice-Hall, 1981.

Liu, C.L., *Elements of Discrete Mathematics.* New York: McGraw-Hill, 1977.

Stanat, Donald, and David McAllister, *Discrete Mathematics in Computer Science.* Englewood Cliffs, N.J.: Prentice-Hall, 1977.

Graph Theory

9

Suppose you join a new class on the first day. You observe that some of the students know each other and some do not. How could you model this situation? One way would be to give a list of all pairs of students who are acquainted—a set of two element sets such as {{Pat, Andre}, {Hank, Diane}, {Hank, Peter}, {Peter, Diane}, {Jean, Frank}, {Frank, Joe}, {Joe, Jean}}. Perhaps a better way would be to represent the list by a diagram such as Figure 9.1. Such lists and diagrams are called **graphs,** or sometimes **undirected graphs,** and the study of them is called **graph theory.** Note that this notion is quite different from the notion of the graph of a function. Graph theory has important applications in computer science, sociology, transportation, chemistry, and other disciplines. It is also widely used in solving games and puzzles.

Figure 9.1 Graph indicating who is acquainted with whom in a class of 10 students.

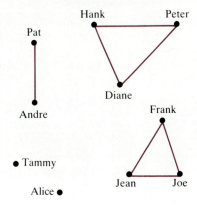

9.1
INTRODUCTION

Historically, graph theory originated in a problem called the **Konigsberg bridge problem.** The Prussian city of Konigsberg (now Kalingrad in the Soviet Union) was built along the Pregel River and occupied both banks and two islands. (See Figure 9.2.) In the eighteenth century, walking was a popular recreation in Konigsberg, and walkers sought a route that would enable them to cross all seven bridges in the city exactly once and return to their starting point. No one could find such a route, so Leonhard Euler, one of the most famous mathematicians of the century, was approached. The concepts he developed in solving this problem were the foundation of graph theory. We will examine Euler's solution below. But first, we need some definitions.

Figure 9.2 The Konigsberg bridge problem.

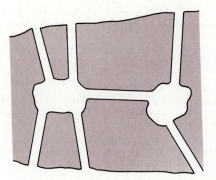

Definition A graph consists of a nonempty set, V, called **vertices,** and another set, E, called **edges,** such that each edge is **associated with** a pair of vertices. Edges are associated with pairs of vertices if there is a function from E to the set of all unordered pairs of vertices. We typically denote a graph by the ordered pair (V,E). If an edge, e, is associated with a particular vertex, v, we say that e is **incident on** v. The number of edges incident on v is the **degree** of v, denoted **deg(v).** A graph, $G' = (V', E')$, is said to be a **subgraph** of another graph, G, if $V' \subset V, E' \subset E$, and each edge in E' is associated with vertices in V'. Note that each edge has one pair of vertices associated with it, but distinct edges may be associated with the same pair of vertices.

EXAMPLE 1 Figure 9.1 represents a graph. The vertices are the circles that denote the 10 students. The edges are the line segments that indicate which students are acquainted. Thus the edge joining the vertices labeled Peter and Diane is incident on those vertices. The degree of the vertex labeled Peter is 2, the degree of the vertex labeled Pat is 1, and the degree of the vertex labeled Alice is 0. The set of vertices {Hank, Peter, Diane} together with the three edges joining them is a subgraph of the entire graph. ∎

Definition A **loop** in a graph is an edge for which the associated vertices are identical. Two edges, e_1 and e_2, are said to be **parallel** if they are both associated with the same pair of vertices. A graph with parallel edges is said to be a **multigraph.** A graph with neither parallel edges nor loops is said to be a **simple graph.**

When we say that G is a graph, we assume that G is not a multigraph unless this possibility is explicitly stated. These terms are summarized in Table 9.1.

TABLE 9.1

Term	Loops?	Parallel Edges?
Multigraph	Allowed	Allowed
Graph	Allowed	No
Simple Graph	No	No

The word *graph* is used by some writers to mean a simple graph. Because there are times when we want to allow loops and times when we do not, we will find it more convenient to keep the distinction. Note that the possibility of multigraphs explains why we say that an edge is *associated with* a pair of vertices rather than *being* the pair of vertices; we may want multiple edges associated with the same pair of vertices.

EXAMPLE 2   Figure 9.3 illustrates a multigraph, a graph, and a simple graph. ∎

Figure 9.3   (a) A multigraph, (b) a graph, and (c) a simple graph.

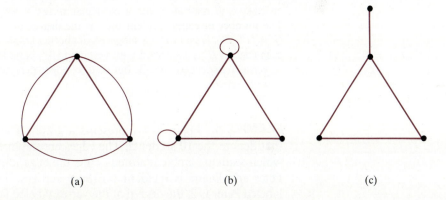

(a)   (b)   (c)

Some Applications of Graphs

EXAMPLE 3   Euler modeled the Konigsberg bridge problem by treating each land mass as a vertex and each bridge as an edge, deriving the multigraph shown in Figure 9.4. In the next section we will see how he used this multigraph to answer the question posed to him. ∎

EXAMPLE 4   The traveling salesman problem can be modeled using graph theory. The vertices are the cities and the edges are the routes joining them. This situation can be modeled using a simple graph, since loops can be ignored. Similarly, if there are two routes between cities, the shorter one can be selected and the other ignored. If one of the streets runs one way, the

Figure 9.4 Euler's model of the Konigsberg bridge problem.

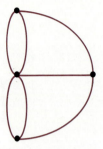

problem must be modeled by a digraph. In its usual formulation, however, it is modeled by a simple graph. ∎

EXAMPLE 5 Electronic circuits can also be modeled using graphs. For instance, the circuit in Figure 9.5(a) can be modeled by the graph in Figure 9.5(b). With a circuit like this, the vertices are the connections and the edges are the components. Such circuits do not normally give rise to graphs with loops. Parallel edges are common, however. Alternating current (AC) circuits are normally modeled with graphs, but direct current (DC) circuits are modeled with digraphs. ∎

Figure 9.5 A graphic representation of an electronic circuit.

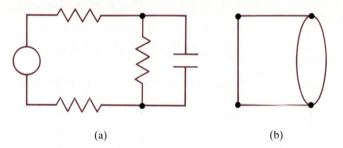

(a)   (b)

EXAMPLE 6 Molecules can be modeled using graphs. For instance, Figure 9.6(a) represents benzene. In Figure 9.6(b), the vertices represent the atoms, the edges represent the chemical bonds between the atoms, and the entire graph represents the molecule. ∎

EXAMPLE 7 In the previous chapter, we examined the structure chart of a computer program as an example of a digraph (Figure 8.3). This program could also be modeled graphically by saying that two modules are related if one calls the other. In this case, the model would not be a multigraph; it could have loops if there were a module called recursively. This situation is more naturally modeled by a digraph than a graph, however, since the statement "*A* calls *B*" does not imply that "*B* calls *A*." ∎

Figure 9.6 Graphic representation of a molecule.

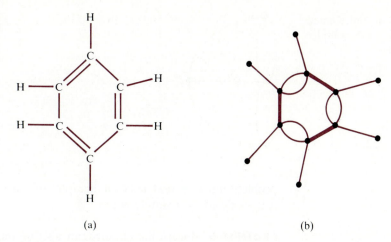

(a)   (b)

Preliminary Results

The traveling salesman problem is a graph theory problem, although it has an additional feature: Weights (or costs) are associated with each edge. The problem is how to travel from vertex to vertex along the edges in such a way that every vertex is visited, while minimizing the total cost and returning home at the end. But there are many other interesting questions one could ask about graphs. Here are four questions that we will address.

1. Is it possible to move from vertex to vertex in a graph and return to the original starting point in such a way that every *edge is crossed* exactly once?

2. Is it possible to move from vertex to vertex in a graph and return to the original starting point in such a way that every *vertex is visited* exactly once?

3. What does it mean for two graphs that appear different to have the same structure, and how can one tell if the structures are the same?

4. Which graphs can be drawn in the plane and which ones cannot?

Answering these questions will occupy Sections 9.2 through 9.4. Before considering them, however, we must present a few preliminary definitions and results.

Definition   A **walk** is a list $u_1, e_1, u_2, e_2, \ldots, u_n, e_n, u_{n+1}$ of alternating vertices and edges such that each e_i is incident on u_i and u_{i+1}. In a graph (not a multigraph), a walk may be denoted $u_1 u_2 \cdots u_{n+1}$. A **path** is a walk with no repeated edge. A **simple path** is a path with no repeated vertex except that u_1 is allowed to equal u_n. The **length** of a path is the number of edges it contains. Any single vertex by itself is considered to be a path of **length zero.**

We summarize the first three of these definitions in Table 9.2.

TABLE 9.2 Some Common Terms

Term	Repeated Edges?	Repeated Vertices?
Walk	Allowed	Allowed
Path	No	Allowed
Simple Path	No	No, except that the first and last vertices may be the same

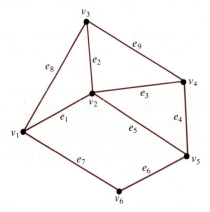

Figure 9.7 A graph with six vertices and nine edges.

EXAMPLE 8 Consider the graph of Figure 9.7. We see that $v_1 \, e_1 \, v_2 \, e_2 \, v_3 \, e_8$ $v_1 \, e_1 \, v_2 \, e_5 \, v_5 \, e_6 \, v_6$ is a walk but not a path and that $v_1 \, e_1 \, v_2 \, e_2 \, v_3 \, e_9 \, v_4 \, e_3 \, v_2 \, e_5$ $v_5 \, e_6 \, v_6$ is a path but not a simple path. However, $v_1 \, e_1 \, v_2 \, e_2 \, v_3 \, e_9 \, v_4 \, e_4 \, v_5$ is a simple path. Note also the degrees of the vertices. Vertices v_1, v_3, v_4, and v_5 all have degree 3. Vertex v_6 has degree 2 and vertex v_2 has degree 4. The sum of these six degrees is 18, exactly double the number of edges. This is true in general, as we shall see. ∎

It is in the definition of the word *path* and its variants that graph theory is the least standardized. The underlying concepts are generally the same as those presented here, although the word *path* is often used to denote what we call *walk* and the term *elementary path* is often introduced to make the distinction we make between a path and a simple path. The point is, whatever book you are reading, be sure to check the author's definitions carefully.

We are now ready to derive some fundamental properties of graphs that will be used throughout the rest of the chapter.

Properties of the Degrees of Vertices

THEOREM 9.1   Let $G = (V,E)$ be a graph or multigraph. Then $\sum_{v \in y} \deg(v) = 2\,|\,E\,|$.

PROOF   Each edge is associated with a pair of vertices $\{u,v\}$. Thus each edge contributes 2 to $\Sigma\deg(v)$. Summing this over all edges, we get $\Sigma\deg(v) = 2\,|\,E\,|$.   □

EXAMPLE 9   Suppose a graph has five vertices. Is it possible for each to have degree 3?

SOLUTION   If each has degree 3, then $\Sigma\deg(v) = 15$. But $\Sigma\deg(v) = 2\,|\,E\,|$ and hence must be even. So it is impossible.   ■

EXAMPLE 10   Suppose a graph, G, is not a multigraph and has five vertices. If all vertices are to have equal degrees, find all possible structures for G.

SOLUTION   Since each vertex can share up to one edge with any of the other four vertices, the maximum degree possible for any vertex is 4. Since the degrees are equal, as in the previous example, they cannot equal 3 or 1. Thus each vertex must have degree 0, 2, or 4. Possible representations for G are shown in Figure 9.8. We will discuss how to verify that these are the only possibilities in Section 9.3   ■

Figure 9.8   Graphs with five vertices and even degree at each vertex.

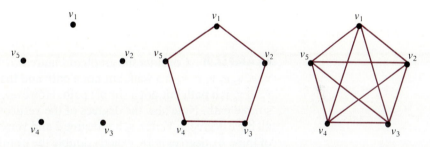

COROLLARY 9.2   In any graph or multigraph, the number of vertices of odd degree is even.

PROOF   If there were an odd number of vertices of odd degree, the sum of the degrees would be odd.   □

Properties of Paths

The following theorem and corollary are used frequently.

THEOREM 9.3   Let $G = (V,E)$ be a graph or multigraph. If two vertices of G are joined by a walk, they are joined by a simple path.

PROOF   The proof is by induction. Let u and v be vertices of G and let P denote a walk joining them. Let n denote the length of P. Then if n equals 1, the walk is a simple path and thus u and v are joined by a simple path. Suppose any two vertices joined by a walk of length $n - 1$ or less are also joined by a simple path. Denote the walk joining u and v by $u_0\, u_1\, u_2 \cdots u_{n-1}\, u_n$, where u_0 denotes u and u_n denotes v. If the walk is a path, we are done. If not, some u_i must appear twice; i.e., the walk must be of the form $u_0\, u_1\, u_2 \cdots u_i\, u_{i+1} \cdots u_j\, u_i\, u_{j+2} \cdots u_{n-1}\, u_n$. But then the vertices $u_{i+1} \cdots u_j$ can be omitted, leaving a walk joining u and v that has a length less than n. Thus there is a simple path joining u and v. (See Figure 9.9.)   □

Figure 9.9   The edges joining u_2, u_3, and u_4 in the walk $u_0\, u_1\, u_2\, u_3\, u_4\, u_2\, u_5\, u_6$ may be omitted, yielding a simple path from u_0 to u_6.

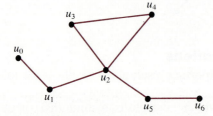

Definition   A graph, G, is **connected** if all pairs of vertices have a path joining them.

COROLLARY 9.4   Let G be a connected graph or multigraph and let $n = |V|$. Then any two vertices are joined by a path of length $n - 1$ or less.

PROOF   Applying the previous theorem, any two vertices are joined by a simple path. Since no vertices are repeated in a simple path, the maximum length of a simple path joining two vertices is $n - 1$.   □

EXAMPLE 11   Consider Figure 9.7 again. Note that there are several simple paths joining v_1 and v_3, for instance e_8; $e_1 e_2$; $e_7 e_6 e_9$; and $e_7 e_6 e_5 e_3 e_9$. Each of these paths is of different length, but the longest is of length 5, which equals $n - 1$.   ∎

EXAMPLE 12   Find a graph with n vertices in which every pair of vertices is joined by a path of length *less* than $n - 1$.

SOLUTION   See Figure 9.10. Note that this example can be generalized as follows: if a graph has a vertex of degree greater than 2, the longest path will have degree less than $n - 1$. The proof of this statement is left for the Exercises.   ∎

Figure 9.10   An eight-vertex graph in which the longest path has length 5.

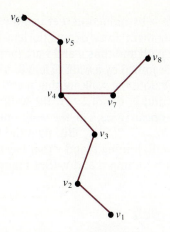

Graphs and Equivalence Relations

Another important result about graphs is the following.

THEOREM 9.5   Establish a relation $R(G)$ on the vertices of a graph, G, as follows: Two vertices, u and v, are related iff there is a path joining them. Then R is an equivalence relation.

PROOF R is reflexive, since a single vertex is a path of length zero. R is symmetric, since a path, $e_1, \ldots, e_n$, joining u to v defines a path $e_n, \ldots, e_1$ joining v to u. Lastly, R is transitive, since a path from u to v can be joined with a path from v to w to form a walk from u to w. This in turn can be replaced by a path joining u and w. Thus R is an equivalence relation.   ☐

Definition   The equivalence classes of $R(G)$ are called the **components** of G. Note that any component of a graph is itself connected.

EXAMPLE 13   The graph of Figure 9.10 has only one component, namely, the entire graph, since there are paths joining any two vertices. The graph of Figure 9.1, the relationships that already exist on the first day of class, has five components: {Tammy}, {Alice}, {Pat, Andre}, {Hank, Peter, Diane}, and {Jean, Frank, Joe}.   ∎

EXAMPLE 14   Suppose G is not connected and $|V| = n$. Show that G has at most $((n-1)(n-2))/2$ edges.

SOLUTION   If G is not connected, it has at least two components. Suppose there are m vertices in one component, G_1, where $1 \le m \le n-1$. Then there are at most $n - m$ in some other component, G_2. Since G is not a multigraph, each vertex can have up to $n - 1$ edges incident on it. Summing over all n vertices gives $n(n-1)$. Since this process counts every edge twice, there is a maximum of $(n(n-1))/2$ edges in G. Of these possible edges, there are $f(m) = m(n-m) = mn - m^2$ edges that cannot be included in G, since otherwise there would be a connection between G_1 and G_2. The equation $f(m) = mn - m^2$ is a concave downward parabola (Figure 9.11) that attains its smallest positive value when $m = 1$ or $m = n-1$. Taking either value for m, $f(m)$, the number of edges excluded, cannot be less than $n-1$. Thus G cannot have more than $(n(n-1))/2 - (n-1)$ edges, which equals $((n-1)(n-2))/2$ edges.

Figure 9.11 Graph of $f(m) = mn - m^2$.

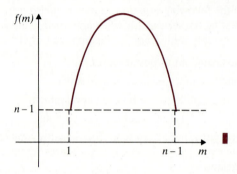

TERMINOLOGY

graph	incident on	walk
undirected graph	degree of a vertex	path
graph theory	deg(v)	simple path
Konigsberg bridge problem	subgraph	length of a path
vertices	loop	zero length path
edges	parallel edges	connected graph
associated with	multigraph	component of a graph
(V,E)	simple graph	

SUMMARY

1. Graphs can be used to model unordered relationships between pairs of elements.
2. If (V,E) is a graph or multigraph, $\Sigma\deg(v) = 2\,|\,E\,|$.
3. If (V,E) is a graph or multigraph, it must have an even number of vertices of odd degree.

4. If two vertices of a graph or multigraph are joined by a path, they are joined by a simple path.

5. In a connected graph or multigraph, any two vertices are joined by a path of length $n - 1$ or less.

6. The property of being joined by a path is an equivalence relation on the set of vertices of a graph. The equivalence classes are called *components*.

EXERCISES 9.1

Each of the following situations can be modeled using graph theory. For each one, indicate what the vertices and edges are; whether the situation would be better modeled by a graph, multigraph, or digraph; and whether loops are present.

1. The laying out of postal routes.

2. Countries that exchange ambassadors.

3. Running utilities to a housing development.

4. Cities that have direct airline connections.

5. Food webs in an ecosystem. (For instance, one part of such a food web is the following chain: hawks eat songbirds which eat grasshoppers which eat vegetation.)

6. A computer network.

7. The state diagram of a digital circuit. (Such circuits are found in one of a finite number of states. When input is given to the circuit, it may jump to some other state, depending on the input and the current state.)

8. An electronic pattern recognition device scans an image and measures various quantities. Two patterns are said to be *similar* if their measurements on the specified quantities do not differ by more than some predetermined value.

9. A matching question on a test.

10. The assignment of employees to jobs.

11. Four factories produce goods which can be shipped to any of three warehouses depending on demand and shipping costs.

12. The lamprey is a parasite that entered Lake Ontario some years ago through the St. Lawrence Seaway and devastated the lake trout population. It then spread through canals and streams to other lakes. Some lakes that had no connection to Lake Ontario were not affected, however.

13. The motion of a knight (or bishop or any other piece) on a chessboard.

14. A round robin tournament.

15. A single elimination tournament.

16. The spread of rumors.

17. Friendship relationships in a large group.

18. A route through an art gallery or museum that would enable one to see every item on display without ever having to retrace one's steps.

Draw graphs with the following characteristics.

19. $|V| \geq 4$, no simple path has a length more than 2.
20. $|V| \geq 3$, every vertex is of odd degree.
21. $|V| \geq 8$, no simple path has a length more than 4.
22. $|V| \geq 6$, G has at least three components.
23. G is connected, but removing any edge makes G disconnected.
24. There is no edge that, when removed, makes the graph disconnected.
25. Removing one particular edge makes the graph disconnected, but only that edge.

In each of the following graphs find a walk that is not a path, a path that is not a simple path, and a simple path of length at least 4.

26.

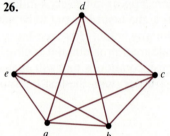

27.

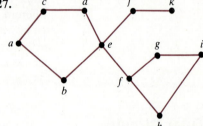

28.

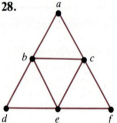

29.

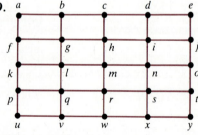

Verify Theorem 9.1 for the following graphs.

30. Exercise 26   **31.** Exercise 27
32. Exercise 28   **33.** Exercise 29
34. Show that $2|E| \leq |V|^2 - |V|$ for simple graphs.
35. How many edges does a graph contain that has four vertices of degree 3 and four vertices of degree 4?
36. A graph with 7 vertices and 15 edges has a vertex of degree 1, a vertex of degree 2, and a vertex of degree 3. Find an equation for the possible degrees of the remaining vertices.
37. Show that a component of a graph is a connected subgraph.
38. Show that the number of possible simple graphs on a set of n vertices is $2^{C(n,2)}$.

9.2

EULER CIRCUITS AND HAMILTONIAN CYCLES

In the last section, we listed four questions about graphs. This section focuses on two of them:

Is it possible to move from vertex to vertex in a graph and return to the original starting point in such a way that every *edge is traversed* exactly once?

Is it possible to move from vertex to vertex in a graph and return to the original starting point in such a way that every *vertex is visited* exactly once?

These two questions appear to be very similar, yet they are extraordinarily different. Answering the first turns out to be quite easy, and we will shortly see how to do it for any graph. However, the second question is as difficult as the traveling salesman problem and is closely related to it. The best we can do is to establish conditions that will enable us to answer the question for *some* graphs. We start with the first question.

Euler Circuits

Definition Let $G = (V, E)$ be a graph or multigraph. A **circuit** is a path in which the initial and terminal vertices are the same. An **Euler circuit** is a circuit that includes every edge in E. An **Euler path** is a path that includes every edge in E but does not have the same initial and terminal vertices. A **cycle** is a *simple* path in which the initial and terminal vertices are the same. A graph that has an Euler circuit is called **Eulerian**.

These terms are summarized in Table 9.3. Note that every cycle is a circuit, but not the converse.

TABLE 9.3

Term	Initial and Terminal Vertices the Same	Must Include Every Edge?	Repeated Vertices Allowed?
Circuit	Yes	No	Yes
Cycle	Yes	No	No
Euler circuit	Yes	Yes	Yes
Euler path	No	Yes	Yes

EXAMPLE 1 In Figure 9.12 *abcdecfa* is a circuit because it starts and ends at the same vertex and because it is a path—a walk with no repeated edges. Since it includes every edge of the graph, it is an Euler circuit. It is not a cycle because it is not a simple path; the vertex *c* is used twice. However, *abcfa* is a cycle.

Figure 9.12 *abcfa* is a cycle; *abcdecfa* is a circuit.

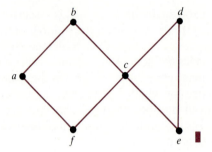

Euler paths and circuits are named after Leonhard Euler, who solved the Konigsberg bridge problem. As mentioned in the last section, the problem was to find a path that could be used to cross each of the seven Konigsberg bridges exactly once and return to one's starting point or to show that such a path was impossible. Euler solved the problem by modeling the situation using a multigraph as shown in Figure 9.13.

Figure 9.13 Euler's model of the Konigsberg bridge problem.

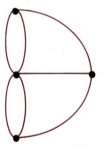

Euler's argument was as follows: Each of the vertices is of odd degree; one is of degree 5 and the others are of degree 3. A path of the desired type would have to enter each vertex and leave it again. If it passed through a vertex only once, that vertex would have to have degree 2. If it passed through twice, the vertex would have to have degree 4, etc. Thus, for an Euler circuit to exist, every vertex would have to be of even degree. Since none of the vertices in the Konigsberg bridge problem are of even degree, the graph does not have an Euler circuit.

This argument of Euler's can be generalized to the following theorem.

THEOREM 9.6   Let $G = (V,E)$ be a graph or multigraph. Then G is Eulerian iff G is connected and each vertex is of even degree.

PROOF   First, suppose that G is Eulerian and denote an Euler circuit by

$$e_1 e_2 \ldots e_{n-1} e_n$$

where e_1 is associated with $\{x_1,x_2\}$, e_2 is associated with $\{x_2,x_3\}$, ..., and e_n is associated with $\{x_n,x_1\}$. Then the number of times a vertex, v, appears in the list $\{x_1,x_2\}$, ..., $\{x_n,x_1\}$ will be its degree, since its degree is precisely the number of edges incident on it and since every edge is included in an Euler circuit. But for every edge for which v is the ending vertex of $\{x_i,x_{i+1}\}$, there is another edge for which it is the starting vertex. Thus every vertex must appear an even number of times, and thus every vertex has even degree. Also, since G is Eulerian, the circuit itself forms a path joining any two vertices. Thus G is connected.

Conversely, suppose that G is connected and every vertex has even degree. Start with any vertex, v. Follow any edge to another vertex. Since this vertex has even degree, there is another edge leaving it. Follow this edge to yet another vertex. Since every vertex entered can also be left, and since there are only finitely many edges, this process will eventually return to v. If all edges have been used, we have an Euler circuit.

If not all edges have been used, remove those edges that have been traversed and any resulting isolated vertices yielding a subgraph. Each vertex in it is of even degree. Since G is connected, at least one of the remaining vertices must have one of the removed edges incident upon it, since otherwise there would be no path joining v to the vertices that remain. (See Figure 9.14 and the discussion in Example 2.) Start from such a vertex and continue until it is returned to. This path can now be inserted into the original path. Since E is finite, continuing in this fashion will generate a path that uses every edge. ☐

Figure 9.14   (*a*) A cycle within a graph. (b) Same graph with five edges and three vertices removed.

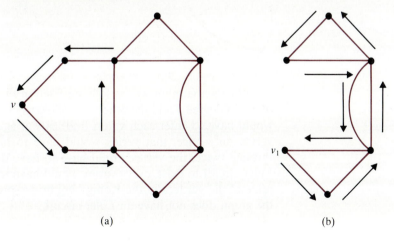

(a)   (b)

EXAMPLE 2   Figure 9.14(a) is a graph with vertex v and one circuit marked. Figure 9.14(b) is the same graph with the circuit removed. Vertex v_1 is common to both the circuit that was removed and the remaining subgraph. A circuit with initial and terminal point v_1 is marked. Inserting this circuit into the first circuit at v_1 will yield an Euler circuit for the entire graph. ∎

COROLLARY 9.7 G has an Euler path if G is connected, two vertices are of odd degree, and all other vertices are of even degree.

PROOF   The proof is left for the Supplementary Exercises.   ⬜

EXAMPLE 3   Figure 9.15 is called *Mohammed's scimitars*. Show that it can be drawn without lifting the pen from the paper.

Figure 9.15

Mohammed's scimitars. Reprinted from *Colloquium Publications,* (1962) "Theory of Graphs," Oystein Ore, Vol. 38, p. 39, by permission of the American Mathematical Society.

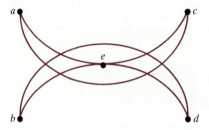

SOLUTION   Each of the four corner vertices has degree 2, and the vertex in the middle has degree 4. Since every vertex is of even degree, there is an Euler circuit. This implies that the figure can be drawn without lifting the pen from the paper.   ∎

Note that Euler's theorem does not *state* an algorithm for finding an Euler circuit; it is an existence theorem. However, the proof of the theorem is constructive; it does, in fact, provide such an algorithm. For instance, suppose we attempted to draw Mohammed's scimitars by drawing from a to e to c and then back to a again. This is a cycle that does not solve the problem. But if its edges are deleted, we can start again from e and form a new cycle such as *edbe.* This can be inserted into the original cycle, giving a solution to our problem—the circuit *aedbeca.*

EXAMPLE 4   Show that the graph in Fig. 9.16 is Eulerian.

SOLUTION   The graph is connected, and each vertex is of degree 4. Thus it is Eulerian. The procedure used in the proof of Euler's theorem could be employed to generate an Euler circuit.

Figure 9.16 An Eulerian graph.

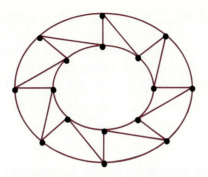

EXAMPLE 5 What additional bridges would be needed in Konigsberg to make an Euler circuit possible?

SOLUTION Each vertex in Figure 9.13 is of odd degree, and there are four vertices. Thus two more bridges would be needed, each joining any two distinct pairs of vertices. For instance, Figure 9.17(a) shows a map of Konigsberg with two bridges added, and Figure 9.17(b) shows a graph representation of the change.

Figure 9.17 The Konigsberg bridge problem with two bridges added.

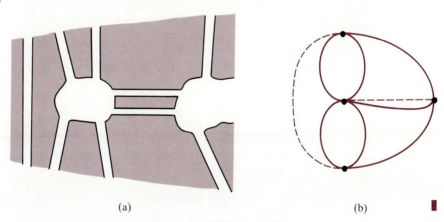

(a)   (b)

Hamilton Cycles

The second question considered in this section is whether, given a graph, it is possible to find a path that visits each vertex exactly once and returns to its starting point. Such a path must be a cycle, not a circuit, since no vertex can be visited twice. A related problem is this: Start at a vertex, visit every other vertex, and stop at a vertex different from the starting point. The solution to the second problem is a simple path. Everything we say in solving these two problems is just as applicable to multigraphs as to graphs and simple graphs, since loops and parallel edges can simply be ignored in producing a path that visits each vertex once. The ideas are formalized as follows.

Definition A **Hamilton cycle** in a graph is a cycle that includes every vertex. A graph that has a Hamilton cycle is called **Hamiltonian**. A **Hamilton path** is a simple path that includes every vertex but is not a cycle.

William Rowan Hamilton was an Irish mathematician who popularized the Hamilton cycle problem with his publication in 1858 of a puzzle he called the *Icosian game*. The game is sketched in Figure 9.18(a). It simulates a trip around the world in which 20 cities are to be visited, returning home without having visited any city twice.

EXAMPLE 6 The graphs in Figure 9.18 are Hamiltonian. Hamilton cycles are illustrated in Figure 9.19. ▮

Figure 9.18 Two Hamiltonian graphs.

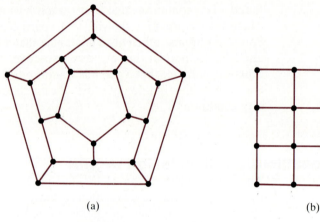

(a)   (b)

Figure 9.19 Hamilton cycles.

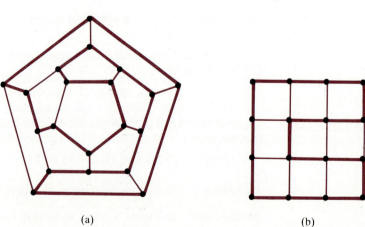

(a)   (b)

Finding Hamilton cycles in graphs has some interesting and diverse applications. For instance, suppose someone giving a dinner party wants to seat a group of guests in such a way that each person is seated next to some-one he or she knows. The relationships among the guests could be modeled by a graph similar to Figure 9.1. If there is such a seating arrangement, it is a Hamilton cycle in the graph. Alternatively, suppose a postman wants to visit each house on his route once without doubling back. Each house could be represented by a vertex, and every possible route joining the houses could be represented by an edge. The desired route is a Hamilton cycle. Finding Hamilton cycles, however, can be very difficult. For the existence of an Euler circuit, there is a simple pair of necessary and sufficient conditions: the graph must be connected, and the degrees of all of its vertices must be even. Given a graph that satisfies these conditions, the proof of the theorem even provides an algorithm with which to construct an Euler cycle. No such necessary and sufficient conditions have been found for the existence of Hamilton cycles. Necessary conditions have been found—i.e., conditions that every graph having such a cycle must satisfy. These are very useful in showing that a particular graph is *not* Hamiltonian. Some conditions sufficient to guarantee that a graph is Hamiltonian have also been found, but they are very restrictive. That is, although the conditions do guarantee that a graph is Hamiltonian, there are many graphs that are Hamiltonian but do not satisfy the conditions. We start with the necessary conditions.

Necessary Conditions for Hamiltonian Graphs

There are a few important elementary properties that every Hamilton cycle must possess, and these are often all that is needed to show that a graph is not Hamiltonian. These properties are as follows:

1. A Hamilton cycle is always of length n. A Hamilton path is always of length $n - 1$.

2. If a vertex has degree 2, both edges incident on it must be included in any Hamilton cycle.

3. In tracing out a possible Hamilton cycle, once a vertex has been reached along some edge and an edge has been chosen to depart from that vertex, all other edges incident on that vertex can be deleted from the graph.

4. Hamilton cycles and Hamilton paths cannot contain any subcycles.

EXAMPLE 7   Show that the graphs of Figure 9.20 are not Hamiltonian.

SOLUTION   In Figure 9.20(a), vertices $a, b, c, g, h,$ and i are all of degree 2. Hence both edges incident on them must be included in any Hamilton cycle. But then the edges joining d to e and e to f cannot be included, by property 3 in the preceding list. Hence e is excluded and the graph does not

Figure 9.20 Two non-Hamiltonian graphs.

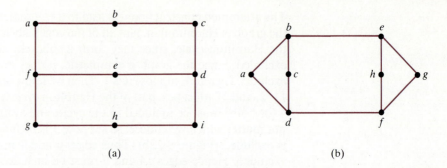

(a)   (b)

have a Hamilton cycle. Note, however, that it does have a Hamilton path —*abcdefghi*. A graph that has a Hamilton cycle automatically has a Hamilton path; deleting any edge from the cycle yields a Hamilton path. But this example shows that the converse is not true.

In Figure 9.20(b), *a* and *c* are both of degree 2. The edges through them must be included, but they form a subcycle, *abcda*. Thus this graph is not Hamiltonian either. ∎

EXAMPLE 8   Figure 9.21(a) is called *Petersen's graph*. Show that it is not Hamiltonian. Show also that Figure 9.21(b) is Hamiltonian.

Figure 9.21 Two graphs, each with 10 vertices and 15 edges, one Hamiltonian, one not.

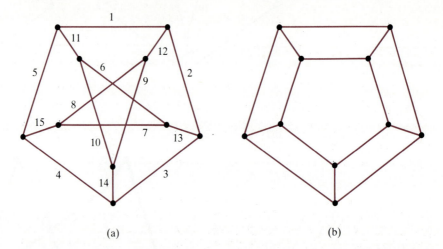

(a)   (b)

SOLUTION   The contrast between the two graphs in Figures 9.21(a) and 9.21(b) shows how graphs that are very similar can behave very differently. By a combination of the elementary properties previously given and by the symmetry of the graph, we can show that Petersen's graph is not Hamiltonian. For simplicity, we have numbered the edges rather than labeling the vertices. Our strategy will be to attempt to construct a Hamilton cycle by exploiting the symmetry of the graph in order to eliminate certain edges.

The attempted construction will lead to a contradiction. First, suppose that the graph is Hamiltonian. Not all of the edges labeled 1 through 5 can be in the Hamilton cycle, since they form a subcycle, so at least one must be excluded. Since the graph is symmetric, it does not matter which one we exclude. Therefore suppose it is 1. Then we know immediately that edges 5, 11, 2, and 12 must be a part of the Hamilton cycle (Figure 9.22a). Edge 3 or 4 (or both) must be included in the cycle, and again, by symmetry, it does not matter which we choose, so suppose 3 is included. This tells us that 13 is excluded (Figure 9.22b). Thus edges 6 and 7 must be included and 10 excluded. Hence edges 14 and 9 must be included and 4 excluded. Thus our construction of a Hamilton cycle breaks down into two subcycles (Figure 9.22c) and a Hamilton cycle does not exist.

Figure 9.22 Attempt to trace out a Hamilton cycle for Petersen's graph.

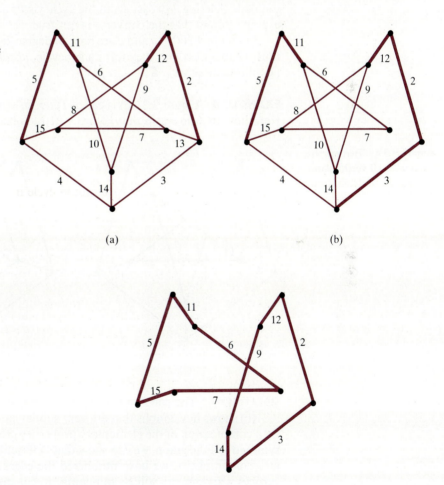

(a)

(b)

(c)

Figure 9.23 illustrates one possible Hamilton cycle for the graph of Figure 9.21(b). Thus this graph is Hamiltonian.

Figure 9.23 Hamilton cycle for the graph of Figure 9.21(b).

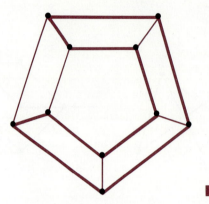

Another useful condition is given by the following theorem. In the theorem and in what follows, whenever we speak of "removing a vertex," we will assume that all edges incident on that vertex are also removed.

THEOREM 9.8 Let G be a graph. Then if the removal of s vertices from G breaks G into a graph with more than s components, G is not Hamiltonian.

PROOF We will prove the contrapositive of the statement. Suppose G is Hamiltonian and consider a Hamilton cycle in G. Let G_H denote the subgraph consisting of the original vertices of G and only those edges in the Hamilton cycle. The proof proceeds by induction on s. Removal of one vertex from G_H will still leave a Hamilton path through the vertices that are left, and hence what is left of G_H has one component. Suppose the removal of $s - 1$ vertices from G_H leaves a subgraph with no more than $s - 1$ components. Removal of an additional vertex will either leave a subgraph with at most $s - 1$ components (if it is the endpoint of a subpath) or will leave a subgraph with s components (if it is not the endpoint of a subpath). Since the subgraph has at most s components, the graph (which has more edges) has at most s components. ☐

EXAMPLE 9 Show that the graphs of Figure 9.24 are not Hamiltonian.

SOLUTION

a) In this graph, removal of the two vertices, a and b (and incident edges), causes the graph to break into three components—the single vertices c, d, and e.

b) This is perhaps the earliest example of a non-Hamiltonian graph. It was devised by the Rev. T.P. Kirkman in 1856. Removal of the six vertices marked in the diagram breaks the graph into seven components. ∎

Figure 9.24 Removal of s vertices yields more than s components.

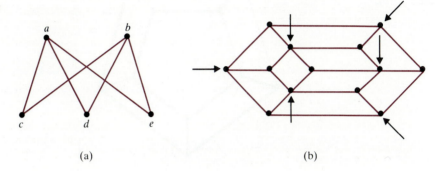

(a)   (b)

Figure 9.24(a) is an example of an important kind of graph known as a bipartite graph. A **bipartite graph** is a graph, $G = (V,E)$, whose vertices consist of two subsets, V_1 and V_2, such that every edge in E is associated with one vertex from V_1 and one from V_2. If every vertex in V_1 is associated with every vertex in V_2, G is called a **complete bipartite graph** and is denoted $K_{n,m}$, where $|V_1| = n$ and $|V_2| = m$. Thus Figure 9.24(a) is $K_{2,3}$.

A number of further necessary conditions for Hamiltonian graphs have been developed. See, for example, *The Traveling Salesman Problem* by Lawler et al., Chapter 11 (see References).

Sufficient Conditions for Hamilton Cycles and Paths

One of the most widely known sufficient conditions for Hamilton paths is the following. It is proven in *Elements of Discrete Mathematics* by Liu, Section 5.7 (see References).

THEOREM 9.9   Suppose $G = (V,E)$ is a graph such that for any two vertices, u and v,

$$\deg(u) + \deg(v) \geq n - 1,$$

where n denotes $|V|$. Then G has a Hamilton path. ▯

A closely related theorem was proven by G.A. Dirac in 1952.

THEOREM 9.10   Let $G = (V,E)$ with $|V| = n > 2$. Suppose G is such that for any vertex v, $\deg(v) \geq n/2$. Then G is Hamiltonian. ▯

In the original description of the traveling salesman problem at the beginning of this book, every city had to be joined by a direct route to every other city. If the cities are vertices and the routes are edges, such a graph is an example of a complete graph. More formally, we define a **complete graph on *n* vertices** to be a simple graph in which every vertex shares an edge with every other vertex. Thus every vertex has degree $n - 1$. The complete graph on n vertices is denoted K_n.

EXAMPLE 10 K_n is Hamiltonian for all $n > 2$.

SOLUTION   This follows directly from Theorem 9.10, since $n > 2$ implies that $n - 1 \geq n/2$ and that every vertex has degree $n - 1$. ∎

There are still a number of problems that cannot be solved by applying the necessary and sufficient conditions presented here. For instance, one famous problem is the **knight's tour.** The legal moves of a knight on a chessboard are an L shape—two spaces horizontally or vertically followed by one space in a line perpendicular to the first direction (see Figure 9.25).

The question is this: Is it possible to start on one square of a chessboard and move a knight in such a way that every square is visited exactly once? If we let each square of the board be a vertex of a graph and associate an edge with every pair of vertices between which a knight can move, the question becomes one of the existence of a certain Hamilton circuit. In fact, a knight's tour does exist; one is presented in Figure 9.26. The numbers indicate the sequence in which the squares are visited. But the sufficient conditions presented here are of no help; the highest degree of any vertex in the knight's tour problem is 8, and there are 64 vertices.

Figure 9.25   Legal moves for a knight on a chessboard.

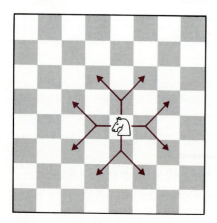

Figure 9.26   One of many solutions to the knight's tour problem. Reprinted from *Colloquium Publications,* (1962) "Theory of Graphs," Oystein Ore, Vol. 38, p. 53, by permission of the American Mathematical Society.

56	41	58	35	50	39	60	33
47	44	55	40	59	34	51	38
42	57	46	49	36	53	32	61
45	48	43	54	31	62	37	52
20	5	30	63	22	11	16	13
29	64	21	4	17	14	25	10
6	19	2	27	8	23	12	15
1	28	7	18	3	26	9	24

Relationship to the Traveling Salesman Problem

The existence of Hamilton cycles is closely related to the traveling salesman problem. For instance, suppose we have a solution technique for the traveling salesman problem. Suppose we also have a graph that we want to test to see if it is Hamiltonian. We could "extend" the graph by adding to it all possible edges that it does not include, i.e., all edges in K_n not already included. If the original edges are all assigned a length of 0 and all the new edges are assigned a length of 1, we have a complete graph with a "distance" assigned to each edge. Our solution technique for the traveling salesman problem can then be applied to this new graph. If the shortest possible route for the salesman to follow has length 0, that route is a Hamilton cycle. If not, a Hamilton cycle does not exist; if one did exist, the shortest route would have 0 length. Note that the extra edges can be added in polynomial time since K_n has $n(n - 1)/2$ edges. Thus, if the traveling salesman problem is tractable, so is the Hamilton cycle problem. The converse is also true; it has been shown that the problem of recognizing Hamilton graphs is *NP*-complete. (See Section 4.4 for a discussion of *NP*-complete problems.) Thus the likelihood of there being necessary and sufficient conditions for the Hamilton cycle problem that can be easily applied, like the necessary and sufficient conditions for the Euler circuit problem, is very slight.

TERMINOLOGY

circuit	Eulerian graph
Euler circuit	complete graph on n vertices
Hamilton path	Hamilton cycle
Euler path	Hamiltonian graph
bipartite graph	K_n
cycle	knight's tour problem
complete bipartite graph	

SUMMARY

1. A graph is Eulerian iff it is connected and each vertex is of even degree.
2. A graph has an Euler path if it is connected, two vertices are of odd degree, and all other vertices are of even degree.

3. Hamilton paths and cycles must have the following properties:

   a) A Hamilton cycle is always of length n. A Hamilton path is always of length $n - 1$.

   b) If a vertex has degree 2, both edges incident on it must be included in a Hamilton cycle.

   c) In tracing out a possible Hamilton cycle, once a vertex has been reached along an edge and another edge has been chosen to depart from that vertex, all other edges incident on that vertex can be deleted from the graph.

   d) Hamilton cycles and Hamilton paths cannot contain any subcycles.

   e) The removal of s vertices from a Hamiltonian graph cannot break the graph into more than s components.

4. If for any two vertices, u and v, of a graph, $\deg(u) + \deg(v) \geq n - 1$, where n denotes $|V|$, G has a Hamilton path.

5. Suppose G is such that for any vertex v, $\deg(v) \geq n/2$. Then G is Hamiltonian.

6. The Hamilton cycle problem can be considered a special case of the traveling salesman problem. Like the traveling salesman problem, it is *NP*-complete.

EXERCISES 9.2

Sketch the following graphs.

1. K_2 2. K_3

3. K_4 4. K_5

5. $K_{3,3}$ 6. $K_{4,4}$

7. $K_{3,4}$ 8. $K_{3,5}$

Test each of the following graphs for Euler circuits and Hamilton cycles.

9. K_4 10. K_5

11. $K_{3,3}$ 12. $K_{4,4}$

13. $K_{3,4}$ 14. $K_{3,5}$

15. Tetrahedron   16. Octahedron   17. Cube   18. Icosahedron

  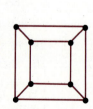 

19. Cube with diagonals

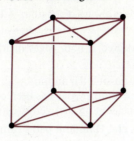

20. Star

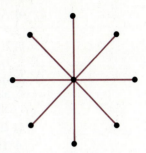

21. Wheel

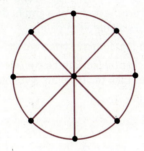

22.[†]

23.

24.[†]

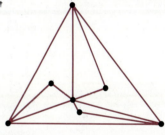

Show that the following graphs are not Hamiltonian.

25. Any graph with a degree 1 vertex.

26. A star graph of any size (see Exercise 20).

27. $K_{n,m}$; $n \neq m$.

28. The $n \times n$ square graph, where n is odd. (See Figure 9.18b for an example of a 4×4 square graph.)

Show that the following are Hamiltonian.

29. $K_{n,n}$ for any n.

30. The wheel graph for any number of spokes greater than two. (See Exercise 21.)

† From *The Traveling Salesman Problem,* Lawler, Lenstra, Rinnooy Kan, and Shmoys. Copyright © 1985. Reprinted by permission of John Wiley & Sons, Ltd.

31. The $n \times n$ square graph, where n is even.

32. Suppose a graph has the property that edges can be removed from it, so that in the resulting graph, every vertex has degree 2. Is the graph necessarily Hamiltonian?

33. Under what conditions is $K_{n,m}$ Eulerian?

34. When is an Euler circuit a Hamilton cycle? Give an example.

35. Find an example of a graph that has neither Euler circuits nor Hamilton cycles.

36. Find an example of a graph that has both Euler circuits and Hamilton cycles.

37. Find an example of a graph that has an Euler circuit but does not have a Hamilton cycle.

38. Find an example of a graph that does not have an Euler circuit but does have a Hamilton cycle.

39. Find another Hamilton circuit for the icosian game.

40. How many distinct Hamilton cycles are there in K_n?

41. Show that the knight's tour problem has no solution on a 3×3 chessboard.

42. Show that the knight's tour problem has no solution on a 4×4 chessboard.

43. Show that the knight's tour problem has no solution on a 5×5 chessboard.

44. Find a knight's tour on a 6×6 chessboard.

9.3

GRAPH ISOMORPHISMS AND REPRESENTATIONS

In this section we will address our third question about graphs:

What does it mean for two graphs that appear different to have the same structure, and how can one tell if the structures are the same?

For instance, do the two graphs in Figure 9.27 have the same structure?

Figure 9.27 Two perspectives on a cube.

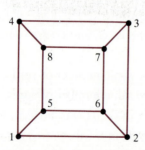 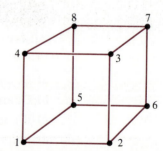

Near the end of this section, we will also apply some of the concepts developed in answering this question to another question: How can we tell how many paths connect two distinct vertices in a graph?

The two graphs are certainly not the same if by *same* we mean identical. But they are the same in the sense that they are both two-dimensional renderings of a cube. The question of what it means for two structures to be the same is extremely important in all branches of mathematics that are concerned with the structure of sets. It is typically addressed by using the concept of the *isomorphism,* a term that can be roughly translated from Greek as *iso,* meaning "same," and *morph,* meaning "form." Although the mathematical definition of isomorphism varies, depending on the type of structure being studied, the underlying concept is always the same: two sets are isomorphic if their structure is the same, whatever that structure might be. For graphs, it is formally defined as follows.

Definition   Two graphs, $G_1 = (V_1, E_1)$ and $G_2 = (V_2, E_2)$, are **isomorphic** if there exists a one-to-one correspondence, f, between V_1 and V_2 such that $\{u, v\}$ is in E_1 iff $\{f(u), f(v)\}$ is in E_2. The function, f, is called an **isomorphism.** Note that this definition applies to graphs but not to multigraphs.

EXAMPLE 1   Consider the two graphs in Figure 9.27. Suppose a function, f, maps each vertex in Figure 9.27(a) to the one with the same number in Figure 9.27(b). Then f is a one-to-one correspondence. Also, each edge, $\{i, j\}$, in Figure 9.27(a) corresponds to $\{i, j\}$, in Figure 9.27(b). Thus the two graphs are isomorphic. If by *same* for graphs we mean isomorphic, the two graphs in Figure 9.27 are the same.   ∎

Verifying that two graphs are isomorphic can be difficult. It may take quite a bit of trial and error before one can find a correspondence between the vertices that yields a consistent correspondence between the edges. Showing that graphs are *not* isomorphic is often easier. This is because there are a number of properties that isomorphic graphs must share. Several of these properties are presented in the next theorem.

THEOREM 9.11   Suppose G_1 and G_2 are isomorphic, with isomorphism f. Then

a) G_1 and G_2 have the same number of vertices.

b) G_1 and G_2 have the same number of edges.

c) if u is adjacent to v in G_1, $f(u)$ is adjacent to $f(v)$ in G_2.

d) if $u_1 u_2 \cdots u_n$ is a cycle in G_1, then $f(u_1) f(u_2) \cdots f(u_n)$ is a cycle in G_2.

e) if u has degree k in G_1, then $f(u)$ has degree k in G_2.

f) G_1 and G_2 have the same number of components.

g) if G_1 is Eulerian, so is G_2.

h) if G_1 is Hamiltonian, so is G_2.

PROOF

a) This is obvious since f is a one-to-one correspondence.

b) Each edge $\{u,v\}$ in E_1 is associated with a unique edge $\{f(u), f(v)\}$ in E_2 and vice versa. Thus $|E_1| = |E_2|$.

c) If u is adjacent to v, then $\{u,v\}$ is an edge in G_1. Thus $\{f(u), f(v)\}$ is an edge in G_2. Thus $f(u)$ is adjacent to $f(v)$.

d) This is left for the Exercises.

e) This is left for the Exercises.

f) Suppose u and v are in G_1 and $u\, u_1 u_2 \cdots u_n\, v$ is a path joining u and v. Then $f(u) f(u_1) f(u_2) \cdots f(u_n) f(v)$ is a path joining $f(u)$ and $f(v)$. Thus, for any two vertices that are connected by a path in G_1, their corresponding vertices are connected by a path in G_2. Similarly, for any two vertices that are connected in G_2, their corresponding vertices in G_1 are connected. Thus the number of components of both graphs must be the same.

g) This follows directly from part (e).

h) This is left for the Exercises. □

EXAMPLE 2 Show that $K_{3,3}$ and K_6 are not isomorphic (Figure 9.28).

Figure 9.28 (a) K_6 and (b) $K_{3,3}$.

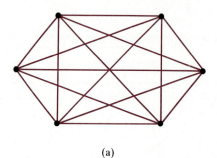

(a)

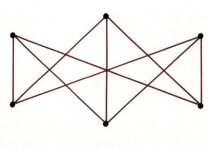

(b)

SOLUTION   Every vertex in $K_{3,3}$ has degree 3, while every vertex in K_6 has degree 5. Hence they cannot be isomorphic. ∎

EXAMPLE 3   Show that the graphs in Figure 9.29 are not isomorphic.

Figure 9.29   Two noniso-morphic graphs.

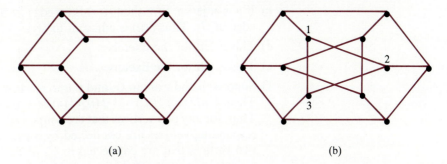

(a)   (b)

SOLUTION   Even though both graphs have the same number of vertices and edges, Figure 9.29(b) has 3-cycles (i.e., three vertices that form a cycle). For instance, the three vertices labeled 1, 2, and 3 form a three-cycle. We can use the symmetry of Figure 9.29(a) to show that it does not have three-cycles. We can select any vertex on the outer or inner hexagon and observe that the smallest cycle including the vertex is at least a four-cycle. Thus, by symmetry, no vertices are contained in three-cycles and the graphs are not isomorphic. ∎

EXAMPLE 4   Show that the graphs in Figure 9.30 are not isomorphic.

Figure 9.30   Nonisomor-phic graphs.

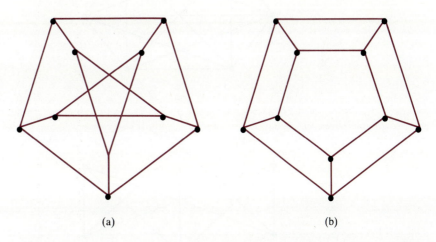

(a)   (b)

SOLUTION   Even though these two graphs look similar to the ones in the previous example, we will show that they are nonisomorphic by a different argument. The first is Petersen's graph, which was shown to be non-Hamiltonian in Section 9.2. The second graph was also shown to be Hamiltonian. Thus they are not isomorphic. ∎

EXAMPLE 5   Show that the graph in Figure 9.31(a) has a subgraph isomorphic to $K_{3,3}$ (Figure 9.31b).

Figure 9.31   A graph with a subgraph isomorphic to $K_{3,3}$.

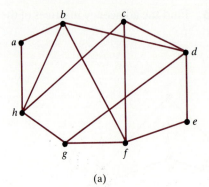

(a)

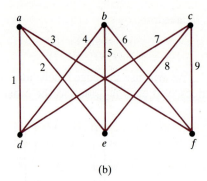

(b)

SOLUTION   We will use the numbers on the edges of $K_{3,3}$ later. Vertices a and e of Figure 9.31(a) have degree 2 and hence cannot correspond to vertices in $K_{3,3}$. The remaining vertices, however, can be partitioned into the subsets $\{b, c, g\}$ and $\{d, f, h\}$. Each vertex in each of these sets is connected to each vertex in the other set and to none of the vertices in its own set. Thus the subgraph generated by deleting a and e from the original graph is isomorphic to $K_{3,3}$. ∎

Matrix Representation of Graphs

Another approach to representing the structure of graphs involves the use of matrices. In Chapter 8, we examined the adjacency matrix for digraphs. Graphs can also be represented by an adjacency matrix.

Definition By the **adjacency matrix** of a graph, $G = (V, E)$, we mean a Boolean matrix whose ijth entry is 1 if there is an edge joining i and j, and 0 if there is no such edge. As in the definition of isomorphism, this is applicable to graphs but not to multigraphs.

EXAMPLE 6 Find the adjacency matrices of the tetrahedron (Figure 9.32) and of $K_{3,3}$.

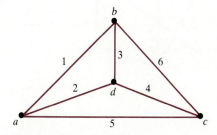

Figure 9.32 Tetrahedron.

SOLUTION We will use the edge labels later. For the tetrahedron, there are four vertices, and each is joined to the other three. Thus the matrix is

$$\begin{bmatrix} 0 & 1 & 1 & 1 \\ 1 & 0 & 1 & 1 \\ 1 & 1 & 0 & 1 \\ 1 & 1 & 1 & 0 \end{bmatrix}.$$

For $K_{3,3}$, if we label the vertices as in Figure 9.31(b) and list them alphabetically, the matrix is

$$\begin{bmatrix} 0 & 0 & 0 & 1 & 1 & 1 \\ 0 & 0 & 0 & 1 & 1 & 1 \\ 0 & 0 & 0 & 1 & 1 & 1 \\ 1 & 1 & 1 & 0 & 0 & 0 \\ 1 & 1 & 1 & 0 & 0 & 0 \\ 1 & 1 & 1 & 0 & 0 & 0 \end{bmatrix}.$$

If we order the vertices differently, we get a different matrix. If the vertices are listed in the order $\{a, d, b, e, c, f\}$, for instance, we get the matrix

$$\begin{bmatrix} 0 & 1 & 0 & 1 & 0 & 1 \\ 1 & 0 & 1 & 0 & 1 & 0 \\ 0 & 1 & 0 & 1 & 0 & 1 \\ 1 & 0 & 1 & 0 & 1 & 0 \\ 0 & 1 & 0 & 1 & 0 & 1 \\ 1 & 0 & 1 & 0 & 1 & 0 \end{bmatrix}.$$

Thus there is no unique adjacency matrix for a particular graph; rather, the matrix depends on the order in which the vertices are listed.

Note that each of the preceding matrices is symmetric. This is true in general for adjacency matrices, since edges are associated with *pairs* of vertices, not ordered pairs. For this reason, the adjacency matrix for a graph is often written in triangular form. If we assume that we are dealing with simple graphs, there are no loops. Thus the main diagonal will be all 0s. Thus the lower triangular form for the preceding two matrices is

$$\begin{bmatrix} 1 & & \\ 1 & 1 & \\ 1 & 1 & 1 \end{bmatrix} \quad \begin{bmatrix} 1 & & & & \\ 0 & 1 & & & \\ 1 & 0 & 1 & & \\ 0 & 1 & 0 & 1 & \\ 1 & 0 & 1 & 0 & 1 \end{bmatrix}. \quad \blacksquare$$

Given the adjacency matrix, it is also possible to find the graph it represents.

EXAMPLE 7   Find a graph represented by the following matrix:

$$\begin{bmatrix} 0 & 1 & 0 & 1 & 1 \\ 1 & 0 & 1 & 0 & 0 \\ 0 & 1 & 0 & 0 & 1 \\ 1 & 0 & 0 & 0 & 0 \\ 1 & 0 & 1 & 0 & 0 \end{bmatrix}.$$

Figure 9.33   Graph drawn
for a given matrix.

SOLUTION   See Figure 9.33.

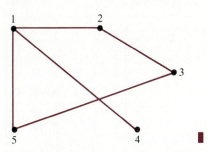

Another approach to representing graphs by matrices is the following.

Definition   The **incidence matrix** of a graph, $G = (V,E)$ with $|V| = n$ and $|E| = m$, is an $n \times m$ Boolean matrix whose ijth entry is 1 if the ith vertex is incident on the jth edge.

Note that the incidence matrix can also be used to represent multigraphs.

EXAMPLE 8   Find the incidence matrices of the tetrahedron and of $K_{3,3}$.

SOLUTION   The vertices and edges of these graphs are labeled in Figures 9.32 and 9.31(b). The incidence matrices are

Tetrahedron

$$
\begin{bmatrix}
1 & 1 & 0 & 0 & 1 & 0 \\
1 & 0 & 1 & 0 & 0 & 1 \\
0 & 0 & 0 & 1 & 1 & 1 \\
0 & 1 & 1 & 1 & 0 & 0
\end{bmatrix}
$$

$K_{3,3}$

$$
\begin{bmatrix}
1 & 1 & 1 & 0 & 0 & 0 & 0 & 0 & 0 \\
0 & 0 & 0 & 1 & 1 & 1 & 0 & 0 & 0 \\
0 & 0 & 0 & 0 & 0 & 0 & 1 & 1 & 1 \\
1 & 0 & 0 & 1 & 0 & 0 & 1 & 0 & 0 \\
0 & 1 & 0 & 0 & 1 & 0 & 0 & 1 & 0 \\
0 & 0 & 1 & 0 & 0 & 1 & 0 & 0 & 1
\end{bmatrix}.
$$

There is a third approach to representing graphs by matrices, called the *edge adjacency matrix*. This will be examined in the Exercises. The adjacency matrix is the most commonly used of the three, however.

Properties of the Adjacency Matrix

We began this section by asking what it means for two graphs to have the same structure. This question was addressed using the concept of isomorphism. We then looked at matrices as a way to represent the structure of a graph. The connection between these two concepts is found in the following theorem.

THEOREM 9.12 Two graphs are isomorphic if and only if, for some permutation of the vertices of one of them, they have the same adjacency matrix.

PROOF Suppose that two graphs, G_1 and G_2, are isomorphic. Now let $\{g_1, g_2, \ldots, g_n\}$ denote the vertices of G_1. Then there is a one-to-one, onto function, f, from $\{g_1, g_2, \ldots, g_n\}$ to the vertices of G_2. Since all vertices and all edges correspond via this function, arranging the vertices of G_2 in the order $\{f(g_1), f(g_2), \ldots, f(g_n)\}$, the adjacency matrices must be identical. But that order is just a permutation of the order of the second set of vertices.

Conversely, suppose that G_1 and G_2 have the same adjacency matrix for some permutation of the vertices of G_2. Then that permutation defines a one-to-one correspondence between the vertices of G_1 and the vertices of G_2 such that all edges correspond as well. Hence the graphs are isomorphic. □

The main benefit of this theorem is that it enriches our understanding of the concepts of isomorphism and of the adjacency matrix. It may occasionally be of some help in deciding whether two graphs are isomorphic, but its value in this situation is limited. Consider the two matrix representations of $K_{3,3}$ examined in Example 6. It is not all obvious from the matrices themselves that some permutation of the vertices of the graphs they represent would make them identical. And since the number of permutations of n vertices is $n!$, attempting to establish isomorphism in this way may prove impractical. The one situation in which this theorem really can be helpful is when the degrees of the vertices vary greatly. Then the vertices of each graph can be sorted from highest degree to lowest, and one can just consider the permutations that preserve that order. This can greatly reduce the number of permutations that must be examined.

One important application of the adjacency matrix of a graph is to connectivity. It enables us to answer two questions: Given any number, m, how many walks of length m join any two vertices, and how can we tell from the adjacency matrix whether a graph is connected? These questions are answered in Theorem 9.13 and Corollary 9.14. In both the theorem and the corollary, A^m denotes matrix A multiplied by itself m times. The multiplication used is standard matrix multiplication and the addition is ordinary integer addition, not the logical OR used in Section 8.4.

THEOREM 9.13 Let $G = (V,E)$ be a graph and let A be its adjacency matrix. Then $(A^m)_{ij}$ is the number of walks of length m that connect vertices i and j.

PROOF   The proof is by induction on m. First, let $m = 1$. Then

$$A_{ij} = 0 \text{ if there is no edge joining } i \text{ and } j,$$
$$= 1 \text{ if there is an edge joining } i \text{ and } j.$$

Since G is a graph, not a multigraph, this 0 or 1 is precisely the number of length 1 walks joining i and j.

Now suppose $m > 1$. Assume that $(A^{m-1})_{ij}$ gives the number of length $m - 1$ walks joining vertex i to vertex j for each i and j. Since $A^m = A \cdot A^{m-1}$, the ijth entry in A^m is the dot product of the ith row of A and the jth column of A^{m-1}. That is, $(A^m)_{ij} = A_{i1}(A^{m-1})_{1j} + \cdots + A_{in}(A^{m-1})_{nj}$. Now the ith row of A consists entirely of 1s and 0s. Suppose $A_{ik} = 1$ for some k. Then there is an edge joining i to k. There also exist $(A^{m-1})_{kj}$ walks of length $m - 1$ joining vertex k to vertex j. Hence there exist $(A^{m-1})_{kj}$ walks of length m that join vertex i to vertex j, passing through vertex k. Summing the $(A^{m-1})_{kj}$ over all k for which $A_{ik} = 1$ gives all length m walks joining vertex i to vertex j that pass through any other vertex on the m-1st step. ⬜

EXAMPLE 9   Find the number of length 1, 2, and 3 walks that join any pair of vertices of the tetrahedron (Figure 9.32).

SOLUTION   For the tetrahedron

$$A = \begin{bmatrix} 0 & 1 & 1 & 1 \\ 1 & 0 & 1 & 1 \\ 1 & 1 & 0 & 1 \\ 1 & 1 & 1 & 0 \end{bmatrix}.$$

This gives the number of length 1 walks joining any pair of vertices. Then

$$A^2 = \begin{bmatrix} 3 & 2 & 2 & 2 \\ 2 & 3 & 2 & 2 \\ 2 & 2 & 3 & 2 \\ 2 & 2 & 2 & 3 \end{bmatrix} \qquad A^3 = \begin{bmatrix} 6 & 7 & 7 & 7 \\ 7 & 6 & 7 & 7 \\ 7 & 7 & 6 & 7 \\ 7 & 7 & 7 & 6 \end{bmatrix}.$$

A^2 gives the number of walks of length 2 joining any pair of vertices; for instance, there are three walks of length 2 joining vertex a to itself in Figure 9.32. These are *ada*, *aca*, and *aba*. Also, there are two walks of length 2 joining vertex a to vertex b. These are *adb* and *acb*. If we consider A^3, we can see that there are six walks of length 3 joining vertex a to itself. These are *abda*, *adba*, *acda*, *adca*, *abca*, and *acba*. Note that these are walks rather than paths, since an edge may be included more than once. ∎

COROLLARY 9.14   Let $G = (V,E)$ be a graph with adjacency matrix A and let $B = I + A + A^2 + \cdots + A^{n-1}$, where $n = |V|$. Then vertices i and j are connected if and only if $B_{ij} \neq 0$.

PROOF   Suppose i and j are connected. Then they are joined by a simple path, and such a path has a maximum length of $n - 1$. Hence at least one of $(A^m)_{ij}$ is nonzero, $1 \le m \le n - 1$. Since none of the $(A^m)_{ij}$ are negative, we can conclude that $B_{ij} \ne 0$. Similarly, if $B_{ij} \ne 0$, there is a walk joining vertex i to vertex j, and thus i and j are connected. The identity matrix, I, is included in the sum since each vertex is connected with itself by a length 0 path.   □

EXAMPLE 10   Consider the graph of Figure 9.34.

Figure 9.34   Graph with one isolated vertex.

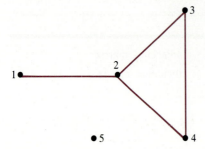

Its adjacency matrix is

$$A = \begin{bmatrix} 0 & 1 & 0 & 0 & 0 \\ 1 & 0 & 1 & 1 & 0 \\ 0 & 1 & 0 & 1 & 0 \\ 0 & 1 & 1 & 0 & 0 \\ 0 & 0 & 0 & 0 & 0 \end{bmatrix}.$$

Then

$$A^2 = \begin{bmatrix} 1 & 0 & 1 & 1 & 0 \\ 0 & 3 & 1 & 1 & 0 \\ 1 & 1 & 2 & 1 & 0 \\ 1 & 1 & 1 & 2 & 0 \\ 0 & 0 & 0 & 0 & 0 \end{bmatrix} \qquad A^3 = \begin{bmatrix} 0 & 3 & 1 & 1 & 0 \\ 3 & 2 & 4 & 4 & 0 \\ 1 & 4 & 2 & 3 & 0 \\ 1 & 4 & 3 & 2 & 0 \\ 0 & 0 & 0 & 0 & 0 \end{bmatrix}$$

$$A^4 = \begin{bmatrix} 3 & 2 & 4 & 4 & 0 \\ 2 & 11 & 6 & 6 & 0 \\ 4 & 6 & 7 & 6 & 0 \\ 4 & 6 & 6 & 7 & 0 \\ 0 & 0 & 0 & 0 & 0 \end{bmatrix} \qquad B = \begin{bmatrix} 5 & 5 & 6 & 6 & 0 \\ 5 & 17 & 11 & 11 & 0 \\ 6 & 11 & 12 & 10 & 0 \\ 6 & 11 & 10 & 12 & 0 \\ 0 & 0 & 0 & 0 & 1 \end{bmatrix}.$$

Note that the fifth row and the fifth column of each A^m consist entirely of 0s. Thus so do the fifth row and column of B, with the exception of B_{55}. This is consistent with the fact that vertex 5 is not connected to any other vertex. Note also that there are three walks of length 3 joining vertex 1 to vertex 2. These are 1232, 1242, and 1212. As in these two examples, the numbers that appear in A^m include many walks that have repeated edges. ∎

TERMINOLOGY

isomorphic	adjacency matrix
isomorphism	incidence matrix

SUMMARY

1. Two graphs are isomorphic if there is a one-to-one correspondence between their vertices that preserves the correspondence between the edges as well.

2. Suppose G_1 and G_2 are isomorphic graphs with isomorphism f. Then

   a) G_1 and G_2 have the same number of vertices.

   b) G_1 and G_2 have the same number of edges.

   c) if u is adjacent to v in G_1, $f(u)$ is adjacent to $f(v)$ in G_2.

   d) if $u_1 u_2 \cdots u_n$ is a cycle in G_1, then $f(u_1) f(u_2) \cdots f(u_n)$ is a cycle in G_2.

   e) if u has degree k in G_1, then $f(u)$ has degree k in G_2.

   f) G_1 and G_2 have the same number of components.

   g) if G_1 is Eulerian, so is G_2.

   h) if G_1 is Hamiltonian, so is G_2.

3. Two ways to represent graphs by matrices are the adjacency matrix and the incidence matrix. These representations are not unique but depend on how the vertices are numbered.

4. Two graphs are isomorphic iff, for some permutation of their vertices, their adjacency matrices are the same.

5. If A is the adjacency matrix of a graph, G, $(A^m)_{ij}$ is the number of length m walks that join vertex i to vertex j.

6. Let $B = I + A + A^2 + \cdots + A^{n-1}$, where $n = |V|$. Then vertices i and j are connected iff $B_{ij} \neq 0$.

EXERCISES 9.3

Decide whether the following pairs of graphs are isomorphic and justify your answer.

1.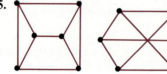

2.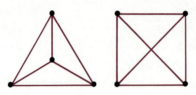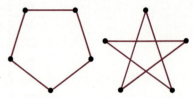

3.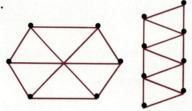

4.

5.

6.

7.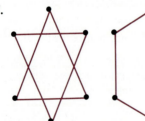

8.

Represent each of the following graphs by adjacency and by incidence matrices.

9. K_5
10. $K_{4,4}$

11. Exercise 3
12. Exercise 4

For each of the following adjacency matrices, draw the graph associated with A.

13. $\begin{bmatrix} 0 & 1 & 1 \\ 1 & 0 & 1 \\ 1 & 1 & 0 \end{bmatrix}$

14. $\begin{bmatrix} 0 & 1 & 0 & 1 \\ 1 & 0 & 1 & 0 \\ 0 & 1 & 0 & 1 \\ 1 & 0 & 1 & 0 \end{bmatrix}$

15.
$$\begin{bmatrix} 0 & 0 & 1 & 0 & 0 \\ 0 & 0 & 0 & 1 & 1 \\ 1 & 0 & 0 & 0 & 0 \\ 0 & 1 & 0 & 0 & 1 \\ 0 & 1 & 0 & 1 & 0 \end{bmatrix}$$

16.
$$\begin{bmatrix} 0 & 0 & 0 & 0 & 1 \\ 0 & 0 & 0 & 0 & 1 \\ 0 & 0 & 0 & 0 & 1 \\ 0 & 0 & 0 & 0 & 1 \\ 1 & 1 & 1 & 1 & 0 \end{bmatrix}$$

For each of the following, find A^2 and A^3 and list all walks of length 2 and length 3 connecting vertex 1 to vertex 3. In Exercises 21–24, number the vertices clockwise, starting at the upper left vertex.

17. Exercise 13

18. Exercise 14

19. Exercise 15

20. Exercise 16

21. Figure 9.28(a)

22. Figure 9.28(b)

23. Exercise 2

24. Exercise 4

25. Find A and A^2 for $K_{m,n}$.

26. Generalize your answer to Exercise 25 to A^j and justify it.

27. Find all nonisomorphic simple graphs with two vertices.

28. Find all nonisomorphic simple graphs with three vertices.

The edge adjacency matrix of a graph is defined as follows. The ith row and the ith column of the matrix are both associated with the ith edge of the graph. A 1 in the ith row and the jth column indicates that edges i and j have a vertex in common. A 1 on the main diagonal indicates that edge i is a loop.

29. Find the edge adjacency matrix for the graph in Figure 9.33.

30. Find the edge adjacency matrix for the graph in Figure 9.32.

31. Show that edge adjacency matrices are symmetric.

32. Generalize the concept of the edge adjacency matrix so that it can represent multigraphs and, using it, represent the graph associated with the Konigsberg bridge problem.

33. Treat the matrix you found in Exercise 29 as an adjacency matrix. Draw the graph associated with it and find its edge adjacency matrix.

Prove the following.

34. K_5 is isomorphic to a subgraph of K_6.

35. K_n is isomorphic to a subgraph of K_{n+1}.

36. Isomorphism is an equivalence relation.

37. Theorem 9.11(d).

38. Theorem 9.11(e).

39. Theorem 9.11(h).

9.4
PLANAR GRAPHS

The fourth question we asked about graphs was, "Which graphs can be drawn in the plane without intersecting edges and which cannot?" For instance, a classic puzzle is as follows:

> Three new houses are being built, and each needs to be connected to the electrical, water, and gas utilities. Is it possible to connect each house to each utility without crossing any of the lines?

Since each utility is to be connected to each house, a relationship between houses and utilities is defined that can be modeled as a graph. The question is whether this graph can be drawn in the plane without the edges intersecting anywhere except at the vertices. One unsuccessful attempt to solve this problem is presented in Figure 9.35.

Figure 9.35 An unsuccessful attempt to join three utilities to three houses without crossing the lines.

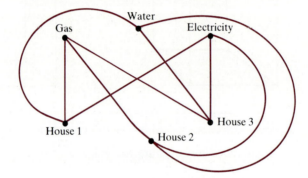

Another variation on the same kind of problem, but with a more geometric flavor, is the following:

> Place five points in the plane and join each point to the other four by curves in such a way that no two curves intersect (other than starting and ending at the five points).

The first problem is modeled by $K_{3,3}$ and the second by K_5. By the end of this section, we will show that neither of these graphs can be drawn in the plane without their edges intersecting. Developing the mathematical tools to prove this and to solve other similar problems is the subject of this section. It has applications to more than just puzzles, however; for instance, it can be applied to the layout of printed circuit boards.

Recall that a graph is defined as a set of vertices and a set of edges. In drawing graphs, we have represented vertices by points in the plane and edges by curves or line segments. There may, in fact, be several different but isomorphic representations of the same graph in the plane, depending on the placement of the edges and vertices. Thus another way to look at our question is to ask which graphs have planar representations in which the edges do not intersect.

Definition   A graph is **planar** if it has a representation in the plane with the edges intersecting only at the vertices on which they are incident.

EXAMPLE 1   Figure 9.36 repeats the two representations of the cube we examined in the last section. The second has no edges intersecting. Hence, the cube is a planar graph.

Figure 9.36   Two representations of a cube.

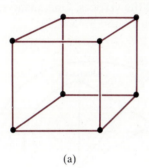 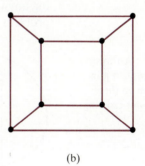

(a)   (b)   ▌

A few more concepts are needed.

Definition   Let G be a graph. Suppose G is represented in the plane by points for the vertices and line segments for the edges. A subset of the plane is **path connected** if any two points in it can be joined by a curve that is entirely contained in the set. The path-connected subsets of the plane that remain after the edges of G are deleted are called **regions**. A region is called an **infinite region** if it cannot be contained within a circle of any radius, no matter how large.

EXAMPLE 2   Consider Figure 9.36(b) again. If the edges of this graph are deleted from the plane, the resulting subset of the plane has six regions, one of which is an infinite region. Note that this graph has 8 vertices, 12 edges, and 6 regions. If we let v denote the number of vertices, e the number of edges, and r the number of regions,

$$v - e + r = 8 - 12 + 6 = 2.$$

As we shall see, this is true for all planar graphs.   ▌

EXAMPLE 3 Consider the graph of Figure 9.37(a). It has four regions—not five, as might first appear. This is because the intersection of the two diagonals is not a vertex. The graph can be redrawn as a planar graph, as in Figure 9.37(b), where it can be seen that it has only four regions. It has four vertices and six edges, so once again, $v - e + r = 4 - 6 + 4 = 2$.

Figure 9.37 Graph with four regions.

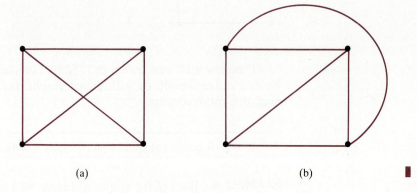

(a)   (b)   ▮

EXAMPLE 4 The graphs in Figure 9.38 have only an infinite region. In Figure 9.38(a), $v = 3$, $e = 2$, and $r = 1$. In Figure 9.38(b), $v = 9$, $e = 8$, and $r = 1$. In both cases, $v - e + r = 2$.

Figure 9.38 Two planar graphs that have only an infinite region.

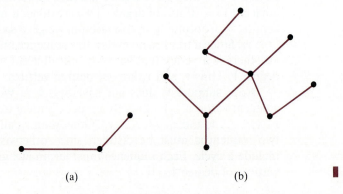

(a)   (b)   ▮

EXAMPLE 5 Consider the graph of Figure 9.39. It has two regions, one infinite, one finite. It has five vertices and five edges, so $v - e + r = 5 - 5 + 2 = 2$.

Figure 9.39　Graph with two regions, one of degree 4, one of degree 6.

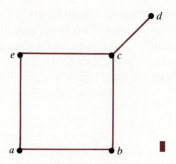

Theorem 9.17 was proven in 1752 by Leonhard Euler and is the basis for everything we will do with planar graphs. Before we can prove it, we need one more concept.

Definition　A connected graph with no cycles is called a **tree**.

EXAMPLE 6　Both of the graphs in Figure 9.38 are trees.　∎

Trees are the principal topic of Chapter 10 and will be extensively discussed there. For now, we need only the following two lemmas.

LEMMA 9.15　Every finite tree has at least two vertices of degree 1.

PROOF　Let T be a tree and let v_1 be any vertex in T. Suppose that v_1 has degree 1. We then have one of the desired vertices. There is a vertex, v_2, adjacent to v_1. If it is of degree 1, we are done. If not, we can select v_3 adjacent to v_2. Continuing in this fashion, we get a sequence of vertices, v_1, v_2, . . . , v_k. Since a tree has no cycles, this sequence must terminate for some k, i.e., there is a vertex that has only one adjacent vertex. Such a vertex has degree 1. Thus v_1 and v_k are our desired vertices.

Now, suppose v_1 does not have degree 1. We can select two distinct adjacent vertices, v_2 and v_2'. By a process similar to the one previously used, v_1, v_2, . . . , v_i and v_1, v_2', . . . , v_j'. Other than v_1, all of the elements of these two sequences must be distinct, since otherwise the sequences would include a cycle. Each sequence must terminate, giving us our two desired vertices of degree 1.　∎

LEMMA 9.16　Let $G = (V, E)$ be a tree with v vertices and e edges. Then $e = v - 1$.

PROOF　The proof is by induction on v. First, we suppose that $v = 1$. Then G has zero edges, so the lemma is proven in this case. Now suppose that any tree with $v - 1$ vertices has $v - 2$ edges. As shown in Lemma 9.15,

G has a degree 1 vertex. If we remove such a vertex and the edge incident on it, the resulting graph is still connected and has no cycles. Thus it is a tree. Since it has $v - 1$ vertices, by the induction hypothesis it has $v - 2$ edges. Thus G has $v - 1$ edges. □

THEOREM 9.17 Let G be a connected planar graph with $|V| = v$, $|E| = e$, and r regions. Then

$$v - e + r = 2.$$

PROOF The proof is by induction on the number of edges. Suppose first that $e = 0$, i.e., has no edges. Since G is connected, $v = 1$ and hence $r = 1$. Thus $v - e + r = 1 - 0 + 1 = 2$. Now suppose that $e > 0$. Assume that for any graph with v vertices, r regions, and $e' < e$ edges,

$$v - e' + r = 2.$$

We split the problem into two cases: G has no cycles and G has at least one cycle.
Case 1: If G has no cycles, it has no regions that are enclosed. Hence it has only an infinite region and thus $r = 1$. Since it is connected, by the previous lemma, $e = v - 1$, and thus $v - e + r = v - (v - 1) + 1 = 2$.
Case 2: If G has cycles, it has finite regions. Select a region that has a common edge with the infinite region and remove such an edge. Then the new graph has one less edge and one less region. We can apply the induction hypothesis to the new graph, getting

$$v - (e - 1) + (r - 1) = 2.$$

Thus $v - e + r = 2.$ □

Note that we have already verified this theorem for the graphs presented in Examples 2, 3, 4, and 5.

Euler's theorem is the principal tool we will use to show that $K_{3,3}$ and K_5 are not planar. Before we can show this, we need the following lemma and two corollaries to the theorem.

LEMMA 9.18 If $e > 2$, the boundary of any region in a simple graph contains at least three edges.

PROOF If a region is finite, its boundary is a cycle. Cycles of length 2 can occur only in multigraphs; cycles of length 1 are loops that cannot occur in simple graphs; cycles of length 0 are impossible. Thus, for finite regions, the degree has to be at least 3. If a region is infinite and is the only region in the graph, its boundary includes all of the edges. Since $e > 2$, the lemma is proven in this case. If there is more than one region, the boundary of the infinite region will surround all of the finite regions. Thus it contains a cycle and hence has a length of at least 3. □

COROLLARY 9.19   Let G be a connected, planar, simple graph with $e \geq 2$. Then $(3/2)r \leq e$.

PROOF   If $e = 2$, then $r = 1$ and we are done. If $e > 2$, by the previous lemma the boundary of each region includes at least three edges. Thus there are at least $3r$ edges if we sum over every region of the graph. But in computing this sum, each edge in the graph is counted twice. Thus the sum must be $2e$. Hence $2e \geq 3r$; i.e., $(3/2)r \leq e$.   ☐

COROLLARY 9.20   Let G be a connected, planar, simple graph with $e \geq 2$. Then $e \leq 3v - 6$.

PROOF   This follows by combining Euler's theorem with the previous inequality. The details are left for the Exercises.   ☐

EXAMPLE 7   Show that K_5 is not planar.

SOLUTION K_5 has 5 vertices and 10 edges. Thus $3v - 6 = 9$, which violates the previous inequality. Since K_5 is a simple, connected graph, we can conclude that it is not planar.   ∎

EXAMPLE 8   Show that $K_{3,3}$ is not planar.

SOLUTION $K_{3,3}$ has no three-cycles. Thus every region in a planar representation of $K_{3,3}$ has at least four edges in its boundary. Thus, if we apply the same argument used in Corollary 9.19, $2e \geq 4r$ for $K_{3,3}$. Combining this with Euler's theorem, we have

$$r = 2 - v + e$$
$$2e \geq 4(2 - v + e)$$
$$e \geq 4 - 2v + 2e$$
$$e \leq 2v - 4.$$

But $v = 6$ and $e = 9$, which violates this inequality. $K_{3,3}$ is simple and connected. Hence it is not planar.   ∎

Kuratowski's Theorem

The results obtained so far provide a way to show that certain graphs are *not* planar. However, we do not yet have a way to show that a particular graph *is* planar other than actually producing a planar representation of it. A theorem proven by Kasimir Kuratowski in 1930 provides such a means. First, we need one concept.

Definition Two graphs, G_1 and G_2, are **homeomorphic** if one can be obtained from the other by (1) one or more replacements of a degree 2 vertex and its incident edges with a single edge or by (2) one or more placements of a vertex on an edge in such a way that the edge becomes two edges that share the new vertex as a degree 2 vertex.

In regard to planarity, a degree 2 vertex and its incident edges are the same as a single edge with no vertex. That is, such vertices can be introduced into a graph by placing a vertex on any edge. Similarly they can be removed from any graph by replacing them and their edges by a single edge. Neither of these operations affects the planarity of a graph. Graphs that are the same except for such replacements are homeomorphic. Note that this is quite different from being isomorphic. Isomorphic graphs must have the same number of vertices, whereas homeomorphic graphs need not.

EXAMPLE 9 The graphs in Figure 9.40 are all homeomorphic. Figure 9.40(b) can be obtained from Figure 9.40(a) by replacing degree 2 vertices and their edges by single edges. Figure 9.40(c) can be obtained from either Figure 9.40(a) or Figure 9.40(b) by adding degree 2 vertices.

Figure 9.40 Homeomorphic graphs.

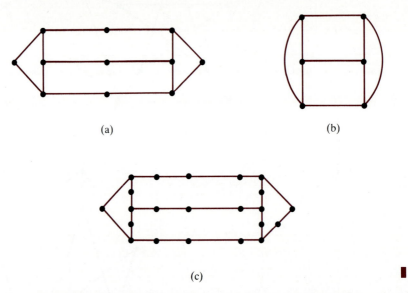

(a)

(b)

(c)

We will prove only one direction of Kurotowski's theorem—that if a graph has a subgraph homeomorphic to K_5 or $K_{3,3}$, it is not planar. For a proof of the converse, see *Introduction to Combinatorial Mathematics* by C.L. Liu (see References).

THEOREM 9.21   A graph is nonplanar iff it contains a subgraph homeomorphic to K_5 or $K_{3,3}$.

PROOF   Two homeomorphic graphs are either both planar or both nonplanar. If a subgraph is homeomorphic to K_5 or $K_{3,3}$, it is not planar, as shown in the previous two examples. But if the subgraph is not planar, neither is the original graph (which has more edges and vertices).   □

EXAMPLE 10   Show that K_n is not planar for $n > 5$.

SOLUTION K_5 is the subgraph obtained from K_n by deleting all except five vertices. Hence K_n is not planar.   ∎

EXAMPLE 11   Show that Petersen's graph (Figure 9.41a) is not planar.

Figure 9.41   Petersen's graph has a subgraph homeomorphic to $K_{3,3}$.

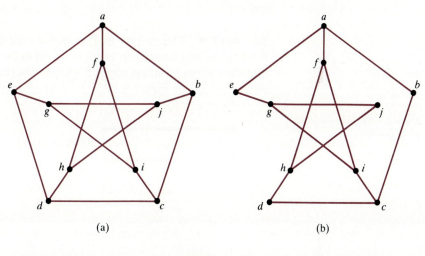

(a)   (b)

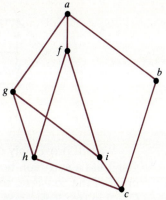

(c)

SOLUTION   Petersen's graph is shown in Figure 9.41(a). Figure 9.41(b) is a subgraph of Petersen's graph formed by deleting edges *de* and *bj*. Figure 9.41(c) is a third graph homeomorphic to Figure 9.41(b). This latter graph is $K_{3,3}$, taking $\{a, i, h\}$ as one set of vertices and $\{f, c, g\}$ as the other. Thus Petersen's graph has a subgraph homeomorphic to $K_{3,3}$. Hence it is not planar. ▋

TERMINOLOGY

planar graph   infinite region

path-connected subset of the plane   tree

region   homeomorphic graphs

SUMMARY

1. Let G be a connected planar graph with $|V| = v$, $|E| = e$, and r regions. Then $v - e + r = 2$.

2. Let G be a connected, planar, simple graph with $e \geq 2$. Then $(3/2)r \leq e \leq 3v - 6$.

3. A graph is nonplanar iff it contains a subgraph homeomorphic to K_5 or $K_{3,3}$.

EXERCISES 9.4

Draw planar representations of the following.

1. K_3 2. K_4

3. $K_{2,4}$ 4. $K_{2,n}$

Verify Euler's formula for each of the following planar graphs.

5. 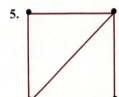   6.

7.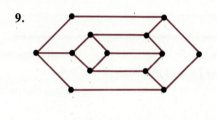

8.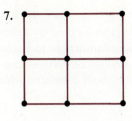

9.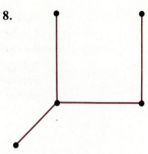

10.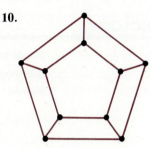

Is it possible to find planar, connected graphs with the following properties? If so, draw such a graph. If not, explain why not.

11. Four regions, each with a boundary consisting of three edges.

12. Six regions, each with a boundary consisting of four edges.

13. Ten regions, each with a boundary consisting of three edges.

14. Five edges, seven vertices.

15. One region, eight edges.

16. More regions than edges.

Find graphs homeomorphic to the following that have the smallest possible number of vertices.

17.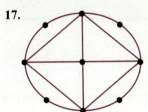

18.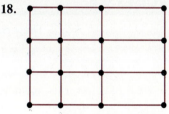

19.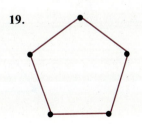

20.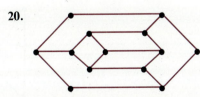

Find subgraphs homeomorphic to K_5 or $K_{3,3}$ in the following.

21. K_6

22. $K_{4,4}$

23.

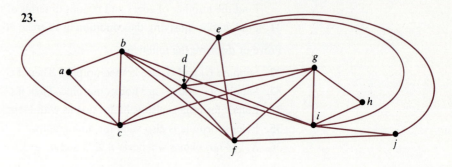

24.

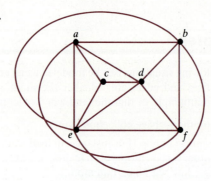

Redraw the following graphs to make them planar or show that they are not planar.

25.

26.

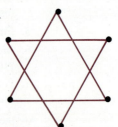

27.

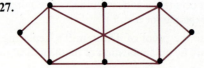

28.

29. $K_{3,3}$ does not violate the inequality $e \leq 3v - 6$ even though it is not planar. For what values of n does $K_{n,n}$ not violate this inequality?

30. Suppose every region in a certain planar graph is bounded by a three-cycle. Find the number of edges and regions in terms of the number of vertices.

31. Find graphs satisfying the condition of Exercise 30 when $|V| = 3, 4,$ and 5.

Prove or disprove the following.

32. Suppose G_1 and G_2 are homeomorphic. Then if G_1 is Hamiltonian, so is G_2.

33. Suppose G_1 and G_2 are homeomorphic. Then if G_1 is Eulerian, so is G_2.

34. Euler's formula is also valid for graphs with loops.

35. Euler's formula is also valid for multigraphs.

36. $K_{n,m}$ is not planar whenever $n \geq 3$ and $m \geq 3$.

REVIEW EXERCISES—Chapter 9

The following situations can be modeled using graphs. For each one, indicate what the vertices and edges are; whether the situation would be better modeled by a graph, multigraph, or digraph; and whether loops are present.

1. Planning garbage collection routes.

2. Chemical bonds. (For instance, in hydrocarbons, a carbon atom may be linked to other carbon atoms or to hydrogen atoms. A hydrogen atom can form only a single link with a carbon atom.)

3. A group of students are entered into a data file. There is a connection between individuals if they are both enrolled in the same course.

4. Two telephone exchanges are related if they are in each other's local calling district.

5. Find a graph G such that $|V| = n$ and $|E| = (n^2 - n)/2$ for $n = 4, 5,$ and 6.

6. Modify the formula of Exercise 34 of Section 9.1 for graphs with loops and prove it.

7. Using Exercise 37 of Section 9.1, show that a component of a graph is a maximal connected subgraph.

Are the following graphs Hamiltonian? Eulerian?

8. A graph with more than two degree 2 vertices adjacent to the same vertex.

9. Two concentric n-gons with their corresponding vertices joined. (Figure 9.21b is an example for $n = 5$.)

10. Rewrite the proof of Theorem 9.6, replacing the phrase "continuing in this fashion" by an application of mathematical induction.

11. Prove Corollary 9.7.

12. Show that if a graph has exactly two vertices of odd degree, they must be in the same component.

13. n vertices are equally spaced around the circumference of a circle. A vertex is selected at random and joined by an edge to the kth vertex from it, moving counterclockwise. This is then joined to the kth vertex from it. This process continues until the original vertex is returned to. Find conditions on k and n that will cause all the vertices to be connected.

Decide whether each of the following pairs of graphs are isomorphic and justify your answer.

14.

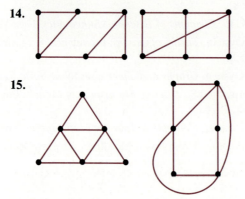

15.

16. Write a definition of isomorphism suitable for use with multigraphs.

17. Generalize the concept of adjacency matrix to multigraphs. Using your generalization, find adjacency and incidence matrices for the graph associated with the Konigsberg bridge problem.

18. Prove: two graphs are isomorphic iff their *incidence* matrices are identical for some permutation of the vertices and of the edges.

19. Prove: let A be an $n \times n$ Boolean matrix. Then $(I + A)^{n-1}$ has no zeroes if the graph associated with A is connected.

20. Generalize Euler's formula to graphs with n components.

21. Draw K_5 with any edge removed. Show that the resulting graph is isomorphic to a planar graph.

22. Repeat Exercise 21 for $K_{3,3}$.

23. A mosaic pattern is built using only hexagons. For such a pattern, show that $r \le e/3 \le (v-2)/2$.

REFERENCES

Garey, Michael R., and David S. Johnson, *Computers and Intractability: A Guide to the Theory of NP-Completeness.* San Francisco: W.H. Freeman, 1985.

Grimaldi, Ralph, *Discrete and Combinatorial Mathematics.* Reading, Mass.: Addison-Wesley, 1985.

Horowitz, Ellis, and Sartaj Sahni, *Fundamentals of Data Structures.* Rockville, Md.: Computer Science Press, 1986.

Knuth, Donald, *The Art of Computer Programming,* Vol. 1., *Fundamental Algorithms.* Reading, Mass.: Addison-Wesley, 1973.

Lawler, E.L., J.K. Lenstra, A.H. Rinnooykan, D.B. Shmoys, *The Traveling Salesman Problem.* New York: Wiley, 1985.

Liu, C.L., *Introduction to Combinatorial Mathematics.* New York: McGraw-Hill, 1968.

Liu, C.L., *Elements of Discrete Mathematics.* New York: McGraw-Hill, 1977.

Ore, Oystein, *Theory of Graphs.* Providence, R.I.: American Mathematical Society, 1962.

Ore, Oystein, *Graphs and Their Uses.* New York: Random House, 1963.

Roberts, Fred S., *Discrete Mathematical Models.* Englewood Cliffs, N.J.: Prentice-Hall, 1976.

Sahni, Sartaj, *Concepts in Discrete Mathematics.* Fridley, Minn.: Camelot, 1981.

Stanat, Donald, and David McAllister, *Discrete Mathematics in Computer Science.* Englewood Cliffs, N.J.: Prentice-Hall, 1977.

Trees

10

Any graph that is connected and has no cycles is a **tree.** For instance, problems such as the following are typical sources of trees.

A building is being wired. Figure 10.1 is an outline of one floor of the building, with the location of each electrical outlet marked by a large dot. Possible connections are marked by line segments. Find a subset of these connections that minimizes the amount of wire needed while still connecting all of the outlets.

Figure 10.1 Floor plan of a building with work stations and possible connections marked.

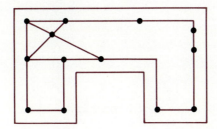

A minimizing subset does not contain any cycles; if two vertices were on a cycle, there would be more than one path joining them. One of the two paths is unnecessary, and thus at least one edge can be deleted. Thus the solution to problems of this sort involves finding a subgraph that is a tree. We will see how to solve this problem and others like it in Section 10.2.

Another source of trees is the hierarchical organization of data. For instance, consider the biological classification scheme in Figure 10.2.

Figure 10.2 Example of a tree arising from a hierarchical relationship.

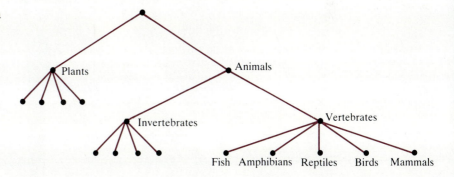

A graph representing such a hierarchy is connected and has no cycles. Hence it is a tree. However, Figure 10.2 is better modeled by a digraph than a graph, since there is a natural order associated with each relationship. In Sections 10.1 and 10.2, we will study trees as previously defined—connected, acyclic graphs. In Sections 10.3 and 10.4, we will study the **rooted tree**—a type of tree that is a digraph.

10.1
INTRODUCTION TO TREES

The Basic Concepts

We have already defined a tree. There are two further concepts that we will need.

Definition   Given the graph $G = (V,E)$, a **spanning tree** is a subgraph of G that is a tree and that contains all of the vertices of V. A **forest** is a collection of trees.

EXAMPLE 1   Figure 10.3 presents a spanning tree for the wiring problem of Figure 10.1. Note that if any edges other than the ones incident on the degree (1) vertices are removed, the resulting graph is no longer a tree, but a forest. The tree diagrammed in Figure 10.3 does not necessarily minimize the amount of wire needed. We will see how to solve the minimization problem in Section 10.2.

Figure 10.3   Spanning tree for the graph in Figure 10.1.

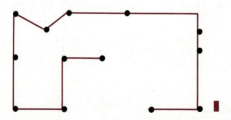

EXAMPLE 2   One individual hears a rumor and tells it to others, who in turn tell still others. At any moment in time, the path along which the rumor has traveled can be modeled by a tree. The tree spans the graph representing the relationships that exist among those individuals who have heard the rumor (Figure 10.4). Trees can also be used to model the spread of infectious diseases.

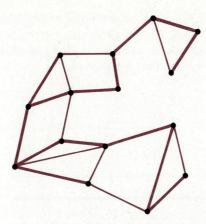

Figure 10.4 Spread of a rumor. Circles denote people; edges indicate people who are in contact; heavy edges show the actual path of communication.

EXAMPLE 3 One of the origins of the mathematical study of trees was in the study of molecular structures in the nineteenth century. Figure 10.5, for instance, represents methane, CH_4, and ethane, C_2H_6, two hydrocarbons. Regarding each atom as a vertex and each bond between the atoms as an edge, these molecules can be represented by trees. These diagrams are drawn as chemists usually draw them, using the symbols for the elements as vertices rather than dots or circles, as is usually done in graph theory. Chemists speak of the *valence* of an element, meaning the number of bonds it can form in a compound. Carbon has valence 4 and hydrogen has valence 1. Thus they are represented by degree 4 and degree 1 vertices, respectively.

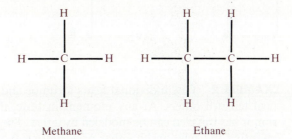

Methane   Ethane

Figure 10.5 Two hydrocarbons whose structure can be modeled by trees.

EXAMPLE 4 Figure 10.6 shows some further examples of trees.

Figure 10.6 Three examples of trees.

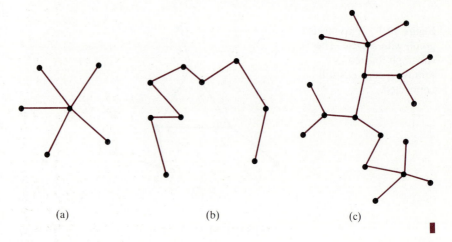

(a)   (b)   (c)

Properties of Trees

Trees have all of the properties common to simple graphs, but they have some special properties as well. These properties are presented in the following results; Lemma 10.1 and one part of Theorem 10.3 were proven in Section 9.4.

LEMMA 10.1 Every tree has at least two vertices of degree 1. ☐

The following lemma tells us that, given a simple graph that includes one or more cycles, we can delete an edge and still have a connected graph. We will use the lemma in Theorem 10.3 and in the construction of spanning trees.

LEMMA 10.2 Let $G = (V,E)$ be a connected simple graph and suppose that G has one or more cycles. Let e denote an edge included in some cycle. The subgraph $G' = (V,E - \{e\})$ is connected.

PROOF   Since G has cycles, it must have at least three vertices. Let u and v denote any two distinct vertices of V; we will show that there is a path in G' joining u and v. Let $e_1 e_2 \ldots e_n$ be any simple path in G joining u and v. If e is not included in $e_1 \ldots e_n$, we have a path joining u and v in G'. Thus suppose e is included in $e_1 \ldots e_n$. Since e is included in a cycle, the vertices it is incident on are connected by a path that can be denoted $e_j \ldots e_k$ and that does not include e. Replacing e by the path $e_j \ldots e_k$ will yield a walk that connects u and v. By Theorem 9.3, there is a simple path joining u and v, and thus they are connected. ☐

EXAMPLE 5   Figure 10.7(a) is a graph with one cycle that includes the edge labeled e. Figure 10.7(b) is a subgraph with the same vertices and e removed.

Figure 10.7   A connected graph with a cycle. The subgraph formed by removing edge e is still connected.

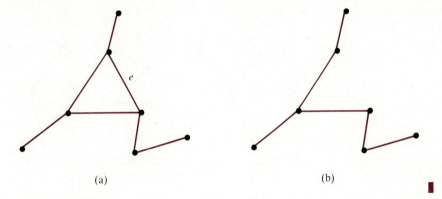

(a)   (b)

THEOREM 10.3   Let $T = (V,E)$ be a graph with $|V| = n$. The following are equivalent.

a) T is a tree.

b) T is connected and has $n - 1$ edges.

c) T has $n - 1$ edges and no cycles.

PROOF   The proof proceeds by showing that $a \Rightarrow b \Rightarrow c \Rightarrow a$.

That $a \Rightarrow b$ was shown in Section 9.4.

We now prove that $b \Rightarrow c$. Suppose T is connected and has $n - 1$ edges. If T has cycles, by Lemma 10.2, one or more edges can be removed without removing any vertices or disconnecting the resulting subgraph. Suppose that after the removal of k such edges, no cycles remain. Then the result is a tree, T', with n vertices and $n - 1 - k$ edges. Since we have shown that part (a) of this theorem implies part (b), we can conclude that T' has $n - 1$ edges; hence k must be zero. Thus T has no cycles.

Lastly, we show that $c \Rightarrow a$. Suppose T has $n - 1$ edges and no cycles. Suppose also that T is not connected and has k components with n_1, n_2, $\cdots$, n_k vertices, respectively. Then each is connected and has no cycles, and hence is a tree. Thus the components have $n_1 - 1$, $n_2 - 1$, $\cdots$, $n_k - 1$ edges, respectively. Adding these together, T has $n_1 + n_2 + \cdots n_k - k = n - k$ edges. But our hypothesis is that T has $n - 1$ edges. Thus $k = 1$ and T is connected. Hence T is a tree. ◻

Theorem 10.3 is concerned with three properties that graphs may have —being connected, having $n - 1$ edges, and being acyclic. The theorem tells us that any graph having two of these properties also has the third and is thus a tree.

EXAMPLE 6   The trees in Figure 10.6 have, respectively, (a) 6 vertices, 5 edges, (b) 9 vertices, 8 edges, and (c) 18 vertices, 17 edges. ∎

EXAMPLE 7   There are 13 outlets in the floor plan of Figure 10.1. Thus any spanning tree must have exactly 12 connecting links. ∎

The following theorem provides another property characterizing those graphs that are trees.

THEOREM 10.4   Let $G = (V, E)$ be a graph. Then G is a tree iff there exists a unique path between every pair of vertices in G.

PROOF   First, we suppose that G is a tree. Let u and v be any two vertices of G. Since G is connected, there exists a path joining u to v. If there is more than one path joining u to v, a subset of the vertices on the two paths will form a cycle. Thus there is a unique path between u and v.

Conversely, suppose that there is a unique path joining every pair of vertices. Then G is connected. If G has a cycle, any two distinct vertices on the cycle are joined by more than one path. Thus G has no cycles and is a tree. ☐

EXAMPLE 8   Figure 10.8 shows the unique path between one pair of vertices in each of the trees in Figure 10.6.

Figure 10.8 Paths joining pairs of vertices in a tree are unique.

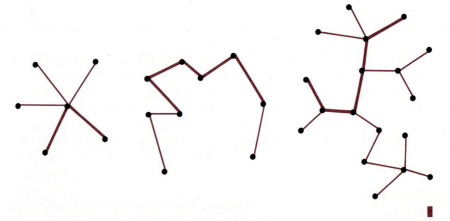

∎

Given a set, V, of size n, there are $2^{C(n,2)}$ simple graphs whose vertices are elements of V (Exercise 38, Section 9.1). However, as we can see from the previous two theorems, not many of those graphs will be trees. It is often possible to list all possible trees with certain characteristics, as is done in the next example.

EXAMPLE 9   Find all nonisomorphic trees with six vertices.

SOLUTION   The sum of the degrees of all of the vertices must be 10, since there are five edges. At least two vertices must have degree 1. Thus the sum of the degrees of the other four vertices must be 8, with each being at least 1. That is, $d_1 + d_2 + d_3 + d_4 = 8$ and each $d_i \geq 1$. There is no need to distinguish solutions such as (1, 1, 1, 5) and (5, 1, 1, 1), as we shall see. All possible solutions of this equation are:

a) (5, 1, 1, 1)

b) (4, 2, 1, 1)

c) (3, 3, 1, 1)

d) (3, 2, 2, 1)

e) (2, 2, 2, 2).

Trees with these degree sequences are given in Figure 10.9.

Figure 10.9   All nonisomorphic trees with six vertices.

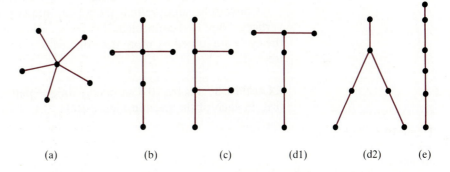

(a)   (b)   (c)   (d1)   (d2)   (e)

These are in fact the only nonisomorphic trees with six vertices. For instance, consider solution (a). Any six-vertex tree with a degree 5 vertex, v, must have five other vertices adjacent to v. This accounts for all five edges; hence all vertices except v must have degree 1. If one had two such trees, a function that maps the degree 5 vertex of one tree to the degree 5 vertex of the other tree and matches the degree 1 vertices of each tree in any way will be an isomorphism; it is one-to-one, onto, and preserves all relationships. Thus Figure 10.9(a) is the only nonisomorphic six-vertex tree with a degree 5 vertex.

Similarly, consider solution (d). It will yield a tree with three vertices of degree 1, two vertices of degree 2, and one vertex of degree 3. If two of the vertices of degree 1 are adjacent to the degree 3 vertex, there is also a vertex of degree 2 adjacent to the degree 3 vertex, and the only possible structure is Figure 10.9(d1). If one of the vertices of degree 1 is adjacent to a degree 3 vertex and the other vertices of degree 1 are adjacent to the vertices of degree 2, we get Figure 10.9(d2).

Similar arguments can be applied to solutions (b), (c), and (e), but they are left for the Exercises.   ∎

We conclude this section by showing that any connected graph has a spanning tree. We will use this result in the next section as we study algorithms to find spanning trees.

THEOREM 10.5 A graph is connected iff it has a subgraph that is a spanning tree.

PROOF If a graph has a spanning tree, it is clearly connected. As for the converse, let $G = (V,E)$ be a connected graph. If G has no cycles, it is a tree and we are done. If it has cycles, we can remove edges without disconnecting the graph or deleting vertices. This process can be continued until no cycles remain. The subgraph that remains is a spanning tree. ☐

EXAMPLE 10 Figures 10.3 and 10.4 show spanning trees in graphs. Figure 10.10 is a graph that is not connected and thus has no spanning tree.

Figure 10.10 A graph that has no spanning tree.

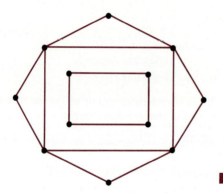

TERMINOLOGY

tree   spanning tree
rooted tree   forest

SUMMARY

1. Trees are acyclic, connected graphs.
2. The following are true of trees:
    T is a tree iff it is connected and has $n - 1$ edges iff it has $n - 1$ edges and no cycles.

T is a tree iff there is a unique path joining every pair of vertices.

Every tree has at least two degree 1 vertices.

3. A graph is connected iff it has a spanning tree.

EXERCISES 10.1

Sketch diagrams representing the following molecules.

1. C_3H_8
2. C_4H_{10}

Find all nonisomorphic trees with the following characteristics ($n = |V|$).

3. $n = 3$
4. $n = 4$
5. $n = 5$
6. $n = 7$ and one vertex has degree 5.
7. $n = 7$ and one vertex has degree 4.

Show that the following subdiagrams of Figure 10.9 are the only nonisomorphic trees with vertices of the specified degrees.

8. Figure 10.9(b)
9. Figure 10.9(c)
10. Figure 10.9(e)

Prove or disprove the following. n denotes $|V|$.

11. Every path in a tree is simple.
12. Adding an edge to a tree without adding vertices creates a cycle.
13. A connected graph with $n + k - 1$ edges has k cycles.
14. A graph with fewer than $n - 1$ edges is a forest.
15. A graph with exactly $n - 1$ edges and at least one cycle has more than one component.
16. A Hamilton path in a graph is a spanning tree.
17. Every tree is planar.
18. If a tree has a vertex of degree k, the longest possible path is of length $n - k + 1$.
19. How many edges are there in a forest with k components?
20. How many edges need to be added to a forest with k components to make it a tree?

10.2
SPANNING TREES

The focus of this section is three algorithms for finding spanning trees. The first two find spanning trees in unweighted graphs; the last finds a minimal spanning tree in a weighted graph.

An important application of spanning trees is graph searching. For instance, suppose you need a particular component for your stereo. The store where you bought the stereo does not have it in stock, so its owners contact their regional headquarters. Each regional headquarters contacts the other stores it is responsible for, as well as other regional headquarters, which in turn contact their stores (Figure 10.11).

Figure 10.11   Large circles denote regional headquarters; smaller circles denote individual stores.

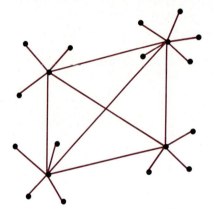

This is a typical example of a graph searching problem. The collection of stores can be modeled by a graph with the edges representing communication links. Each vertex has a unique name (typically called a **key**) but also has other information associated with it (in this case, its inventory). The problem is to use the keys to move from vertex to vertex in the graph, stopping at each vertex to search the inventory for your part. The sequence in which the stores are visited provides a unique path to each vertex. Thus the sequence of visitation generates a spanning tree, and algorithms for finding spanning trees are typically used to solve the graph searching problem.

Algorithms for Finding Spanning Trees

We will consider two algorithms for finding spanning trees—depth-first search and breadth-first search. Although the word *search* is often used in the names of these algorithms, keep in mind that they are actually algo-rithms for generating spanning trees and that graph searching is only one application of spanning trees. Each includes the phrase "visit v." This is where the actual searching, if searching is to be done, takes place. Since what happens at each vertex depends on the particular application, no details of that visit will be spelled out in the algorithms. Two typical mean-ings of "visit" other than searching are displaying the vertex just visited and displaying the edge used to reach the current vertex.

The strategy of the two algorithms is somewhat different. Each uses an initial vertex, v_0. Depth-first search starts from v_0 and follows one path as far as possible. It then backs up from the end of that path until it can follow another path, and again follows it as far as possible. It continues in this fashion until all vertices have been visited. Breadth-first search starts from v_0 and visits every vertex adjacent to v_0. It then selects one of these and visits every vertex adjacent to it. It continues to visit all vertices adjacent to ones already visited in the order of original visitation until all vertices have been visited.

EXAMPLE 1 Find depth-first and breadth-first spanning trees for the graph of Figure 10.12.

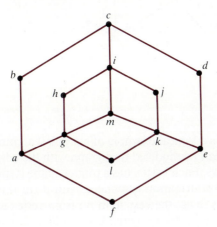

Figure 10.12

SOLUTION One depth-first tree starts at a and goes as far as possible along one path, such as the one to f (Figure 10.13a). It then backs up to e, and goes as far as possible, namely, to l (Figure 10.13b). It then backs up to g and goes to m, completing the depth-first spanning tree (Figure 10.13c).

Figure 10.13 Stages in the execution of depth-first search.

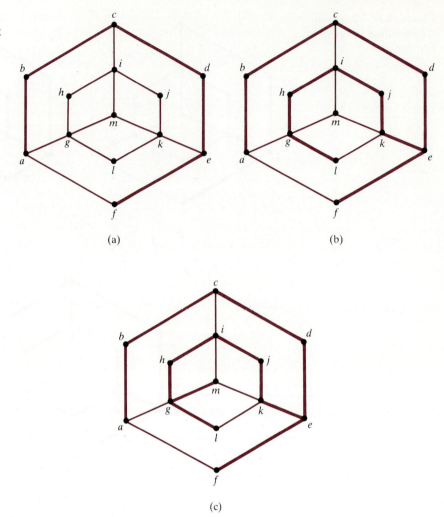

(a)

(b)

(c)

The breadth-first approach first visits all vertices adjacent to *a* (Figure 10.14a). It then visits all vertices adjacent to them (Figure 10.14b). At each stage, breadth-first search visits all vertices adjacent to those visited in the previous round of visitations. We will adopt the convention that if there are several vertices that can be used as a "base" for further visits at a particular stage, we will use them in lexicographic order. For instance, in Figure 10.14(b), any of *c, m, l,* or *e* could be used for visiting adjacent vertices. By using *c* before *m,* for example, *i* is visited along the path *abci* rather than *agmi.* Figure 10.14(c) is the resulting spanning tree. ▮

Figure 10.14 Stages in the execution of breadth-first search.

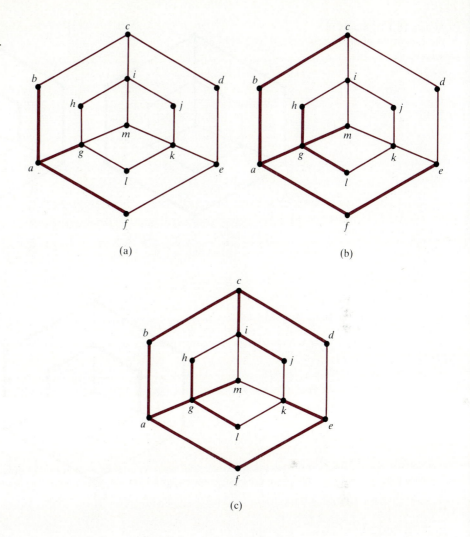

(a)

(b)

(c)

The formal versions of the depth-first and breadth-first spanning tree algorithms are as follows. Note that depth-first search is recursive.

ALGORITHM 35   Finding a Depth-First Spanning Tree

Procedure Depth_first(G, v; var visited);

{G is the graph for which a spanning tree is to be found; v is the initial vertex; visited is an n-element boolean array where $n = |V|$. The algorithm assumes that visited has been initialized to zero before the procedure is invoked.}

```
    begin
1.    visit v;
2.    visited[v] := 1;
3.    for each vertex w adjacent to v do
4.        if visited[w] = 0
5.        then Depth_first(G,w,visited)
    end;
```

ALGORITHM 36 Finding a Breadth-First Spanning Tree

Procedure Breadth_first(G,v; var visited);

{G is the graph for which a spanning tree is to be found; v is the initial vertex; visited is an n-element boolean array where n = |V|. Queue is an ordered set to which the names of vertices will be added or deleted. The algorithm assumes that visited has been initialized to zero and queue to the empty set before the procedure is invoked. u and w denote vertices.}

```
    begin
1.    visit v;
2.    visited[v] := 1;
3.    add v to queue;
4.    while queue is not empty do begin
5.        remove the first entry from queue and call it u;
6.        for all vertices w adjacent to u do
7.            if visited[w] = 0
                then begin
8.                visit w;
9.                visited[w] := 1;
10.               add w to queue
                end {for}
            end {while}
    end;
```

Note that queue is used in a "FIFO" manner—first in, first out. That is, when a vertex is added to queue, it is placed at the *end* of the list; when a vertex is removed from queue, it is taken from the *beginning* of the list.

Note also that queue and visited could be initialized *within* breadth-first search rather than *before* it is involved.

EXAMPLE 2   Find *depth-first* and *breadth-first* spanning trees for the graph of Figure 10.15, using Algorithms 35 and 36 and using *a* as the initial vertex.

Figure 10.15

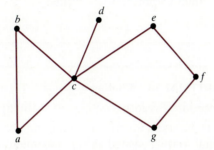

SOLUTION   First, we find the depth-first spanning tree. In lines 1 and 2, the algorithm visits *a* and records it as visited. Thus visited becomes

$$[1, 0, 0, 0, 0, 0, 0].$$

Line 3 begins a loop that will examine, in succession, *b* and *c*. Suppose it takes *b* first. Since *b* has not been visited, *depth-first* is called again with *b* as the initial vertex. In lines 1 and 2, *b* is visited. Thus visited becomes

$$[1, 1, 0, 0, 0, 0, 0].$$

Line 3 starts a loop that examines *a* and *c*. *a* has been visited, so the test in line 4 fails on the first pass; *c* has not been visited, however, so the test in line 4 succeeds for *c* and *depth-first* is called again with *c* as the initial vertex. Visited becomes

$$[1, 1, 1, 0, 0, 0, 0].$$

Thus, at this point, the algorithm has followed the path *abc*. *a, d, e,* and *g* are all adjacent to *c*. Following our convention of using lexicographic order, *a* is examined first. It has been visited. Then *d* is examined. Depth-first is called again with *d* as the initial vertex. Visited becomes

$$[1, 1, 1, 1, 0, 0, 0].$$

This time the only vertex adjacent to *d* is *c*, which has been visited, so depth-first search terminates and the algorithm continues its loop through the vertices adjacent to *c*. *e* has not been visited. Continuing in the same fashion, *e, f,* and *g* are visited in succession. Thus visited finally becomes

$$[1, 1, 1, 1, 1, 1, 1].$$

The depth-first spanning tree produced can be seen in Figure 10.16(a).

Figure 10.16   Depth-first and breadth-first spanning trees.

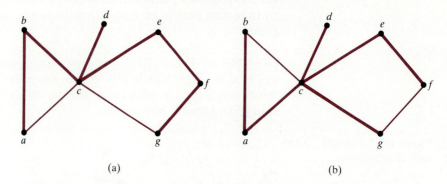

(a)   (b)

We now examine breadth-first search. In lines 1 and 2, *a* is visited, and in line 3, *a* is added to queue. In line 5, *a* is then removed from queue and each vertex adjacent to *a*, namely, *b* and *c*, is successively visited and added to queue. Thus at the first completion of the for loop

> *a, b,* and *c* have been visited
>
> queue = *b, c.*

b is now removed from queue, and since the vertices adjacent to *b*, namely, *a* and *c*, have been visited, nothing new is added to queue. Thus, at the second completion of the for loop,

> *a, b,* and *c* have been visited
>
> queue = *c.*

c is now removed from queue, and *d, e,* and *g* are visited and added to queue. Thus, after the third completion of the for loop,

> *a, b, c, d, e,* and *g* have been visited
>
> queue = *d, e, g.*

d is now removed from the list, and no changes occur. *e* is removed and *f* is visited and added to the end of queue. At this point,

> *a, b, c, d, e, g,* and *f* have been visited in that order
>
> queue = *g, f.*

g and *f* are now removed from queue, one at a time. Since all vertices adjacent to them have been visited, no changes occur and the algorithm halts. The spanning tree it generates is seen in Figure 10.16(b). ∎

The worst-case complexity of each of these algorithms is $O(|E| + |V|)$. While we won't go through the computation in detail, the worst-case occurs when the graph being searched is K_n. (See the Exercises.)

The verification involves showing that each algorithm visits every vertex and produces a unique path from v_0 to each vertex. The proofs proceed by induction path from v_0 to any given vertex. The details are left for the Exercises.

EXAMPLE 3   Figure 10.17 represents the graph-searching problem presented at the beginning of this section, but with the vertices labeled. If the initial vertex is a, in what order will the vertices be visited by depth-first search? By breadth-first search?

Figure 10.17 Graph-searching problem.

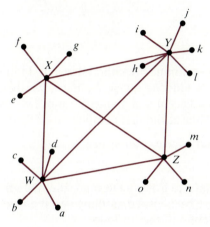

SOLUTION   We will solve this problem only for depth-first search; breadth-first search will be left for the Exercises. We will continue to follow the convention that if more than one vertex is acceptable at any stage in the algorithm, we will select one using lexicographic order. Thus, starting at a, the first vertex visited must be W. b, c, and d will be visited next, in that order. Continuing in this fashion, the order of visitation will be X, e, f, g, Y, h, i, j, k, l, Z, m, n, o. ∎

Minimal Spanning Trees

Suppose G is a weighted graph. For instance, G might be the graph that arose in the wiring problem of the last section. Figure 10.18 presents that

Figure 10.18 Wiring problem. Numbers indicate meters of wire needed for each connection.

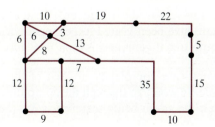

graph, with weights added to the edges to indicate the lengths of each section of wire. The problem is to find the spanning tree that minimizes the total length of wire used.

An algorithm to solve this problem was developed by Joseph Kruskal in 1956. It is an example of a greedy algorithm that works. The idea behind it is very simple. We start by selecting the edge with the smallest weight. We then keep on selecting the edge with the smallest weight as long as we do not form a cycle. Once we have selected the $(n-1)$th edge, we have a tree and are done.

ALGORITHM 37   Finding a Minimal Spanning Tree

Procedure Min_span(n, V, E, W; var T);

{$G = V,E$) is a graph; W is the set of weights associated with each edge in E; $n = |V|$; T will become the minimal spanning tree.}

begin

1.   initialize T to the empty graph;

2.   while there are less than $n - 1$ edges in T do begin

3.   find an edge, e, in E that has minimal weight and that has not previously been selected;

4.   if e does not create a cycle in T

5.   then add e to T

end;

EXAMPLE 4   Use Kruskal's algorithm to find a minimal set of connections for Figure 10.18.

SOLUTION   After six edges have been selected, T is as in Figure 10.19(a). The final spanning tree is presented in Figure 10.19(b).

Figure 10.19 Stages in building a minimal spanning tree using Kruskal's algorithm.

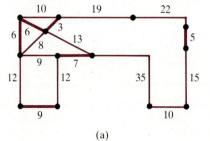

(a)

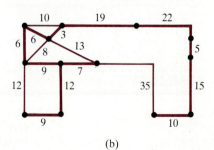

(b)

Verification of Kruskal's Algorithm

First, we must show that the algorithm produces a spanning tree. Whenever this algorithm is carried out on graph G, at its completion T will have $n - 1$ edges and no cycles. T must include all n vertices of G, since, if it included fewer than n vertices, it would have to have cycles. Thus, with $n - 1$ edges and no cycles, T must be a tree. Hence it is a spanning tree.

To show that T is a minimal spanning tree, let S denote that spanning tree of G whose edges have a minimum total weight of $w(S)$. Let the edges of T be $\{e_1, e_2, \ldots, e_{n-1}\}$ in the order in which they were selected, and let e_k denote the first edge in the list $\{e_1, e_2, \ldots, e_{n-1}\}$ that is not in S. If e_k joins vertices u and v, then since S is a spanning tree, there is a path in S that joins u and v. This path, together with e_k, must form a cycle. Thus there must exist at least one edge, e_s, in this cycle that is not in T. If we remove e_s from S and insert e_k, we get a connected graph, S', with $n - 1$ edges. Thus S' is a tree. We know that

$$w(S') = w(S) + w(e_k) - w(e_s).$$

Since S is a minimal spanning tree,

$$w(S) \leq w(S').$$

But since $e_1, \ldots, e_{k-1}$ are common to both T and S, and since e_k is the smallest weight edge that could have been selected without creating a cycle,

$$w(e_k) \leq w(e_s).$$

Hence

$$w(e_k) - w(e_s) \leq 0$$

and thus

$$w(S') \geq w(S).$$

Hence $w(S') = w(S)$. Continuing in this fashion, $e_{k+1}, \ldots, e_{n-1}$ could all be inserted in S without changing its weight. Thus $w(T) = w(S)$ and T is a minimal spanning tree. ☐

TERMINOLOGY

key	breadth-first spanning tree
depth-first spanning tree	minimal spanning tree

SUMMARY

1. Two commonly used algorithms for finding spanning trees in unweighted graphs are depth-first search and breadth-first search. Both have worst-case complexity $O(|E| + |V|)$.

2. The principal applications examined here for these algorithms are to connectivity problems and to graph searching.

3. Kruskal's algorithm can be used to find minimal spanning trees in weighted graphs. Kruskal's algorithm is an example of a greedy algorithm that works.

EXERCISES 10.2

1. Find the breadth-first spanning tree for Figure 10.17 in the text if a is the initial vertex.

Find depth-first and breadth-first spanning trees for the specified graphs.

2. Exercise 15, Section 9.2   3. Exercise 16, Section 9.2

4. Exercise 22, Section 9.2   5. Exercise 23, Section 9.2

Answer the following questions.

6. Show that K_n has at least two spanning trees with no edges in common for $n \geq 4$.

7. Show that if $|V| \leq 3$ and V is not a multigraph, any two spanning trees must share an edge.

8. Show that the depth-first spanning tree for K_n is a chain; the breadth-first spanning tree for K_n is a star.

9. Show that any graph for which the depth-first and breadth-first spanning trees are the same is itself a tree. (Hint: what do the algorithms do when they reach a vertex that is part of a cycle?)

10. Give arguments to justify the claim that line 4 of depth-first search will be executed $O(n^2)$ times when the graph being searched is K_n.

11. Give arguments to justify the claim that line 7 of breadth-first search will be executed $O(n^2)$ times when the graph being searched is K_n.

12. Verify that depth-first search produces a spanning tree.

13. Verify that breadth-first search produces a spanning tree.

Find minimal spanning trees for Exercises 14–15.

14.

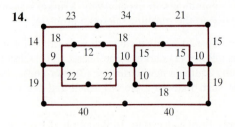

15.

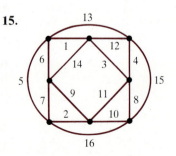

16. The graph given by the following adjacency matrix:

$$
\begin{matrix}
\infty & 5 & 4 & \infty & \infty & \infty \\
5 & \infty & \infty & 3 & 2 & \infty \\
4 & \infty & \infty & 3 & \infty & 6 \\
\infty & 3 & 3 & \infty & \infty & 2 \\
\infty & 2 & \infty & \infty & \infty & 3 \\
\infty & \infty & 6 & 2 & 3 & \infty
\end{matrix}
$$

17. A cable TV company wants to connect a group of small towns to its system. The following table presents the distances between the towns. Find a set of connections that minimizes the total length of cable for the company, starting from Braxton.

Braxton	0					
Gates	10	0				
Alison	12	9	0			
Lewis	25	20	11	0		
Elba	25	22	13	8	0	
Canton	30	20	15	18	14	0
	Braxton	**Gates**	**Alison**	**Lewis**	**Elba**	**Canton**

Answer the following questions.

18. Rewrite Kruskal's algorithm so that it finds maximal spanning trees.

19. Explain how Kruskal's algorithm could be used to find a spanning tree in a connected but unweighted graph.

20. Suppose a graph is known to have k components. How could Kruskal's algorithm be modified to find a minimal spanning forest?

21. Greedy algorithms based on repeatedly "selecting the cheapest" element typically have a counterpart based on "eliminating the most expensive" element. Write an algorithm like Kruskal's based on this idea.

22. Verify the algorithm you developed in Exercise 21.

23. Give arguments to explain why Kruskal's algorithm cannot be successfully applied to the traveling salesman problem.

An alternative to Kruskal's algorithm is called *Prim's algorithm.* It is informally stated as follows: (1) start at some vertex, v_1; (2) select the minimal edge incident on v_1, thereby adding a new vertex, v_2; (3) after j edges have been selected, choose the $(j + 1)$st such that no cycle is formed and the new edge is incident on one of $\{v_1, \ldots, v_j\}$.

24. Apply Prim's algorithm to the graph of Exercise 14.

25. Apply Prim's algorithm to the graph of Exercise 15.

26. Apply Prim's algorithm to the graph of Exercise 16.

27. Apply Prim's algorithm to solve Exercise 17.

28. Write Prim's algorithm in pseudo-code.

29. Verify Prim's algorithm.

10.3
ROOTED TREES

In this section, we examine trees that are a special kind of digraph. Such trees can be used to model a variety of hierarchical relationships—for example, administrative organization charts, biological classification systems, library classification systems, genealogies, and tournaments. They also have a number of important applications in computer science, notably to sorting and searching, writing assemblers and compilers, and in the study of formal and natural languages. Our basic definition will seem quite different from the definition of a tree used in Sections 10.1 and 10.2, although it is closely related. We will still think of trees as graphs without cycles, but with digraphs we will be concerned with both directed and undirected cycles. Thus we use an unambiguous approach that does not mention cycles in the definition.

Basic Concepts

Definition   Let G be a digraph and let v be a vertex of G. The **in-degree** of v is the number of distinct edges of the form (x, v). The **out-degree** of v is the number of distinct edges of the form (v, y). A **rooted tree** is a digraph that has the following properties:

1. It has one vertex whose in-degree is 0, called the **root**.

2. Every other vertex has in-degree 1.

3. Every vertex is joined to the root by a directed path.

Rooted trees are more similar to the trees studied in Sections 10.1 and 10.2 than they might appear to be in the definition. If we treat each edge as undirected rather than directed, a rooted tree becomes a tree. We will prove this in Theorem 10.6.

EXAMPLE 1   Figure 10.20 gives three examples of rooted trees. When they are drawn without arrows, it is assumed that the root is at the top and the direction of all edges is down. Note that the out-degrees of the vertices vary from 0 to 3, while the root has in-degree 0 and all other vertices have in-degree 1.   ∎

Figure 10.20 Three examples of rooted trees.

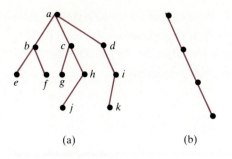

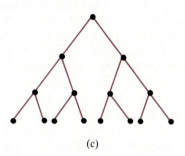

(a)   (b)

(c)

Definition   Let $T = (V,E)$ be a rooted tree and suppose that (a,b) is an edge in T. Then a is called the **parent** of b and b is a **child** of a. Two or more children of the same parent are called **siblings**. If there is a directed path starting at c and ending at d, c is an **ancestor** of d and d is a **descendant** of c. A vertex with no children is called a **leaf**. A vertex with children is called an **internal vertex**. $T' = (V',E')$ is a **subtree** of T if $V' \subset V$, $E' \subset E$, and T' is itself a rooted tree. The **level** of a vertex in a rooted tree is the length of the directed path connecting it to the root. The **height** of a rooted tree is the greatest level of any leaf.

EXAMPLE 2   In Figure 10.20(a), a is the parent of b, c, and d. b is the parent of e and f. Similarly, a and b are ancestors of e and e is a descendant of a and b. e, f, g, j, and k are the leaves, while all of the other vertices are internal. b, c, and d are at level 1, e, f, g, h, and i are at level 2, and j and k are at level 3. The height of the tree is 3. ∎

EXAMPLE 3   The portion of an organizational hierarchy chart in Figure 10.21 is a rooted tree. If the chart were changed so that an employee at some level reported to two or more supervisors, the structure would no longer be a tree, since that employee would be represented by a vertex with in-degree greater than or equal to 2. It is not unusual for major American corporations to have organizational charts with a height as much as 8. A

Figure 10.21 Organizational chart.

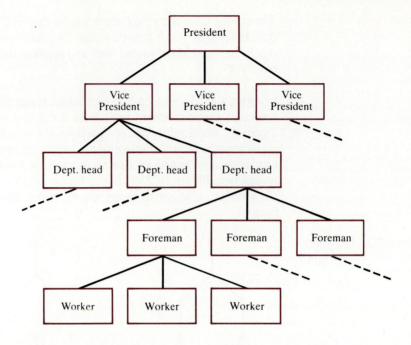

current trend is to regard such structures as "management heavy" and to
reduce their height. ∎

EXAMPLE 4   The prerequisite structure for the set of computer science
courses illustrated in Figure 10.22 is not a rooted tree.

Figure 10.22 Portion of a
prerequisite chart.

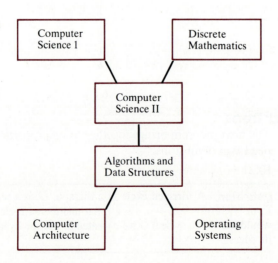

The course Computer Science 2, has in-degree 2, so the structure is not a tree. In spite of this, it is not unusual for a faculty to borrow some terminology from the study of rooted trees and speak of the "height of a prerequisite tree."   ∎

EXAMPLE 5   In Section 8.3, we studied Hasse diagrams, a way to visually represent a partial order $(A, \preceq)$. Such a diagram may or may not be a tree. Typically, when we represent a partial order by a Hasse diagram, the smallest element is at the bottom and the largest element is at the top. Following our usual convention of having the root at the top, Figure 10.23(a) is a rooted tree; Figure 10.23(b) is not. Figure 10.23(c) is also a tree if we regard the smallest element as being the root and allow the root to be at the bottom.

Figure 10.23   Hasse diagrams: *a* and *c* are rooted trees; *b* is not.

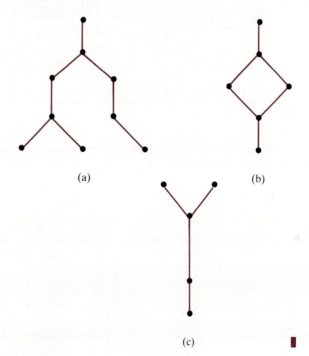

(a)   (b)

(c)   ∎

Properties of Rooted Trees

The next theorem brings together some properties of rooted trees. First, we need two definitions.

Definition   A **directed cycle** in a digraph, G, is a sequence of edges in G of the form (v_1, v_2), (v_2, v_3), $\cdots$, (v_n, v_1). An **undirected cycle** is a sequence of edges that would become a cycle if G were treated as a graph by ignoring the direction of all of its edges.

EXAMPLE 6 Figure 10.24(a) is a digraph that includes a directed cycle. Figure 10.24(b) includes an undirected cycle.

Figure 10.24 Directed and undirected cycles.

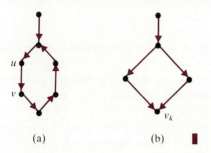

(a)   (b)

THEOREM 10.6 Let T be a rooted tree. Then

a) T has no loops.

b) T has no cycles, either directed or undirected.

c) There is a unique directed path from the root to each vertex.

d) Every directed path in T is simple.

PROOF

a) The root cannot have a loop, since it has in-degree 0. But neither can the other vertices; by definition, there is a path from the root to each vertex. That path plus a loop would give a vertex in-degree 2.

b) As for directed cycles, suppose that there is one. Choose any vertex on that cycle, say v. There exists a unique path from the root to v. The intersection of the set of vertices along this path and the cycle is non-empty, since it contains at least v. Thus we can let u be the vertex of T of the lowest level that is both on the cycle and on the path from the root to v (see Figure 10.24a). u cannot be the root, since the root has in-degree 0 and every vertex on a cycle has in-degree at least 1. Since u is of the lowest level on both the cycle and the path and is not the root, it has in-degree 2—one for the cycle and one for the path. This is impossible, and hence T has no directed cycles.

If T has an undirected cycle, this cycle must be a walk of the form $\{v_1,v_2\}, \{v_2,v_3\}, \ldots, \{v_n,v_1\}$, which has the property that if the direction of each of the edges is considered, it is not a directed cycle. That is, if we start from some vertex, v_i, on the cycle and follow the directed edges $(v_i,v_{i+1}), (v_{i+1},v_{i+2}), \ldots$, we must come to a first vertex, v_k, for which $\{v_{k-1},v_k\}$ and $\{v_k,v_{k+1}\}$ are part of the undirected cycle beginning and ending at v_1, but for which (v_{k-1},v_k) and (v_k,v_{k+1}) are not part of a directed cycle from v_1 to itself. (If $k = 1$, v_{k-1} is v_n.) That is, the direction of the edge $\{v_k,v_{k+1}\}$ must be (v_{k+1},v_k). Hence v_k has in-degree 2, which is impossible (see Figure 10.24b).

c) Suppose there is a vertex that has two distinct paths to it from the root. Such paths would form an undirected cycle.

d) If a directed path were to cross itself, it would include a directed cycle. ◻

EXAMPLE 7 Consider a rooted tree such as the one in Figure 10.20(a). It is easy to see that each property of Theorem 10.6 is satisfied, i.e., it has no loops and no cycles, there is a directed path from the root to each vertex, and every directed path is simple. ∎

EXAMPLE 8 Structure charts used in the design of computer programs are not usually trees, as they frequently contain undirected cycles (see Figure 10.25).

Figure 10.25 Structure charts are not usually rooted trees.

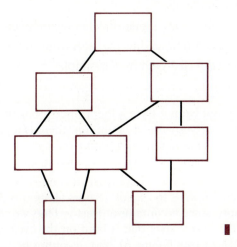

The following definitions are very important in the applications of rooted trees.

Definition An *m*-ary tree is either the empty set or a rooted tree in which the out-degree of every vertex is less than or equal to *m*. An **ordered *m*-ary tree** is an *m*-ary tree in which the edges leaving each vertex are ordered from 0 to $m - 1$.

EXAMPLE 9 In Figure 10.26, there are four *m*-ary trees. For Figure 10.26(a); $m = 3$; $m = 2$ for the others. If we order the vertices from left to right, each is an ordered *m*-ary tree. Figure 10.26(d) is a subtree of Figure 10.26(c). ∎

Figure 10.26 *m*-ary trees.

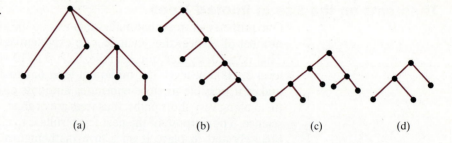

(a)   (b)   (c)   (d)

Definition A **binary tree** is an *m*-ary tree with $m = 2$ and with the property that for each parent, each of its children is either designated as a **left child** or a **right child.**

EXAMPLE 10 Figure 10.26(b–d) consists of binary trees. ∎

The difference between binary trees and ordered *m*-ary trees with $m = 2$ is that ordered *m*-ary trees do not distinguish a left and a right child. Thus the two binary trees in Figure 10.27 are different binary trees; as ordered *m*-ary trees, they would not be distinguished.

Figure 10.27 Two distinct binary trees.

Note that if we remove the root from each of the trees in Figure 10.26(b–d), each binary tree breaks down into a collection of binary trees. It is this property that has prompted the formulation of the following definition of binary tree that is especially popular in textbooks on data structures: *A binary tree is a finite set, V, of vertices that is either the empty set or a vertex, v, (the root) together with two other disjoint subsets of V − {v} that are themselves binary trees and are called the left subtree and the right subtree.* The recursive nature of this definition is very attractive in data structures because it is useful in writing recursive algorithms. Both definitions are equivalent, although we will not prove this here. The advantages of the rooted tree definition are that it enables us to study binary trees in the context of graphs and digraphs and it is easier to apply to the kinds of theorems we want to prove.

Theorems on the Size of Rooted Trees

One property of trees that makes them so useful in applications is that the number of vertices they include is an exponential function of their height. That is, a 10-ary tree, for example, has up to 10 vertices at level 1, 100 at level 2, 1000 at level 3, 1 million at level 6, and 1 billion at level 9. This makes it possible to store enormous amounts of data in trees and reach them along very short paths; thus trees are of great importance in computer science. The purpose of the next few results is to clarify this property more precisely and to place it on a firm mathematical foundation. All of the results proven here are for binary trees; we will leave the generalization to m-ary trees for the Exercises.

THEOREM 10.7   Let T be a binary tree. Then T has between 0 and 2^L vertices at any level, L.

PROOF   The proof is by induction on L. At level 0, T has at most one vertex, the root. Thus the theorem holds at that level. Now assume that there are at most 2^{L-1} vertices at level $L - 1$. Since T is a binary tree, each vertex has at most two children. Thus there are at most $2 \cdot (2^{L-1}) = 2^L$ vertices at level L.   ☐

THEOREM 10.8   A binary tree of height h has between $h + 1$ and $2^{h+1} - 1$ vertices.

PROOF   Since it has height h, there must be at least one vertex at each level from 0 to h. Thus there are at least $h + 1$ vertices. By Theorem 10.7, there are at most 2^L vertices at each level. Thus there are at most $1 + 2 + 4 + \cdots + 2^h = 2^{h+1} - 1$ vertices in the entire tree.   ☐

EXAMPLE 11   Both the upper and lower bounds given in the previous theorem for the number of vertices in a binary tree are attainable. A tree of height h in which every internal vertex has exactly one child has a total of $h + 1$ vertices. A tree in which every internal vertex has exactly two children has $2^{h+1} - 1$ vertices (see Figure 10.28).   ■

Figure 10.28   Two binary trees, one with maximum vertices and one with minimum vertices.

(a)   (b)

THEOREM 10.9   If a binary tree has L leaves and height h, then $\log_2 L \leq h$.

PROOF   Let T be a binary tree and let h denote its height. The proof is by induction on the height of T. Suppose $h = 1$. Then there are either one or two leaves, and thus $\log_2 L \leq h$. Now suppose $h > 1$ and that for binary trees of height $h - 1$ or less, the number, x, of leaves is such that $\log_2 x \leq h - 1$. Remove all leaves that are at level h from T. Then the number, N, of leaves in the resulting tree will satisfy the inequalities

$$L \geq N \geq (1/2)\,L,$$

since every parent has at most two children.

Hence

$$\log_2 N \geq \log_2(L/2).$$

Thus

$$\log_2 N \geq \log_2 L - \log_2 2 = \log_2 L - 1.$$

Since the height of the resulting tree is $h - 1$, by the induction hypothesis, $h - 1 \geq \log_2 N$. Therefore

$$\log_2 L - 1 \leq h - 1$$

and thus

$$\log_2 L \leq h. \quad \square$$

Definition   A **full** (or **regular**) binary tree is one in which every internal vertex has both a left and a right child.

EXAMPLE 12   The binary trees in Figure 10.26(b–d) are all full binary trees.   ∎

THEOREM 10.10   If T is a full binary tree with L leaves, height h, and i internal vertices, then

a) T has $2i + 1$ vertices and $i + 1$ leaves.

b) $L \geq h + 1$.

PROOF

a) Since T is full, every internal vertex has exactly two children. Thus there are $2i$ children. The only vertex that is not a child is the root. Thus there is a total of $2i + 1$ vertices. Since i are internal, $i + 1$ are leaves.

b) Let $v_0, v_1, \ldots, v_h$ be a path from the root to some vertex, v_h, at level h. Since T is full, all of the v_i except v_h have exactly two children. Thus there is a sequence $T_0, T_1, \ldots, T_{h-2}$ of subtrees of T formed by taking the child other than v_{i+1} at each v_i as the root of T_i. For instance, the root of T_0 is the sibling of v_1. Each of these subtrees is disjoint from the others and has at least one leaf. v_{h-1} has two children, both of which are leaves. Thus there are at least $(h - 1) + 2 = h + 1$ leaves. ∎

Although the number of vertices *possible* at any level of a binary tree grows exponentially with level, we can see from Theorem 10.8 and Example 11 that it is also possible for the actual number of vertices to grow linearly with height if each parent has only one child. The following concept is introduced to distinguish a class of trees whose number of vertices does grow exponentially with height.

Definition A rooted tree with height h is **balanced** if all of its leaves are at level h or level $h - 1$ and every vertex at a level less than $h - 1$ has out-degree 2.

THEOREM 10.11 A balanced binary tree of height h has between 2^h and $2^{h+1} - 1$ vertices.

PROOF We have already shown that a binary tree has a maximum of $2^{h+1} - 1$ vertices. The tree for which this maximum is attained has all of its leaves at level h and every vertex has out-degree 2. Thus it is balanced. On the other hand, a balanced binary tree must have at least $1 + 2 + 4 + \cdots + 2^{h-1} = 2^h - 1$ vertices at levels 0 through $h - 1$. (See the Exercises.) Since there must be at least one vertex at level h, the result follows. ∎

COROLLARY 10.12 If T is a balanced binary tree with height h, L leaves, and n vertices, then

$$h = \lceil \log_2 L \rceil = \lfloor \log_2 n \rfloor.$$

PROOF By the previous theorem,

$$2^h \leq n < 2^{h+1}.$$

Thus $h \leq \log_2 n < h + 1$. Since h is an integer,

$$h = \lfloor \log_2 n \rfloor.$$

By Theorem 10.10, $n = 2i + 1 = (i + 1) + (i + 1) - 1 = 2L - 1$.

Thus

$$2^h + 1 \leq 2L < 2^{h+1} + 1.$$

Hence,

$$2^h < 2L \leq 2^{h+1}$$

and

$$2^{h-1} < L \le 2^h.$$

Thus $h - 1 < \log_2 L \le h$. Again, since h is an integer,

$$h = \lceil \log_2 L \rceil. \quad \square$$

Corollary 10.12 makes precise the notion stated at the beginning of this subsection: that the number of vertices in a rooted tree grows exponentially with its height or, equivalently, that the height is a logarithmic function of the number of vertices. It also shows that the same relationship holds between the leaves and the height. This is very important in computer science, as it enables us to construct algorithms using binary trees that have a complexity of $O(\log_2 n)$; this will be discussed more fully in the next section. But Corollary 10.12 imposes a major restriction—that this relationship holds only when the tree is balanced. In fact, both the number of vertices and the number of leaves can grow linearly with height when every parent has just one child.

TERMINOLOGY

in-degree	descendant	*m*-ary tree
out-degree	leaf	ordered *m*-ary tree
rooted tree	internal vertex	binary tree
root	subtree	left child
parent	level	right child
child	height	full binary tree
siblings	directed cycle	regular binary tree
ancestor	undirected cycle	balanced binary tree

SUMMARY

1. Rooted trees are digraphs with one vertex, the root, having in-degree 0, and all other vertices having in-degree 1. Rooted trees are useful in modeling a variety of hierarchical relationships.

2. Rooted trees have no loops, no directed or undirected cycles, have a unique path joining every vertex to the root, and have the property that all paths in them are simple.

3. A binary tree has

   a) between 0 and 2^L vertices at each level, L.

    b) between $h + 1$ and 2^{h+1} vertices, where h is its height.

    c) the property that $\log_2 L \leq h$.

4. A full binary tree with i internal vertices has $i + 1$ leaves and $2i + 1$ vertices. Also, $L \geq h + 1$.

5. For a balanced binary tree

    a) the number of vertices is between 2^h and $2^{h+1} - 1$.

    b) $h = \lceil \log_2 L \rceil = \lfloor \log_2 n \rfloor$.

EXERCISES 10.3

As defined at the beginning of this section, a rooted tree must satisfy three conditions. Find examples of digraphs that satisfy the specified two conditions but are not rooted trees.

1. Conditions 1 and 2 but not 3   **2.** Conditions 1 and 3 but not 2

3. Conditions 2 and 3 but not 1

Add labels to the vertices of Figure 10.20(c) and find vertices or pairs of vertices that satisfy the following conditions.

4. One is the parent of the other.

5. One is the ancestor of the other but not its parents.

6. Two siblings.   **7.** A vertex at level 2.

8. Find the height of the tree.

Draw examples of two binary trees, one with $h + 1$ and one with $2^{h+1} - 1$ vertices when

9. $h = 3$.   **10.** $h = 4$.

Sketch binary trees having the following characteristics or state why this cannot be done.

11. height 4, four vertices   **12.** height 4, 15 vertices

13. height 4, regular, 15 vertices   **14.** height 4, balanced, 10 vertices

Are each of the following trees full? Balanced? Why?

15.   **16.**

17. 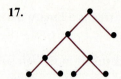   18.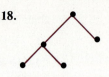

If i denotes the number of internal vertices and L denotes the number of leaves of a binary tree, find examples of binary trees with

19. $i > L$.   20. $i < L$.

21. $i = L$.

Generalize the following results of this section to m-ary trees.

22. Theorem 10.7   23. Theorem 10.8

24. Theorem 10.9   25. Theorem 10.10(a)

26. Theorem 10.11

Prove the following.

27. If a binary tree has height h and every vertex at a level less than h has out-degree 2, it has 2^L vertices for each level $L < h$.

28. Show that the examples of the trees in Figure 10.26(b–d) satisfy the recursive definition of a binary tree given right after Figure 10.27.

29. Show that a single elimination tournament is a binary tree.

30. Show that in a single elimination tournament the number of matches is one less than the number of players.

31. The level of a vertex in a binary tree can also be defined recursively: the root is at level 0; all other vertices are at a level one greater than that of their parent. Show that this definition is equivalent to the one given in this section.

10.4

APPLICATIONS OF BINARY TREES IN COMPUTER SCIENCE

In this final section, we shall examine three applications of binary trees, all of which are important in computer science.

Search Trees

One important problem in computer science is how to store large quantities of data in ways that allow rapid access to particular items. Search trees provide one solution to that problem. Data are typically organized into records, and each record has a key. The records are then stored by means of their keys. A **binary search tree** is a binary tree in which

1. each vertex is a record.

2. the key of every record stored in the left subtree of a parent precedes the key of the parent.

3. the key of every record stored in the right subtree follows the key of its parent, where *precedes* and *follows* typically indicate lexicographic order.

Note that the preceding definition assumes that binary search trees have no duplicate keys. In Algorithm 39, we will see how to construct binary search trees; even if the data from which the tree is constructed have duplicates, our construction will not place duplicate keys in the tree. One way to handle duplicates is to store the frequency of their occurrence along with each key. A typical search for a record stored in a search tree starts at the root, compares the key to the root, and then searches the left or right subtree. Thus it follows a path until either the key is found or a leaf is reached that is not the key being searched for. Since there is a unique path in any binary tree from the root to each vertex, this process either leads directly to the key being searched for or proves that it is not included in the tree. If the tree is balanced and has n elements, an algorithm using this type of search is in $O(\log_2 n)$ and is thus very efficient. For instance, the search tree in Figure 10.29 uses the names of certain English poets as keys.

Figure 10.29 A binary search tree.

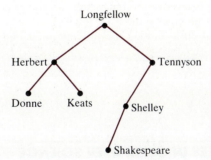

Suppose we wanted to search for the record whose key is "Shelley." Shelley follows Longfellow in alphabetical order, so if Shelley is present, he must be in the right subtree. We then find that Shelley precedes Tennyson. We search the left subtree of Tennyson and find Shelley. If we were looking for Yeats, we would again start at the root, find that Yeats follows Longfellow, and search the right subtree. Yeats follows Tennyson, so we again search the right subtree. But Tennyson's right subtree is empty, so we conclude that Yeats is not in the tree. This process is formalized in the following algorithm. For simplicity, it is set up merely to indicate whether the key, X, being searched for was found. It can be modified to select other data from a record or to modify a record once that record has been located.

ALGORITHM 38 **Binary Tree Search**

Procedure Tree_search(T,X; var found);

{T is a binary search tree; X is the item being searched for. Found is a boolean variable that indicates whether X was located in T.}

begin

1. If T is empty
2. then found := false
3. else if X is the key of the root of T
4. then found := true
5. else if X precedes the key of the root of T
6. then Tree_search(left subtree of T, X, found)
7. else Tree_search(right subtree of T, X, found)

end;

EXAMPLE 1 List the vertices of Figure 10.29 encountered in their order of visitation when searching for Hopkins.

SOLUTION Longfellow is visited first, then Herbert, followed by Keats. The left subtree of Keats is empty, so the algorithm halts after visiting these two. ∎

If a search tree is balanced, as mentioned previously, such a search will have a complexity of $O(\log_2 n)$. If a search tree is completely unbalanced, i.e., if every vertex has only one child, the complexity degenerates to that of a linear search—$O(n)$. Algorithm 39 can be used to insert records into a binary search tree. If the records are already sorted or close to being sorted by their keys before being inserted, the algorithm will yield a tree that is very unbalanced. However, if the records arrive in random order, the tree will show features of a balanced tree and thus yield searches that are more efficient than linear searches. It has been proven that the average height, h, of a binary tree generated by random insertions is such that $h \leq 1.4 \log_2 n$. (See Aho, et al., *Data Structures and Algorithms*, Section 5.2; see References.) Variations of the algorithm such as including the frequency of duplicate data items or adding data in addition to the key to each record can be done by altering the line "visit the root."

ALGORITHM 39   Binary Tree Insertion

Procedure Insert(*X;* var *T*);

{*T* is a binary tree; *X* is the key of an item to be inserted.}

begin

1.   if *T* is empty

then begin

2.   create a root node for *T* with key value of *X;*

3.   visit the root;

end;

4.   else

5.   if *X* is the key of the root of *T*

6.   then visit the root

else

7.   if *X* precedes the key of the root of *T*

8.   then insert (*X,*left subtree of *T*)

9.   else insert (*X,*right subtree of *T*)

end;

EXAMPLE 2   Use Algorithm 39 to place the letters of the word *procedure* into a search tree.

SOLUTION   *p* becomes the root. Since *r* follows *p,* it is placed in the right subtree of *p.* Since *o* precedes *p,* it is placed in the left subtree of *p.* Continuing in this fashion, we get the tree of Figure 10.30.

Figure 10.30   The letters of the word *procedure* placed in a binary search tree.

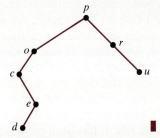

One of the more common applications of search trees in computer science is in the construction of symbol tables in assemblers and compilers. Two-pass assemblers typically construct a symbol table on their first pass. Thus a binary search tree provides an efficient way of looking up the symbols on the second pass. In some languages, like Pascal, all variables are declared before the body of the program. These variables can be placed in a search tree, and the tree provides an efficient tool to test the validity of symbols and their usage during compilation. There are algorithms for balancing binary trees, but they are beyond the range of this discussion.

Representing Algebraic Expressions

Without parentheses the algebraic expression $x + y * z$ is usually interpreted as meaning $x + (y * z)$, since we normally adopt a precedence rule that gives multiplication higher priority than addition. However, without the precedence rule, it is ambiguous; it is not clear if it means $(x + y) * z$ or $x + (y * z)$. Also, the evaluation of expressions with parentheses, even when they are unambiguous, can be very inefficient for a compiler, since the compiler has to scan back and forth to evaluate and assemble the components of the expression. Binary trees provide an effective way to represent expressions unambiguously without parentheses and provide a means to evaluate expressions efficiently.

Algebraic expressions such as $-3(xy + z) + 2w$ are built up using binary and unary operators. **Binary operators** allow the combination of two symbols using connectives such as $+$, $*$, $/$, $-$, and exp. For instance, a product involving three symbols such as xyz is the result of two binary operations: $x * y$ and that product times z. **Unary operators** modify a single symbol or expression, as does the minus sign in algebra or the "not" symbol in propositional logic. We will consider only binary operators here and leave unary operators for the Exercises. A valid algebraic expression using a binary operator and our conventional notation is of the form

expression operator expression

where *expression* means either a symbol, a number, or another expression. (Note the recursive definition.)

We place valid expressions in binary trees by placing the operator at the root and the two subexpressions in the subtrees. For instance, $x + (y * z)$ is diagrammed in Figure 10.31(a) and $(x + y) * z$ is diagrammed in Figure 10.31(b). Note that $x + y * z$ is not ambiguous when placed in a binary tree; either the $+$ or the $*$ is at the root.

Figure 10.31 Trees representing (a) $x + (y * z)$ and (b) $(x + y) * z$.

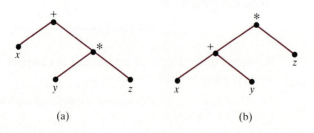

(a)   (b)

EXAMPLE 3   Represent the expression $3(x^2y + y^2x)(z - 1)$ in a binary tree.

SOLUTION   This expression involves the product of three subexpressions: 3, $(x^2y + y^2x)$, and $(z - 1)$. We first add parentheses around all binary operations. The expression then becomes

$$((3*(((x \exp 2) * y) + ((y \exp 2) * x))) * (z - 1)).$$

That is, we multiply 3 times $(x^2y + y^2x)$ first and then multiply the result by $z - 1$. The resulting tree is in Figure 10.32. If we had parenthesized so as to multiply 3 times the result of multiplying $(x^2y + y^2x)$ by $z - 1$, we would have gotten a different tree.

Figure 10.32
Tree representing
$3(x^2y + y^2x)(z - 1)$.

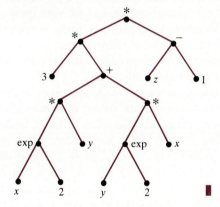

Thus binary trees provide a scheme for representing algebraic expressions unambiguously. Also, two notational schemes for representing expressions unambiguously without parentheses have been developed. They are called **Polish** and **reverse Polish notation** after the mathematician Jan Lukasiewicz (1878–1956). These notations have a close connection with binary tree representations of expressions in that they can be viewed as the result of visiting each of the vertices in the binary tree representing an expression in a certain way. Before examining this connection, however, we will examine the notational schemes themselves.

Polish notation, also called the **prefix form** of an expression, is written using the pattern

operator expression expression.

Reverse Polish notation, also called the **postfix form** of an expression, is written using a different pattern:

expression expression operator.

Thus, for instance, $(x + y)$ is written $+xy$ in prefix form and $xy+$ in postfix form. Similarly, $(x + y) * z$ is written $* + xyz$ in prefix form and $xy + z*$ in postfix form. That is, since prefix expressions are written *operator expression expression,* when writing $(x + y) * z$, the operator, $*$, is written first, followed by the prefix form for $(x + y)$, namely, $+xy$, followed by z. For the postfix form, the notation for $(x + y)$ is written first, followed by z, followed by the operator, $*$. Evaluation of prefix and postfix expressions follows a similar pattern.

EXAMPLE 4 Let $x = 2$, $y = 3$, and $z = 4$.

a) Evaluate the postorder expressions

$$xyz*+$$
$$xy + z*.$$

b) Evaluate the preorder expressions

$$+ x * yz$$
$$* + xyz.$$

SOLUTION

a) If we replace each symbol by its value, we get

$$2\ 3\ 4\ *\ +.$$

Reading left to right, when we get to $3\ 4*$, we have a sequence of symbols that fits the pattern. Hence we replace $3\ 4*$ by its value, namely, 12, yielding

$$2\ 12 +$$
$$= 14.$$

For $xy + z*$, we replace all symbols by their value, giving

$$2\ 3 + 4\ *.$$

Since this begins with $2\ 3 +$, we can replace this portion of the expression by its value, 5, yielding

$$5\ 4\ * = 20.$$

b) With preorder notation, binary expressions are of the form

$$\text{operator expression expression.}$$

Substituting for x, y, and z, $+ x * yz$ becomes

$$+ 2 * 3\ 4.$$

Reading from left to right, we cannot do any computation until we get to * 3 4. Thus we get

$$+ 2\ 12 = 14.$$

For * + xyz, we first get

$$* + 2\ 3\ 4$$
$$= * 5\ 4$$
$$= 20. \quad \blacksquare$$

EXAMPLE 5   Evaluate the preorder expression $+ * a - bcd$ when $a = b = c = d = 3$.

SOLUTION   Proceeding as in Example 4, we substitute values and read the expression from left to right until we find a sequence of symbols that fits the pattern *operator expression expression.* Thus we get

$$+ * 3 - 3\ 3\ 3$$
$$= + * 3\ 0\ 3$$
$$= + 0\ 3$$
$$= 3. \quad \blacksquare$$

The process of visiting all of the vertices of a tree is called **traversing** a tree. The three principal algorithms for tree traversal are Algorithms 40, 41, and 42. The different notational schemes for algebraic expressions we have just examined are the result of applying these three algorithms to the binary tree representing an expression. These algorithms also have applications to any situation in which one must visit every vertex in a binary tree.

ALGORITHM 40   Preorder Traversal

Procedure Preorder(T);

{T is a binary tree.}

begin

1.   If T is not empty then

2.   visit the root;

3.   Preorder(left subtree);

4.   Preorder(right subtree)

end;

ALGORITHM 41   Inorder Traversal

Procedure Inorder(T);

{T is a binary tree.}

begin

1.   if T is not empty then

2.   Inorder(left subtree);

3.   visit the root;

4.   Inorder(right subtree);

  end;

ALGORITHM 42   Postorder Traversal

Procedure Postorder(T);

{T is a binary tree.}

begin

1.   if T is not empty then

2.   Postorder(left subtree);

3.   Postorder(right subtree);

4.   visit the root;

  end;

EXAMPLE 6   List the vertices of the tree of Figures 10.32(a) and 10.32(b) in the order of their visitation for the preorder, inorder, and postorder traversals.

SOLUTION   For Figure 10.32(a) the traversals are

$$\text{preorder: } + x * yz$$
$$\text{inorder: } x + y * z$$
$$\text{postorder: } xyz * + .$$

For Figure 10.32(b), the traversals are

$$\text{preorder: } * + xyz$$
$$\text{inorder: } x + y * z$$
$$\text{postorder: } xy + z * . \quad \blacksquare$$

The key point to note in the previous example is that the preorder and postorder traversals of the two trees in Figure 10.32 are different, although they have the same inorder traversal. That is, the inorder traversal yields the conventional notation, which is ambiguous without the use of parentheses. The pre- and postorder traversals do distinguish them and, in fact, give a unique way to denote any expression without the need of parentheses. The result of a preorder traversal is the prefix form. The postorder traversal yields the postfix form. The result of an inorder traversal is called the **infix form.**

EXAMPLE 7   Find an infix expression that is equivalent to the prefix expression $+ * a - bcd$ and parenthesize it appropriately.

SOLUTION   First, construct the binary tree representing the expression; then traverse it using an inorder traversal, inserting parentheses wherever needed. The tree is shown in Figure 10.33. It is constructed using the pattern that prefix expressions follow:

$$\text{operator expression expression.}$$

Thus the $+$ goes at the root. We begin the construction of a left subtree by placing the $*$ at the root. When we get to a symbol that is not an operator (a, b, c, or d), we have gotten to a leaf. Thus a is placed to the left of $*$ and $-bc$ to its right. The remains to form the right subtree of $+$. The infix expression is $a * b - c + d$. The parenthesized expression is $(a * (b - c)) + d$.

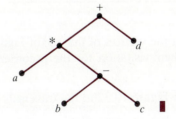

Figure 10.33   Binary tree formed from the prefix expression $+ * a - bcd$.

Expressions written in prefix or postfix notation are particularly attractive to writers of compilers, since evaluation can be done very efficiently; no scanning of the expression back and forth is required. Thus compilers typically handle infix expressions by placing them in a binary tree and then generating equivalent postfix or prefix expressions by an appropriate traversal.

Huffman Codes

Another important computer science topic to which binary trees can be applied is coding. Since information is stored and transmitted using only 1's and 0's, binary codes for every character that might be used in any particular application must be developed. The most commonly used system for coding alphanumeric data is the ASCII coding scheme. In this scheme, every character or letter is given a unique 7-bit code; an eighth bit is usually reserved as a parity bit or used to expand the set of characters that can be coded. Thus the storage or transmission of n characters requires $8n$ bits. If, however, we use variable-length codes, we can assign the most frequently used characters the shorter codes and substantially reduce the length of messages. This brings up an immediate problem, however: If the length of the code for a character is not fixed, how does the receiver know where one character ends and the next one begins? One solution is to use special separator characters. However, we then have the problem of determining whether a sequence of bits being received is the end of the previous message or the beginning of a separator. Also, the separator wastes space. Morse code, used for telegraphy in the late nineteenth and early twentieth centuries, used a kind of separator—a pause between characters (see Figure 10.34).

Figure 10.34 Morse code. Dots are short clicks; dashes are long clicks.

A . _	N _ .	1 . _ _ _ _
B _ . . .	O _ _ _	2 . . _ _ _
C _ . _ .	P . _ _ .	3 . . . _ _
D _ . .	Q _ _ . _	4 . . . . _
E .	R . _ .	5 . . . . .
F . . _ .	S . . .	6 _ . . . .
G _ _ .	T _	7 _ _ . . .
H	U . . _	8 _ _ _ . .
I . .	V . . . _	9 _ _ _ _ .
J . _ _ _	W . _ _	10 _ _ _ _ _
K _ . _	X _ . . _	period . _ . _ . _
L . _ . .	Y _ . _ _	comma _ _ . . _ _
M _ _	Z _ _ . .	? . . _ _ . .

The code for an *e* is one dot and for an *i* is two dots; to distinguish two *e*'s from one *i*, a longer pause is left between the two dots for the two *e*'s than would be left between the two dots for an *i*. This is really not satisfactory, though, since we want to transmit the message as rapidly as possible and would like to eliminate the separator entirely.

Another solution is to look for codes that have the **prefix property**, i.e., codes in which no character has a code that is the prefix for any other code. Morse code does not have the prefix property. The code for an *e*, for instance, begins the code for i, h, j, and many other characters as well. The scheme

00—a	110—n
1110—d	011—p
100—e	1111—y
101—l	010—blank

is a coding scheme for the eight characters a, d, e, l, n, p, y and blank, which has the prefix property. That is, if we receive, for instance, a transmission beginning with 00, we know that we have received an a; there is no need to wonder if subsequent bits will tell us that 00 was the beginning of some other character. A message and its decoding are presented in Figure 10.35.

Figure 10.35 A message decoded from a variable-length code.

00110010000110111011000100001011110001111
00/110/010/00/011/011/101/100/010/00/010/1110/00/1111
a n a p p l e a d a y

Binary trees provide a simple way to construct and decipher codes with the prefix property. We draw a binary tree having the same number of leaves as the characters needed for our messages. We then assign each character to a leaf. As a character's code, we use a sequence of 0's and 1's. Starting from the root, if a character is in its left subtree, we begin with an 0; if it is in its right subtree, we begin with a 1. We continue to follow the (unique) path to each leaf, adding a 0 or a 1 each time we descend a level. Such codes have the prefix property, since the code for a character, say a, could be the prefix of the code for another, say b, only if a were the ancestor of b. Since all characters are associated with leaves, that cannot happen. The binary tree from which the code for the previous example was constructed is given in Figure 10.36.

Figure 10.36 Binary tree used for constructing a prefix code.

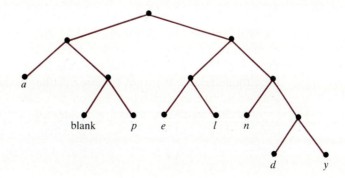

Decoding can also be done from the tree. Starting at the root and at the beginning of the message, follow the branches specified by the sequence of 0's and 1's until reaching a leaf. Then return to the root.

A more challenging problem is to find the optimal coding scheme for a set of characters, given the expected frequency with which they occur in messages. For instance, Table 10.1 shows the average frequency of occurrence of the letters in the English language.

TABLE 10.1 Average Frequency of Occurrence of English Letters

Letter	Frequency in 1000 Letters	Letter	Frequency in 1000 Letters	Letter	Frequency in 1000 Letters
a	82	j	1	s	61
b	14	k	4	t	105
c	28	l	34	u	25
d	38	m	25	v	9
e	131	n	71	w	15
f	29	o	80	x	2
g	20	p	20	y	20
h	53	q	1	z	1
i	63	r	68		

From C. L. Liu, *Elements of Discrete Mathematics*, 2nd ed., p. 199, © 1985. Reprinted by permission of McGraw-Hill Book Company, New York.

Thus the problem is to find the code that will give, on the average, the shortest messages for the set of characters needed in a particular application. Let $\{c_1, \ldots, c_n\}$ denote the set of characters to be coded and let $\{f_1, \ldots, f_n\}$ denote the corresponding frequencies of those characters. Also, denote the length in bits of the code for each character by $\{l_1, \ldots, l_n\}$. Then, by the **weight** of a coding scheme, we mean

$$\sum_{i=1}^{n} f_i l_i.$$

The weight, then, is the number of bits required to represent 1000 letters, including f_1 occurrences of c_1, f_2 occurrences of c_2, etc. It is a measure of the average number of bits in 1000 characters of text using the given coding scheme. By the **optimal code** we mean one that has the prefix property and has minimal weight for these characters and corresponding frequencies over all codes that have the prefix property. Such an optimal code is called a **Huffman code**, after its inventor, David Huffman. Huffman developed the following algorithm for finding the tree from which one can read the optimal code.

The algorithm proceeds by arranging the characters to be coded in order by decreasing frequency. A binary tree is then built, one step at a time. Suppose n characters are to be added. First, a new symbol, v_n, is introduced and made the key of the root of a tree. The least frequent character, c_n, is made its left child and the second least frequent character, c_{n-1}, is made its right child. c_n and its frequency are then deleted from the list. The character c_{n-1} is replaced by the symbol v_n, and the frequency assigned v_n is the sum of the original frequencies of c_n and c_{n-1}. Thus the two characters are, in effect, replaced by the one symbol, v_n. The list is now sorted in order of decreasing frequency and the process is repeated. When it is time to insert the symbol v_n as a left or right child in a tree, the entire tree of which it is the key of the root is inserted. In this way, a tree is built that has the c_i's as leaves and the symbols $v_2, \ldots, v_n$ as internal vertices. This process is formally presented in Algorithm 43.

ALGORITHM 43   Generating Huffman Codes

Procedure Huffman(n, c, f; var T);

{n is the number of characters to be coded; c is an n-element array consisting of those characters; f is an array consisting of the frequencies of the corresponding characters; T is the binary tree from which the Huffman code will be read.}

begin

1.   sort c and f in order of decreasing frequency;

2.   For $i := n$ down to 2 do begin

3.   let T_i be a tree that has only a root and let the symbol v_i be its key;

4.   let the left subtree of T_i be a single node containing just $c[i]$ or the tree of which $c[i]$ is the root;

5.   let the right subtree of T_i be a single node containing just $c[i-1]$ or the tree of which $c[i-1]$ is the root;

6.   replace the current character $c[i-1]$ by the symbol v_i and replace $f[i-1]$ by $f[i-1] + f[i]$

7.   delete $c[i]$ and $f[i]$

8.   reorder c and f in such a way that c is still arranged by decreasing frequency

     end;

9. $T := T_2$

end;

EXAMPLE 8 Use Huffman's algorithm to find a Huffman code for messages, using the letters *a* through *g* and the frequencies given in Table 10.1.

SOLUTION The characters, arranged in decreasing order by their corresponding frequencies are as follows:

e	131
a	82
d	38
f	29
c	28
g	20
b	14

Figure 10.37 presents the results of the first three iterations of the algorithm and the final tree. There are seven characters; in the first iteration, for instance, the frequencies of *b* and *g* are added and assigned as the frequency of v_7. v_7 is then moved up the list to its appropriate place and a tree with root v_7 and *b* and *g* as leaves is constructed. In the second iteration, a tree with root v_6 and leaves *f* and *c* is constructed. The process is repeated until only v_2 is left in the list of characters.

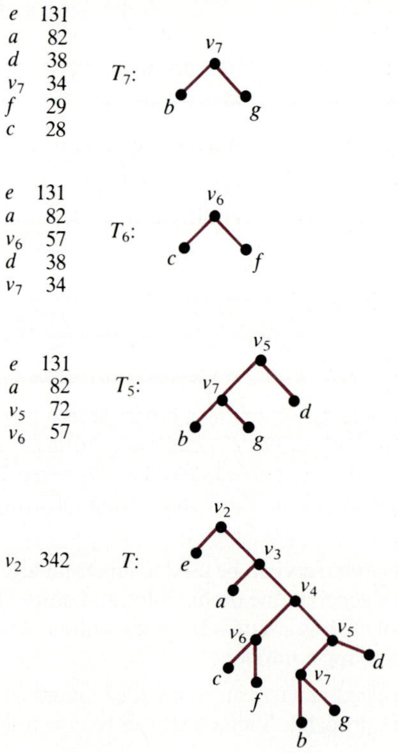

Figure 10.37 Construction of a tree for a Huffman code.

Thus an optimal code can then be read directly from the tree:

b	11100
g	11101
c	1100
f	1101
d	1111
a	10
e	0. ∎

For a proof that Huffman's algorithm generates the minimal tree with the prefix property for any given set of characters and frequencies, see Grimaldi, *Discrete and Combinatorial Mathematics,* Section 15.3 (see References).

TERMINOLOGY

binary search tree	preorder traversal
binary operator	inorder traversal
unary operator	postorder traversal
Polish notation	infix form
reverse Polish notation	prefix property
prefix form	weight of a code
postfix form	optimal code
traversing	Huffman code

SUMMARY

1. Algorithm 38 for searching a binary search tree is in $O(\log_2 n)$ if the tree is balanced. Algorithm 39 provides a way to insert data into a binary search tree, although it does not guarantee a balanced tree. Binary search trees have important applications in the construction of symbol tables for assemblers and compilers.

2. Binary search trees can be used to represent algebraic expressions. Traversals of such trees generate the prefix, infix, and postfix form for expressions; the prefix and postfix forms are unambiguous without parentheses and have applications in constructing compilers.

3. Variable-length codes can be decoded unambiguously as long as the codes have the prefix property. Such codes can be constructed by assigning the characters

to be coded to the leaves of any binary tree and using a sequence of 0's and 1's as the code. Optimal codes with the prefix property can be constructed by means of Algorithm 43.

EXERCISES 10.4

Using the binary search tree of Figure 10.29, list the keys visited in their order of visitation when searching for the following.

1. Swinburne

2. Keats

3. Shakespeare

4. Eliot

Using the binary tree insertion algorithm, add the following poets to the tree of Figure 10.29.

5. Byron

6. Eliot

7. Pound

8. Swinburne

Using the binary tree insertion algorithm, construct binary search trees for the following lists.

9. The letters of "Four score and seven years ago"

10. The numbers 19, 38, 27, 10, 104, 92, 107, 21, 37, 2

11. The symbols point1, point2, *x, y,* entry, eval, total, header

12. The symbols alpha, exit, getdata, testy, num2, num1, format, testx, getflag

Answer the following questions.

13. Suppose that in a particular application of binary search trees, records can only be stored at the leaves. What effect would this have on the size of the tree?

14. Answer Exercise 13 for the height of the tree.

For the trees in Figure 10.38, give the order of visitation of the vertices by the specified traversal.

Figure 10.38

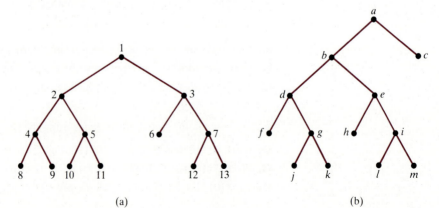

(a) (b)

15. Figure 10.38(a), preorder

16. Figure 10.38(a), postorder

17. Figure 10.38(a), inorder

18. Figure 10.38(b), preorder

19. Figure 10.38(b), postorder

20. Figure 10.38(b), inorder

Represent the following algebraic expressions by binary trees:

21. $\dfrac{a + bc}{2d}$

22. $\dfrac{xy + xz + yz}{x + y + z}$

Write the specified expressions in the indicated notation.

23. Exercise 21; prefix

24. Exercise 22; prefix

25. Exercise 21; postfix

26. Exercise 22; postfix

Change each of the following expressions from the notation indicated to parenthesized infix notation.

27. $+ * a - bcd$; preorder

28. $* - a - bx/yz$; preorder

29. $ab * c * d *$; postorder

30. $pqr + * st/ +$; postorder

Evaluate each of the following expressions as was done in Examples 4 and 5.

31. Exercise 27; $a = 10$, $b = 2$, $c = 2$, $d = 6$

32. Exercise 28; $a = 6$, $b = 5$, $x = 4$, $y = 10$, $z = 2$

33. Exercise 29; $a = 1$, $b = 2$, $c = 3$, $d = 4$

34. Exercise 30; $p = 1$, $q = 2$, $r = 3$, $s = 6$, $t = 3$

Answer the following questions.

35. Are trees that represent expressions that have no unary operators necessarily full? Why or why not?

36. Are trees that represent expressions that have no unary operators necessarily balanced? Why or why not?

37. Show how binary trees can be used to represent expressions with unary operators by representing $(-ab)/(a - b)$.

38. This is the same question as for Exercise 37, but with the expression $-(ab/(a^2 + b^2))$.

39. Would the presence of unary operators affect the answers to Exercises 35 and 36?

40. Can variables ever be represented by an internal vertex? Why or why not?

41. Can operators ever be represented by a leaf? Why or why not?

42. Find three distinct trees, all of which can be represented by the infix expression $a - b * c/d$.

In Exercises 43–46, decode the message using the specified prefix code.

43. 000100111010011110100111100110; Figure 10.36.

44. 00110111101010110000110010011001101010101; Figure 10.36.

45. 11011011110101100101110101; Figure 10.37.

46. 11010011111011100101101; Figure 10.37.

47. Compute the weight of the code in Figure 10.37.

48. Compute the weight of the following code, using the frequencies of Table 10.1.

a—1

b—110

c—100

d—01

e—0

f—10

g—101.

Compare this weight to that of the code constructed in Example 8. Which is heavier? Why?

Using the frequencies in Table 10.1, construct Huffman codes for the following character sets.

49. $\{e, i, s, h, t, m, o\}$

50. $\{x, k, z, q, j, v\}$

51. $\{m, n, o, p, q, r\}$

REVIEW EXERCISES—CHAPTER 10

1. Prove that if a tree with n vertices has a path of length $n - 1$, the largest degree of any vertex is 2.

2. Prove that if a tree has a vertex of degree 3, there exist at least three vertices of degree 1.

3. Generalize Exercise 2 to a statement about vertices of degree k and prove it.

4. Find two other examples of hierarchies that can be represented by trees.

5. Does the depth-first spanning tree always include the longest possible path in a graph? Why or why not?

6. Is the length of the path from the initial vertex, v_0, to another vertex, v, in a breadth-first spanning tree necessarily equal to the distance from v_0 to v? Why or why not?

7. Generalize Kruskal's algorithm to handle multigraphs.

8. Show that if all edges in a weighted graph have different weights, Kruskal's algorithm finds the unique minimal spanning tree.

Generalize the following results to m-ary trees and prove them.

9. Theorem 10.10(b). (*Hint:* the generalization is $L \geq m + (m - 1)(h - 1)$.)

10. Corollary 10.12

The *skew* of a full binary tree can be defined as the difference between the lengths of the longest and shortest paths to its leaves.

11. Show that a balanced tree has a skew of 0 or 1.

12. Find the maximum possible value of the skew and an example of a full binary tree with such a skew.

13. Find the upper and lower bounds on the number of vertices in a full binary tree with skew equal to 2 and height h.

14. Suppose the concept of skew were applied to trees that are not full. Find examples of trees with skew 0 and skew $h - 1$.

15. Modify the binary search tree algorithm so that it is iterative rather than recursive.

16. Argue that the algorithm you developed in Exercise 15 is in $O(\log_2 n)$ if the tree is balanced.

17. Represent the boolean expression $(a \vee b) \wedge (b \rightarrow c)$ by a binary tree.

18. Represent the boolean expression $((a \vee \neg b) \rightarrow b) \vee c)$ by a binary tree.

19. Using the frequencies in Table 10.1, construct a Huffman code for $\{a, e, i, o, u, y, h, s, t\}$.

20. Show that any code with the prefix property can be represented by a binary tree.

REFERENCES

Aho, Alfred V., John E. Hopcroft, and Jeffrey D. Ullman, *Data Structures and Algorithms.* Reading, Mass.: Addison-Wesley, 1983.

Grimaldi, Ralph, *Discrete and Combinatorial Mathematics.* Reading, Mass.: Addison-Wesley, 1985.

Horowitz, Ellis, and Sartaj Sahni, *Fundamentals of Data Structures.* Rockville, Md.: Computer Science Press, 1976.

Knuth, Donald, *The Art of Computer Programming,* Vol. 1., *Fundamental Algorithms.* Reading, Mass.: Addison-Wesley, 1973.

Liu, C.L., *Elements of Discrete Mathematics.* New York: McGraw-Hill, 1977.

Stanat, Donald, and David McAllister, *Discrete Mathematics in Computer Science.* Englewood Cliffs, N.J.: Prentice-Hall, 1977.

Appendix A
Arrays

A great deal of scientific and commercial work involves the management of data—often a large amount of data. These data typically exist in the form of lists and tables. For instance, a physics experiment might involve taking a measurement every second. The resulting numbers can form a substantial list. Alternatively, a stock market report might include the performance of several stocks over many days. The resulting data can be organized into a table.

While lists and tables may not appear very exciting, they give rise to one of the most useful and interesting mathematical models—the array. One-dimensional arrays are also called *vectors* and two or more dimensional arrays are called *matrices*. Virtually any situation involving information about a finite set gives rise to a vector or a matrix.

The Concept of the Array

Definition   An **array** (or **vector**) is a ordered n-tuple $[x_1, \ldots, x_n]$, where each x_i is an element of some specified set.

Alternatively and more precisely, an array can be defined as a function with domain $\{1, \ldots, n\}$ and any codomain. We will use the ordered n-tuple idea as the principal definition because we can visualize 2-tuples as points in planes and 3-tuples as points in space. Hence we can visualize an n-tuple as a point in n-dimensional space.

Subscripts

Note particularly the use of **subscripts**—the 1, i, and n below the x in the definition. Unlike the concept of a set, in which the order of the elements is not specified (the set {3, 1, 2} is exactly the same as the set {1, 2, 3}), it is essential in an array to know which is first, which is second, etc. The subscripts enable us to do this. Just as with sigma notation, the i is called the **index**.

EXAMPLE 1   The following are all arrays:

$$[1, 4, 5, 3, 7]   [8, -1, 2]   [16, 0]   [17.5, 67.5, 91, 31.2].$$ ▮

The arrays in Example 1 are called **row vectors.**

EXAMPLE 2   The following are **column vectors.** We regard them as different from the vectors in Example 1, even though the numerical values in each are the same.

$$\begin{bmatrix} 1 \\ 4 \\ 5 \\ 3 \\ 7 \end{bmatrix} \quad \begin{bmatrix} 8 \\ -1 \\ 2 \end{bmatrix} \quad \begin{bmatrix} 16 \\ 0 \end{bmatrix} \quad \begin{bmatrix} 17.5 \\ 67.5 \\ 91 \\ 31.2 \end{bmatrix}$$ ▮

Historically, vectors originated in problems of geometry and physics in which scientists needed to represent quantities such as position, velocity, acceleration, and force in terms of their two- or three-dimensional components. Thus, of the four vectors presented in Example 1, only $[8, -1, 2]$ and $[16, 0]$ would have been recognized as vectors. More recently, the concept of vector has been broadened to include any type of list, not just vectors that are basically geometric. Thus, for instance, a grocery list is a vector. The words *array* and *vector* are used almost interchangeably, although *vector* has a more geometric connotation and is preferred by mathematicians. Computer scientists speak more frequently of *arrays.*

Applications of Arrays

In the following examples, we will examine several typical situations to which the concept of the array can be applied.

EXAMPLE 3 Suppose there are 10 students. Their scores and names can be collected into arrays such as

$$G = \begin{bmatrix} 65 \\ 69 \\ 71 \\ 75 \\ 82 \\ 82 \\ 83 \\ 90 \\ 93 \\ 98 \end{bmatrix} \qquad N = \begin{bmatrix} \text{B. Smith} \\ \text{H. Jackson} \\ \text{S. DeVries} \\ \text{P. Correa} \\ \text{T. Volger} \\ \text{H. Antesa} \\ \text{P. Gluchowski} \\ \text{N. Chu} \\ \text{J. Scholz} \\ \text{E. Liddel} \end{bmatrix}$$

If the arrays have been set up properly, the names and scores should be in corresponding positions. Thus, for example, $G_1 = 65$ and $N_1 = $ B. Smith; i.e., B. Smith got a 65 on the test. Similarly, $G_7 = 83$ and $N_7 = $ P. Gluchowski. Mathematicians normally use single-letter names and subscripts, whereas computer scientists use longer names and brackets. Thus, G_1 would be written in a computer language as perhaps Grades[1] and N_1 as Names[1]. ∎

EXAMPLE 4 Suppose a heart monitoring device measures the pulse rate every 15 seconds. Successive observations can be collected in an array. For instance, if D denotes the name of the array, we could write

$$D = \begin{bmatrix} 68.5 \\ 68.9 \\ 69.0 \\ 68.7 \\ 69.1 \\ \cdot \\ \cdot \\ \cdot \end{bmatrix}$$

Then $D_1 = 68.5$, $D_2 = 68.9$, $D_3 = 69.0$, etc. Often arrays used in the collection of scientific data are very large, sometimes including tens of thousands of observations. ∎

EXAMPLE 5 Represent the coefficients of the polynomials $x^2 + 2x + 2$ and $3x^3 + 2x^2 + 2x + 1$ by arrays.

SOLUTION The polynomial $x^2 + 2x + 2$ can be denoted $[2,2,1]$, putting the a_0 coeffecent on the left. Similarly, the polynomial $3x^3 + 2x^2 + 2x + 1$ can be denoted $[1, 2, 2, 3]$. ∎

EXAMPLE 6 Represent the frequencies of the letters available in a Scrabble game by an array.

SOLUTION For instances, the array

$$F = \begin{bmatrix} 9 \\ 2 \\ 2 \\ 4 \\ 12 \\ 2 \\ 3 \\ 2 \\ 9 \\ 1 \\ 1 \end{bmatrix}$$

represents the frequency of the letters A through J in a complete Scrabble set. Thus $F[1] = 9$, $F[2] = 2$, etc. If one of the A's were lost, for instance, $F[1]$ would drop to 8.

We have said that the concept of array can be defined as a function from $\{1, \ldots, n\}$ into some codomain. This can be easily generalized to other domains. For instance, in this example the domain can be taken as the letters A through Z plus the word *blank,* since there are two blank disks in a Scrabble game. Using this domain, we would write $F['A'] = 9$, $F['B'] = 2$. $F['Blank'] = 2$, etc. ∎

EXAMPLE 7 Suppose a four-unit by nine-unit rectangle is located in the first quadrant of a coordinate plane with one corner at the origin. Represent all four corners of the rectangle as arrays.

SOLUTION First, sketch the rectangle (Figure A.1).

Figure A.1 A four-unit by nine-unit rectangle with one corner at the origin.

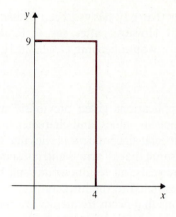

The four corners can be represented by the vectors [0,0], [0,9], [4,9], [4,0] if we start at the origin and move around the rectangle counterclockwise. ▮

Boolean Arrays

Definition A **Boolean array** is an array in which each x_i is either a 0 or a 1.

EXAMPLE 8 Represent the subsets of the set $S = \{a, b, c, d, e, f\}$ by means of a Boolean array.

SOLUTION Use a 1 to indicate that an element belongs to a particular subset and a 0 to indicate that it doesn't belong. Suppose

$$A = \begin{bmatrix} 1 \\ 0 \\ 0 \\ 1 \\ 1 \\ 0 \end{bmatrix} \quad \text{and } B = \begin{bmatrix} 0 \\ 0 \\ 1 \\ 0 \\ 0 \\ 1 \end{bmatrix}$$

Then A represents the set $\{a,d,e\}$ and B represents $\{c,f\}$. ▮

EXAMPLE 9 Consider the traveling salesman problem once again. Suppose we want to start at city A and visit six other cities, B, C, D, E, F, and G. We can use a six-element array called V (or *Visited*) in which each element indicates whether that city has yet been visited. Thus, initially, V would equal [0,0,0,0,0,0]. However, after, say, B, E, and G had been visited but not C, D, and F, V would become [1,0,0,1,0,1]. ∎

Algebraic Properties of Arrays

Note that the applications given previously used codomains that were integers, reals, boolean values, and character strings. Arrays of integers or reals have some special algebraic properties that we must examine. For simplicity, we will assume that all arrays and constants to be used in the following discussion are reals; all results would still be true even if arrays and constants were restricted to the integers.

Two operations that occur frequently are adding vectors and multiplying a vector by a constant. For instance, suppose a company sells four products—widgets, gidgets, gadgets, and hammers. Four units of widgets, three units of gidgets, one unit of gadgets, and six units of hammers are shipped to customer A. This situation is represented by the vector [4,3,1,6]. If customer B buys one unit of widgets, two units of gidgets, five units of gadgets, and four units of hammers, this is represented by the vector [1,2,5,4]. Combining these to get the total sales to the two customers we add the vector [4,3,1,6] to the vector [1,2,5,4]. The resulting vector is [5,5,6,10], the result of adding each of the items (or **components**) individually. Similarly, suppose company A submits the preceding request and later asks to double its order. This new order can be represented by [8,6,2,12], with each component doubled. If a customer does not order anything in a particular month, this situation is represented by the vector [0,0,0,0], called the **zero vector**. This example illustrates the concepts underlying the following definitions.

Addition and Scalar Multiplication of Vectors

Definition Let $u = [u_1, u_2, \ldots, u_n]$, let $v = [v_1, v_2, \ldots, v_n]$, and let α be any real number, also called a **scalar.** Then

$$u + v = [u_1 + v_1, u_2 + v_2, \ldots, u_n + v_n]$$
$$\alpha u = [\alpha u_1, \alpha u_2, \ldots, \alpha u_n]$$
$$-u = [-u_1, -u_2, \ldots, -u_n]$$
$$u - v = u + (-v).$$

EXAMPLE 10 Let

$$u = [1,4,3,6], \quad v = [8,5,7,1], \text{ and } w = [2,2,3,1]$$

Find: **a)** $u + v$ **b)** $v + w$ **c)** $3u$ **d)** $-w$ **e)** $u - 2w$
f) $2u + 3w.$

SOLUTION

a) $u + v = [1,4,3,6] + [8,5,7,1] = [9,9,10,7]$

b) $v + w = [8,5,7,1] + [2,2,3,1] = [10,7,10,2]$

c) $3u = 3[1,4,3,6] = [3,12,9,18]$

d) $-w = [-2,-2,-3,-1]$

e) $u - 2w = [1,4,3,6] + [-4,-4,-6,-2] = [-3,0,-3,4]$

f) $2u + 3w = [2,8,6,12] + [6,6,9,3] = [8,14,15,15].$ ∎

Properties of Addition

The following theorem summarizes the basic algebraic properties of **vector addition.**

THEOREM A.1 Let $u = [u_1, u_2, \ldots, u_n]$, $v = [v_1, v_2, \ldots, v_n]$, $w = [w_1, w_2, \ldots, w_n]$, and let $\mathbf{0}$ denote the zero vector. Then

a) $u + v = v + u$

b) $u + (v + w) = (u + v) + w$

c) $u + \mathbf{0} = u$

d) $u + (-u) = \mathbf{0}.$

PROOF

a) $u + v = [u_1 + v_1, u_2 + v_2, \ldots, u_n + v_n]$
$= [v_1 + u_1, v_2 + u_2, \ldots, v_n + u_n]$
$= v + u$

b) (Left for the Exercises.)

c) $u + \mathbf{0} = [u_1 + 0, u_2 + 0, \ldots, u_n + 0]$
$= [u_1, u_2, \ldots, u_n]$
$= u$

d) (Left for the Exercises.) ☐

Properties of Scalar Multiplication

THEOREM A.2 Let $u = [u_1, u_2, \ldots, u_n]$ and let $v = [v_1, v_2, \ldots, v_n]$. Also, let α and β be any real numbers. Then

a) $(\alpha + \beta)u = \alpha u + \beta u$

b) $\alpha(u + v) = \alpha u + \alpha v$

c) $(\alpha\beta)u = \alpha(\beta u)$

d) $1 \cdot u = u.$

PROOF

a) $(\alpha + \beta)u = [(\alpha + \beta)\, u_1, (\alpha + \beta)\, u_2, \ldots, (\alpha + \beta)\, u_n]$
$= [\alpha u_1 + \beta u_1, \alpha u_2 + \beta u_2, \ldots, \alpha u_n + \beta u_n]$
$= [\alpha u_1, \alpha u_2, \ldots, \alpha u_n] + [\beta u_1, \beta u_2, \ldots, \beta u_n]$
$= \alpha[u_1, u_2, \ldots, u_n] + \beta[u_1, u_2, \ldots, u_n]$
$= \alpha u + \beta.$

b) (Left for the Exercises.)

c) $(\alpha\beta)u = (\alpha\beta)[u_1, u_2, \ldots, u_n]$
$= [(\alpha\beta)u_1, (\alpha\beta)u_2, \ldots, (\alpha\beta)u_n]$
$= [\alpha(\beta u_1), \alpha(\beta u_2), \ldots, \alpha(\beta u_n)]$
$= \alpha[\beta u_1, \beta u_2, \ldots, \beta u_n)]$
$= \alpha(\beta u)$

d) (Left for the Exercises.)   ☐

The Dot Product

We have now examined the process of adding two vectors and the process of multiplying a vector by a scalar. But we have not discussed the process of multiplying a vector by another vector. There are several ways to do this. However, there is one that is of particular interest to us here: the **dot product.** It has many applications in economics as well as in the natural sciences. Perhaps the easiest way to understand it is to return to the small business we looked at earlier. If you recall, we assumed that customer A ordered four units of widgets, three units of gidgets, one unit of gadgets, and six units of hammers. This formed a vector [4,3,1,6] that is called the **quantity vector.** Suppose these items were sold at the following unit prices: widgets, \$4.50; gidgets, \$13.50; gadgets, \$2.00; and hammers, \$14.00. These amounts can be collected into the following **price vector:** [4.5,13.5,2,14]. The total value of the sale is the sum of the price times the quantity of each item. That is, the total value $= (4 \cdot 4.50) + (3 \cdot 13.5) + (1 \cdot 2) + (6 \cdot 14) = 18 + 40.5 + 2 + 84 = \$144.50$. In other words, the two vectors, price and quantity, are multiplied, component by component, and then the results are added. This process is more formally stated as follows:

Definition   Let $u = [u_1, u_2, \ldots, u_n]$ and let $v = [v_1, v_2, \ldots, v_n]$. Then we denote the **dot product** of u and v by $u \cdot v$ and define it to be

$$u \cdot v = u_1 v_1 + u_2 v_2 + \cdots + u_n v_n.$$

The dot product is also sometimes called the **inner product** or **scalar product.** We will avoid the term *scalar product* because it so similar to the term *scalar multiplication,* previously discussed. Note that the dot product of two vectors is a scalar but that the product of a scalar and a vector is a vector.

EXAMPLE 11   Again, let

$$u = [1,4,3,6]$$
$$v = [8,5,7,1]$$
$$w = [2,2,3,1].$$

Find $u \cdot v$, $u \cdot w$, $v \cdot w$, and $w \cdot v$.

SOLUTION

$$u \cdot v = 8 + 20 + 21 + 6 = 55$$
$$u \cdot w = 2 + 8 + 9 + 6 = 25$$
$$v \cdot w = 16 + 10 + 21 + 1 = 48$$
$$w \cdot v = 16 + 10 + 21 + 1 = 48. \quad \blacksquare$$

THEOREM A.3   Let $u = [u_1, u_2, \ldots, u_n]$, $v = [v_1, v_2, \ldots, v_n]$, and $w = [w_1, w_2, \ldots, w_n]$. Also, let α be any real number. Then

a) $u \cdot v = v \cdot u$

b) $u \cdot (v + w) = u \cdot v + u \cdot w$

c) $(\alpha u) \cdot v = \alpha(u \cdot v) = u \cdot (\alpha v)$

d) $0 \cdot u = 0.$

PROOF

a) $u \cdot v = u_1 v_1 + u_2 v_2 + \cdots + u_n v_n$
$ = v_1 u_1 + v_2 u_2 + \cdots + v_n u_n$
$ = v \cdot u$

b) (Left for the Exercises.)

c) $(\alpha u) \cdot v = [\alpha u_1, \alpha u_2, \ldots, \alpha u_n] \cdot [v_1, v_2, \ldots, v_n]$
$ = (\alpha u_1) v_1 + (\alpha u_2) v_2 + \cdots + (\alpha u_n) v_n$
$ = \alpha(u_1 v_1) + \alpha(u_2 v_2) + \cdots + \alpha(u_n v_n)$
$ = \alpha(u_1 v_1 + u_2 v_2 + \cdots + u_n v_n)$
$ = \alpha(u \cdot v)$
(Remainder is left for the Exercises.)

d) (Left for the Exercises.) $\square$

TERMINOLOGY

array	scalar
vector	vector addition
subscript	vector subtraction
index	scalar multiplication
row vector	dot product
column vector	quantity vector
Boolean array	price vector
components	inner product
zero vector	sclaar product

SUMMARY

1. Arrays can be used to model many situations, such as tabulating data, representing polynomials and subsets, and tabulating frequencies.

2. The major properties of numerical arrays are as follows:

Theorem 1
   a) $u + v = v + u$
   b) $u + (v + w) = (u + v) + w$
   c) $u + \mathbf{0} = u$
   d) $u + (-u) = \mathbf{0}.$

Theorem 2
   a) $(\alpha + \beta)u = \alpha u + \beta u$
   b) $\alpha(u + v) = \alpha u + \alpha v$
   c) $(\alpha \beta)u = \alpha(\beta u)$
   d) $1 \cdot u = u.$

Theorem 3
   a) $u \cdot v = v \cdot u$
   b) $u \cdot (v + w) = u \cdot v + u \cdot w$
   c) $(\alpha u) \cdot v = \alpha(u \cdot v) = u \cdot (\alpha v)$
   d) $\mathbf{0} \cdot u = 0.$

EXERCISES

Represent each of the following polynomials using an array.

1. $3x^2 + 2x - 1$ **2.** $1 - 3x - x^3$

3. $x^4 - 2x^2 + 9$ **4.** $2x^3 - x^5 + 1$

Let $S = \{1, 2, 3, 4, 5, 6, 7\}$. Represent each of the following subsets of S using an array.

5. $\{1, 3, 5, 7\}$ **6.** $\{2, 4, 6\}$

7. $\{1, 2, 3\}$ **8.** $\{1, 7\}$

9. $\emptyset$ **10.** $\{1, 2, 3, 4, 5, 6, 7\}$

11. Suppose a three-unit by seven-unit rectangle is located in the first quadrant of a coordinate plane with one corner at the origin. Represent all four corners of the rectangle as arrays.

12. Suppose the rectangle in Exercise 11 is moved so that its sides remain parallel to both axes but the corner that was at the origin is moved to the point (2,3). Now represent all four corners as arrays.

13. Suppose a rectangular solid has one corner at the origin, has sides of 2, 3, and 4, and is located in the first octant. Represent all of its corners by arrays.

14. Suppose the rectangle in Exercise 13 is shifted parallel to each axis so that the corner that was at the origin is moved to the point (2,1,3). Now represent all of the corners by arrays.

15. Represent the frequencies of each letter in the sentence "The quick brown fox jumped over the lazy dog's back" by an array with domain 'A' $\cdots$ 'Z'.

16. Represent the frequencies of the digits 0 to 9 as they occur in the statements of Exercises 1 to 10 by an array. (Count the question number, as well as the digits that appear in the question.)

For Exercises 17 through 26, let $u = [4, 0, 8, 3]$, $v = [1, -1, 2, 6]$, and $w = [2, 1, 3, 6]$. Find:

17. $u + v$ **18.** $u - v$

19. $2w$ **20.** $-w$

21. $u + 2w$ **22.** $3u - 2v$

23. $u + v + w$ **24.** $v - w - u$

25. x such that $u + v = x + w$ **26.** y such that $y + v = 2w$

For Exercises 27 through 34, verify the following for $u = [4, 0, 8, 3]$, $v = [1, -1, 2, 6]$, and $w = [2, 1, 3, 6]$. Let α and β be any real numbers.

27. $u + v = v + u$ **28.** $u + (v + w) = (u + v) + w$

29. $u + 0 = u$ **30.** $u + (-u) = 0$

31. $(\alpha + \beta) u = \alpha u + \beta u$ **32.** $\alpha(u + v) = \alpha u + \alpha v$

33. $(\alpha\beta)u = \alpha(\beta u)$ **34.** $1 \cdot u = u$

For Exercises 35 through 38, verify the following properties for $u = [u_1, \ldots u_n]$, $v = [v_1, \ldots, v_n]$, and $w = [w_1, \ldots, w_n]$. Let α and β be any real numbers.

35. $u + (v + w) = (u + v) + w$ **36.** $u + (-u) = 0$

37. $\alpha(u + v) = \alpha u + \alpha v$ **38.** $1 \cdot u = u$

39. Let $u = [u_1, \ldots u_n]$, $v = [v_1, \ldots, v_n]$. If $u + v = v$, show that $u = 0$.

40. Let $u = [u_1, u_2]$ and let $v = [v_1, v_2]$. Find examples of a u and a v such that it is impossible for αu to equal v for any α.

In each of the following situations, there are products and prices. Represent each by a quantity vector and a price vector and compute the requested item.

41. A publisher produces four different magazines. *Modern Living* has a monthly circulation of 120,000 and sells for $1.50 per copy; *Northern Lights* has a circulation of 18,000 and sells for $3.50 per copy; *Wine and Cheese Tasting* has a circulation of 75,000 and sells for $1.95; *Space Digest* has a circulation of 95,000 and sells for $0.95. Find the gross income.

42. An electric power generating plant burns coal to produce steam and uses the steam to drive its generators. In an hour of operation, it uses 30 tons of coal A, twice as much coal B, and one and one-half times as much of coal C. Coal A costs $14 per ton, B costs $10 per ton, and C costs $12 per ton. Find the total cost of coal for a hour's operation.

Let $u = [5, 2, 1, 0, 7]$, $v = [-1, 2, 3, 1, 2]$, and $w = [0, 6, 2, 1, -2]$. Compute the following.

43. $u \cdot v$ **44.** $v \cdot u$

45. $v \cdot w$ **46.** $w \cdot u$

For u, v, and w as previously given and α and β representing any real numbers, verify the following laws.

47. $u + v = v + u$ **48.** $u + (v + w) = (u + v) + w$

49. $u + 0 = u$ **50.** $u + (-u) = 0$

51. $(\alpha + \beta)u = \alpha u + \beta u$ **52.** $\alpha(u + v) = \alpha u + \alpha v$

53. $(\alpha\beta)u = \alpha(\beta u)$ **54.** $1 \cdot u = u$

For Exercises 55 through 57, verify the following properties for $u = [u_1, \ldots u_n]$, $v = [v_1, \ldots, v_n]$, and $w = [w_1, \ldots, w_n]$. Let α and β be any real numbers.

55. $u \cdot (v + w) = u \cdot v + u \cdot w$ **56.** $\alpha(u \cdot v) = u \cdot (\alpha v)$

57. $0 \cdot u = 0$

58. Let u^2 denote $u \cdot u$. Show that u^2 is always greater than or equal to zero.

59. Show that $u^2 = 0$ if and only if $u = 0$.

60. Show that $(u + v)^2 = u^2 + 2u \cdot v + v^2$.

61. Show that $(u - v)^2 = u^2 - 2u \cdot v + v^2$.

The vectors u and v are orthogonal if $u \cdot v = 0$. Test whether the following vectors are orthogonal.

62. $[3,1,7]$ and $[1,1,-1]$ **63.** $[4,0,-1,2]$ and $[0,7,2,1]$

64. $[6,3,-3,-1]$ and $[1,-1,0,3]$ **65.** $[18,4,-7]$ and $[1,-3,1]$

66. Find α such that $[3,-1,7]$ and $[2,0,\alpha]$ are orthogonal.

67. Find any values of α and β such that $[\beta,-1,7,2]$ and $[3,1,\alpha,6]$ are orthogonal.

Appendix B
Introduction to Matrices

In the previous appendix we examined arrays. We noted that arrays can be looked upon as functions whose domain is the set $\{1, \cdots, n\}$. But data often exist not just as a function of one variable but as a function of two or more variables. Tables of data are typical examples. Such situations are frequently modeled by **matrices,** also called **rectangular arrays,** which we will soon define formally. We will discuss only two-dimensional matrices for simplicity and ease of visualization and because they have so many significant applications. But keep in mind that the principles developed here can also be generalized to higher dimensions.

Definition   A **matrix,** A, is a rectangular display of numbers:

$$A = \begin{bmatrix} a_{11} \cdots a_{1n} \\ a_{m1} \cdots a_{mn} \end{bmatrix}$$

Any domain may be chosen for the numbers, most often $\mathbf{R}$, $\mathbf{Z}$, or $\mathbf{Z}_2$, although there are times when we even drop the restriction that the matrix must consist of numbers and allow matrices of characters, character strings, etc. Thus, for instance,

$$\begin{bmatrix} 1 & 2 \\ 3 & 5 \end{bmatrix} \quad \begin{bmatrix} 2 & 5 & 6 \\ 1 & 8 & 9 \\ 9 & 5 & 3 \\ 8 & 1 & 2 \end{bmatrix} \quad \begin{bmatrix} 3 & 3 & 6 \\ 8 & 6 & -1 \\ 1 & 2 & 4 \end{bmatrix}$$

are all matrices of integers.

A **row** or **row vector** is the set of all horizontal entries for some value of *i*. Thus, for instance, (3, 5) is the row vector given by the second row of the matrix.

$$\begin{bmatrix} 1 & 2 \\ 3 & 5 \end{bmatrix}$$

Similarly, a **column,** or **column vector** is the set of all vertical entries for some value of *j*. For instance,

$$\begin{bmatrix} 1 \\ 3 \end{bmatrix}$$

is the column vector given by the first column of the preceding matrix.

The notation

$$(a_{ij})\ 1 \leq i \leq m,\ 1 \leq j \leq n$$

is often used to denote a matrix. *m* stands for the number of rows and *n* for the number of columns. *A* is called an **m × n** matrix. Similarly, the notation a_{ij} (with no parentheses) stands for the number that occurs in the *i*th row and *j*th column. Note that in both notations the row is always specified first and the column second. Thus, in the matrix

$$\begin{bmatrix} 1 & 2 \\ 3 & 5 \end{bmatrix}$$

$a_{11} = 1$, $a_{12} = 2$, $a_{21} = 3$, and $a_{22} = 5$.

EXAMPLE 1   For the matrix

$$\begin{bmatrix} 2 & 5 & 6 \\ 1 & 8 & 9 \\ 9 & 5 & 3 \\ 8 & 1 & 2 \end{bmatrix}$$

specify the values of *m* and *n* and identify a_{12}, a_{21}, a_{23}, and a_{32}.

SOLUTION $m = 4$, $n = 3$, $a_{12} = 5$, $a_{21} = 1$, $a_{23} = 9$, and $a_{32} = 5$.   ∎

EXAMPLE 2   Matrices can be used to represent systems of simultaneous equations. For instance, the left-hand sides of the system of equations

$$2x + 3y - z = 0$$
$$x + 2y + 2z = 7$$
$$-2x - 3y + z = 16$$

can be represented by the following **coefficient matrix:**

$$\begin{bmatrix} 2 & 3 & -1 \\ 1 & 2 & 2 \\ -2 & -3 & 1 \end{bmatrix}$$

The right-hand side can be represented by the following column vector:

$$\begin{bmatrix} 0 \\ 7 \\ 16 \end{bmatrix}. \quad \blacksquare$$

EXAMPLE 3 Matrices can be used to represent weighted graphs—for instance, the following traveling salesman problem from Chapter 1 (Figure B.1):

Figure B.1 Five-city traveling salesman problem.

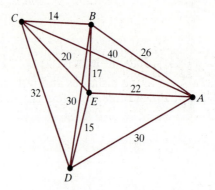

Let the rows stand for the starting city, the columns for the terminating city, and the entries for the distance between them:

	A	B	C	D	E
A	0	26	40	30	22
B	26	0	14	30	17
C	40	14	0	32	20
D	30	30	32	0	15
E	22	17	20	15	0

$\blacksquare$

Figure B.1 is a **complete graph,** i.e., there is a direct route from every city to every other city.

EXAMPLE 4   In Example 3, we used the entries in the matrix to represent the distances between cities. In the case when a graph may or may not be complete, we can also use matrices to indicate whether there is a direct route between cities (Figure B.2).

Figure B.2 Five cities, not all connected by direct routes.

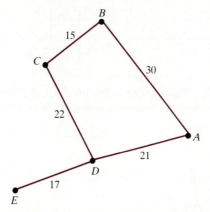

In this case, we ignore the distances and just enter a 0 or a 1 to indicate whether there is a direct route from one city to another.

	A	B	C	D	E
A	1	1	0	1	0
B	1	1	1	0	0
C	0	1	1	1	0
D	1	0	1	1	1
E	0	0	0	1	1

Any matrix such as this, in which the entries are elements of $\mathbf{Z}_2$, is called a **Boolean matrix.** ∎

Some Special Matrices

There are a few patterns in the arrangement of the entries in a matrix that occur frequently enough to be given special names. Five of the most common patterns are as follows.

The **zero matrix,** denoted **0,** is a matrix all of whose entries are zero. For instance,

$$\begin{bmatrix} 0 & 0 & 0 \\ 0 & 0 & 0 \\ 0 & 0 & 0 \end{bmatrix}$$

is the 3×3 zero matrix.

A **square matrix** is one with $m = n$, i.e, with the same number of rows as columns. For example,

$$\begin{bmatrix} 1 & 3 & 5 \\ 3 & 0 & 7 \\ 2 & 4 & 6 \end{bmatrix}$$

is a square matrix, as are the coefficient matrix in Example 3 and the matrices in Examples 3 and 4.

A **diagonal matrix** is a square matrix in which all entries are zero except possibly a_{11}, a_{22}, etc. That is, $a_{ij} = 0$ whenever $i \neq j$. For instance,

$$\begin{bmatrix} 1 & 0 \\ 0 & 2 \end{bmatrix} \qquad \begin{bmatrix} 3 & 0 & 0 \\ 0 & 0 & 0 \\ 0 & 0 & -1 \end{bmatrix}$$

are both diagonal matrices. The set of entries for which $i = j$ (a_{11}, a_{22}, etc.) is called the **main diagonal.** Note that having zero entries on the main diagonal is permissible.

A **symmetric matrix** is a matrix for which the corresponding entries on opposite sides of the main diagonal are equal, i.e., $a_{ij} = a_{ji}$ for all i and j. For example,

$$\begin{bmatrix} 1 & 3 & 2 \\ 3 & 4 & 6 \\ 2 & 6 & 7 \end{bmatrix} \qquad \begin{bmatrix} 0 & 1 & 7 \\ 1 & 9 & 8 \\ 7 & 8 & 3 \end{bmatrix}$$

are symmetric matrices, as are the matrices in Examples 3 and 4. Note that symmetric matrices are necessarily square.

An **upper triangular matrix** is a matrix in which all of the entries below the main diagonal are zero, i.e., $a_{ij} = 0$ whenever $i > j$. For instance,

$$\begin{bmatrix} 2 & 3 & 6 \\ 0 & 1 & 8 \\ 0 & 0 & 1 \end{bmatrix}$$

is an upper triangular matrix. The **lower triangular matrix** is defined in a similar fashion (see the Exercises). A matrix is called **triangular** if it is either upper or lower triangular.

An **identity matrix** is a diagonal matrix in which all of the entries on the main diagonal are 1, i.e., $a_{ij} = 1$ if $i = j$ and $a_{ij} = 0$ if $i \neq j$. For instance,

$$\begin{bmatrix} 1 & 0 & 0 \\ 0 & 1 & 0 \\ 0 & 0 & 1 \end{bmatrix}$$

is an identity matrix.

We must present one more concept before proceeding to matrix operations.

Definition Two matrices, A and B, are **equal** if they are the same size and if all of their corresponding entries are equal, i.e., both must be $m \times n$ and $a_{ij} = b_{ij}$ $\forall i, j$.

Addition of Matrices

Definition Suppose we have two $m \times n$ matrices, A and B, where

$$A = \begin{bmatrix} a_{11} & \cdots & a_{1n} \\ \cdot & & \cdot \\ \cdot & & \cdot \\ \cdot & & \cdot \\ a_{m1} & \cdots & a_{mn} \end{bmatrix} \qquad B = \begin{bmatrix} b_{11} & \cdots & b_{1n} \\ \cdot & & \cdot \\ \cdot & & \cdot \\ \cdot & & \cdot \\ b_{m1} & \cdots & b_{mn} \end{bmatrix}$$

By $A + B$ we mean the matrix formed by adding each of the individual entries of A and B. That is,

$$A + B = \begin{bmatrix} a_{11} + b_{11} & \cdots & a_{1n} + b_{1n} \\ \cdot & & \cdot \\ \cdot & & \cdot \\ \cdot & & \cdot \\ a_{m1} + b_{m1} & \cdots & a_{mn} + b_{mn} \end{bmatrix}$$

Alternatively, we can write $A + B = (a_{ij} + b_{ij})$; i.e., $A + B$ is the matrix whose ijth entry is $a_{ij} + b_{ij}$. Note that matrix addition is undefined if the two matrices are not the same size.

Properties of Matrix Addition

There are two major properties of matrix addition that we must prove. These properties are stated in the following theorem.

THEOREM B.1 Let A and B be $m \times n$ matrices. Then

a) $A + B = B + A$ (addition of matrices is commutative)

b) $A + (B + C) = (A + B) + C$ (addition of matrices is associative)

PROOF

Let $A = (a_{ij})$ $1 \leq i \leq m$ and $1 \leq j \leq n$. Let $B = (b_{ij})$ $1 \leq i \leq m$ and $1 \leq j \leq n$. Then

a) $A + B = (a_{ij} + b_{ij})$ $1 \leq i \leq m$ and $1 \leq j \leq n$.

$= (b_{ij} + a_{ij})$ $1 \leq i \leq m$ and $1 \leq j \leq n$, since addition of real numbers is commutative.

$= B + A$.

b) $A + (B + C) = (a_{ij} + (b_{ij} + c_{ij}))$ $1 \leq i \leq m$ and $1 \leq j \leq n$.

$= ((a_{ij} + b_{ij}) + c_{ij})$ $1 \leq i \leq m$ and $1 \leq j \leq n$, since addition of real numbers is associative.

$= (A + B) + C$. ☐

Multiplication of a Matrix by a Constant

Definition Let A be the matrix

$$A = \begin{bmatrix} a_{11} & \cdots & a_{1n} \\ & & \\ \cdot & & \cdot \\ \cdot & & \cdot \\ \cdot & & \cdot \\ & & \\ a_{m1} & \cdots & a_{mn} \end{bmatrix}$$

and let c be any constant. By cA we mean the matrix

$$A = \begin{bmatrix} ca_{11} & \cdots & ca_{1n} \\ & & \\ \cdot & & \cdot \\ \cdot & & \cdot \\ \cdot & & \cdot \\ & & \\ ca_{m1} & \cdots & ca_{mn} \end{bmatrix}$$

In other words, $cA = (ca_{ij})$.

There are two properties of multiplication by a constant that are commonly used and that are proven in the following theorem.

THEOREM B.2

a) $c(A + B) = cA + cB$

b) $(c + d) A = cA + dA$

PROOF

a) $c(A + B) = c(a_{ij} + b_{ij})$
$= (c(a_{ij} + b_{ij}))$
$= (ca_{ij} + cb_{ij})$
$= (ca_{ij}) + (cb_{ij})$
$= cA + cB$

b) (Left for the Exercises.) ▯

These are both forms of the distributive law as applied to multiplication of matrices by scalars.

Matrix Multiplication

Unlike multiplication of a matrix by a constant, matrix multiplication involves the multiplication of two matrices. Its definition is a generalization of the concept of the dot product of two vectors discussed in Appendix A.

Definition   Suppose we have two matrices, A and B, where A is $m \times p$ and B is $p \times n$:

$$A = \begin{bmatrix} a_{11} & \cdots & a_{1p} \\ \cdot & & \cdot \\ \cdot & & \cdot \\ \cdot & & \cdot \\ a_{m1} & \cdots & a_{mp} \end{bmatrix} \qquad B = \begin{bmatrix} b_{11} & \cdots & b_{1n} \\ \cdot & & \cdot \\ \cdot & & \cdot \\ \cdot & & \cdot \\ b_{p1} & \cdots & b_{pn} \end{bmatrix}$$

By the **product of A and B,** denoted AB, we mean the $m \times n$ matrix whose ijth entry is the dot product of the ith row of A and the jth column of B. $AB = (c_{ij})$, where

$$c_{ij} = \sum_{k=1}^{p} a_{ik} b_{kj}.$$

Thus, just as the dot product of two vectors is not defined unless the two vectors are of the same length, matrix multiplication is not defined unless the two matrices are such that the number of columns of the one on the left is the same as the number of rows of the one on the right.

EXAMPLE 5 Compute AB where

$$A = \begin{bmatrix} 2 & 0 & 7 \\ 1 & -1 & 6 \\ -2 & 1 & 3 \end{bmatrix} \qquad B = \begin{bmatrix} -1 & 3 & 4 \\ 3 & 4 & 2 \\ 0 & 2 & 1 \end{bmatrix}$$

Denote the product by C. Two entries of C are computed in detail; the rest are left for you to verify.

c_{11} is the dot product of $(2, 0, 7)$ and $(-1, 3, 0)$, i.e.,

$$C_{11} = (2 \cdot -1) + (0 \cdot 3) + (7 \cdot 0) = -2.$$

c_{23} is the dot product of $(1, -1, 6)$ and $(4, 2, 1)$, i.e.,

$$C_{23} = 4 - 2 + 6 = 8.$$

Hence

$$C = \begin{bmatrix} -2 & 20 & 15 \\ -4 & 11 & 8 \\ 5 & 4 & -3 \end{bmatrix}. \quad \blacksquare$$

Properties of Matrix Multiplication

The major properties of matrix multiplication are summarized in the following theorem.

THEOREM B.3 Let A, B, and C be $n \times n$ matrices. Then

a) $A(B + C) = AB + AC$
b) $A(BC) = (AB)C$
c) AB does not necessarily equal BA.
d) If $A \neq 0$ and $AB = AC$, we cannot conclude that $B = C$.

Part (c) indicates that the commutative law does not hold for matrix multiplication; part (d) indicates that cancellation does not apply to matrix multiplication. Note that we have restricted the four statements to $n \times n$ matrices for clarity of presentation. They can be generalized to matrices that are not square. (See the Exercises.)

PROOF

a) Let $A = (a_{ij})$, $B = (b_{ij})$, and $C = (c_{ij})$. Then

$$B + C = (b_{ij} + c_{ij}).$$

Hence

$$A(B + C) = \left(\sum_{k=1}^{n} a_{ik}(b_{kj} + c_{kj}) \right)$$

$$= \left(\sum_{k=1}^{n} (a_{ik}b_{kj} + a_{ik}c_{kj}) \right)$$

$$= \left(\sum_{k=1}^{n} a_{ik}b_{kj} \right) + \left(\sum_{k=1}^{n} a_{ik}c_{kj} \right)$$

$$= AB + AC.$$

b) We will use the same notation as in part (a). Then

$$BC = \left(\sum_{k=1}^{n} b_{ik}c_{kj} \right)$$

Hence

$$A(BC) = \left(\sum_{p=1}^{n} a_{ip} \left(\sum_{k=1}^{n} b_{pk}c_{kj} \right) \right)$$

$$= \left(\sum_{p=1}^{n} \sum_{k=1}^{n} a_{ip}b_{pk}c_{kj} \right)$$

$$= (a_{i1}b_{11}c_{1j} + a_{i1}b_{12}c_{2j} + \cdots + a_{i1}b_{1n}c_{nj}$$
$$+ a_{i2}b_{21}c_{1j} + a_{i2}b_{22}c_{2j} + \cdots + a_{i2}b_{2n}c_{nj}$$
$$+ \cdots$$
$$+ a_{in}b_{n1}c_{1j} + a_{in}b_{n2}c_{2j} + \cdots + a_{in}b_{nn}c_{nj})$$

$$= (a_{i1}b_{11}c_{1j} + a_{i2}b_{21}c_{1j} + \cdots + a_{in}b_{n1}c_{1j}$$
$$+ a_{i1}b_{12}c_{2j} + a_{i2}b_{22}c_{2j} + \cdots + a_{in}b_{n2}c_{2j}$$
$$+ \cdots$$
$$+ a_{i1}b_{1n}c_{nj} + a_{i2}b_{2n}c_{nj} + \cdots + a_{in}b_{nn}c_{nj})$$

$$= \left(\left(\sum_{p=1}^{n} a_{ip}b_{p1} \right)c_{1j} + \left(\sum_{p=1}^{n} a_{ip}b_{p2} \right)c_{2j} + \cdots + \left(\sum_{p=1}^{n} a_{ip}b_{pn} \right)c_{nj} \right)$$

$$= \left(\sum_{k=1}^{n} \left(\sum_{p=1}^{n} a_{ip}b_{pk} \right)c_{kj} \right)$$

$$= (AB)C.$$

c) Since this involves disproving the statement $AB = BA$, only a counter-example is necessary.

$$\text{Let } A = \begin{bmatrix} 1 & 2 \\ 0 & 1 \end{bmatrix} \text{ and let } B = \begin{bmatrix} 2 & 1 \\ 1 & 1 \end{bmatrix}$$

On your own, verify that

$$AB = \begin{bmatrix} 4 & 3 \\ 1 & 1 \end{bmatrix} \text{ and } BA = \begin{bmatrix} 2 & 5 \\ 1 & 3 \end{bmatrix}$$

d) This is another disproof, so again we only need a counterexample.

$$\text{Let } A = \begin{bmatrix} 0 & 0 \\ 0 & 2 \end{bmatrix}, B = \begin{bmatrix} 1 & 1 \\ 2 & 1 \end{bmatrix}, \text{ and } C = \begin{bmatrix} 3 & 7 \\ 2 & 1 \end{bmatrix}$$

Then it is easy to verify that $AB = AC$ but $B \neq C$. □

TERMINOLOGY

matrix	rectangular array
row	row vector
column	column vector
$m \times n$ matrix	coefficient matrix
complete graph	boolean matrix
zero matrix	square matrix
diagonal matrix	main diagonal
symmetric matrix	upper triangular matrix
lower triangular matrix	triangular matrix
identity matrix	equal matrices
addition of matrices	multiplication by a constant
matrix multiplication	product of matrices A and B

SUMMARY

1. Matrices are rectangular arrays of numbers and are one of the most commonly used tools in discrete modeling.

2. Some important common matrices are the zero matrix, the square matrix, the diagonal matrix, the symmetric matrix, the upper and lower triangular matrices, and the identity matrix.

3. Addition of matrices is commutative and associative.

4. Multiplication of matrices by a scalar obeys the two distributive laws

$$c(A + B) = cA + cB \text{ and } (c + d)A = cA + cD.$$

5. Matrix multiplication is a generalization of the dot product for vectors. It is distributive and associative but not necessarily commutative. Cancellation (if $AB = AC$ then $B = C$) may not be valid, either.

EXERCISES

Let A be the matrix

$$\begin{bmatrix} 1 & 2 & 3 \\ 4 & 5 & 6 \\ 7 & 8 & 9 \end{bmatrix}$$

For Exercises 1 through 12, find all entries, a_{ij}, for which the stated condition is true.

1. $i = 1, j = 2$
2. $i = 2, j = 3$
3. $i = 3, j \geq 2$
4. $j = 1, i \geq 2$
5. $i = j$
6. $i > j$
7. $j > i$
8. $j \geq i$
9. $j \geq i + 1$
10. $i \leq j - 1$
11. $a_{ij} < a_{ji}$
12. $a_{ij} = a_{ji} + 2$
13. Write the column vectors for the preceding matrix.
14. Write the row vectors for the preceding matrix.

Represent the following systems of linear equations using matrices.

15. $2x + 3y = 16$
    $3x + y = 5$

16. $x - y = 7$
    $2x + 3y = -1$

17. $x - y + z = 7$
    $2x - y - 3z = 1$
    $-3x + 2y + 2z = 13$

18. $2x + 3y - 2z = 1$
    $-x - 2y - 3z = 14$
    $-3x + y + 2z = -7$

Represent the following weighted graphs using a matrix.

19.

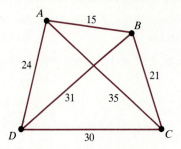

20.

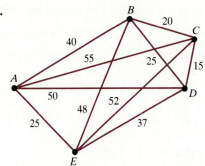

In Exercises 21 through 28, compute the values of the given expressions for the following matrices.

$$A = \begin{bmatrix} 2 & 1 \\ 1 & 3 \end{bmatrix} \quad B = \begin{bmatrix} -1 & 3 \\ 1 & -3 \end{bmatrix} \quad C = \begin{bmatrix} 6 & 2 \\ -1 & 2 \end{bmatrix} \quad I = \begin{bmatrix} 1 & 0 \\ 0 & 1 \end{bmatrix}$$

21. $A + B$

22. $A - B$

23. $2C$

24. $3I$

25. $A - 2C$

26. $B + 4I$

27. $A + 2I - C$

28. $A - 2C - 2B$

Using A, B, C, and I as just given, solve the following equations for the matrix X.

29. $X + I = B - C$

30. $2X + B = 3C - I$

31. $C - X = -5C + 2X$

32. $aC + X = aB + I$

For each of the following conditions, create an example of a matrix satisfying it.

33. $a_{ij} = 0$ if $i \neq j$

34. $a_{ij} = a_{ji}$

35. $a_{ij} = 0$ if $i > j$

36. $a_{ij} = 0$ if $i < j$

Suppose A is required to be an $n \times n$ Boolean matrix. How many different A's are possible for each of the following additional restrictions on A?

37. A is diagonal.   **38.** A is upper triangular.

39. A is symmetric.   **40.** no additional restrictions

For the matrices A, B, and C previously specified, verify the following properties.

41. $A + B = B + A$ **42.** $A + (B + C) = (A + B) + C$

43. $c(A + B) = cA + cB$ **44.** $(c + d)A = cA + dA$

45. If A is an $n \times n$ matrix and c and d are constants, show that

$$(c + d)\, A = cA + dA.$$

46. If A is an $n \times n$ matrix and c and d are constants, show that

$$(c - d)A = cA - dA.$$

For the matrices of Exercises 21 through 28, compute the following.

47. AB **48.** AC

49. A^2 **50.** $B^2 + 2B - I$

51. $A^3 - I$ **52.** $(A - B)(A + B)$

53. Show that $A(B + C) = AB + AC$.

54. Show that $A(BC) = (AB)C$.

In addition to the preceding A, B, C, and I, suppose that

$$D = \begin{bmatrix} 1 & 0 \\ 1 & 1 \end{bmatrix} \quad E = \begin{bmatrix} 4 & 3 \\ 2 & 2 \end{bmatrix} \quad V = \begin{bmatrix} 2 \\ 1 \end{bmatrix}.$$

55. Find $AV + IV$.

56. Find $DV - EV$.

57. Show that the cancellation law does not hold for matrices by showing that $BD = BE$ even though $D \neq E$.

58. Show that the commutative law does not hold for matrix multiplication by showing that $DE \neq ED$.

59. Verify the noncommutativity of the matrices in the proof of Theorem B.3, part (c).

60. Verify that cancellation does not hold for the matrices in the proof of Theorem B.3, part (d).

Evaluate the given expressions for the following matrices.

$$A = \begin{bmatrix} 2 & 3 & 1 \\ -1 & 2 & 1 \\ -4 & 7 & 1 \end{bmatrix} \quad B = \begin{bmatrix} -1 & 1 & 7 \\ 2 & 2 & 3 \\ 3 & 1 & -2 \end{bmatrix} \quad I = \begin{bmatrix} 1 & 0 & 0 \\ 0 & 1 & 0 \\ 0 & 0 & 1 \end{bmatrix}$$

61. BI

62. IB

63. $A^2 - AI$

64. $IA - 2I$

65. $AB - 7I$

66. $BA - 2AI$

Suppose A, B, and C are $n \times n$ matrices and α and β are real numbers. Prove the following properties of matrix multiplication.

67. $(A + B)C = AB + AC$

68. $\alpha(B + C) = \alpha B + \alpha C$

69. $AI = A$

70. $IA = A$

71. $(\alpha A)B = \alpha(AB)$

72. $\alpha(AB) = A(\alpha B)$

73. $(\alpha + \beta)C = \alpha C + \beta C$

74. $(\alpha A)(\beta B) = \alpha\beta(AB)$

Let D, E, and V be as specified in Exercises 55 through 58. Also let

$$M = \begin{bmatrix} a & b \\ c & d \end{bmatrix} \qquad W = \begin{bmatrix} y \\ z \end{bmatrix}$$

Solve the equations.

75. $MD = I$

76. $DM = I$

77. $DW = V$

78. $EW = V$

79. $EM = I$

80. $ME = I$

Each of the following expressions is, in general, false for matrices. For each one, indicate why it is false and state a condition under which it would be true.

81. $(A + B)^2 = A^2 + 2AB + B^2$

82. $(A + B)(A - B) = A^2 - B^2$

83. $(AB)^2 = A^2B^2$

Appendix C
Simultaneous Linear Equations and Gaussian Elimination

Mathematical models often involve simultaneous linear equations. Such models arise in any situation in which there is more than one variable and the dynamics of the underlying situation are *linear*—that is, variables can be multiplied by constants and added but never multiplied by themselves or by other variables. These models occur frequently in economics—for instance, in equations of the form $c_1x_1 + \cdots + c_nx_n = b$, where the c_i are constant prices and the x_i are variable quantities. Additional linear equations arise from the *constraints*—various restrictions placed on the quantities available.

Because simultaneous linear equations arise so often, they have been widely studied. Perhaps the best-known systematic method for solving them is an algorithm called **Gaussian elimination,** developed by Carl Friedrich Gauss, whose name we encountered in connection with finite series. Our main objective in this section is to examine this algorithm. Before we can do this, though, we must present a few preliminary concepts.

A Brief Review

EXAMPLE 1 Suppose we have a pair of simultaneous equations such as

$$x + y = 2$$
$$2x - y = 3.$$

The method for solving these equations typically taught in high school is as follows:

Multiply the first equation by -2, giving

$$-2x - 2y = -4.$$

Add this to the second equation. The pair of equations is then

$$x + y = 2$$
$$-3y = -1.$$

607

Now solve the second equation for y, giving

$$y = 1/3.$$

Substitute this into the first equation, giving

$$x + (1/3) = 2.$$

Hence we conclude that $x = 5/3$. ∎

The Use of Matrices

Gaussian elimination is a direct generalization of the technique just applied. Recall that in the last appendix, we noted that a system of simultaneous equations can be represented by matrices. For instance, the preceding system can be written

$$\begin{bmatrix} 1 & 1 \\ 2 & -1 \end{bmatrix} \begin{bmatrix} x \\ y \end{bmatrix} = \begin{bmatrix} 2 \\ 3 \end{bmatrix}.$$

Suppose we have n equations in n unknowns. Gaussian elimination is based on the **augmented coefficient matrix,** which is the $n \times n$ coefficient matrix with an $n + 1$th column added. The additional column is the vector containing the right-hand sides of the system of equations. Thus the augmented coefficient matrix for the preceding system is

$$\begin{bmatrix} 1 & 1 & 2 \\ 2 & -1 & 3 \end{bmatrix}.$$

Now let's apply the same sequence of steps that we applied to the preceding system of equations to the augmented coefficient matrix. We start with

$$\begin{bmatrix} 1 & 1 & 2 \\ 2 & -1 & 3 \end{bmatrix}.$$

Multiply the first row by -2, giving

$$-2 \quad -2 \quad -4.$$

Adding this to the second row, the matrix becomes

$$\begin{bmatrix} 1 & 1 & 2 \\ 0 & -3 & -1 \end{bmatrix}.$$

Dividing the second row by -3, we get

$$\begin{bmatrix} 1 & 1 & 2 \\ 0 & 1 & \frac{1}{3} \end{bmatrix}.$$

Hence $y = 1/3$. Rewriting the first row as an equation and substituting for y (called **back substitution**), we get

$$x + (1/3) = 2.$$

Thus $x = 5/3$.

EXAMPLE 2   Use this method to solve the following system of equations:

$$x + y - z = 2$$
$$2x - y - 3z = 3$$
$$-x + 3y + 2z = 0.$$

SOLUTION   First, we form the augmented coefficient matrix:

$$\begin{bmatrix} 1 & 1 & -1 & 2 \\ 2 & -1 & -3 & 3 \\ -1 & 3 & 2 & 0 \end{bmatrix}.$$

We then proceed to transform the matrix as follows:

$$\begin{bmatrix} 1 & 1 & -1 & 2 \\ 2 & -1 & -3 & 3 \\ -1 & 3 & 2 & 0 \end{bmatrix} \quad \xrightarrow{\text{row2} + (-2) * \text{row1}}$$

$$\begin{bmatrix} 1 & 1 & -1 & 2 \\ 0 & -3 & -1 & -1 \\ -1 & 3 & 2 & 0 \end{bmatrix} \quad \xrightarrow{\text{row3} + (1) * \text{row1}}$$

$$\begin{bmatrix} 1 & 1 & -1 & 2 \\ 0 & -3 & -1 & -1 \\ 0 & 4 & 1 & 2 \end{bmatrix} \quad \xrightarrow{\text{row2} * (-\frac{1}{3})}$$

$$\begin{bmatrix} 1 & 1 & -1 & 2 \\ 0 & 1 & \frac{1}{3} & \frac{1}{3} \\ 0 & 4 & 1 & 2 \end{bmatrix} \quad \xrightarrow{\text{row4} + (-4) * \text{row2}}$$

$$\begin{bmatrix} 1 & 1 & -1 & 2 \\ 0 & 1 & \frac{1}{3} & \frac{1}{3} \\ 0 & 0 & -\frac{1}{3} & \frac{2}{3} \end{bmatrix}.$$

At this point, we can see immediately that $z = -2$. Rewriting the second row as an equation, we get

$$y + (1/3) \cdot (-2) = 1/3; \text{ hence } y = 1.$$

Rewriting the first row as an equation, we get

$$x + y - z = 2.$$

Substituting for y and z, we have

$$x + 1 + 2 = 2, \text{ which implies that } x = -1. \quad \blacksquare$$

Elementary Row Operations

There are three types of operations that can be used to solve simultaneous equations by Gaussian elimination. We used two of them in Example 1. They are called **elementary row operations (ERO's)** and may be applied to any augmented coefficient matrix. They have the effect of transforming a system of equations into an **equivalent system**—another system that has the same solution set. They are as follows:

1. Any two rows may be interchanged.

2. Any row may be multiplied by a constant.

3. A constant multiple of any row may be added to another row.

Later we will see examples that will illustrate why ERO1 is needed. The objective in applying the ERO's to an augmented coefficient matrix is to produce a matrix that is upper triangular and in which the diagonal entries starting from the upper left are a series of 1's possibly followed by a series of 0's.

Geometric Interpretation

Before examining the algorithm for Gaussian elimination in detail, we must understand why some systems of simultaneous equations have no solutions and others have infinitely many. Consider the following geometric interpretation of systems of simultaneous equations.

Examine again the first system we looked at:

$$x + y = 2$$
$$2x - y = 3.$$

Graphically, these equations form a pair of straight lines (Figure C.1). Their point of intersection is on both lines. Thus the system has coordinates that satisfy both equations. Hence these coordinates are the solution of the system of equations.

Figure C.1 The lines $x + y = 2$ and $2x - y = 3$.

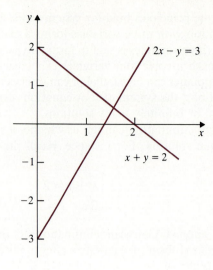

If, however, the lines are parallel, they will never intersect. Thus the system of equations

$$x + y = 2$$
$$x + y = 3$$

has no solution (Figure C.2).

If the two equations represent the same line, there will be infinitely many points satisfying both equations. For instance, the equations

$$x + y = 2$$
$$3x = 6 - 3y$$

are a system with infinitely many solutions.

Figure C.2 The lines $x + y = 2$ and $x + y = 3$ do not intersect.

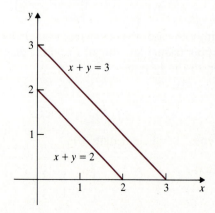

The same principles hold for systems with three or more variables. A linear equation with three variables forms a plane; a linear equation with more than three variables is said to form a **hyperplane.** Suppose four linear equations each involve four variables. Then each will form a hyperplane. These hyperplanes can be parallel or can intersect. Thus, just as with two or three equations, the system of four equations can have a unique solution, no solution, or infinitely many solutions.

Gaussian elimination can tell us whether a system of linear equations has a unique solution. For instance, recall the first set of equations we examined:

$$x + y = 2$$
$$2x - y = 3.$$

When we applied Gaussian elimination to the augmented coefficient matrix, the final form of the matrix after application of the ERO's was

$$\begin{bmatrix} 1 & 1 & 2 \\ 0 & 1 & \frac{1}{3} \end{bmatrix}.$$

In Example 2, we solved the system

$$x + y - z = 2$$
$$2x - y - 2z = 1$$
$$-x + 2y + z = 1.$$

The final form of that augmented coefficient matrix after application of the ERO's was

$$\begin{bmatrix} 1 & 1 & -1 & 2 \\ 0 & 1 & 0 & 1 \\ 0 & 0 & 1 & -2 \end{bmatrix}.$$

Note that in both cases the coefficient portion of the augmented coefficient matrix is upper triangular with all 1's on the diagonal. Now apply Gaussian elimination to the system

$$x + y = 2$$
$$x + y = 3.$$

The augmented coefficient matrix is

$$\begin{bmatrix} 1 & 1 & 2 \\ 1 & 1 & 3 \end{bmatrix}.$$

Applying the third ERO to this matrix it becomes

$$\begin{bmatrix} 1 & 1 & 2 \\ 0 & 0 & 1 \end{bmatrix}.$$

The coefficient portion is upper triangular, but the diagonal entries include a zero. Note that the last row indicates that $0x + 0y = 1$, i.e., $0 = 1$. Since this is impossible, this system has no solution. Systems whose equations reduce to a contradiction are said to be **inconsistent.**

Consider again the system

$$x + y = 2$$
$$3x = 6 - 3y.$$

Rewriting this with the variables on the left and the constant on the right, we get

$$x + y = 2$$
$$3x + 3y = 6.$$

This yields the augmented coefficient matrix

$$\begin{bmatrix} 1 & 1 & 2 \\ 3 & 3 & 6 \end{bmatrix}.$$

Applying the third ERO to this matrix, we get

$$\begin{bmatrix} 1 & 1 & 2 \\ 0 & 0 & 0 \end{bmatrix}.$$

The last row indicates that $0x + 0y = 0$. This is not inconsistent, like the previous example, but it doesn't tell us anything either. A system of equations that has a row reduced augmented coefficient matrix containing a row of all zeroes is called **redundant.** Such a system (if no other rows are inconsistent) has infinitely many solutions.

Gaussian Elimination

We are now ready to examine a formal algorithm for Gaussian elimination. The formal algorithm is needed for several reasons: It helps to clarify the process of Gaussian elimination better than examples alone; it serves as a jumping-off point if we code Gaussian elimination as a computer program; and it enables us to analyze the complexity of the algorithm. This algorithm does not distinguish between redundant and inconsistent systems. Modification of the algorithm to include this feature is left for the Exercises. The algorithm assumes n equations in n inknowns.

ALGORITHM 44   Solving Simultaneous Equations by Gaussian Elimination

Procedure Gaussian (n, A; var X);

 {n, i, j, and k are integers;

 A is an n by $n + 1$ augmented coefficient matrix;

 X is a 1 by n array, which will be used to hold the solution of the system of equations;

 OK_to_proceed is Boolean.}

begin

1. OK_to_proceed := true;

2. $i := 1$;   {start at first row}

  Repeat

3.   let k denote the row number of that entry in column i between row i and row n with the largest magnitude;

4.   if $A[k,i] = 0$

5.   then OK_to_proceed := false   {detects inconsistency or redundancy}

6.   else if $k \neq i$ then switch   {apply ERO1} row k and row i;

7.   if OK_to_proceed

   then begin

8.   divide the new row i by $A[i,i]$; {apply ERO2}

9.   for $j := i + 1$ to n do

10.   add $-A[j,i]$ times row i to   {apply ERO3} row j;

11. $i := i + 1$;   {move to next row}

   end;

12. Until not OK_to_proceed OR ($i > n$);

13. if OK_to_proceed

  then begin

14. $X[n] := A[n,n + 1]$;

15.   for $i := n - 1$ downto 1 do

16. $X[i] := A[i,n + 1] -$

$$\sum_{j=i+1}^{n} X[j] * A[i,j]$$ {back substitution}

else
17.   write ('This system of equations does not have a unique solution')
    end;

Examples

EXAMPLE 3   Trace the algorithm for Gaussian elimination for the following systems of equations:

$$2x - y - z = -6$$
$$x + 2y - 4z = -5$$
$$-x + y + 2z = 7.$$

SOLUTION   We first form the augmented coefficient matrix:

$$\begin{bmatrix} 2 & -1 & -1 & -6 \\ 1 & 2 & -4 & -5 \\ -1 & 1 & 2 & 7 \end{bmatrix}.$$

For this example, a detailed trace is provided.

Step	Action
1	*OK_to_proceed* is set to true.
2	$i := 1$.
3	$k := 1$, since 2 is the number with the largest magnitude in the first column.
4	$A[1,1] \neq 0$, so jump to step 6.
6	$1 = 1$, so do not switch any rows.
7	*OK_to_proceed* is true, so continue to step 8.
8	Divide row 1 by 2, giving the new matrix.

$$\begin{bmatrix} 1 & -0.5 & -0.5 & -3 \\ 1 & 2 & -4 & -5 \\ -1 & 1 & 2 & 7 \end{bmatrix}$$

9	$j := 2$, which is less than 3, so proceed to step 10.

10   Add $-A[2,1]$, which is -1, times row 1 to row 2, giving

$$\begin{bmatrix} 1 & -0.5 & -0.5 & -3 \\ 0 & 2.5 & -3.5 & -2 \\ -1 & 1 & 2 & 7 \end{bmatrix}.$$

9    $j := 3$ is still less than or equal to 3, so proceed to step 10 again.

10   Add $-A[3,1]$, which is $+1$, times row 1 to row 3, giving

$$\begin{bmatrix} 1 & -0.5 & -0.5 & -3 \\ 0 & 2.5 & -3.5 & -2 \\ 0 & 0.5 & 1.5 & 4 \end{bmatrix}.$$

9    $j := 4$, greater than 3, so proceed to step 11.

11    $i := 2$.

12   not Ok_to_proceed = false; $(2 > 3)$ = false, so return to step 3.

3    $k := 2$, since 2.5 is the largest entry in column 2 between row 2 and row 3.

4    $A[2,2] = 2.5 \neq 0$. Therefore proceed to step 6.

6    $2 = 2$, so don't switch—continue to step 7.

7   Ok_to_proceed = true, so continue to step 8.

8   Divide row 2 by 2.5, yielding

$$\begin{bmatrix} 1 & -0.5 & -0.5 & -3 \\ 0 & 1 & -1.4 & -0.8 \\ 0 & 0.5 & 1.5 & 4 \end{bmatrix}.$$

9    $j := 3$, which is less than or equal to 3, so proceed to step 10.

10   Add $-A[3,2]$, namely, -0.5, times row 2 to row 3, giving

$$\begin{bmatrix} 1 & -0.5 & -0.5 & -3 \\ 0 & 1 & -1.4 & -0.8 \\ 0 & 0 & 2.2 & 4.4 \end{bmatrix}.$$

9    $j := 4$; hence proceed to step 11.

11    $i := 3$.

12   Not Ok_to_proceed = false; $(3 > 3)$ = false, so proceed to step 3.

3    $k := 3$, since there is only one entry in column 3 between row and row 3, namely, $A[3,3]$.

4    $A[3,3] \neq 0$, so proceed to step 6.

6    $3 = 3$, so don't switch; proceed to step 7.

7   Ok_to_proceed = true, so continue to step 8.

8   Divide row 3 by 2.2, yielding

$$\begin{bmatrix} 1 & -0.5 & -0.5 & -3 \\ 0 & 1 & -1.4 & -0.8 \\ 0 & 0 & 1 & 2 \end{bmatrix}.$$

9    $j := 4$, which exceeds 3, so jump to step 11.

11    $i := 4$.

12   Not Ok_to_proceed = false but $(4 > 3)$ is true, so continue to step 13.

13   Ok_to_proceed = true, so continue to step 14.

14    $X[3] := 2$—this is the solution for z.

15    $i := 2$.

16   This step begins the process of back substitution—$X[2] := A[2,4] - (X[3] * A[2,3]) = -0.8 - (2 * (-1.4)) = 2$.

15    $i := 1$.

16    $X[1] := A[1,4] - X[2] * A[1,2] - X[3] * A[1,3] = -3 - (2 * (-0.5) + 2 * (-0.5)) = -1$. ∎

EXAMPLE 4   Use the Gaussian elimination algorithm to solve the following system of equations.

$$2x + 2y - z = 3$$
$$-4x - 4y + 3z = -3$$
$$x - y + 2z = 5.$$

SOLUTION   This time only the steps that modify the augmented coefficient matrix are included.

$$\begin{bmatrix} 2 & 2 & -1 & 3 \\ -4 & -4 & 3 & -3 \\ 1 & -1 & 2 & 5 \end{bmatrix} \qquad \xrightarrow{\begin{array}{c} \text{step 6} \\ \text{switch row1 and row2} \end{array}}$$

$$\begin{bmatrix} -4 & -4 & 3 & -3 \\ 2 & 2 & -1 & 3 \\ 1 & -1 & 2 & 5 \end{bmatrix} \qquad \xrightarrow{\begin{array}{c} \text{step 8} \\ \text{divide row1 by } -4 \end{array}}$$

$$\begin{bmatrix} 1 & 1 & -0.75 & 0.75 \\ 2 & 2 & -1 & 3 \\ 1 & -1 & 2 & 5 \end{bmatrix} \quad \xrightarrow{\begin{array}{c} \text{step 10} \\ \hline \end{array}}$$
add −2 * row1 to row2

$$\begin{bmatrix} 1 & 1 & -0.75 & 0.75 \\ 0 & 0 & 0.5 & 1.5 \\ 1 & -1 & 2 & 5 \end{bmatrix} \quad \xrightarrow{\begin{array}{c} \text{step 10} \\ \hline \end{array}}$$
add −1 * row1 to row3

$$\begin{bmatrix} 1 & 1 & -0.75 & 0.75 \\ 0 & 0 & 0.5 & 1.5 \\ 0 & -2 & 2.75 & 4.25 \end{bmatrix} \quad \xrightarrow{\begin{array}{c} \text{step 6} \\ \hline \end{array}}$$
switch row2 and row3

$$\begin{bmatrix} 1 & 1 & -0.75 & 0.75 \\ 0 & -2 & 2.75 & 4.25 \\ 0 & 0 & 0.5 & 1.5 \end{bmatrix} \quad \xrightarrow{\begin{array}{c} \text{step 8} \\ \hline \end{array}}$$
divide row2 by −2

$$\begin{bmatrix} 1 & 1 & -0.75 & 0.75 \\ 0 & 1 & -1.375 & -2.125 \\ 0 & 0 & .5 & 1.5 \end{bmatrix} \quad \xrightarrow{\begin{array}{c} \text{step 10} \\ \hline \end{array}}$$
add 0 * row2 to row3

$$\begin{bmatrix} 1 & 1 & -0.75 & 0.75 \\ 0 & 1 & -1.375 & -2.125 \\ 0 & 0 & 0.5 & 1.5 \end{bmatrix} \quad \xrightarrow{\begin{array}{c} \text{step 8} \\ \hline \end{array}}$$
divide row3 by 0.5

$$\begin{bmatrix} 1 & 1 & -0.75 & 0.75 \\ 0 & 1 & -1.375 & -2.125 \\ 0 & 0 & 1 & 3 \end{bmatrix}$$

We then begin the back substitution steps and get

Step 14: $z = x[3] := 3$

Step 16: $y = x[2] := A[2,4] - x[3] * A[2,3]$
$$:= -2.125 - 3 * (-1.375) := 2$$

Step 16: $x = x[1] := A[1,4] - x[2] * A[1,2] - x[3] * A[1,3]$
$$:= 0.75 - 2 - (3 * (-0.75)) := 1. \quad \blacksquare$$

The most important thing to notice in this example is the two uses of step 6: ERO1—switching the order of the rows. Consider the matrix just before the second application of step 6. This was the second use of ERO1. We want $A[3,2]$ to become zero. But we cannot do this with ERO2 or ERO3. The only way to do it is to switch equations 2 and 3.

In the first use, we switched rows 1 and 2 in order to move the largest coefficient in the column to the first row and the first column. The algorithm could have proceeded without this step; however, when it is implemented on a computer, this step improves the accuracy of the answer. That is, real numbers are stored in computers approximately and hence are accurate only to a certain number of decimal places. An algorithm such as this, which involves many computations can gradually cause those inaccuracies to increase in size to the point where the algorithm computes seriously inaccurate answers. Always keeping the largest coefficient in a column in the $A[i,i]$ position, also called the **pivot position,** prevents this from happening. For a discussion of why this procedure improves the accuracy of Gaussian elimination, see *Numerical Analysis* by Johnson and Reiss or *Elementary Numerical Analysis—An Algorithmic Approach* by Conte and deBoor (see References following Appendix D).

EXAMPLE 5   Apply Gaussian elimination to solve the system

$$x + y - z = 3$$
$$2x - y + 2z = 2$$
$$x - 2y + 3z = 4.$$

SOLUTION   We again form the augmented coefficient matrix and carry out the algorithm:

$$\begin{bmatrix} 1 & 1 & -1 & 3 \\ 2 & -1 & 2 & 2 \\ 1 & -2 & 3 & 4 \end{bmatrix} \xrightarrow[\text{switch row1 and row2}]{\text{step 6}}$$

$$\begin{bmatrix} 2 & -1 & 2 & 2 \\ 1 & 1 & -1 & 3 \\ 1 & -2 & 3 & 4 \end{bmatrix} \xrightarrow[\text{divide row1 by 2}]{\text{step 8}}$$

$$\begin{bmatrix} 1 & -0.5 & 1 & 1 \\ 1 & 1 & -1 & 3 \\ 1 & -2 & 3 & 4 \end{bmatrix} \xrightarrow[\text{add } -1 * \text{row1 to row2}]{\text{step 10}}$$

$$\begin{bmatrix} 1 & -0.5 & 1 & 1 \\ 0 & 1.5 & -2 & 2 \\ 1 & -2 & 3 & 4 \end{bmatrix}$$

$\xrightarrow{\text{step 10}}$

add $-1 *$ row1 to row3

$$\begin{bmatrix} 1 & -0.5 & 1 & 1 \\ 0 & 1.5 & -2 & 2 \\ 0 & -1.5 & 2 & 3 \end{bmatrix}$$

$\xrightarrow{\text{step 8}}$

divide row2 by 1.5

$$\begin{bmatrix} 1 & -0.5 & 1 & 1 \\ 0 & 1 & -1.33 & 1.33 \\ 0 & -1.5 & 2 & 3 \end{bmatrix}$$

$\xrightarrow{\text{step 10}}$

add $1.5 *$ row2 to row3

$$\begin{bmatrix} 1 & -0.5 & 1 & 1 \\ 0 & 1 & -1.33 & 1.33 \\ 0 & 0 & 0 & 1 \end{bmatrix}$$

Now, when the algorithm returns to step 3, k becomes 3. Hence, in step 4, $A[3,3]$ is zero and *Ok_to_proceed* becomes false. The algorithm then drops down to line 17 and writes the message "This system of equations does not have a unique solution." ▮

The important thing to notice in the previous example is that the system of equations is inconsistent. If the third equation had had a different right-hand side, so that the 1 in the lower left-hand corner had become a zero instead, the system would have been redundant.

Fewer or More Equations Than Unknowns

Suppose we have a system of simultaneous linear equations in which there are m equations and n unknowns and $m < n$. As long as the equations are consistent, this is the same situation as that in which there are n equations and n unknowns but one or more of the equations is redundant. In this case, it is possible to find solutions for some of the variables in terms of the other variables. For instance, consider

$$x + 2y - z = 1$$
$$2x - y + 3z = 0.$$

If we multiply the first equation by -2 and add it to the second, we get

$$-5y + 5z = -2.$$

We can then solve for either y or z in terms of the other. Choosing z,

$$z = (5y - 2)/5.$$

Substituting this into the first equation, we get

$$x + 2y - (5y - 2)/5 = 1,$$

which implies that

$$x = (3 - 5y)/5.$$

The Gaussian elimination algorithm we have examined assumed that $m = n$; if there are redundant equations, it will only tell us that the solution is not unique. The algorithm can be easily modified to find at least one solution in the case where $m < n$. It cannot be easily modified, though, to find a solution for some of the variables in terms of the others.

Gaussian elimination can also handle the situation where there are more equations than unknowns. The augmented coefficient matrix is formed for such a system and Gaussian elimination is applied. If any of the equations turn out to be inconsistent, there is, of course, no solution. If the algorithm (which may rearrange the order of the equations) reveals all equations after the nth to be redundant, those equations may be ignored and there will be either a unique solution (if none of the remaining equations are redundant) or infinitely many solutions (if any of the remaining equations are redundant).

Complexity of Gaussian Elimination

Following is an analysis of the maximum number of computations that this version of Gaussian elimination requires. Each addition, subtraction, multiplication, and division is again counted as one computation; also, each assignment (such as i:=1) is counted as a computation. Thus switching two rows requires $3(n + 1)$ computations, since switching each element in the rows requires three assignment statements. Similarly, evaluating the conditional in "Until not Ok_to_proceed or $(i > n)$" requires three computations —one to evaluate each of the predicates and one to combine them using OR. For this analysis, we will assume that the equations are neither inconsistent nor redundant.

Step	Number of computations	Total
1	1	1
2	1	1
3	$(4n-1)+(4n-5)+\cdots+3=\sum_{i=1}^{n}(4i-1)$	$2n^2+n$
4	n	n
5	0	0
6	$3(n+1)+3n+\cdots+6=\sum_{i=1}^{n}3(i+1)$	$\frac{3}{2}n^2+\frac{9}{2}n$
7	n	n
8	$n(n+1)$	n^2+n
9	$2(n+(n-1)+\cdots+1)$	n^2+n
10	$n(n-1)/2*(2n+2)$	n^3-n
11	n	n
12	$3n$	$3n$
13	1	1
14	1	1
15	$2n$	$2n$
16	$3+5+\cdots+(2n-1)=\sum_{i=1}^{n-1}(2i+1)$	n^2-1
17	0	0
Total		$n^3+\frac{13}{2}n^2+\frac{29}{2}n+3$

A few of these steps require explanation:

Step 3: This step finds the location of the largest entry in column i. When $i=1$, this involves examining the entries in rows 1 through n; when $i=2$, it involves examining the entries in rows 2 through n; etc. If this step were carried out using Algorithm 2 of Section 4.1, with n items to examine, it would take at most $4n-1$ computations; with $n-1$ items, at most $4(n-1)-1$ computations, etc.

Step 9: When $i=1$, $j=2$ and thus this step will be executed n times, leading to $2n$ computations. Similarly, if $i=2$, then $j=3$, leading to $n-1$ executions and $2(n-1)$ computations, etc.

Step 10: This line will be executed $(n-1)+(n-2)+\cdots+1=n(n-1)/2$ times. However, it is a compound step, i.e., each step involves $n+1$ additions and $n+1$ multiplications.

Step 16: When $i = n - 1$, there are three computations—a multiplication, a subtraction, and an assignment. When $i = n - 2$, there are two more —one more addition and one more multiplication. This pattern continues until the finite sequence has $n - 1$ terms.

Thus Gaussian elimination is an $O(n^3)$ algorithm. Note that the n^3 term appears only in step 10; hence, when this algorithm is carried out, most of the time will be spent executing step 10. Thus, if the algorithm is to be made significantly more efficient, step 10 is the logical focal point. Such improvements are definitely possible; the algorithm is written to be understandable by people rather than to have maximum efficiency. In particular, many entries in the reduced matrix will be found to be zero. It is thus unnecessary to add $-A[j,i]$ times row i to row j for those entries; they could simply be set to zero or even ignored. With such an enhancement, the n^3 term in the total could be replaced by $n^3/3$. Although still $O(n^3)$, this can represent a substantial savings in time when n is large. For the details of this argument, see the Exercises.

Gauss-Jordan Elimination

There is a variant of Gaussian elimination called *Gauss-Jordan elimination* that eliminates the back substitution phase of Gaussian elimination. It involves continuing the process of row reduction so that instead of the matrix being reduced to upper triangular form, it is reduced to the $n \times n$ identity matrix. For instance, suppose the reduced form arrived at by Gaussian elimination is

$$\begin{bmatrix} 1 & 1 & -0.75 & 0.75 \\ 0 & 1 & -1.375 & -2.125 \\ 0 & 0 & 1 & 3 \end{bmatrix}$$

Gauss-Jordan elimination would then continue to use elementary row operations to eliminate the nonzero entries in row 1, columns 2 and 3, and row 2, column 3. That is, row 3 would be multiplied by 1.375 and added to row 2, etc. The resulting matrix would be

$$\begin{bmatrix} 1 & 0 & 0 & 1 \\ 0 & 1 & 0 & 2 \\ 0 & 0 & 1 & 3 \end{bmatrix}$$

and the solutions can be read directly from the last column without back substitution. Gaussian elimination is generally preferred to Gauss-Jordan elimination because the row reduction involved in Gauss-Jordan elimination requires further applications of a step like step 10, which is $O(n^3)$; the back substitution portion of Gaussian elimination (steps 13 to 16) is $O(n^2)$. Hence Gaussian elimination is more efficient than Gauss-Jordan elimination.

TERMINOLOGY

Gaussian elimination   augmented coefficient matrix

back substitution   elementary row operation (ERO)

equivalent systems of equations   hyperplane

inconsistent system   redundant system

pivot position

SUMMARY

1. Gaussian elimination can be used both to solve systems of simultaneous linear equations and to determine when they have a unique solution.

2. Let m = the number of equations in a system of simultaneous linear equations and let n = the number of variables. Then:

	If Equations Are:	The Number of Solutions is:
$m < n$	Consistent	∞
	Inconsistent	0
$m = n$	Consistent and not redundant	1
	Consistent and redundant	∞
	Inconsistent	0
$m > n$	Consistent with n nonredundant	1
	Consistent with less than n nonredundant	∞
	Inconsistent	0

3. Gaussian elimination proceeds by attempting to reduce the coefficient matrix portion of the augmented coefficient matrix to an upper triangular matrix. Another algorithm, Gauss-Jordan elimination, proceeds by attempting to reduce the coefficient matrix portion of the augmented coefficient matrix to an identity matrix.

4. The number of computations required by Gaussian elimination can be reduced to $n^3/3$. Gauss-Jordan elimination cannot be made this efficient.

EXERCISES

Each of the following systems of equations has a unique solution. Find it by forming the augmented coefficient matrix and applying the Gaussian elimination algorithm.

1. $2x + y = 2$
$3x - y = -6$

2. $x - y = 6$
$-x + 3y = -4$

3. $-2x + y = 0$
$4x - 7y = -10$

4. $3x - y = 16$
$-2x + 3y = -13$

5. $x + y + z = 1$
$x - y - z = 3$
$2x - y + 3z = 9$

6. $x + 2y - z = -2$
$-x - 2y + 3z = 7$
$-x + 3y + z = -3$

7. $-x + 4y - z = 3$
$2x - 8y + 3z = -5$
$x - 2z = -2$

8. $3x + y + z = 2$
$x + 2y - z = -3$
$x + 2y - 3z = -7$

9. $x + y + z + w = 4$
$x - y - z - 2w = -6$
$2x + 3y - z - w = 0$
$x - 2z + w = 0$

10. $2x + 3y - 4z + w = 7$
$2y - 3z + w = 4$
$x + 3z - 3w = -5$
$x + y + z = 0$

11. $u + v - w - x - y = 3$
$2u - 2v + w + x + 2y = -1$
$-u + v + 2w + 3x - y = -1$
$u + 2v - w - x + y = 4$
$-2u + 2v - x - y = -1$

12. $u + 2y = 4$
$w + x + y = 3$
$v - x - y = 0$
$v - 3w = -5$
$u + v + 2x = 3$

For each of the following systems of equations, decide whether the system has a unique solution. If it does not, indicate whether the equations are inconsistent or redundant.

13. $2x + 3y = 2$
$2x - y = -6$

14. $x - y = 6$
$-x + y = -4$

15. $-2x + y = 0$
$2y = 4x - 10$

16. $3x - y = 16$
$-2x + y = 9$

17. $x + y + 2z = 1$
$x - 2y - z = 3$
$2x - y - z = 9$

18. $x + y - z = -2$
$-x - 2y + 3z = 7$
$-2y + 4z = 10$

19. $-x + 4y - z = 3$
$2x - 8y + 3z = -5$
$x - 2z = -2$

20. $3x + y + z = 2$
$x + 2y - z = -3$
$x + 2y - 3z = -7$

21. $\begin{aligned} x + y + z + w &= 4 \\ x - y - z - 2w &= -6 \\ 2x + 3y - z - w &= 0 \\ -3y + z &= 0 \end{aligned}$

22. $\begin{aligned} 2x + 3y - 4z + w &= 7 \\ 2y - 3z + w &= 4 \\ 2x + 7y - 10z + 3w &= 15 \\ 2x + 5y - 7z + 2w &= 11 \end{aligned}$

23. $\begin{aligned} u + v - w - x - y &= 3 \\ 2u - 2v + w + x + 2y &= 2 \\ -u + v + 2w + 3x - y &= 6 \\ u + 2v - w - x + y &= -1 \\ -2u + 2v - x - y &= 3 \end{aligned}$

24. $\begin{aligned} u + 2y &= 4 \\ w + x + y &= 3 \\ v - x - y &= 0 \\ v + w &= 1 \\ u + v + 2x &= 3 \end{aligned}$

For what values of a, b, and c do the following equations have a unique solution?

25. $\begin{aligned} x + 2y &= 3 \\ -x + ay &= 13 \end{aligned}$

26. $\begin{aligned} x + by &= 7/b \\ bx + y &= 7 \end{aligned}$

27. $\begin{aligned} x + y + z &= 1 \\ -x + 2y - z &= 4 \\ x - 3y + cz &= 7 \end{aligned}$

28. $\begin{aligned} ax + y - 3z &= 0 \\ 2ax + 3y - 2z &= 12 \\ y + 4z &= 6 \end{aligned}$

For what values of a, b, and c are the following equations inconsistent? For what values are they redundant?

29. $\begin{aligned} -x + 3y &= 6 \\ 2x - 6y &= a \end{aligned}$

30. $\begin{aligned} x &= 3 - y \\ y &= b - x \end{aligned}$

31. $\begin{aligned} x + y + z &= c \\ x - y - 2z &= 1 \\ 2x - z &= 17 \end{aligned}$

32. $\begin{aligned} x + 2y - z &= a \\ -x - 3y + 2z &= b \\ x + y &= c \end{aligned}$

For the following systems, solve in terms of one or more of the variables.

33. $\begin{aligned} x + 2y - z &= 6 \\ 3x - y - z &= 1 \end{aligned}$

34. $\begin{aligned} -x - 2y - z &= 3 \\ x + y - 3z &= 4 \end{aligned}$

35. $\begin{aligned} x + y + z + w &= 2 \\ -x - 2y - z + 3w &= 4 \\ 2x - 3y + 4z &= 6 \end{aligned}$

36. $\begin{aligned} 2x - 3y + 2z - w &= 17 \\ -x - y - 3z + 4w &= 4 \end{aligned}$

Decide whether the following systems are inconsistent or redundant.

37. $\begin{aligned} x + 2y &= 4 \\ -3x + y &= 6 \\ -3x + 8y &= 24 \end{aligned}$

38. $\begin{aligned} x - 3y &= 7 \\ x - 4y &= 12 \\ x - 2y &= 3 \end{aligned}$

39. $\begin{aligned} x + y - z &= 6 \\ -x - y - 12z &= 4 \\ 2x + y - z &= 3 \\ x + 2y - 15z &= -1 \end{aligned}$

40. $\begin{aligned} x - y - 2z &= 1 \\ 2x - 3y - z &= 4 \\ 3x - y - 4z &= 6 \\ 3x - 4y - 3z &= 2 \\ 5y - 7z &= -11 \end{aligned}$

Modify the algorithm for Gaussian elimination so that it does the following.

41. If a system does not have a unique solution, it indicates whether the equations are inconsistent or redundant.

42. If a system of equations has more equations than unknowns, it will either solve them or indicate that the system does not have a unique solution.

43. It carries out the process of Gauss-Jordan elimination.

44. It carries out the process of Gauss-Jordan elimination when the number of equations is greater than the number of unknowns.

45. Modify step 10 in the Gaussian elimination algorithm so that it ignores all computations involving columns 1 through i.

46. In the modification suggested in the previous exercise, explain why the computations involving columns 1 through i can be ignored.

47. Show that the new step 10 written in Exercise 45 is executed $\sum_{i=1}^{n} i^2$ times and hence is $O(n^3/3)$.

48. Compute the number of computations required to carry out the Gauss-Jordan elimination algorithm from Exercise 43.

Appendix D
Matrix Inversion

Definition Suppose A is an $n \times n$ matrix. By the **inverse** of A we mean another $n \times n$ matrix, denoted A^{-1}, with the property that

$$A \cdot A^{-1} = I \qquad \text{and} \qquad A^{-1} \cdot A = I,$$

where I denotes the $n \times n$ identity matrix.

Recall that for any $n \times n$ matrix B, $BI = B$ and $IB = B$. It is also true that for any real number b, $b \cdot 1 = 1 \cdot b = b$. In other words, I serves the same role for $n \times n$ matrices that the number 1 does for real numbers. Similarly, every real number except zero has a *multiplicative inverse*—a number by which it can be multiplied to get a product of 1. Thus, for instance, the multiplicative inverse of 2 is 1/2. We denote the multiplicative inverse of b by $1/b$ or b^{-1} and call it the **reciprocal** of b. The matrix inverse, then, serves the same role for $n \times n$ matrices that the reciprocal does for real numbers.

The concept of the matrix inverse is very useful in at least two ways. First, suppose we have a system of n simultaneous equations in n unknowns. These can be denoted

$$AX = B,$$

where A is the $n \times n$ coefficient matrix, X is an $n \times 1$ column vector whose entries are the unknowns, and B is an $n \times 1$ column vector—the right-hand sides of the equations. Suppose also that we know A^{-1}. Then we can multiply both sides of the equation by A^{-1} and get

$$A^{-1}(AX) = A^{-1}B$$
$$(A^{-1}A)X = A^{-1}B \qquad \text{(since matrix multiplication is associative)}$$
$$IX = A^{-1}B$$
$$X = A^{-1}B.$$

In other words we can solve the system of equations simply by multiplying vector B by A^{-1}.

Second, similar operations can often be carried out on more complex matrix equations as well.

There are, however, two problems that must be faced in using the matrix inverse. The first is that not every matrix has an inverse. A matrix that has an inverse is called **invertible.** With real numbers, the situation is simpler; 0 is the only real number that does not have an inverse. However, there are infinitely many matrices that do not have inverses. So we will need a means to distinguish matrices that have inverses from those that do not. The second problem is a practical one. Suppose we have the system of equations $AX = B$. The time required to find A^{-1} in order to solve the system by computing $A^{-1}B$ is greater than the time required to solve the system directly using Gaussian elimination. Because of this difficulty, interest in the matrix inverse as a practical tool has declined greatly in recent years. Even so, the matrix inverse is still an important theoretical tool when dealing with matrix equations and, as such, is worth understanding. Thus, in this section, we will first examine some matrices that have inverses and some that do not. Then we will examine an algorithm to compute the matrix inverse. The algorithm will also tell us whether the matrix has an inverse.

Examples

EXAMPLE 1   Show that the inverse of

$$\begin{bmatrix} 1 & 2 \\ 0 & 4 \end{bmatrix} \text{ is } \begin{bmatrix} 1 & -0.5 \\ 0 & 0.25 \end{bmatrix}$$

SOLUTION

$$\begin{bmatrix} 1 & 2 \\ 0 & 4 \end{bmatrix} \begin{bmatrix} 1 & -0.5 \\ 0 & 0.25 \end{bmatrix} = \begin{bmatrix} 1 & 0 \\ 0 & 1 \end{bmatrix}$$

Similarly

$$\begin{bmatrix} 1 & -0.5 \\ 0 & 0.25 \end{bmatrix} \begin{bmatrix} 1 & 2 \\ 0 & 4 \end{bmatrix} = \begin{bmatrix} 1 & 0 \\ 0 & 1 \end{bmatrix} \quad \blacksquare$$

Suppose A is invertible. It can be shown in general that if $AB = I$, it is necessarily true that $BA = I$*. Thus for the next example, we will show that $AA^{-1} = I$, but omit showing that $A^{-1}A = I$.

---

* If $AB = I$, then $ABA = A$. Hence $A^{-1}ABA = A^{-1}A$. Thus $(A^{-1}A)BA = I$ and thus $BA = I$.

EXAMPLE 2   Show that the inverse of

$$\begin{bmatrix} 1 & 1 & 0 \\ -1 & 0 & 0.5 \\ 0 & 1 & 1 \end{bmatrix} \text{ is } \begin{bmatrix} -1 & -2 & -1 \\ 2 & 2 & 1 \\ -2 & -2 & 2 \end{bmatrix}$$

SOLUTION

$$\begin{bmatrix} 1 & 1 & 0 \\ -1 & 0 & 0.5 \\ 0 & 1 & 1 \end{bmatrix} \begin{bmatrix} -1 & -2 & 1 \\ 2 & 2 & -1 \\ -2 & -2 & 2 \end{bmatrix} =$$

$$\begin{bmatrix} -1+2 & -2+2 & 1-1 \\ 1-1 & 2-1 & -1+1 \\ 2-2 & 2-2 & -1+2 \end{bmatrix} = \begin{bmatrix} 1 & 0 & 0 \\ 0 & 1 & 0 \\ 0 & 0 & 1 \end{bmatrix} \blacksquare$$

EXAMPLE 3   Show that the matrix

$$A = \begin{bmatrix} 1 & 0 \\ 0 & 0 \end{bmatrix}$$

does not have an inverse.

SOLUTION   Let

$$B = \begin{bmatrix} a & b \\ c & d \end{bmatrix}$$

be an arbitrary 2×2 matrix. We shall multiply A by B and see if it is possible to solve for a, b, c, and d in such a way that $AB = I$. We have

$$AB = \begin{bmatrix} a & b \\ 0 & 0 \end{bmatrix}$$

Thus, no matter what values we choose for a, b, c, and d, the second row of AB will be 0, not 1. Hence it is impossible to find a matrix B such that $AB = I$, and hence A does not have an inverse. $\blacksquare$

EXAMPLE 4   Show that the matrix

$$C = \begin{bmatrix} 1 & 1 \\ 2 & 2 \end{bmatrix}$$

does not have an inverse.

SOLUTION   Using the same technique as in the last problem, we multiply C by

$$\begin{bmatrix} a & b \\ c & d \end{bmatrix}$$

and get

$$\begin{bmatrix} a+c & b+d \\ 2a+2c & 2b+2d \end{bmatrix}.$$

In order for this product to be the identity matrix, we must find a, b, c, and d that solve the equations

$$a + c = 1$$
$$2a + 2c = 0$$
$$b + d = 0$$
$$2b + 2d = 1.$$

But this system is inconsistent! (See the Exercises.) Hence they cannot be solved, and hence C does not have an inverse. ∎

Finding the Inverse

The inverse of a matrix can be found by a variation of Gauss-Jordan elimination. Furthermore, the algorithm will not only find the inverse when there is one, but will also tell us when no inverse exists. We will first discuss the mechanics of the algorithm, then work through a couple of examples, and finally, explain why it works. We will leave the actual writing of it as a formal algorithm and the analysis of its complexity for the Exercises.

Recall that in Gaussian elimination we formed the augmented coefficient matrix by adding a column vector formed by the right-hand sides of the system of equations. We then proceeded to apply elementary row operations to the augmented coefficient matrix until the coefficient portion of it became upper triangular, with only 1's and possibly 0's on the diagonal. Gauss-Jordan elimination simply continued to apply more ERO's until the coefficient portion became the identity matrix or until the last row of the coefficient portion became all zeroes.

To find the matrix inverse, we augment the coefficient matrix by placing the entire identity matrix to its right. We then apply Gauss-Jordan elimination. If the original coefficient matrix reduces to the identity, the matrix has an inverse and the matrix on the right will be it. If the original coefficient matrix cannot be reduced to the identity, then it did not have an inverse.

EXAMPLE 5 Find the inverse of

$$A = \begin{bmatrix} -1 & 2 \\ 1 & 0.5 \end{bmatrix}$$

SOLUTION We form the augmented coefficient matrix and then apply Gauss-Jordan elimination:

$$\begin{bmatrix} -1 & 2 & 1 & 0 \\ 1 & 0.5 & 0 & 1 \end{bmatrix} \xrightarrow{}$$
$$\text{row2} + 1 * \text{row1}$$

$$\begin{bmatrix} -1 & 2 & 1 & 0 \\ 0 & 2.5 & 1 & 1 \end{bmatrix} \xrightarrow{}$$
$$\text{row2} * (1/2.5)$$

$$\begin{bmatrix} -1 & 2 & 1 & 0 \\ 0 & 1 & 0.4 & 0.4 \end{bmatrix} \xrightarrow{}$$
$$\text{row1} + (-2) * \text{row2}$$

$$\begin{bmatrix} -1 & 0 & 0.2 & -0.8 \\ 0 & 1 & 0.4 & 0.4 \end{bmatrix} \xrightarrow{}$$
$$\text{row1} * (-1)$$

$$\begin{bmatrix} 1 & 0 & -0.2 & 0.8 \\ 0 & 1 & 0.4 & 0.4 \end{bmatrix}$$

Thus

$$A^{-1} = \begin{bmatrix} -0.2 & 0.8 \\ 0.4 & 0.4 \end{bmatrix} \blacksquare$$

EXAMPLE 6 Find the inverse of

$$B = \begin{bmatrix} 1 & 0 & 2 \\ 2 & -1 & 3 \\ -1 & 0.5 & 6 \end{bmatrix}$$

SOLUTION We proceed as before except that, to save space, some steps have been combined.

$$\begin{bmatrix} 1 & 0 & 2 & 1 & 0 & 0 \\ 2 & -1 & 3 & 0 & 1 & 0 \\ -1 & 0.5 & 6 & 0 & 0 & 1 \end{bmatrix} \xrightarrow{}$$
$$\text{row 2} + (-2) * \text{row1}$$
$$\text{and row3} + (1) * \text{row1}$$

$$\begin{bmatrix} 1 & 0 & 2 & 1 & 0 & 0 \\ 0 & -1 & -1 & -2 & 1 & 0 \\ 0 & 0.5 & 8 & 1 & 0 & 1 \end{bmatrix}$$

$\Longrightarrow$ row3 + (.5) * row2
and row3 * (1/7.5)

$$\begin{bmatrix} 1 & 0 & 2 & 1 & 0 & 0 \\ 0 & -1 & -1 & -2 & 1 & 0 \\ 0 & 0 & 1 & 0 & 0.067 & 0.133 \end{bmatrix}$$

$\Longrightarrow$ row2 + (1) * row3
row1 + (−2) * row3

$$\begin{bmatrix} 1 & 0 & 0 & 1 & -0.133 & -0.266 \\ 0 & -1 & 0 & -2 & 1.067 & 0.133 \\ 0 & 0 & 1 & 0 & 0.067 & 0.133 \end{bmatrix}$$

$\Longrightarrow$ row2 * (−1)

$$\begin{bmatrix} 1 & 0 & 0 & 1 & -0.133 & -0.266 \\ 0 & 1 & 0 & 2 & -1.067 & -0.133 \\ 0 & 0 & 1 & 0 & 0.067 & 0.133 \end{bmatrix}$$

Hence

$$B^{-1} = \begin{bmatrix} 1 & -0.133 & -0.266 \\ 2 & -1.067 & -0.133 \\ 0 & 0.067 & 0.133 \end{bmatrix} \quad \blacksquare$$

Why Does This Algorithm Work?

Let I_1 denote the first column of the $n \times n$ identity matrix, I_2 the second column, etc. Thus

$$I_1 = \begin{bmatrix} 1 \\ 0 \\ \cdot \\ \cdot \\ \cdot \\ 0 \end{bmatrix} \quad I_2 = \begin{bmatrix} 0 \\ 1 \\ 0 \\ \cdot \\ \cdot \\ 0 \end{bmatrix} \quad \cdots \quad I_n = \begin{bmatrix} 0 \\ 0 \\ \cdot \\ \cdot \\ \cdot \\ 0 \end{bmatrix}$$

Suppose A is an $n \times n$ coefficient matrix. If we form an augmented coefficient matrix by appending I_1 to A and applying Gauss-Jordan elimination, we can solve the system of equations $AX = I_1$. That solution would appear in the rightmost column of the augmented matrix in the position where I_1 was located. Similarly, applying Gauss-Jordan eliminiation with I_2 appended to A gives us a solution to $AX = I_2$ that appears where I_2 was located. Thus, by applying Gauss-Jordan elimination to the matrix A augmented by adding the entire identity matrix, I, we are, in effect, solving n

systems of equations all at once. The n systems are $AX = I_1$, $AX = I_2$, ..., $AX = I_n$. The solution to each will appear in the location formerly occupied by the corresponding column vector, I_j. But if we apply the following theorem, we can conclude that the solutions to these systems are the columns of a matrix, B, with the property that $AB = I$. That is, B is the inverse of A.

THEOREM D.1 Let A, B, and C be $n \times n$ matrices such that $AB = C$. Denote B by $[B_1, \cdots, B_n]$ and C by $[C_1, \cdots, C_n]$, where B_j and C_j are the jth columns of B and C, respectively. Then, for each j, B_j is the solution of the system of equations $AX = C_j$.

PROOF Using the usual notation for A, B, and C, we have

$$
\begin{bmatrix} C_{11} & \cdots & C_{1n} \\ \cdot & & \cdot \\ \cdot & & \cdot \\ \cdot & & \cdot \\ C_{n1} & \cdots & C_{nn} \end{bmatrix} = \begin{bmatrix} A_{11}B_{11} + \cdots + A_{1n}B_{n1} & \cdots & A_{11}B_{1n} + \cdots + A_{1n}B_{nn} \\ \cdot & & \cdot \\ \cdot & & \cdot \\ \cdot & & \cdot \\ A_{n1}B_{11} + \cdots + A_{nn}B_{n1} & \cdots & A_{n1}B_{1n} + \cdots + A_{nn}B_{nn} \end{bmatrix}.
$$

Thus, in general, we can say that

$$
C_j = \begin{bmatrix} C_{1j} \\ \cdot \\ \cdot \\ \cdot \\ C_{nj} \end{bmatrix} = \begin{bmatrix} A_{11}B_{1j} + \cdots + A_{1n}B_{nj} \\ \cdot & \cdot \\ \cdot & \cdot \\ \cdot & \cdot \\ A_{11}B_{1j} + \cdots + A_{1n}B_{nj} \end{bmatrix}.
$$

Note also that

$$
B_j = \begin{bmatrix} B_{1j} \\ \cdot \\ \cdot \\ \cdot \\ B_{nj} \end{bmatrix}
$$

and hence

$$
AB_j = \begin{bmatrix} A_{11}B_{1j} + \cdots + A_{1n}B_{nj} \\ \cdot & \cdot \\ \cdot & \cdot \\ \cdot & \cdot \\ A_{11}B_{1j} + \cdots + A_{1n}B_{nj} \end{bmatrix}.
$$

Thus $AB_j = C_j$ and hence B_j is the solution of the system of equations $AX = C_j$. ☐

Complexity of Matrix Inversion

Since Gauss-Jordan elimination has already been shown to be less efficient than Gaussian elimination, and since the matrix inversion algorithm depends on an extended version of Gauss-Jordan elimination, it is obvious that this algorithm for matrix inversion is less efficient than Gaussian elimination. Hence solving $AX = B$ by finding A^{-1} by this algorithm would not be an efficient way to proceed. But, as pointed out earlier, the inverse of a matrix is important for other reasons.

TERMINOLOGY

inverse of a matrix   invertible matrix

reciprocal of a number

SUMMARY

1. Some matrices have inverses; others do not.
2. The matrix inverse can be found, if it exists, by augmenting the matrix by the $n \times n$ identity matrix and applying Gauss-Jordan elimination.
3. If a matrix cannot be reduced to the identity matrix by applying Gauss-Jordan elimination, it does not have an inverse.

EXERCISES

1. Verify that the matrix computed in Example 5 is in fact the inverse of A in that example.
2. Verify that the matrix computed in Example 6 is the inverse of B in that example.

Let matrices A, B, C, and D be as given.

$$A = \begin{bmatrix} 1 & 2 \\ 3 & 4 \end{bmatrix} \qquad B = \begin{bmatrix} -2 & 1 \\ 1.5 & -0.5 \end{bmatrix}$$

$$C = \begin{bmatrix} 1 & 2 & 1 \\ 0 & -1 & 2 \\ 1 & 0 & -1 \end{bmatrix} \qquad D = \begin{bmatrix} \frac{1}{6} & \frac{1}{3} & \frac{5}{6} \\ \frac{1}{3} & -\frac{1}{3} & -\frac{1}{3} \\ \frac{1}{6} & \frac{1}{3} & -\frac{1}{6} \end{bmatrix}$$

Verify that:

3. $AB = I$ 4. $BA = I$ 5. $CD = I$ 6. $DC = I$

Compute the inverses of the following matrices.

7. $\begin{bmatrix} 3 & 1 \\ 1 & 3 \end{bmatrix}$

8. $\begin{bmatrix} 4 & 2 \\ 1 & 2 \end{bmatrix}$

9. $\begin{bmatrix} 4 & 3 \\ 2 & 1 \end{bmatrix}$

10. $\begin{bmatrix} -1 & 6 \\ 2 & -3 \end{bmatrix}$

11. $\begin{bmatrix} 2 & 1 & 0 \\ -1 & 2 & 2 \\ 0 & 1 & 1 \end{bmatrix}$

12. $\begin{bmatrix} 1 & 2 & 3 \\ -1 & 1 & 2 \\ 2 & 1 & 5 \end{bmatrix}$

13. Show that the system of equations obtained in Example 4 is inconsistent.

Show by the method used in Examples 3 and 4 that the following matrices do not have inverses.

14. $\begin{bmatrix} 2 & 0 \\ 1 & 0 \end{bmatrix}$

15. $\begin{bmatrix} 1 & 3 \\ 3 & 9 \end{bmatrix}$

Show by Gauss-Jordan elimination that the following matrices do not have inverses.

16. The matrix of Exercise 14

17. The matrix of Exercise 15

18. $\begin{bmatrix} 3 & 1 & 0 \\ 2 & -1 & 3 \\ 1 & -1 & -3 \end{bmatrix}$

19. $\begin{bmatrix} -1 & 2 & 4 \\ 2 & -1 & 7 \\ 0 & 5 & 1 \end{bmatrix}$

20. $\begin{bmatrix} 1 & 4 & 3 & -1 \\ 0 & 2 & 3 & 1 \\ 4 & 1 & 2 & 8 \\ -2 & 9 & 7 & -9 \end{bmatrix}$

Let C and D be as given in Exercises 3 through 6. Let

$$I_1 = \begin{bmatrix} 1 \\ 0 \\ 0 \end{bmatrix} \qquad I_2 = \begin{bmatrix} 0 \\ 1 \\ 0 \end{bmatrix} \qquad I_3 = \begin{bmatrix} 0 \\ 0 \\ 1 \end{bmatrix}$$

Show that:

21. $CD_1 = I_1$

22. $CD_2 = I_2$

23. $DC_3 = I_3$

24. $DC_1 = I_1$

25. Suppose that A and B are $n \times n$ invertible matrices. Show that $(AB)^{-1} = B^{-1}A^{-1}$.

26. Show that for any size n, $I^{-1} = I$.

27. Let

$$A = \begin{bmatrix} a & b \\ c & d \end{bmatrix}$$

and suppose that $A^2 = I$. What can be said about a, b, c, and d?

28. Calculate the inverse of

$$\begin{bmatrix} 1 & 0 & 0 \\ 0 & 2 & 0 \\ 0 & 0 & 3 \end{bmatrix}.$$

29. Generalize the result of Exercise 28 and prove that your generalization is correct.

30. Show that the inverse of any invertible 3×3 upper triangular matrix is also upper triangular.

31. Suppose A is an $n \times n$ matrix. One way to find the inverse of A would be to denote B by

$$\begin{bmatrix} b_{11} \cdots b_{1n} \\ \cdot \quad \cdot \\ b_{n1} \cdots b_{nn} \end{bmatrix}.$$

One can then multiply A times B, set the result equal to I, and solve the resulting system of equations for b_{11}, b_{12}, etc. Show that the complexity of this approach is $O(n^6)$. (Hint: use what you already know about the complexity of Gaussian elimination.)

32. Write the algorithm presented in this section for finding the inverse of a matrix as a formal algorithm.

33. Compute the complexity of the algorithm for matrix inversion used in this section.

REFERENCES (APPENDICES A–D)

Anton, Howard, *Elementary Linear Algebra.* 2nd ed. New York: Wiley, 1977.

Conte, S.D., and C. deBoor, *Elementary Numerical Analysis—An Algorithmic Approach.* New York: McGraw-Hill, 1980.

Florey, Francis G., *Elementary Linear Algebra with Applications.* Englewood Cliffs, N.J.: Prentice-Hall, 1979.

Forsythe, G.E., M.A. Malcolm, and C.B. Moler, *Computer Methods for Mathematical Computations.* Englewood Cliffs, N.J.: Prentice-Hall, 1977.

Grossman, S.I., *Elementary Linear Algebra.* Belmont, Calif.: Wadsworth, 1980.

Isaak, S., and M.N. Manougian, *Basic Concepts of Linear Algebra.* New York: Norton, 1976.

Johnson, L., and R.D. Reiss, *Numerical Analysis,* 2nd ed. Reading, Mass.: Addison-Wesley, 1982.

Lang, S., *Linear Algebra.* Reading, Mass.: Addison-Wesley, 1966.

Noble, Ben, *Applied Linear Algebra.* Englewood Cliffs, N.J.: Prentice-Hall, 1969.

O'Nan, M., *Linear Algebra.* New York: Harcourt Brace Jovanovich, 1971.

Strang, G., *Linear Algebra and Its Applications.* New York: Academic Press, 1976.

Williams, G., *Linear Algebra with Applications.* Boston: Allyn & Bacon, 1984.

Appendix E
Complex Numbers and Recurrence Equations

The quadratic equation,

$$at^2 + bt + c = 0$$

has two solutions:

$$t = \frac{-b \pm \sqrt{(b^2 - 4ac)}}{2a}.$$

The latter equation is called the **quadratic formula.** The term $b^2 - 4ac$ is called the **discriminant** and whenever it is negative, the equation will have complex roots, that is, roots that involve i. Thus it is not unusual for a second-order, linear, homogeneous recurrence equation with constant coefficients to have a characteristic equation that has complex roots. A **complex number** is any number of the form

$$a + bi$$

where we use i to denote $\sqrt{-1}$. Such numbers can be represented geometrically as points in the **complex plane** as shown in Figure E.1. This representation turns out to be the key to resolving a paradox—that recurrence

Figure E.1 The complex plane with a complex number represented in rectangular coordinates.

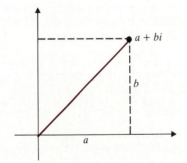

Figure E.2 The complex plane with a complex number represented in rectangular and polar coordinates.

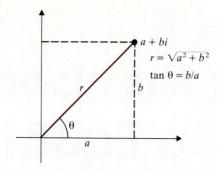

equations with real coefficients that model real world situations have solutions that involve complex numbers. When we represent a complex number by the notation $a + bi$ we are representing it in rectangular coordinates in the complex plane, a being the horizontal position of a point in the plane and b being the vertical position. But we can also represent the point in **polar coordinates.** Suppose we draw a line segment from the origin to the point as in Figure E.2. We can represent the point as an ordered pair (r, θ), where r is the distance from the origin to the point and θ is the angle that line segment makes with the positive x-axis. Thus we can write

$$a + bi = r \cos \theta + ir \sin \theta = r (\cos \theta + i \sin \theta).$$

The reason this is helpful is that

$$(\cos \theta + i \sin \theta)^n = (\cos n\theta + i \sin n\theta).$$

We won't prove this identity here, but its proof can be found in almost any Calculus book. For instance, we can apply the identity to the equation we looked at above (see Figure E.3):

$$\frac{-1 + i\sqrt{3}}{2} = \left(\cos \frac{2\Pi}{3} + i \sin \frac{2\Pi}{3}\right)$$

and

$$\frac{-1 - i\sqrt{3}}{2} = \left(\cos \frac{4\Pi}{3} + i \sin \frac{4\Pi}{3}\right).$$

Thus

$$f(n) = c_1 \left(\frac{-1 + i\sqrt{3}}{2}\right)^n + c_2 \left(\frac{(-1 - i\sqrt{3})}{2}\right)^n$$

$$= c_1 \left(\cos \frac{2\Pi}{3} + i \sin \frac{2\Pi}{3}\right)^n + c_2 \left(\cos \frac{4\Pi}{3} + i \sin \frac{4\Pi}{3}\right)^n$$

$$= c_1 \left(\cos \frac{2n\Pi}{3} + i \sin \frac{2n\Pi}{3}\right) + c_2 \left(\cos \frac{4n\Pi}{3} + i \sin \frac{4n\Pi}{3}\right).$$

Figure E.3 $\dfrac{-1 \pm \sqrt{3}}{2}$ represented in both rectangular and polar coordinates.

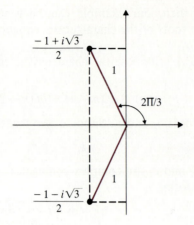

Thus

$$f(0) = c_1 + c_2$$

$$f(1) = c_1\left(\frac{-1}{2} + \frac{i\sqrt{3}}{2}\right) + c_2\left(\frac{-1}{2} - \frac{i\sqrt{3}}{2}\right) = \frac{-1}{2}(c_1 + c_2) + \frac{i\sqrt{3}}{2}(c_1 - c_2),$$

$$f(2) = c_1\left(\frac{-1}{2} - \frac{i\sqrt{3}}{2}\right) + c_2\left(\frac{-1}{2} + \frac{i\sqrt{3}}{2}\right) = \frac{-1}{2}(c_1 + c_2) - \frac{i\sqrt{3}}{2}(c_1 - c_2),$$

$$f(3) = c_1 + c_2,$$

etc.

Now observe what happens if c_1 and c_2 are complex conjugates, that is, if

$$c_1 = k_1 + k_2 i \qquad \text{and} \qquad c_2 = k_1 - k_2 i.$$

Then

$$f(0) = 2k_1$$

$$f(1) = -\frac{1}{2}\, 2k_1 + \frac{i\sqrt{3}}{2}\, 2k_2 i = -k_1 - \sqrt{3}k_2$$

$$f(2) = -\frac{1}{2}\, 2k_1 + \frac{i\sqrt{3}}{2}\, 2k_2 i = -k_1 + \sqrt{3}k_2$$

$$f(3) = 2k_1$$

etc.

Thus

$$f(n) = \begin{cases} 2k_1 & n = 3k \\ -k_1 - \sqrt{3}k_2 & n = 3k + 1 \\ -k_1 + \sqrt{3}k_2 & n = 3k + 2 \end{cases}$$

and the values of $f(n)$ are not complex but are real for every value of n!

In summary, then, our example can be generalized to the following theorem if the roots of the characteristic equation are complex:

THEOREM E.1 Suppose the characteristic equation of the recurrence equation

$$f(n) + af(n - 1) + bf(n - 2) = 0$$

has complex roots, t_1 and t_2. Suppose also that the solution

$$c_1 t_1{}^n + c_2 t_2{}^n$$

is such that c_1 and c_2 are complex conjugates. Then the solution can be written in the form

$$f(n) = r^n(k_1 \cos n\theta + k_2 \sin n\theta).$$

PROOF   We first write the solution

$$f(n) = c_1(a + bi)^n + c_2(a - bi)^n.$$

We can then write $a + bi$ in the form $r(\cos \theta + i \sin \theta)$ and $a - bi$ in the form $r(\cos \theta - i \sin \theta)$. Substituting into the expression for $f(n)$ gives

$$f(n) = c_1 r^n(\cos n\theta + i \sin n\theta) + c_2 r^n(\cos n\theta - i \sin n\theta).$$

Now if we assume c_1 and c_2 are complex conjugates, we can write $c_1 = k_1 + k_2 i$ and $c_2 = k_1 - k_2 i$. Substituting and simplifying, we get

$$f(n) = r^n(k_1 \cos n\theta + k_2 \sin n\theta).$$

We will leave the details of the simplification for the Exercises.   ☐

The key assumption in this proof is that c_1 and c_2 are complex conjugates. This is, in fact, quite a reasonable assumption. For instance, consider the class of recurrence equations of the form

$$f(n) + af(n - 1) + bf(n - 2) = 0, \quad f(0) = v_0 \text{ and } f(1) = v_1,$$

that is, with two initial values specified. Suppose that v_0 and v_1 are real and also suppose that the characteristic equation has complex roots, $a + bi$ and $a - bi$. Then the solution could be written in the form

$$f(n) = c_1(a + bi)^n + c_2(a - bi)^n.$$

Hence

$$f(0) = c_1 + c_2 = v_0,$$
$$f(1) = c_1(a + bi) + c_2(a - bi) = v_1.$$

Thus

$$(c_1 + c_2)a + (c_1 - c_2)bi = v_1,$$
$$(c_1 - c_2)bi = -(c_1 + c_2)a + v_1.$$

Since the right-hand side is a real number, so must the left side be also. Thus $(c_1 - c_2)$ must be a multiple of i.

$$(c_1 - c_2) = ki, \ (c_1 + c_2) = v_0.$$

Hence

$$c_1 = \frac{1}{2}(v_0 + ki) \quad \text{and} \quad c_2 = \frac{1}{2}(v_0 - ki)$$

and thus they are complex conjugates.

The point of the latter argument is this: Under some very reasonable assumptions, namely that the coefficients of the recurrence equation and the initial values for $f(0)$ and $f(1)$ are themselves real, the constants c_1 and c_2 *must be* complex conjugates.

EXAMPLE 1   Solve the initial value problem:

$$f(n) + 2f(n-1) + 2f(n-2) = 0, \quad f(0) = 0, f(1) = 1.$$

SOLUTION   This solution involves evaluating $\cos \theta$ and $\sin \theta$ for different values of θ. (See Table E.1 for $\cos \theta$ and $\sin \theta$ for several common values of θ.) The characteristic equation is

$$t^2 + 2t + 2 = 0,$$

which has roots $-1 \pm i$. These roots can be written in the form

$$r(\cos \theta + i \sin \theta) \quad \text{and} \quad r(\cos \theta - i \sin \theta)$$

where $r = \sqrt{2}$ and $\theta = \dfrac{3\Pi}{4}$. (See Figure E.4.) Thus we have

$$f(n) = r^n(k_1 \cos n\theta + k_2 \sin n\theta).$$
$$= \sqrt{2}^n \left(k_1 \cos \frac{3n\Pi}{4} + k_2 \sin \frac{3n\Pi}{4}\right).$$

Figure E.4 $-1 \pm i$ represented in both rectangular and polar coordinates.

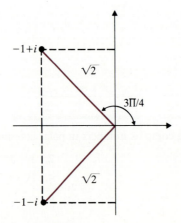

Hence

$$f(0) = k_1 = 0 \quad \text{and} \quad f(1) = \sqrt{2}\left(k_1\left(\frac{-1}{\sqrt{2}}\right) + k_2\left(\frac{1}{\sqrt{2}}\right)\right) = 1.$$

Thus $k_2 = 1$, and $f(n) = \sqrt{2}^n \sin\frac{3n\Pi}{4}$. ∎

TABLE E.1   cos θ and sin θ for some common values of θ

θ	cos θ	sin θ
0	1	0
$\frac{\Pi}{6}$	$\frac{\sqrt{3}}{2}$	$\frac{1}{2}$
$\frac{\Pi}{4}$	$\frac{1}{\sqrt{2}}$	$\frac{1}{\sqrt{2}}$
$\frac{\Pi}{3}$	$\frac{1}{2}$	$\frac{\sqrt{3}}{2}$
$\frac{\Pi}{2}$	0	1
$\frac{3\Pi}{4}$	$\frac{-1}{\sqrt{2}}$	$\frac{1}{\sqrt{2}}$
$\frac{2\Pi}{3}$	$\frac{\sqrt{3}}{2}$	$\frac{-1}{2}$
Π	−1	0
$\frac{5\Pi}{4}$	$\frac{-1}{\sqrt{2}}$	$\frac{-1}{\sqrt{2}}$
$\frac{4\Pi}{3}$	$\frac{-1}{2}$	$\frac{-\sqrt{3}}{2}$
$\frac{3\Pi}{2}$	0	−1
2Π	1	0

EXERCISES

Write the following complex numbers in polar coordinates, in the form (r, θ).

1. $3 + 3i$ 2. $3 - 3i$

3. $-1 + i\sqrt{3}$ 4. $2\sqrt{3} + 2i$

5. i 6. -2

For the following recurrence equations, write the solution in the form

$$f(n) = c_1(a + bi)^n + c_2(a - bi)^n,$$

and using the initial conditions, find c_1 and c_2, thereby proving them to be complex conjugates. Also write the solution in the form

$$f(n) = r^n(k_1 \cos n\theta + k_2 \sin n\theta).$$

7. $f(n + 2) + 2f(n + 1) + 2f(n) = 0; f(0) = 0, f(1) = 1$
8. $f(n + 2) + 2f(n + 1) + 2f(n) = 0; f(0) = 1, f(1) = 0$
9. $f(n) = -3f(n - 2); f(0) = 1, f(1) = 2$
10. $f(n) = -3f(n - 2); f(0) = 2, f(1) = 0$
11. $f(n) = f(n - 1) - f(n - 2); f(0) = 2, f(1) = 0$
12. $f(n) = f(n - 1) - f(n - 2); f(0) = 1, f(1) = 1$
13. In the proof of Theorem E.1, show that $c_1 r^n(\cos n\theta + i \sin n\theta) + c_2 r^n(\cos n\theta - i \sin n\theta)$ can be simplified to $r^n(k_1 \cos n\theta + k_2 \sin n\theta)$ if c_1 and c_2 are complex conjugates.

Appendix F
Counting Equivalence
Relations

Theorem 8.2 enables us to answer the question "How many equivalence relations are there on a set, A, if $|A| = n$?" If we can count the number of partitions, we know the number of equivalence relations.

When counting the number of partitions of A, we first pick any r, $1 \le r \le n$, and compute the number of partitions that contain r subsets of A. One way to visualize the problem of counting these partitions is as follows: Think of r cells; each element of A is to be assigned to one cell in such a way that every cell receives at least one element of A. At first, this problem appears very similar to a problem discussed in Section 7.3: In how many ways can n customers in a restaurant choose one of r desserts? However, it is very different. In the restaurant problem, the customers are not distinguishable (at least as far as the number of each type of dessert is concerned), but the desserts are. Thus n indistinguishable objects are being placed in r distinguishable cells. With partitions, the situation is reversed: n distinguishable elements of A are being assigned to r indistinguishable cells. For a more extensive discussion of these problems, see Hillman et. al., *Discrete and Combinatorial Mathematics* (see References following Appendix F). We will be following an approach similar to theirs. We start with an example.

EXAMPLE 1 List all partitions of $A = \{a, b, c, d\}$ of sizes 1, 2, and 3.

SOLUTION   The number of partitions when $r = 1$ is easy to determine; $\{a, b, c, d\}$ itself is the only partition.

In order to solve the problem of systematically listing all partitions when $r = 2$, we remove one element, say d, from A, and then list the partitions of the subset $\{a, b, c\}$ that is left. We get

$\{a, b, c\}$ (the only one-subset partition)

$\{\{a\}, \{b,c\}\}$ $\{\{b\}, \{a,c\}\}$ $\{\{c\}, \{a,b\}\}$ (all possible as two-subset partitions).

There are then two ways to make two-subset partitions of the original set A. We can either add $\{d\}$ as a separate subset to $\{a, b, c\}$, giving

$$\{\{d\}, \{a, b, c\}\},$$

or we can insert d in every possible place where it can go in each possible two-subset partition. This gives

$$\{\{a, d\}, \{b, c\}\} \ \{\{a\}, \{b, c, d\}\} \ \{\{b,d\}, \{a,c\}\} \ \{\{b\}, \{a,c,d\}\}$$
$$\{\{c,d\}, \{a,b\}\} \ \{\{c\}, \{a,c,d\}\}.$$

Thus we have seven possible two-subset partitions of $\{a, b, c, d\}$.

With three sets, we follow a similar procedure. The only possible three-subset partition for $\{a, b, c\}$ is $\{\{a\}, \{b\}, \{c\}\}$. The possibilities for two-subset partitions of $\{a, b, c\}$ have been previously listed. Thus inserting d in each possible place in $\{\{a\}, \{b\}, \{c\}\}$ gives

$$\{\{a,d\}, \{b\}, \{c\}\} \ \{\{a\}, \{b,d\}, \{c\}\} \ \{\{a\}, \{b\}, \{c,d\}\}$$

and adding $\{d\}$ as an additional element to each possible two-subset partition of $\{a, b, c\}$ gives

$$\{\{d\}, \{a\}, \{b,c\}\} \ \{\{d\}, \{b\}, \{a,c\}\} \ \{\{d\}, \{c\}, \{a,b\}\}.$$

Thus there are six three-subset partitions of $\{a, b, c, d\}$. ∎

Definition Let $n > 0$ and let $1 \le r \le n$. We let **$S(n,r)$** denote the number of distinct r-subset partitions of an n-element set. $S(n,r)$ is called the **Stirling number of the second kind.**

In the previous example, we showed that

$$S(4,1) = 1$$
$$S(4,2) = 7$$

and

$$S(4,3) = 6.$$

Example 1 can be generalized to the following theorem.

THEOREM F.1 If $n, r \in \mathbf{N}$,

a) $S(n,1) = S(n,n) = 1$

b) If $n \ge 3$ and $2 \le r \le n - 1$, $S(n,r) = S(n - 1, r - 1) + rS(n - 1, r)$.

PROOF

a) $S(n,1)$ is the number of possible one-subset partitions. There is clearly only one such partition, namely, the set itself. $S(n,n)$ is the number of

n-subset partitions. Since each set in the partition must be nonempty, there is again only one—the n one-element subsets.

b) Consider $A = \{a_1, a_2, \ldots, a_n\}$. Then if we remove a_n from A, we get $A' = \{a_1, a_2, \ldots, a_{n-1}\}$. We consider two types of partitions of A—those in which $\{a_n\}$ is one of the sets in the partition and those in which $\{a_n\}$ is not one of the sets. Each partition that has $\{a_n\}$ as an element corresponds to an $r-1$-subset partition of A'. Thus there are $S(n-1, r-1)$ partitions of this type. Also, A' itself has $S(n-1, r)$ r-subset partitions. In each of these partitions, a_n can be added to any of the r subsets of which it consists, giving an r-subset partition for A; in fact, any r-subset partition in which $\{a_n\}$ is not included can be produced this way. Thus there are $rS(n-1, r)$ partitions in which $\{a_n\}$ is not included. Hence

$$S(n,r) = rS(n-1,r) + S(n-1,r-1). \quad \square$$

Stirling's numbers of the second kind can be computed by the recursion formula found in Theorem F.1. For instance, using the values of $S(4,1)$, $S(4,2)$, and $S(4,3)$ we computed earlier, we conclude that

$$S(5,2) = 2 \cdot S(4,2) + S(4,1) = 15$$
$$S(5,3) = 3 \cdot S(4,3) + S(4,2) = 25$$
$$S(5,4) = 4 \cdot S(4,4) + S(4,3) = 10.$$

These numbers can be arranged into a table similar to Pascal's triangle. Note that unlike Pascal's triangle, Stirling's triangle is not symmetric (see Figure F.1). Also note that the formula presented in Theorem F.1 is a two-parameter recurrence equation. Because of the two parameters, we cannot solve it by the methods of Chapter 6. In the case where $r = 2$, however, the equation does reduce to a one-parameter recurrence equation that we can solve. (See the Exercises.)

These numbers provide the answer to our original question. The number of equivalence relations on a set, A, with $|A| = n$ is

$$\sum_{r=1}^{n} S(n,r).$$

Figure F.1 Part of Stirling's second triangle.

		1			
	1		1		
	1	3		1	
1	7		6		1
1	15	25	10	1	

EXAMPLE 2   How many equivalence relations are there on the set {a, b, c, d, e}?

SOLUTION   From Figure 8.10, we can compute that there are 52 such relations. ∎

TERMINOLOGY

$S(n,r)$

Stirling number of the second kind

EXERCISES

1. Add the next two lines to Stirling's triangle.

2. List all two-subset partitions of a set, A, with $|A| = 5$.

How many equivalence relations are there on a set, A, with $|A| =$

3. 2   **4.** 3   **5.** 4   **6.** 6

7. Prove that if $n \geq 1$ and $1 \leq r \leq n/2$, then $S(n,r) \geq S(n,n-r+1)$ and interpret this result in terms of Stirling's triangle.

8. See Exercise 7. Prove that if $r \neq 1$, $S(n,r) > S(n,n-r+1)$.

9. Write an algorithm for computing $S(n,r)$ for any $n \geq 1$, r such that $1 \leq r \leq n$.

10. Show that if $r = 2$, the equation in Theorem F.1 reduces to a one-parameter recurrence equation and solve it.

11. Show that if $r \geq 3$, the equation in Theorem F.1 does not reduce to a one-parameter recurrence equation.

Glossary of Key Terms

Addition principle   The principle that the number of elements in two or more disjoint sets is the sum of the number of elements in each individual set. (7.1)

Adjacency matrix   A Boolean matrix indicating which pairs of vertices in a graph or digraph are joined by edges. (8.1, 9.3)

Algorithm   Can be informally described as a step-by-step process for solving a problem from a particular class of problems. (4.1)

Antisymmetric   A relation, R, is antisymmetric if aRb and bRa implies $a = b$. (8.2)

Arithmetic series   A sum of the form $a + (a + d) + (a + 2d) + \cdots + (a + nd)$ for some n, a, and d. (2.6)

Array   A function from the set $\{1, \ldots, n\}$ into some codomain. It is simply an ordered list of elements of the codomain. (App. A)

Asymptotic domination   A function, f, asymptotically dominates a function, g, if there are positive integers m and N such that for every $n > N$, $|f(n)| \le m|g(n)|$. (4.2)

Axiom   A statement that asserts a relationship between the fundamental terms of a mathematical system. Axioms are not proven but rather are taken as the starting points for establishing the theory of that system. (3.1)

Axiomatic method   An approach to mathematics based on using primitive terms and axioms. Theorems can only be added if proven from the axioms, definitions, and/or previously proven theorems. (3.1)

Back substitution   The last step in the solution of simultaneous equations by Gaussian elimination. (App. C)

Binomial theorem   A formula for expanding the expression $(x + y)^n$ using the binomial coefficients and terms of the form $x^i y^{n-i}$. (7.4)

Binary search   An algorithm for locating an element in a sorted set. Based on a divide-and-conquer strategy. (6.6)

Binary tree   A 2-ary tree such that for every parent, each child is designated as left child or right child. (10.3)

653

Bipartite graph   A graph in which the vertices can be partitioned into two subsets, V_1 and V_2, such that every edge is associated with one element of V_1 and one element of V_2. (9.2)

Breadth-first spanning tree   A spanning tree generated by an algorithm that first visits all neighbors of a specified initial vertex, then visits their neighbors, etc. (10.2)

Brute force algorithm   An approach to solving a problem based on exhaustively listing all possible solutions, then selecting the best one. (1.1)

Circuit   A path in which the initial and terminal vertices are the same. (9.2)

Combination   Any subset of a given finite set. (7.3)

Combinational circuit   An electronic circuit whose output can be completely described by a truth table. (3.8)

Complete graph   A graph in which there is one and only one edge associated with every pair of vertices. (9.2)

Complexity of an algorithm   An approximate measure of the numbers of computations an algorithm requires for an input set of size n, typically presented in the form $O(g(n))$ for the appropriate function g. (4.3)

Composite number   A positive integer that has factors other than one and itself. (5.2)

Conditional probability   The conditional probability of an event A given an event B is $p(A \cap B)/p(B)$. It represents the probability of A occurring if it is known that B occurs.

Congruent integers   Two integers, a and b, are congruent mod n for some positive integer, n, if n divides $b - a$. (5.5)

Constructive proof   A proof that does not use proof by contradiction. (3.6)

Cross-product of two sets A and B   The set of all ordered pairs (a,b) such that $a \in A$ and $b \in B$. (2.3)

Cycle   A simple path in which the initial and terminal vertices are the same. (9.2)

Depth-first spanning tree   A spanning tree generated by visiting one neighbor of an initial vertex, then an unvisited neighbor of the previously visited vertex, etc. When it cannot proceed any further, it backtracks until an unvisited neighbor can be found. (9.2)

Diagonal matrix   An $n \times n$ matrix in which all entries off of the main diagonal are zero. (App. B)

Difference equation   An equation in which $f(n + 1) - f(n)$ is expressed in terms of n and $f(n), \ldots, f(0)$. (6.1)

Digraph   A set V, of vertices and a set, E, of directed edges. Each directed edge is associated with an ordered pair of vertices. (8.1)

Divide-and-conquer algorithm   A problem solving strategy based on dividing a problem into subproblems of the same form. (4.5)

Division algorithm   The theorem that if a and b are integers, $b > 0$, then there exist unique integers q and r, $0 \le r < b$, such that $a = bq + r$. (5.3)

Edge   If (V, E) is a graph or digraph, each element of E is an edge. (8.1, 9.1)

Element   A primitive undefined term of set theory. An element can be described as a member of a set. (2.1)

Equiprobable measure   That probability measure in which each simple event has an equal probability. (7.5)

Equivalence class   A subset of elements of a set, A, that are related to each other under an equivalence relation, R. (8.2)

Equivalence relation   A relation on a set, A, that is reflexive, symmetric, and transitive. (8.2)

Euclidean algorithm   A method for finding the greatest common divisor of two integers. (5.3)

Euler circuit   A circuit in a graph that traverses every edge once. (9.2)

Event   Any subset of a sample space. (7.5)

Existential quantifier   The phrase "there is" or "for some" in sentences such as "There is a real number x such that $x - 1 = 0$" and "For some real numbers x, $x > 10$." (3.5)

Expectation   In an experiment with numerical outcomes, the sum of each numerical value times the probability of that value occurring. It represents the "average" outcome. Also called "expected value." (7.5)

Experimental probability   The average frequency of occurrence of an outcome. (7.5)

Function   A subset of a set $A \times B$ such that every $a \in A$ appears in one and only one ordered pair. (2.3)

Fundamental theorem of arithmetic   The theorem that any positive integer can be uniquely factored into primes up to the order in which the primes are written. (5.3)

Gate   An elementary circuit that serves as a fundamental building block of combinational circuits. (3.8)

Gaussian elimination   An algorithm for solving systems of simultaneous linear equations. (App. C)

General solution of a linear recurrence equation   A solution that includes all solutions to the reduced equation plus a term satisfying the non-homogeneous equation. Includes arbitrary constants that can be selected to satisfy the initial conditions. (6.3)

Geometric series   The sum $ax^i + ax^{i+1} + \cdots + ax^n$ for some a, i, x, and n. (2.6)

Graph   A set (V, E) of vertices and edges such that there is a function from E to the set of pairs of elements of V. Such a function is often described by saying that each edge is "associated with" a pair of vertices. (9.1)

Hamilton cycle   A cycle that includes every vertex. (9.2)

Hoffman code   A variable-length code that provides the minimal expected length for messages given the requisite characters and their expected frequencies. (10.4)

Homeomorphic graphs   Two graphs are homeomorphic if one can be transformed into the other by a process of placing degree two vertices on edges and replacing degree two vertices and their incident edges by an edge. (9.4)

Identity matrix   An $n \times n$ matrix with 1's on the main diagonal and all other entries 0. (App. B)

Incidence matrix   A Boolean matrix used to represent graphs and multigraphs. Each row corresponds to a vertex, each column to an edge. A 1 is entered in each column in the row corresponding to its incident vertices. (9.3)

Inclusion-exclusion principle   A formula for counting the number of elements in $A_1 \cup \ldots \cup A_n$ based on the number of elements in each A_i and in each intersection. (7.1, 7.4)

Independent events   Events A and B are independent if $p(A \cap B) = p(A) \cdot p(B)$. (7.7)

Indirect proof   Proving $p \Rightarrow q$ by proving its contrapositive. (3.6)

Infix form   The standard form for presenting algebraic expressions; for example, $x + (y + z)$. (10.4)

Inverse of a function f   A function, f^{-1}, such that $f^{-1}(f(x)) = x$ for all x in the domain and $f(f^{-1}(y)) = y$ for all y in the codomain. (2.4)

Inverse of a matrix   Given an $n \times n$ matrix, A, its inverse is another $n \times n$ matrix A^{-1} such that $AA^{-1} = I$. (App. D)

Invertible matrix   A matrix that has an inverse. (App. D)

Isomorphic graphs   Two graphs are isomorphic if there is a one-to-one correspondence between their vertices V_1 and V_2 such that $\{u, v\}$ is an edge in V_1 iff $\{u', v'\}$ is an edge in V_2 for the corresponding vertices u' and v'.

Knapsack problem   Any of a class of problems that involve selecting the subset of a finite set that best satisfies some specified constraints. (4.4)

Konigsberg bridge problem   A famous problem generally regarded as the origin of graph theory. It was solved by Leonhard Euler and involved proving the non-existence of a route through the city of Konigsberg that crosses every bridge in the city once and only once and returns to the start. (9.1, 9.2)

m-ary tree   Either the empty set or a rooted tree in which every vertex has out-degree at most m. (10.3)

Matrix   A rectangular display of elements from some domain. (App. B)

Mathematical induction   A technique for proving statements of the form $p(n)$ where n is any natural number. There are two forms of induction; each uses a basis step and an induction step. (3.7)

Mathematical model   Representation of an idealized real world situation using symbols to denote both the entities and relationships involved. (1.1)

Method of characteristic roots   A technique for solving linear, second-order, homogeneous recurrence equations based on finding the roots of a characteristic polynomial. (6.4)

Method of substitution   A technique for solving linear, first-order recurrence equations based on matching the coefficients with a standard form. (6.3)

Minimal spanning tree   In a weighted graph, the spanning tree that has minimal weight. (10.2)

Modus ponens A rule of inference. For propositions, it follows the pattern

$$p \Rightarrow q$$
$$\frac{p}{q} \tag{3.6}$$

Modus tollens A rule of inference. For propositions, it follows the pattern

$$p \Rightarrow q$$
$$\frac{\neg q}{\neg p} \tag{3.6}$$

Multinomial theorem A formula for expanding expressions of the form $(x_1 + \cdots + x_k)^n$. (7.4)

Multiplication principle Given k successive events that can occur in $n_1, \ldots, n_k$ ways, respectively, the number of ways the entire sequence can occur is $n_1 \ldots n_k$. (7.2)

Nearest neighbor algorithm A problem solving method for the traveling salesman problem based on selecting the closest adjacent vertex at each step but not looking further ahead than that next step. Is an example of a greedy algorithm. (1.1)

One-to-one correspondence A function that is both one-to-one and onto. (2.4)

One-to-one function A function from a set A to a set B such that every element of B comes from at most one element of A. (2.4)

Onto function A function whose range equals its codomain. (2.4)

Partial order A reflexive, transitive, antisymmetric relation on a set, A. (8.3)

Path A walk in which no edges are repeated. (9.1)

Permutation A one-to-one function from a set onto itself. It is often thought of as an arrangement of the elements of the set. (7.2)

Planar graph A graph that can be represented by points and curves in the plane in such a way that no edges intersect except at the vertices upon which they are incident. (9.4)

Positional representation of an integer A notation $d_n \ldots d_0$ for an integer in which $0 \le d_i < b$ for some integer b and each d_i. The value of the integer is $d_n b^n + d_{n-1} b^{n-1} + \cdots + d_0 b^0$. (5.4)

Postfix form A form for presenting algebraic expressions based on the pattern expression expression operator; for example, $xyz++$.

Predicate A statement that involves one or more variables from some domain and that becomes a proposition when values are assigned to the variables. (3.2)

Prefix form A form for presenting algebraic expressions based on the pattern operator expression expression; for example, $+x+yz$.

Prime number A positive integer that has no factors other than one and itself. (5.2)

Primitive term An item of mathematical vocabulary that is not defined but rather is used as the starting point for defining further terms. Primitive terms are needed to avoid circularity in definition. (3.1)

Principle of superposition If both f and g are solutions of a linear, homogeneous recurrence equation, then so is $c_1 f + c_2 g$ for any c_1 and c_2.

Probability measure　A function defined on the set of all subsets of a sample space. It must satisfy four properties listed in the body of this text. (7.5)

Proof by contradiction　A proof technique based on making an assumption and reasoning to a logical contradiction, thereby concluding that the assumption was false. (3.6)

Proposition　A statement that can be unambiguously assigned a truth value. (3.2)

Quicksort　An algorithm for sorting a list of data. It is extremely efficient in most cases. (6.6)

Recurrence equation　An equation in which $f(n + 1)$ is written in terms of n and $f(n), \ldots, f(0)$. (6.1)

Recursive algorithm　An algorithm that calls (a simpler version of) itself. (6.5)

Recursive definition　A type of definition in which something is defined in terms of simpler versions of itself. In order to avoid circularity, such a definition includes an elementary case in which the definition does not depend on itself. (5.1)

Reflexive relation　A relation on a set A that includes (a, a) for all $a \in A$. (8.2)

Relation　A subset of some cross-product set, $A \times B$. (8.1)

Rooted tree　A digraph with one vertex with in-degree 0, all other vertices with in-degree 1, and every vertex joined to the root by a directed path. (10.3)

Russell's paradox　An apparent contradiction that arose early in the study of set theory. The contradiction is that the set of all sets that are not members of themselves can neither be a member of itself nor not be a member of itself. (2.2)

Sample space　The set of possible outcomes of some experiment. (7.5)

Sequence　A function from N into some codomain. (6.2)

Set　A primitive undefined term of set theory. A set can be described as a collection of objects. (2.1)

Simple path　A path in which no vertices are repeated. (9.1)

Solution of a recurrence equation　A sequence all of whose elements satisfy the recurrence equation and its initial conditions, if any. (6.2)

Spanning tree　A subgraph of a given connected graph that is also a tree. (10.1)

Square matrix　A matrix with the same number of rows as columns. (App. B)

Stack　A means of organizing data used in a number of computer applications. It is based on a first in, last out principle. (7.5)

Stirling number of the second kind　Gives the number of r-element partitions of an n-element set. Thus it can be used to count the number of equivalence relations on a given set. (App. F)

Symmetric relation　A relation, R, on a set, A, such that aRb implies bRa. (8.2)

Tautology　A compound proposition that is true for all assignments of values to its simple propositions. (3.4)

Total order　A partial order in which every pair of elements is comparable. (8.3)

Transitive closure of a relation　The smallest transitive relation containing the given relation. (8.4)

Transitive relation   A relation, R, on a set, A, such that aRb and bRc implies aRc. (8.2)

Traveling salesman problem   A famous problem in theoretical computer science. It involves finding the shortest circuit in a weighted graph that visits all vertices. (1.1 et al.)

Traversal algorithm   An algorithm for visiting all elements of a particular data structure. (10.4)

Tree   A connected, acyclic graph. (10.1)

Triangular matrix   A matrix in which all of the entries above the main diagonal are zero or all of the entries below the main diagonal are zero. (App. B)

Truth table   A tabulation of the values a compound proposition can assume when values are assigned to its simple propositions. (3.3)

Universal quantifier   The phrase "for all" in statements like "For all real numbers x, $x^2 \geq 0$." (3.5)

Venn diagram   A visual representation of a collection of sets and the relationships that exist among them. (2.1)

Verification of an algorithm   Proving that an algorithm is correct. This typically involves a precise statement of the conditions that hold both before its execution (pre-conditions) and after its execution (post-conditions) and a demonstration that the algorithm transforms the pre-conditions into the post-conditions. This typically involves the use of mathematical induction. (4.3)

Vertex   If a graph or digraph is denoted (V, E), any element of V is a vertex. (8.1, 9.1)

Walk   An alternating sequence of vertices and edges of the form $v_1 e_1 v_2 e_2 \ldots$ such that each edge is incident on the vertices written on either side of it. (9.1)

Well-ordering principle   The assertion that every non-empty subset of the natural numbers has a least element. (5.1)

Answers to Odd-Numbered Exercises

Exercises 1.1

1. *ABCDA*   90

5. *AB*   15
    ABC   36
    ABCD   66
    ABCDA   90

11. $(n-1)^2$

17. $n(n-1)(n-2)$

31. Fulfills both conditions.

41. Discrete, stochastic

43. Real world: It is tropical fruit—some will spoil, some will be stolen; the statement, "She has $120 to spend; she can buy mangoes for 30 cents each and pineapples for 45 cents each."
Real model: All the fruit will be sold, neither spoils faster
Math model: $.30m + .45p = 120$ and $p = 2m$
Conclusions: buy 100 mangoes, 20 pineapples
Real world: she makes a profit.

3. *ACBDA*   66

7. *AD*   12
    ADC   28
    ADCB   52
    ADCBA   77

13. *AB*   15
    ABC   36
    ABCDA   90

23. Vertices: *A,B,C,D*
    Edges: *AB* *AC* *AD* *BC* *BD* *CD*
    Weights: 15   35   24   21   31   30

39. Continuous, deterministic

Exercises 2.1

1. $\{x \mid 0 \le x \le 1\}$

5. {0000, 0001, 0010, 0011,
    0100, 0101, 0110, 0111,
    1000, 1001, 1010, 1011,
    1100, 1101, 1110, 1111}

3. $\Sigma = \{0,1\}$
    $\{w \in \Sigma^* \mid \text{length } (w) = 4\}$

9. $\{x \mid x^2 - 4 = 0\}$

661

13. Equal

17. {8,9,10,11}

21. {0,1,2,3}

29. 8

33. 1

37. False—D/L is still a non-empty set of dogs, which is not a subset of cats.

41. True

15. Unequal

19. {4,5,6,7}

25. No

31. 4

35. False—cats are not dogs.

39. True

43.

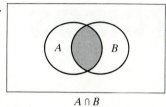

$A \cap B$

47.

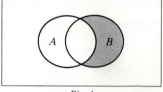

$B \setminus A$

51.

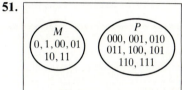

53.

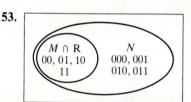

59.

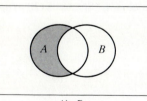

$A \setminus B$ $B \setminus A$

Exercises 2.3

1. Next value: 63
Function: $x^3 - 1$

5. $\varnothing$

9. Is not a function—each x has two different y values.

3. {(1,1), (1,2), (2,1), (2,2), (3,1), (3,2) }

7. Is a function—each x has a unique y associated with it.

11. a) -1 b) 0 c) 0 d) $a^2 - 1$ e) $x^2 + 2x$
f) $x^4 - 1$ g) $x^2 + y^2 - 2$ h) $4x^4 - 4x^2$

13. Range: **R**
Independent var: x
Dependent var: y

21. Is not a function.

y	x
-2	4
-1	1
0	0
1	1
2	4

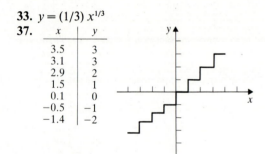

17. Domain: all strings of 0's and 1's
Codomain: all strings of 0's and 1's
Range: all strings of 0's and 1's beginning with
at least 1 zero

25. Domain: **R**
Range: {100}

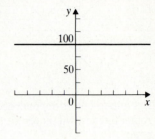

29. $d + kt^2$

31. $y = \sqrt{(x-3)}$

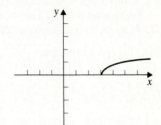

33. $y = (1/3)\, x^{1/3}$
37.

x	y
3.5	3
3.1	3
2.9	2
1.5	1
0.1	0
-0.5	-1
-1.4	-2

35. $y = |x-1|$
39. Range: **R**
Domain: **R**

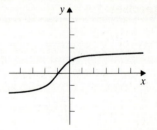

43. Range: **Z** (the set of integers)
Domain: **R**

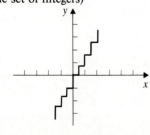

Exercises 2.4

1. $g \circ f = |x-3|$
$f \circ g = |x| - 3$

3. $g \circ f = (x/2)^2$
$f \circ g = x^2/2$

7.

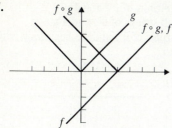

11.

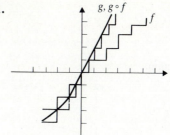

13. Is a one-to-one correpondence since the horizontal line test will never hit more than one point.

17. Is not a one-to-one correspondence since the horizontal line test fails at each integer > 0.

21. Is a one-to-one correspondence.

25. $f(x)$ has an inverse: $\begin{cases} -\sqrt{x} & x > 0, \\ -x & x \le 0. \end{cases}$

29. Inverse is $(1/x)$.

33. 13

15. Is not a one-to-one correspondence since the horizontal line test fails at an infinite number of points.

19. Is a one-to-one correspondence.

23. Has an inverse: $\sqrt[3]{(3x)}$.

27. Has an inverse: $\begin{cases} z + 3 & z < 3, \\ z - 3 & z > 3. \end{cases}$

31. Inverse is $(x + 3)/2$.

35. 4

Exercises 2.5

1.

x	y
-2	0.0625
-1	0.25
0	1
1	4
2	16
3	64

5. $2^8 = 256$

7. $3^0 = 1$

11. $2^{-1} = 0.5$

9. $2^6 = 64$

13.

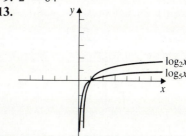

15. -3

19. 3.807

23. 11

17. $x + 1$

21. 7.451

25. 3.011

Exercises 2.6

1. $2 + 3 + 4 + 5 + 6 \cdots + (n + 1)$

3. $4 + 9 + 25 + 36 + 49 + 64 + 81 = 268$

5. $3 + 3 + 3 + \cdots 3 + = 3n$

7. $\displaystyle\sum_{i=1}^{n} (2i - 1)$

9. $\displaystyle\sum_{i=1}^{n} (2^i)$

11. $\displaystyle\sum_{i=1}^{n} (i + 1) = 2 + 3 + 4 + 5 + 6 + 7 + 8 + \cdots + (n + 1)$

$\displaystyle\sum_{i=2}^{n+1} i = 2 + 3 + 4 + 5 + 6 + 7 + 8 + \cdots + (n + 1)$

15. $4\left(n + \dfrac{n(n + 1)}{2}\right)$

17. $2^{n+1} - (n + 2)$

19. $n^2(n + 1)^2/4 + \dfrac{2(n(n + 1))}{2} - n$

21. $\dfrac{n(n + 1)}{2} + n$

23. $n^2 + 2n$

Exercises 3.2

1. Is a proposition.

3. Is a proposition.

5. Is not a proposition.

7. This could be a proposition—given that everyone accepts the same set of letters as being the set of vowels.

15. $\{(x,y) \mid y \le 2 - x\}$

17. The truth set is empty.

Exercises 3.3

1. p and q

3. $p \rightarrow q$

5. q and p

7. Misers like money and my cousins are not greedy.

9. If misers like money, then my cousins are greedy.

11. Misers do not like money and my cousins are greedy.

13. False

15. True

17. True

19.

p	q	p NAND q
False	False	True
False	True	True
True	False	True
True	True	False

Exercises 3.4

1.

p	q	r	p OR q	$(p$ OR $q) \to r$
T	T	T	T	T
T	T	F	T	F
T	F	T	T	T
T	F	F	T	F
F	T	T	T	T
F	T	F	T	F
F	F	T	F	T
F	F	F	F	T

3.

p	q	NOT q	p AND NOT q
T	T	F	F
T	F	T	T
F	T	F	F
F	F	T	F

5. T

7. F

9.

p	p AND p
T	T
F	F

13.

p	F	p AND F
T	F	F
F	F	F

17.

p	q	$(p$ AND $q)$	$((p \to q$ AND $p) \to q$
T	T	T	T
T	F	F	T
F	T	F	T
F	F	F	T

21. p OR q

23. p AND q

25. p

27. (hours > 40) OR (day > 8) OR (Sat OR Sun)

Exercises 3.5

1. False

3. True

5. False

7. False

9. True

11. False

13. U = set of all fish, S is the set of skates
$\exists\, x \in \Sigma \,|\, (x \in U)$

15. U = set of tiresome things, R is the set of rainy days
$\forall\, x \in R \,|\, (X \in U)$

19. U = set of tedious things, S is set of songs of at least one hour
$\forall\, x \in S \,|\, (x \in U)$

21. 6

Exercises 3.6

1. Alphonse is rich; modus ponens

3. Salt is not sugar; modus tollens

5. Some pigs are not eagles; modus tollens

7. Your presents are not tin.

9. My friends do not dine at the higher table.

23. Converse: If you have strep throat the throat culture will come back positive.
Inverse: If the throat culture comes back negative, then you do not have strep throat.
Contrapositive: If you do not have strep throat, the throat culture will come back negative.

29. Let p denote people who fit in well,
j denote belief in justice

$$p(x) \rightarrow j(x)$$
$$\frac{j(\text{you})}{p(\text{you})} \quad \text{Fallacy of the converse}$$

33. p or q
$$\frac{\neg p}{q}$$

Exercises 3.7

1. Let $n = 1$
LHS $= 2$
RHS $= \dfrac{1(2)(3)}{3} = 2$

Assume true for n, then we must prove for $n + 1$.

LHS: $\displaystyle\sum_{i=1}^{n+1} i(i+1) = \sum_{i=1}^{n} i(i+1) + (n+1)(n+2)$

$$= \frac{n(n+1)(n+2)}{3} + (n+1)(n+2) \qquad \text{by induction}$$

$$= \frac{n(n+1)(n+2)}{3} + \frac{3(n+1)(n+2)}{3}$$

$$= \frac{(n+3)(n+1)(n+2)}{3} = \frac{(n+1)(n+2)(n+3)}{3}$$

RHS: By simple substitution we get the following:
$\dfrac{(n+1)(n+2)(n+3)}{3}$, which is equal to the LHS.

9. $n = 1$
LHS $= a^{m+1}$
RHS $= a^m(a^1) = a^{m+1}$
Assume for n, prove for $n + 1$.
LHS: a^{m+n+1}
RHS: $a^m a^{n+1}$

$$
\begin{aligned}
&= a^m a^n a && \text{by given definition}\\
&= a^{m+n} a && \text{by induction hypothesis}\\
&= a^{m+n+1} && \text{by given definition}
\end{aligned}
$$

Exercises 3.8

1.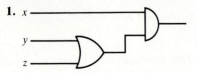

5.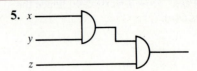

11. $(p \wedge q \wedge \neg r) \vee (p \wedge \neg q \wedge \neg r) \vee (\neg p \wedge q \wedge r) \vee (\neg p \wedge \neg q \wedge \neg r)$

Exercises 4.1

5. $i = 0,1,2,3,4,5$
largest $= -\infty$, $5,5,5,5,5$
location $= ?,1,1,1,1,1$

9. $i := 0$
while more people in list
  read next person:
    $i := i + 1$;
end; {while}
write (i, 'people in list');

15. Not finite

7. $i := 1,2,3,4,5,6,7$
largest $:= (-\infty$ at init) $1,17,17,17,32,32,32$
location $:= 1,2,2,2,5,5,5$

11. Procedure Commission (contact_list);
{contact_list is a list of sales, items, costs}
rate := .08;
total := 0;
while there are more contacts do begin
  get next contract;
  total := total + cost
  write (contact, item, cost);
  if item is on special then rate := .09;
end; {while}
if total > 1000 then rate $= .10$;
commission := rate $*$ total;
write ('commission earned is', commission);
end; {commission}

17. Not definite

Exercise 4.2

5. $f(n) > g(n)$ for $0 < n < 150$
$g(n) > f(n)$ for $n > 150$, $n < 0$

9. Let $N = 1$, $m = 1$

15. This was shown in the book in Example 2(b) if
$c = 2$.

7. Let $N = 1$, $m = 1$
$n > 1$
Multiplying both sides by n^3 we get $n^4 > n^3$

11. $f(n) = \sin n$, $g(n) = 1/n$

Exercises 4.3

1. $10{:}10^{-6}$
$100{:}10^{-6}$
$1000{:}10^{-6}$

5. $10{:}10^{-4}$
$100{:}10^{-2}$
$1000{:}1$

9. $5n - 1$ best case, $7n - 3$ worst case; hence the
algorithm is $O(n)$.

13. $O(n)$

3. $10{:}10^{-5}$
$100{:}10^{-4}$
$1000{:}10^{-3}$

7. $O(n)$

11. $O(n)$

15. Poly 1: $3n$
Poly 2: $2n$
(counting only $+$, $*$, $-$,$/$)

Exercises 4.4

3. 39

7. Items 2 and 3

Exercises 4.5

3. Counter: 0,1
Location: 0, 1
9. Counter: 0, 1
Found f, t
15. $O(n)$

5. Counter: 0, 1, 2, 3, 4, 5
Location: 0, 0, 0, 0, 0, 0
11. Counter: 0, 1, 2, 3, 4, 5
Found: f, f, f, f, f, f
17. The pre-condition is that we have an unordered list of length n. The post-condition is that the variable "location" will be 0 if not found; the actual location in the list otherwise. The loop invariant is that location = counter if counter is the first place in which x is found; otherwise it is 0.

19. Location: ?, 4, 2, 3
High: 8, 4, 4, 4
Low: 1, 1, 2, 2
Found: f, f, f, t
25. 7
29. Until approximately $n = 32$ the linear search could be better; after that the binary search will be faster on average.
33. Average linear: 0.5 sec
Worst linear: 1 sec
Worst binary: 2×10^{-5}

21. Location: ?, 4, 2, 1
High: 8, 4, 2, 2
Low: 1, 1, 1, 1
Found: f, f, f, t
27. $\lceil \log_2 n \rceil$
31. Average linear: 2.5×10^{-5}
Worst linear: 5×10^{-5}
Worst binary: 6×10^{-6}

Exercises 5.1

1. $\text{Succ}(a)$
5. $\text{Pred}^2(a)$
9. $\forall a \in N,\ \text{Succ}(a) \in N$

3. $\text{Succ}^3(a)$
7. If $a \neq 1$, then $\text{Succ}(\text{Pred}(a)) = a$.
19. Assume that a "character" has been defined. Then String^1 = string of length 1 = character. String^{n+1} = character "followed by" String^n.

Exercises 5.2

3. Prime
7. Prime
25. 1.17
29. About 16%

5. Composite
9. Composite
27. 1.17
31. About 6.2%

Exercises 5.3

1. $q = 15, r = 26$

3. $q = 72, r = 67$

5. $q = 0, r = 600$

7. 2

9. 2

11. Not relatively prime

13. Relatively prime

15. $x = 5, y = -4$

17. $x = -6, y = 19$

19. $2 \times 3 \times 5 \times 7$

21. $2^3 \times 3^2 \times 5$

23. $2^4 \times 17 \times 23$

25. n

27. n

29. 1

Exercises 5.4

1. 11

3. 94

5. Every number is the base to the first power.

7. 1000_5: 125
1000_8: 512
1000_{16}: 4096

9. They are the same number between $n(10)$ and $n(10) + 4$ for any integer $n \geq 0$.

13. 53_8

15. 1113_5

17. 644_8

19. 100000010001_2

29. 351_8

31. 1042_5

33. $7_8 R46_8$

35. $103_5 R23_5$

37. 473

39. 270

41. 75_8

43. 100000_2

45. 645_8

47. 10001011_2

49. 444_8
124_{16}

51. 724_8
$1D4_{16}$

53. 11111110100_2

55. 11111000001111_2

Exercises 5.5

1. 3

3. 0

5. 0

7. True

9. True

11. 2

13. -1

15. Complete

17. Not complete

19. 4

21. 10

23. 3

25. 11

Exercises 6.1

1. $f(n) = f(n-1) + c$
5. $f(n) = f(n-1) + cn$

11. $P_{n+1} = P_n + (c/2P_n - r)P_n$
15. $S_n = S_{n-1} + n^2, S_0 = 0$
19. $h(n) = 0.6h(n-1), h(0) = h$

23. $P_{n+1} = (1+r)P_n - (R + nc)$ where c is the amount of increase in each payment.

3. $f(n) = f(n-1) - c/n^2$
7. $\Delta f(n) = -1$
$f(n) = f(n-1) - 1, f(1) = 6$
13. $S_n = S_{n-1} + n, S_1 = 1$
17. $D(n) = 2D(n-1), D(1) = 2$
21. $S(1) = 2$
$S(n) = S(n-1) + n$

Exercises 6.2

1. Is a solution.
5. $f(1)$ is undefined.
9. Non-linear

13. Linear; constant coefficients; non-homogeneous, first order;
17. Has a unique solution.
23. $f(n) = 2n + 1$
27. $f(n) = \dfrac{n(n+1)(2n+1)}{6}$

3. Is a solution.
7. $f(2)$ is undefined.
11. Linear; constant coefficients; non-homogeneous, first order;
15. Has a unique solution.
19. Has a unique solution.
25. $f(n) = n!$

Exercises 6.3

1. $f(n = 3^n$
5. $f(n) = 0$ for all n
9. $f(n) = 3^{n+1}/2 - 1/2$
13. $f(n) = 3n - 2$
17. $S(n) = 2^n$
27. $f(n) = n!$

3. $f(n) = (1/2)^{n-1}$
7. $f(n) = 3(n+1)$
11. $f(n+1) = 2f(n), f(0) = 1$
15. $f(n) = \dfrac{n(n-1)(2n-1)}{6}$
19. $S(n) = \dfrac{n^2 + n + 2}{2}$
29. $f(n) = n! + n + \sum\limits_{i=1}^{n-1} (i) \left(\prod\limits_{j=i}^{n-1} (j+1) \right)$

Exercises 6.4

1. $c_1 2^n + c_2$
5. $f(n) = c_1(3/2)^n + c_2(-2)^n$

3. $f(n) = c_1(5.83)^n + c_2(0.17)^n$
7. $f(n) = (k_1\cos(n\pi/6) + k_2\sin(n\pi/6))$

9. $f(n) = 1$

11. $f(n) = (-0.176)(5.83)^n + (0.176)(0.17)^n$

13. $f(n) = (2/7)(3/2)^n + (-2/7)(-2)^n$

15. $f(n) = (2/\sqrt{3})(\sin(n\pi/6))$

17. $g(n) = -1/2$

19. $f(n) = -(1/2)n - 11/8$

21. $g)n) = (3/2)3^n$

23. $g(n) = -(1/4)n^2 - (11/8)n - 75/32$

25. $g(n) = (1/4)n + 1$

27. $g(n) = (4/7)2^n$

29. $f(n) = (3/5)2^n - (1/10)(-3)^n - 1/2$

31. $f(n) = (7/5)2^n - (1/40)(-3)^n - (1/2)n - 11/8$

33. $f(n) = (-16/10)2^n + (1/10)(-3)^n + (3/2)3^n$

35. $f(n) = (12/5)2^n - (9/160)(-3)^n - (1/4)n^2 - (11/8)n - 75/32$

37. $f(n) = (-1)^{n+1} - (5/4)n(-1)^n + (1/4)n + 1$

Exercises 6.5

1.

n	largest
1	2
2	3
3	3
4	14
5	14

7.

value of n	line	action	stack
$n = 3$	1	test fails go to line 3	
	3	recursive call, $n=2$	3
$n = 3$	1	test fails go to line 3	2
	3	recursive call, $n=1$	3
$n - 1$	1	test succeeds	
	2	longest := 3	

Algorithm halts, 2 is popped from the stack

$n = 2$	3	largest := 3	
	4	$-1 > 3$ fails	3

Algorithm halts, 3 is popped from the stack

$n = 3$	3	largest := 3	
	4	$6 > 2$ so go to line 5	
	5	largest := 6	

Algorithm halts. The stack is empty, so stop.

19. This can be proved by induction.

23. $O(n)$

25. $O(n)$

27. If the variable largest is renamed smallest and line 4 is changed to "if A [n] < smallest," the algorithm will find the smallest.

Exercises 6.6

1. Test fails, algorithm halts. $N = 1$

3. $N = 2$

5. $N = 1$

7. $N = 1$

9. Test fails, algorithm ends. $N = 1$

11. Test fails, algorithm ends. $N = 2$

13. Test fails, algorithm ends. $N = 1$ **15.** $N = 5$
17. $N = 3$ **19.** $N = 2$

Exercises 7.1

1. 8

3. 6

5. 16

7. 3: 2
     4: 3
     5 or less: 9

9. 1296

11. 26×27^6

13. 64000

15. 4096

17. 4810

19. 960

21. The claim is not reasonable.

25. 56 had an excess of both; 95 had no excesses.

27. 333

Exercises 7.2

1. 3628800

3. 1

5. 4200

7. 40320

9. 20!

11. *abc, acb, cab, bac, bca, cba*

13. $19! \times 2$

15. 19!

17. 1294920

19. 17!

21. 336

23. 6!6!6!/18

25. 20160

27. 5040

29. 1663200

31. 540540

33. 70

35. 1324

37. 41253

39. 243165

Exercises 7.3

5. $1 + 4 + 6 + 4 + 1 = 16$

7. $1 + 7 + 21 + 35 + 35 + 21 + 7 + 1 = 128$

9. 220

11. 52!/10!42!

13. 52!/10!42! − 48!/10!38!

15. $m!/k!(m - k)!$

17. 495

19. There are 35 possibilities for each possible winner. There are 4 possible winners, so there are 140 possible majority distributions.

21. 252

23. 75582

25. 91

27. 23751

29. 1246

31. 1345

33. 345

35. 1234

Exercises 7.4

1. $1 + 4x + 6x^2 + 4x^3 + x^4$ **5.** $x^3 - 3x^2y + 3xy^2 - y^3$
7. $x^3 + 3x^2y + 3x^2z + 3xy^2 + 3y^2z + 3xz^2 + 3yz^2 + y^3 + z^3 + 6xyz$
23.

```
        1    7    21    35     35     21    7    1
     1    8    28    56     70     56    28    8    1
  1    9    36    84    126    126    84    36    9    1
```

Exercises 7.5

1. Each of the numbers 0 to 37 plus the number 00
9. A positive integer (or zero)
15. $S = \{dd, dg, gd, gg\}$

19. 0.6

3. Any number 1 to 6
11. Let each space have a probability of 1/38.
17. Assign probabilities to each outcome such that $p_1 + p_2 + p_3 + p_4 + p_5 + p_6 = 1$, and each p_i = proportion of times i was rolled.
21. 7/8

Exercises 7.6

1. 9/37
5. 18/37
9. 3/8
13. 1/6
17. 1/4
21. 1/504
25. 70/256
31. 2/5
35. 1/924
39. 420/924
43. .0035

3. 19/37
7. 1/8
11. 1/4
15. 1/128
19. 1/2
23. 2/132
27. 28/256
33. .237
37. 20/924
41. 1.23×10^{-5}
45. .095

Exercises 7.7

1. 3/51, not independent
5. 1/6
9. 0.129
13. Independent
17. 0.0001

3. 1/2, independent
7. 0.253
11. 0.987
15. Not independent
19. 0.2916

Exercises 8.1

1. $\{(1,a), (1,b), (1,c), (1,d), (2,a), (2,b), (2,c), (2,d), (3,a), (3,b), (3,c), (3,d)\}$
5. $\{(n,1), (n,2), (n,3), (m,1), (m,2), (m,3)\}$ **7.** Is a function.

9. Is not a function.

11. Domain: Set of subsets of S
Codomain: Set of subsets of S
The set of ordered pairs would be (s_1, s_2) if s_1 was a superset of s_2.

13. Domain: set of 10 cities
Codomain: set of 10 cities
The set of pairs would be

17.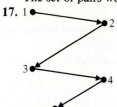

21.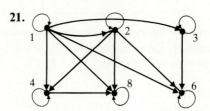

23. $\begin{bmatrix} 0 & 1 \\ 1 & 0 \end{bmatrix}$

25. $\begin{bmatrix} 0 & 1 & 0 & 0 \\ 1 & 0 & 0 & 1 \\ 1 & 0 & 0 & 1 \\ 0 & 0 & 1 & 0 \end{bmatrix}$

27. $\begin{bmatrix} 0 & 0 & 1 & 0 \\ 0 & 0 & 1 & 0 \\ 0 & 0 & 0 & 1 \\ 0 & 0 & 0 & 0 \end{bmatrix}$

29.

35.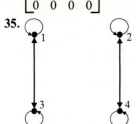

37. This is a relation over $N \times N \times N$.

39. Let I be the set of possible ID numbers, A be the set of possible names, and C be the set of possible codes. Then this is a relation over $I \times A \times C$.

Exercises 8.2

1. Reflexive, symmetric, antisymmetric, transitive
3. Reflexive, symmetric, transitive
5. Antisymmetric, transitive
7. Symmetric, transitive
9. Reflexive, antisymmetric, transitive
11. Reflexive, symmetric, transitive
13. Reflexive, symmetric, transitive
15. Symmetric
27. The matrix of Exercise 1 is such a relation.
29. The identity relation
31. The entire graph is one equivalence class.
35. The sets $O(f)$, for real valued functions on $N \cup \{0\}$.
37. Is a partition.
39. Is a partition.

41. The quotient set will consist of all groups of siblings.

43. $\{\varnothing, \text{ sets of size 1, sets of size 2, } \ldots\}$

45. True

47. False

49. 2^{n^2-n}

51. $2^n 3^k$ where $k = \dfrac{n^2 - n}{2}$.

Exercises 8.3

1. Is not a poset—is not transitive.

3. Is not a poset—is not transitive.

5. Is not a poset—is not reflexive.

7. Is not a poset—is not antisymmetric.

9. Is a poset.

11. Is a poset.

13. Is not a poset.

19. Is a total order.

21. Is not a total order.

23. Maximal: $\{1,2\}$
Minimal: $\varnothing$

25. Maximal: 8,7,6,5

27. Maximal: can not be found
Minimal: can not be found

29. $a, \ldots, k$

37. N

Exercises 8.4

1.

$R_3 \cup R_4 \qquad R_3 \cap R_4$

21. R^3 is the great-grandparent/great-grandchild relationship.

27. $\begin{bmatrix} 1 & 1 & 0 \\ 1 & 1 & 1 \\ 1 & 1 & 0 \end{bmatrix}$

29. $\begin{bmatrix} 1 & 1 & 1 \\ 0 & 0 & 1 \\ 0 & 0 & 1 \end{bmatrix}$

31. $\begin{bmatrix} 1 & 0 & 1 \\ 1 & 1 & 1 \\ 0 & 0 & 1 \end{bmatrix}$

33. $\begin{bmatrix} 1 & 1 & 1 & 1 \\ 1 & 1 & 1 & 1 \\ 1 & 1 & 1 & 1 \\ 1 & 1 & 1 & 1 \end{bmatrix}$

45. Yes

Exercises 9.1

1. Vertices: stops in the route
Edges: ways to get between stops
Multigraph with no loops (digraph if there are one-way streets)

5. Vertices: animals and vegetation
Edges: the "eater" to the "eaten"
Digraph with loops

9. Vertices: elements to be matched
Edges: matched elements
Graph with no loops

13. Vertices: squares on the board
Edges: possible next moves
Graph with no loops

17. Vertices: people
Edges: existing relationships
Digraph with no loops (graph if we assume
"friendship" means a reciprocated relationship)
31. 24
35. 14

27. Not a path: *aca*
Not a simple path: *edcabej*
Simple path: *kjeba*

33. 80

Exercises 9.2

9. No Euler circuit, is Hamiltonian
13. No Euler circuit, no Hamiltonian circuit
17. No Euler circuit, is Hamiltonian
21. No Euler circuit, is Hamiltonian

11. No Euler circuit, is Hamiltonian
15. No Euler circuit, is Hamiltonian
19. Has an Euler circuit, is Hamiltonian
23. No Euler circuit, no Hamiltonian circuit

Exercises 9.3

1. Are not
5. Are

3. Are
7. Are not

9. Adjacency matrix
$$\begin{bmatrix} 0 & 1 & 1 & 1 & 1 \\ 1 & 0 & 1 & 1 & 1 \\ 1 & 1 & 0 & 1 & 1 \\ 1 & 1 & 1 & 0 & 1 \\ 1 & 1 & 1 & 1 & 0 \end{bmatrix}$$

For the incidence matrix, count the edges as follows: 1–2, 1–3, 1–4, 1–5, 2–3, ... Then the matrix is

$$\begin{bmatrix} 1 & 1 & 1 & 1 & 0 & 0 & 0 & 0 & 0 & 0 \\ 1 & 0 & 0 & 0 & 1 & 1 & 1 & 0 & 0 & 0 \\ 0 & 1 & 0 & 0 & 1 & 0 & 0 & 1 & 1 & 0 \\ 0 & 0 & 1 & 0 & 0 & 1 & 0 & 1 & 0 & 1 \\ 0 & 0 & 0 & 1 & 0 & 0 & 1 & 0 & 1 & 1 \end{bmatrix}$$

13.

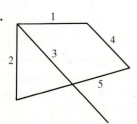

17. $A^2 = \begin{bmatrix} 2 & 1 & 1 \\ 1 & 2 & 1 \\ 1 & 1 & 2 \end{bmatrix}$ $\quad A^3 = \begin{bmatrix} 2 & 3 & 3 \\ 3 & 2 & 3 \\ 3 & 3 & 2 \end{bmatrix}$

Edge sequences are 123 for length two, and
1213, 1313, 1323 for length three.

29.

Numbering edges as shown, we get $\begin{bmatrix} 0 & 1 & 1 & 1 & 0 \\ 1 & 0 & 1 & 0 & 1 \\ 1 & 1 & 0 & 0 & 0 \\ 1 & 0 & 0 & 0 & 1 \\ 0 & 1 & 0 & 1 & 0 \end{bmatrix}$

Exercises 9.4

5. $4 - 5 + 3 = 2$

21. Homeomorphic to any subgraph of 5 vertices in K_6

23. Homeomorphic to $K_{3,3}$

29. $n \leq 4$

9. $13 - 18 + 7 = 2$

27. Is not planar.

Exercises 10.1

1.
$$
\begin{array}{ccccc}
 & H & H & H & \\
 & | & | & | & \\
H- & C- & C- & C- & H \\
 & | & | & | & \\
 & H & H & H &
\end{array}
$$

13. False

17. True

11. True

15. True

19. $n - k$

Exercises 10.2

1. $aWbcdXYZefghijklmno$

Exercises 10.3

1. Let labeling be $a, b, \ldots$, starting at the root and progressing left to right down the levels.

5. a and d

15. Not full, balanced

23. An m-ary tree of height h has between $h + 1$ and
$$\sum_{i=0}^{h} (m^i) \text{ vertices.}$$

7. d

17. Full, not balanced

25. If T is a full m-ary tree with L levels, height h, and i internal vertices, then T has $mi + 1$ vertices and $(m - 1)i + 1$ leaves.

Exercises 10.4

1. Longfellow, Tennyson, Shelley

5. Byron would go as a left child of Donne.

13. It would approximately double the size (amount of memory) needed.

3. Longfellow, Tennyson, Shelley, Shakespeare

7. Pound would go as the left child of Shakespeare.

15. 1, 2, 4, 8, 9, 5, 10, 11, 3, 6, 7, 12, 13

17. 8, 4, 9, 2, 10, 5, 11, 1, 6, 3, 12, 7, 13
23. $/ + a * bc * 2d$
27. $a * (b - c) + d$
35. Yes
41. No
45. *fadeacage*

19. *f, j, k, g, d, h, l, m, i, e, b, c, a*
25. $abc * + 2d */$
29. $(a * b) * c * d$
39. Yes
43. *A play pen*
47. 845

Exercises Appendix A

1. [3, 2, −1]
11. [0,0], [0,3], [3,0], [3,7]
25. [3, −2, 7, 3]
41. 317500
63. Orthogonal

5. [1, 0, 1, 0, 1, 0, 1]
17. [5, −1, 10, 9]
27. Both are [5, −1, 10, 9].
43. 16
65. Not orthogonal

Exercises Appendix B

1. 2
11. 2,3,6

7. 2,3,6
13. [1 4 7], [2 5 8], [3 6 9]

15. $\begin{bmatrix} 2 & 3 \\ 3 & 1 \end{bmatrix} \begin{bmatrix} x \\ y \end{bmatrix} = \begin{bmatrix} 16 \\ 5 \end{bmatrix}$

19. $\begin{bmatrix} 0 & 15 & 35 & 24 \\ 15 & 0 & 21 & 31 \\ 35 & 21 & 0 & 30 \\ 24 & 31 & 30 & 0 \end{bmatrix}$

21. $\begin{bmatrix} 1 & 4 \\ 2 & 0 \end{bmatrix}$

29. $\begin{bmatrix} -6 & 1 \\ 2 & 4 \end{bmatrix}$

33. $\begin{bmatrix} 1 & 0 \\ 0 & 1 \end{bmatrix}$

37. 2^n

39. $2^{(n(n+1)/2)}$

47. $\begin{bmatrix} -1 & 3 \\ 2 & -6 \end{bmatrix}$

51. $\begin{bmatrix} 14 & 20 \\ 20 & 34 \end{bmatrix}$

55. [7 6]

57. Both matrices are the following: $\begin{bmatrix} 2 & 3 \\ -2 & -3 \end{bmatrix}$

63. $\begin{bmatrix} -5 & 16 & 5 \\ -7 & 6 & 1 \\ -15 & 2 & 3 \end{bmatrix}$

75. $\begin{bmatrix} 0 & 1 \\ -1 & 1 \end{bmatrix}$

81. False

Exercises Appendix C

1. $x = -4/5$, $y = 18/5$
3. $x = 1$, $y = 2$
5. $x = 2$, $y = -2$, $z = 1$
7. $x = 0$, $y = z = 1$
9. $x = 0$, $y = 1$, $z = 1$, $w = 2$
11. $u = 1$, $v = 1$, $w = -2$, $x = 1$, $y = 0$
13. Yes
15. No, inconsistent
17. Yes
19. Yes
21. No—first + second − third = fourth
23. Yes
25. $a \neq -2$
27. $c \neq 1$
29. $a = -12$ redundant, else inconsistent
31. $c = 16$ redundant, else inconsistent
33. $z = (7x - 8)/3$, $y = (2x + 5)/3$
35. $x = 22 - 16w$, $y = 4w - 6$, $z = 11w - 14$
37. Redundant
39. Inconsistent
41. If $X[k] \ldots X[n]$ are all 0's, we have redundancy; otherwise we have consistency.
43. Change line 9 so that instead of modifying rows $i + 1$ to n, it modifies all rows except the ith.
45. The addition needs to be done only for columns $i + 1$ to $n + 1$.
47. For each value of i, we are carrying out a multiplication and addition for every entry in the lower rectangle from column $i + 1$ to row $n + 1$ and from row $i + 1$ to row n. Thus this involves $O(i^2)$ computations. Since i runs from 1 to n, we get the sum indicated.

Exercises Appendix D

7. $\begin{bmatrix} 3/8 & -1/8 \\ -1/8 & 3/8 \end{bmatrix}$

9. $\begin{bmatrix} -1/2 & 3/2 \\ 1 & -2 \end{bmatrix}$

11. $\begin{bmatrix} 1 & -1 & 2 \\ 1 & 2 & -4 \\ -1 & -2 & 5 \end{bmatrix}$

27. If $b = 0$ or $c = 0$, $a = \pm 1$ and $d = \pm 1$. If $b \neq 0$ and $c \neq 0$, $a = -d$ and $bc = 1 - a^2$.

29. The inverse is a diagonal matrix with the entries $1, 1/2, 1/3, \ldots, 1/n$ along the diagonal.

33. Same as Gauss-Jordan elimination—$O(n^3)$.

Exercises Appendix E

1. $(3\sqrt{2}, \pi/4)$
3. $(2, \pi/3)$
5. $(1, \pi/2)$
7. $f(n) = (\sqrt{2})^n \sin 3n\pi/4$.
9. $f(n) = (\sqrt{3})^n(\cos n\pi/2 + (2\sqrt{3}/3) \sin n\pi/2)$
11. $f(n) = 2(\cos n\pi/6 - (2\sqrt{3}/3) \sin n\pi/6$

Exercises Appendix F

1.

	1		31		90		65		15		1	
1		63		301		350		140		21		1

3. 2

5. 15

Index